NEUROMETHODS

Series Editor
Wolfgang Walz
University of Saskatchewan
Saskatoon, SK, Canada

For further volumes:
http://www.springer.com/series/7657

Neuromethods publishes cutting-edge methods and protocols in all areas of neuroscience as well as translational neurological and mental research. Each volume in the series offers tested laboratory protocols, step-by-step methods for reproducible lab experiments and addresses methodological controversies and pitfalls in order to aid neuroscientists in experimentation. *Neuromethods* focuses on traditional and emerging topics with wide-ranging implications to brain function, such as electrophysiology, neuroimaging, behavioral analysis, genomics, neurodegeneration, translational research and clinical trials. *Neuromethods* provides investigators and trainees with highly useful compendiums of key strategies and approaches for successful research in animal and human brain function including translational "bench to bedside" approaches to mental and neurological diseases.

Translational Methods for PTSD Research

Edited by

Graziano Pinna

Psychiatric Institute, University of Chicago, Chicago, IL, USA

Editor
Graziano Pinna
Psychiatric Institute
University of Chicago
Chicago, IL, USA

ISSN 0893-2336   ISSN 1940-6045 (electronic)
Neuromethods
ISBN 978-1-0716-3217-8   ISBN 978-1-0716-3218-5 (eBook)
https://doi.org/10.1007/978-1-0716-3218-5

© The Editor(s) (if applicable) and The Author(s), under exclusive license to Springer Science+Business Media, LLC, part of Springer Nature 2023

This work is subject to copyright. All rights are solely and exclusively licensed by the Publisher, whether the whole or part of the material is concerned, specifically the rights of translation, reprinting, reuse of illustrations, recitation, broadcasting, reproduction on microfilms or in any other physical way, and transmission or information storage and retrieval, electronic adaptation, computer software, or by similar or dissimilar methodology now known or hereafter developed.

The use of general descriptive names, registered names, trademarks, service marks, etc. in this publication does not imply, even in the absence of a specific statement, that such names are exempt from the relevant protective laws and regulations and therefore free for general use.

The publisher, the authors, and the editors are safe to assume that the advice and information in this book are believed to be true and accurate at the date of publication. Neither the publisher nor the authors or the editors give a warranty, expressed or implied, with respect to the material contained herein or for any errors or omissions that may have been made. The publisher remains neutral with regard to jurisdictional claims in published maps and institutional affiliations.

"There's always sun just over the horizon" - Delores E. Thompson

This Humana imprint is published by the registered company Springer Science+Business Media, LLC, part of Springer Nature.
The registered company address is: 1 New York Plaza, New York, NY 10004, U.S.A.

Preface to the Series

Experimental life sciences have two basic foundations: concepts and tools. The Neuromethods series focuses on the tools and techniques unique to the investigation of the nervous system and excitable cells. It will not, however, shortchange the concept side of things as care has been taken to integrate these tools within the context of the concepts and questions under investigation. In this way, the series is unique in that it not only collects protocols but also includes theoretical background information and critiques which led to the methods and their development. Thus it gives the reader a better understanding of the origin of the techniques and their potential future development. The Neuromethods publishing program strikes a balance between recent and exciting developments like those concerning new animal models of disease, imaging, in vivo methods, and more established techniques, including, for example, immunocytochemistry and electrophysiological technologies. New trainees in neurosciences still need a sound footing in these older methods in order to apply a critical approach to their results.

Under the guidance of its founders, Alan Boulton and Glen Baker, the Neuromethods series has been a success since its first volume published through Humana Press in 1985. The series continues to flourish through many changes over the years. It is now published under the umbrella of Springer Protocols. While methods involving brain research have changed a lot since the series started, the publishing environment and technology have changed even more radically. Neuromethods has the distinct layout and style of the Springer Protocols program, designed specifically for readability and ease of reference in a laboratory setting.

The careful application of methods is potentially the most important step in the process of scientific inquiry. In the past, new methodologies led the way in developing new disciplines in the biological and medical sciences. For example, Physiology emerged out of Anatomy in the nineteenth century by harnessing new methods based on the newly discovered phenomenon of electricity. Nowadays, the relationships between disciplines and methods are more complex. Methods are now widely shared between disciplines and research areas. New developments in electronic publishing make it possible for scientists that encounter new methods to quickly find sources of information electronically. The design of individual volumes and chapters in this series takes this new access technology into account. Springer Protocols makes it possible to download single protocols separately. In addition, Springer makes its print-on-demand technology available globally. A print copy can therefore be acquired quickly and for a competitive price anywhere in the world.

Saskatoon, SK, Canada   *Wolfgang Walz*

Preface: Research Methods for clinical and preclinical post-traumatic stress disorder (PTSD)

Adverse life events, including highly threatening experiences, affect people's psychological integrity and tend to occur frequently in the lifespan. The lack of developing resilience to these traumatic experiences may lead to post-traumatic stress disorder (PTSD). A hallmark of PTSD is the exaggerated fear that develops following exposure to a traumatic event and the inability by vulnerable individuals to extinguish it over time. Hence, there is a growing interest in the study of extinction of conditioned fear as a strategy to resolve PTSD symptoms [1]. While human studies have shown that individuals with or without PTSD generally show similar acquisition and extinction of fear, individuals with PTSD typically fail to maintain the gains of a successful extinction training, suggesting deficits in retaining extinction learning is potentially important to the development and maintenance of PTSD. Sex differences in PTSD neurobiology suggest that hormonal factors may contribute to women's greater risk for PTSD. However, neurohormonal signaling in response to acute and chronic stress and their effects on retention of fear extinction learning remain poorly investigated.

Fear acquisition and fear extinction are generally studied by experimental procedures mainly based on the Pavlovian conditioning and extinction methods that rely on exposing conditioned experimental animals or humans to fear-eliciting cues in absence of aversive stimuli. In animal models of PTSD, fear is determined by the freezing behavior of the animal in response to exposure to the fear conditioned stimulus (CS), and it is defined by the absence of any movement except for those related to respiration. Increased amplitude of an acoustically elicited startle response as well as emission of ultrasonic calls and changes in respiration and blood pressure are all signs of increased fear in response to exposure to CS [1].

The mechanisms underlying fear extinction and retention of extinction learning are not fully understood; investigating how extinction can be facilitated and better retained may contribute to the development of new therapeutic strategies relevant to enhancing treatment of PTSD. Generally, in animal models, fear extinction is studied by exposing repeatedly or for extended times the fear-conditioned experimental rodents to the CS in absence of the aversive unconditioned stimulus (US). Translationally, prolonged exposure (PE) therapy, a first-line PTSD treatment with an extensive evidence base, was developed based on these principles of extinction learning. PE is a trauma-focused cognitive behavioral therapy involving repeated exposure to trauma memories/reminders. Although PE is one of the most well validated treatments for PTSD, the reduction in PTSD symptom severity remains modest, which underscores the necessity to develop cognition-enhancing drugs to facilitate this process and maintain the therapeutic gains overtime.

While a lot of progress has been achieved in understanding the behavioral and neuronal underpinnings regulating the ability of traumatized individuals to develop resilience, or lack thereof, this branch of neuropsychopharmacology still lacks efficient pharmacological treatments. Development of biomarkers is hampered by the paucity of valid animal models and lack of sophisticated investigative technologies. Knowledge of the molecular brain mechanisms underlying fear extinction and fear extinction retention and other PTSD symptoms will enhance biomarker discovery and may help develop novel pharmacotherapeutic strategies.

In this issue, Zanca and colleagues [2] discuss early life trauma in causing PTSD and laying the foundation for later-life vulnerability to PTSD. They present infant rodent fear conditioning methods for young rats to understand the age-specific effects on PTSD vulnerability and its utility in understanding how infant traumatic experiences affect neuronal processes that can precipitate symptoms. Their methodology explores experimental framework including the infant's developmental environment with the maternal presence. They also describe infant handling procedures for learning in young infants and present the Scarcity-Adversity model of low bedding, an infant maltreatment method that introduces harsh maternal treatment of pups, thereby mimicking a pivotal predictor of human PTSD. Fear conditioning included in this protocol after typical and atypical maternal care provides an animal model that enables to study age-specific features of PTSD and defines a link between infant trauma and later-life vulnerability to PTSD.

Adverse early-life experiences can also be modelled in preclinical models of PTSD by post-weaning social isolation, which entails reasonable face validity, as early social experiences play a critical role in PTSD risks. Drummond and Kim [3] present a rodent model of early life adversity that leads to impaired extinction of conditioned fear. They describe and discuss the various nuances in the different methodologies currently used and suggest how to maximize reliability and reproducibility of fear extinction experimental procedure, for example, by avoiding creating a sensory deprivation condition that may lead to schizophrenic-like behavior rather than PTSD-like behaviors. Their protocol provides important guidelines on how to study early-life stress in a manner useful for translational studies in humans.

Indeed, given that PTSD and other anxiety disorders often begin during childhood and adolescence, this requires a developmental approach as well as the use of age-appropriate methodologies to investigate fear extinction in developmental populations. Marusak and collaborators [4] detail a novel virtual reality paradigm that they developed and validated to investigate fear learning and fear extinction in children and adolescents. Necessary equipment and software, considerations, and guidelines are also provided.

Sleep disturbance and trauma nightmares are among the most common PTSD symptoms, but evidence that sleep disturbance is also involved in the development and persistence of PTSD is accruing. Kobayashi and collaborators [5] present methodological factors important for planning sleep studies and psychophysiological variables that are associated with sleep in PTSD populations. They describe traditional as well as novel methodologies, including traditional and quantitative polysomnography, actigraphy, subjective sleep and dream measures, fear conditioning and extinction, and brain imaging during sleep, for assessing sleep and sleep-associated phenomena. Finally, they discuss the unique methodological challenges that are often encountered with studying sleep in individuals with PTSD.

Miller et al. [6] describe a human threat conditioning and extinction paradigm, which was designed based on rodent threat conditioning paradigms and can be used to investigate PTSD and other psychiatric disorders. This paradigm is based on a passive viewing of emotional learning and memory tasks composed of four phases. The first phase is constituted by the threat conditioning, which is followed by a threat extinction learning. The third phase is represented by a memory for extinction learning, which is evaluated during the extinction recall after a 24-h delay following extinction learning. Finally, the fourth phase is the renewal of threat responding after extinction. During this procedure, participants are in an MRI scanner to identify changes in brain correlates during these emotional learning phases.

Phenomenological features of fear encompassing physiological arousal, defensive and behavioral activation, distress, associative learning, and adaptation are within neurobehavioral responses that should be captured methodologically in PTSD studies. The psychophysiological study of traumatized populations includes sensitive indices that include Pavlovian conditioning, skin conductance response, and fear-potentiated startle. While describing common methodologies for collecting psychophysiological data in PTSD populations, Norrholm and colleagues [7] also discuss the convergence and divergence of these physiological indices and emphasize on the advantages provided by each to inform development of future innovations.

Physiological responses to trauma and stressors, including autonomic nervous system responses, the hypothalamic-pituitary-adrenal (HPA) axis responses, as well as the immune response, have been implicated in PTSD risk and symptom development. Understanding the mechanisms underlying these neurobiological responses may help develop new strategies for prevention and PTSD treatment. Individual variability in neurophysiological response to trauma is believed to play a pivotal role in determining stress vulnerability or, alternatively, resilience. Thus, understanding mechanisms underlying individual variability in stress responses promises to enhance the understanding of mechanisms of PTSD vulnerability. Andersen and collaborators [8] describe an inescapable stress exposure model in rats that has contributed to the understanding of the mechanisms underlying stress vulnerability and stress resilience.

In humans, the apolipoprotein E (APOE) genotype may influence the susceptibility to develop PTSD as well as ensuing symptom severity. APOE is interesting for its association with cognitive health and the role played in cholesterol transport and metabolism that may underlie a link with PTSD. Torres et al. [9] detail liquid chromatography and gas chromatography methods to quantify cholesterol and derivatives in discrete mouse brain areas. They describe methods to measure cholesterol and its metabolites in brains of targeted replacement mice expressing human E2, E3, or E4 under control of the murine apoE promotor. These mice were exposed to the PTSD chronic variable stress model. Understanding of cholesterol metabolism in individuals with PTSD may stimulate novel therapeutic strategies for this condition.

While most of animal models of fear extinction have mainly used male subjects, studies in fear extinction are now more often including sex as a biological variable and more studies are now being conducted with female rodents. Velasco and colleagues [10] detail how to induce fear extinction impairment in both sexes, with male and female mice that have been previously exposed to acute stress, including immobilization.

Fear memory, negative operant conditioning, cognitive impairment, mood instability, and hypervigilance are among the most common PTSD symptoms, but it is rather difficult to reproduce them within a single animal model. Shikanai and collaborators [11] describe the 3WFS model in rats, which includes the exposure of rats to aversive stimuli, such as footshock during the third postnatal week. 3WFS rats show delayed extinction of contextual fear conditioning, which can be reversed by the administration of D-cycloserine. This treatment is also promising to promote extinction of fear memory in individuals with PTSD, which underscores the validity of the 3WFS model in reproducing PTSD symptoms. These investigators suggest that abnormalities in 5-HT system in the hippocampus underlies impaired fear extinction in 3WFS rats. In their chapter, they describe how to create a 3WFS model and they introduce their research finding on fear memory related to the 5-HT neurotransmission in the hippocampus.

Individual variability constitutes a major PTSD feature; therefore, it is crucial to understand factors, both psychological and biological, that underlie mechanisms of vulnerability and resilience to PTSD. Rodent models that incorporate either individual variability or sex differences in response to conditioning fear are valuable to enhance the translation and validity of testing potential pharmacological and non-pharmacological treatments. Careaga and collaborators [12] describe behavioral and neurobiological features of fear conditioning that may be involved in developing vulnerability and resilience to PTSD. They also present literature gaps and propose future directions.

Studies on the therapeutic efficacy of virtual reality exposure therapy (VRET) combined with non-invasive brain stimulation (NIBS) as stand-alone treatment approach for PTSD are limited. In their chapter, Vicario and colleagues [13] discuss limitations and future directions of research exploring the effect of these two PTSD treatments combined, which are emerging research field with potential in improving the treatment of PTSD.

Fear conditioning and avoidance protocols that are commonly implemented in laboratory procedures induce moderate-intensity and specific aversive memories. However, overconsolidated traumatic events are quantitative and qualitative different. After discussing the mnemonic basis of PTSD, its memory-based symptoms, and the neurobiology underlying traumatic memories, Gazarini et al. [14] review and discuss investigations addressing the potential translational value of aversive tasks with single or combined post-training pharmacological interventions and susceptibility to attenuation by extinction and administration of drug to block reconsolidation.

Selective serotonin reuptake inhibitors (SSRIs) are the only approved and most prescribed medications for individuals who suffer PTSD. Many clients fail to respond to this pharmacological treatment and, as of today, there are no ways to determine who will be more likely to respond and benefit from these medications. Loudness Dependence of Auditory Evoked Potentials (LDAEP), a measure derived from auditory event-related potentials (ERPs) to a series of increasingly loud tones, is affected by levels of brain serotonin and promises to serve as a biomarker of SSRI responsiveness. Pineles and colleagues [15] describe the rationale supporting implementing LDAEP tasks in scientific studies aimed at advancing precision medicine strategies for PTSD. They also provide guidance to enable interested investigators to administer the LDAEP task in their labs, including information on the equipment needed and detailed instruction on procedures for data collection, cleaning, and scoring.

To improve pharmacotherapy for PTSD, validated PTSD animal models could enhance the understanding of PTSD neurobiology via discovery of novel pharmacologic targets. However, the complexity of PTSD makes the development of valid rodent models a challenge. Santovito and Pinna [16] utilize protracted social isolation in mice as a stress model to induce a time-dependent downregulation of neurosteroid biosynthesis, including the GABAergic allopregnanolone, which has been previously observed in individuals with PTSD. Social isolation in rodents also changes $GABA_A$ receptor sensitivity and conformation in corticolimbic circuitry, which together with decreased allopregnanolone levels may provide a potential biomarker axis to advance drug discovery and -possibly- future precision medicine for PTSD. Recently, drugs that mimic the pharmacology of neurosteroids is of high interest following clinical trials showing the efficacy of these molecules in treating mood disorders. In a translational approach, these authors discuss the methodology of studying neurosteroids' effects on reconsolidation blockade a mechanism that effectively reduces PTSD-like behavioral deficits, including impaired fear extinction and fear extinction retention in rodent stress models and thereby provides new insights for suitable agents for clinical testing.

Thus, this book aims to fill the gap in experimental procedures in animal models of PTSD and humans affected by this debilitating disorder by describing new methodologies that may be useful for translational research; addressing sex differences; highlighting the state-of-the-art of biomarker discovery in the development and maintenance of PTSD; and featuring new promising agents to enhance fear extinction retention that may help millions of individuals that suffer from PTSD worldwide.

Chicago, IL, USA *Graziano Pinna*

References

1. Raber J, et al. (2019) Current understanding of fear learning and memory in humans and animal models and the value of a linguistic approach for analyzing fear learning and memory in humans. Neurosci Biobehav Rev 105:136–177, ISSN 0149-7634, https://doi.org/10.1016/j.neubiorev.2019.03.015
2. Zanca RM, Stanciu S, Ahmed I, Cain CK, Sullivan RM (2023) Translational model of infant PTSD induction: methods for infant fear conditioning. In: Pinna G (ed) Translational methods for PTSD research. Springer Nature, Berlin
3. Drummond K, Kim JH (2023) Periadolescent social isolation effects on extinction of conditioned fear. In: Pinna G (ed) Translational methods for PTSD research. Springer Nature, Berlin
4. Marusak HA, Peters C, Rabinak CA (2023) Using virtual reality to study fear and extinction in children and adolescents. In: Pinna G (ed) Translational methods for PTSD research. Springer Nature, Berlin
5. Kobayashi I, Pereira ME, Jenkins KD, Johnson FL III, Pace-Schott EF (2023) Assessing the role of sleep in the regulation of emotion in PTSD. In: Pinna G (ed) Translational methods for PTSD research. Springer Nature, Berlin
6. Miller DB, Rassaby MM, Wen Z, Milad MR (2023) Pavlovian conditioning and extinction methods for studying the neurobiology of fear learning in PTSD. In: Pinna G (ed) Translational methods for PTSD research. Springer Nature, Berlin
7. Norrholm SD, Cilley TJ, and Jovanovic T (2023) Reconciling translational disparities between empirical approaches to better understand PTSD. In: Pinna G (ed) Translational methods for PTSD research. Springer Nature, Berlin
8. Andersen ND, Sterrett JD, Costanza-Chavez GW, Zambrano CA, Baratta MV, Frank MG, Maier SF, Lowry CA (2023) An integrative model for endophenotypes relevant to posttraumatic stress disorder (PTSD): detailed methodology for inescapable tail shock stress (IS) and juvenile social exploration (JSE). In: Pinna G (ed) Translational methods for PTSD research. Springer Nature, Berlin
9. Torres ERS, DeBarber AE, Raber J (2023) Apolipoprotein E isoform-related translational measures in PTSD research. In: Pinna G (ed) Translational methods for PTSD research. Springer Nature, Berlin
10. Velasco ER, Florido A, Marin I, Molina P, Perez-Caballero L, Andero R (2023) Stress immobilization inducing fear extinction deficits in male and female mice. In: Pinna G (ed) Translational methods for PTSD research. Springer Nature, Berlin
11. Shikanai H, Matsuzaki H, Kasai R, Kusaka S, Shindo T, Izumi T (2023) 5-HT neural system abnormalities in PTSD model rats. In: Pinna G (ed) Translational methods for PTSD research. Springer Nature, Berlin
12. Careaga L, Girardi CEN, Suchecki D. (2023) Animal models of PTSD: the role of fear conditioning. In: Pinna G (ed) Translational methods for PTSD research. Springer Nature, Berlin

13. Vicario CM, Nitsche MA, Salehinejad MA, Tortora F, Lucifora C, Grasso GM, Lakmehsari AH (2023) Combining virtual reality exposure therapy with non-invasive brain stimulation for the treatment of post-traumatic stress disorder and related syndromes. A perspective. In: Pinna G (ed) Translational methods for PTSD research. Springer Nature, Berlin
14. Gazarini L, Stern CAJ, Bertoglio LJ (2023) Associating aversive task exposure with pharmacological intervention to model traumatic memories in laboratory rodents. In: Pinna G (ed) Translational methods for PTSD research. Springer Nature, Berlin
15. Pineles SL, Pandey S, Shor R, Abi-Raad RF, Kimble MO, Orr SP (2023) Loudness Dependence of Auditory Evoked Potentials (LDAEP): a promising pre-treatment predictor of selective serotonin reuptake inhibitor (SSRI) response. In: Pinna G (ed) Translational methods for PTSD research. Springer Nature, Berlin
16. Santovito LS, Pinna G (2023) Preclinical methods of neurosteroid-induced facilitation of fear extinction and fear extinction retention. In: Pinna G (ed) Translational methods for PTSD research. Springer Nature, Berlin

Contents

Preface to the Series		v
Preface		vii
Contributors		xv
1	Translational Model of Infant PTSD Induction: Methods for Infant Fear Conditioning *Roseanna M. Zanca, Sara Stanciu, Islam Ahmed, Christopher K. Cain, and Regina M. Sullivan*	1
2	Periadolescent Social Isolation Effects on Extinction of Conditioned Fear *Katherine Drummond and Jee Hyun Kim*	23
3	Using Virtual Reality to Study Fear and Extinction in Children and Adolescents *Hilary A. Marusak, Craig Peters, and Christine A. Rabinak*	37
4	Assessing the Role of Sleep in the Regulation of Emotion in PTSD *Ihori Kobayashi, Mariana E. Pereira, Kilana D. Jenkins, Fred L. Johnson III, and Edward F. Pace-Schott*	51
5	Pavlovian Conditioning and Extinction Methods for Studying the Neurobiology of Fear Learning in PTSD *Dylan B. Miller, Madeleine M. Rassaby, Zhenfu Wen, and Mohammed R. Milad*	97
6	Reconciling Translational Disparities Between Empirical Approaches to Better Understand PTSD *Seth D. Norrholm, Timothy J. Cilley Jr., and Tanja Jovanovic*	117
7	An Integrative Model for Endophenotypes Relevant to Posttraumatic Stress Disorder (PTSD): Detailed Methodology for Inescapable Tail Shock Stress (IS) and Juvenile Social Exploration (JSE) *Nathan D. Andersen, John D. Sterrett, Gabriel W. Costanza-Chavez, Cristian A. Zambrano, Michael V. Baratta, Matthew G. Frank, Steven F. Maier, and Christopher A. Lowry*	135
8	Apolipoprotein E Isoform-Related Translational Measures in PTSD Research *Eileen Ruth Samson Torres, Andrea E. DeBarber, and Jacob Raber*	169
9	Stress Immobilization Inducing Fear Extinction Deficits in Male and Female Mice *Eric Raul Velasco, Antonio Florido, Ignacio Javier Marin-Blasco, Patricia Molina, Laura Perez-Caballero, and Raul Andero*	191
10	5-HT Neural System Abnormalities in PTSD Model Rats *Hiroki Shikanai, Hirokazu Matsuzaki, Rina Kasai, Shota Kusaka, Tsugumi Shindo, and Takeshi Izumi*	203

11	Animal Models of PTSD: The Role of Fear Conditioning.................... *Mariella B. L. Careaga, Carlos Eduardo Neves Girardi, and Deborah Suchecki*	215
12	Combining Virtual Reality Exposure Therapy with Non-invasive Brain Stimulation for the Treatment of Post-traumatic Stress Disorder and Related Syndromes: A Perspective................................... *Carmelo M. Vicario, Mohammad A. Salehinejad, Chiara Lucifora, Gabriella Martino, Alessandra M. Falzone, G. Grasso, and Michael A. Nitsche*	231
13	Associating Aversive Task Exposure with Pharmacological Intervention to Model Traumatic Memories in Laboratory Rodents .... *Lucas Gazarini, Cristina A. J. Stern, and Leandro J. Bertoglio*	247
14	Loudness Dependence of Auditory Evoked Potentials: A Promising Pre-treatment Predictor of Selective Serotonin Reuptake Inhibitor Response .... *Suzanne L. Pineles, Shivani Pandey, Rachel Shor, Ronnie F. Abi-Raad, Matthew O. Kimble, and Scott P. Orr*	305
15	Preclinical Methods of Neurosteroid-Induced Facilitation of Fear Extinction and Fear Extinction Retention................................ *Luca Spiro Santovito and Graziano Pinna*	325

Index... 349

Contributors

RONNIE F. ABI-RAAD • *Compumedics USA Inc, Compumedics-Neuroscan, Charlotte, NC, USA*

ISLAM AHMED • *Emotional Brain Institute, The Nathan S. Kline Institute for Psychiatric Research, New York, NY, USA; Child and Adolescent Psychiatry, New York University Langone Medical Center, New York, NY, USA*

RAUL ANDERO • *Institut de Neurociències, Universitat Autònoma de Barcelona, Barcelona, Spain; Departament de Psicobiologia i de Metodologia de les Ciències de la Salut, Universitat Autònoma de Barcelona, Barcelona, Spain; Centro de Investigación Biomédica en Red de Salud Mental (CIBERSAM), Instituto de Salud Carlos III, Madrid, Spain; Unitat de Neurociència Traslacional, Parc Taulí Hospital Universitari, Institut d'Investigació i Innovació Parc Taulí (I3PT), Institut de Neurociències, Universitat Autònoma de Barcelona, Bellaterra, Spain; ICREA, Barcelona, Spain*

NATHAN D. ANDERSEN • *Department of Integrative Physiology, University of Colorado Boulder, Boulder, CO, USA; Center for Neuroscience, University of Colorado Boulder, Boulder, CO, USA*

MICHAEL V. BARATTA • *Center for Neuroscience, University of Colorado Boulder, Boulder, CO, USA; Department of Psychology and Neuroscience, University of Colorado Boulder, Boulder, CO, USA*

LEANDRO J. BERTOGLIO • *Departamento de Farmacologia, Universidade Federal de Santa Catarina, Florianópolis, SC, Brazil*

CHRISTOPHER K. CAIN • *Emotional Brain Institute, The Nathan S. Kline Institute for Psychiatric Research, New York, NY, USA; Child and Adolescent Psychiatry, New York University Langone Medical Center, New York, NY, USA*

MARIELLA B. L. CAREAGA • *Department of Psychobiology, Escola Paulista de Medicina, Universidade Federal de São Paulo, São Paulo, Brazil*

TIMOTHY J. CILLEY JR. • *Neuroscience Center for Anxiety, Stress, and Trauma (NeuroCAST), Department of Psychiatry and Behavioral Neurosciences, Wayne State University School of Medicine, Detroit, MI, USA*

GABRIEL W. COSTANZA-CHAVEZ • *Center for Neuroscience, University of Colorado Boulder, Boulder, CO, USA; Department of Psychology and Neuroscience, University of Colorado Boulder, Boulder, CO, USA*

ANDREA E. DEBARBER • *Department of Chemical Physiology & Biochemistry, Oregon Health & Science University, Portland, OR, USA*

KATHERINE DRUMMOND • *Florey Institute for Neuroscience and Mental Health, Parkville, VIC, Australia*

ALESSANDRA M. FALZONE • *Dipartimento di Scienze Cognitive, Psicologiche, Pedagogiche e degli studi culturali, Università di Messina, Messina, Italy*

ANTONIO FLORIDO • *Institut de Neurociències, Universitat Autònoma de Barcelona, Barcelona, Spain; Departament de Psicobiologia i de Metodologia de les Ciències de la Salut, Universitat Autònoma de Barcelona, Barcelona, Spain*

MATTHEW G. FRANK • *Center for Neuroscience, University of Colorado Boulder, Boulder, CO, USA; Department of Psychology and Neuroscience, University of Colorado Boulder, Boulder, CO, USA*

LUCAS GAZARINI • *Universidade Federal de Mato Grosso do Sul, Três Lagoas, MS, Brazil*

CARLOS EDUARDO NEVES GIRARDI • *Department of Psychobiology, Escola Paulista de Medicina, Universidade Federal de São Paulo, São Paulo, Brazil*

G. GRASSO • *Dipartimento di Scienze Cognitive, Psicologiche, Pedagogiche e degli studi culturali, Università di Messina, Messina, Italy*

TAKESHI IZUMI • *Department of Pharmacology, Faculty of Pharmaceutical Sciences, Health Science University of Hokkaido, Hokkaido, Japan; Advanced Research Promotion Center, Health Science University of Hokkaido, Hokkaido, Japan*

KILANA D. JENKINS • *Behavioral Biology Branch, Center for Military Psychiatry and Neuroscience, Walter Reed Army Institute of Research, Silver Spring, MD, USA*

FRED L. JOHNSON III • *Behavioral Biology Branch, Center for Military Psychiatry and Neuroscience, Walter Reed Army Institute of Research, Silver Spring, MD, USA*

TANJA JOVANOVIC • *Neuroscience Center for Anxiety, Stress, and Trauma (NeuroCAST), Department of Psychiatry and Behavioral Neurosciences, Wayne State University School of Medicine, Detroit, MI, USA*

RINA KASAI • *Department of Pharmacology, Faculty of Pharmaceutical Sciences, Health Science University of Hokkaido, Hokkaido, Japan*

JEE HYUN KIM • *IMPACT – The Institute for Mental and Physical Health and Clinical Translation, School of Medicine, Deakin University, Geelong, VIC, Australia*

MATTHEW O. KIMBLE • *Department of Psychology, Middlebury College, Middlebury, VT, USA*

IHORI KOBAYASHI • *Behavioral Biology Branch, Center for Military Psychiatry and Neuroscience, Walter Reed Army Institute of Research, Silver Spring, MD, USA*

SHOTA KUSAKA • *Department of Pharmacology, Faculty of Pharmaceutical Sciences, Health Science University of Hokkaido, Hokkaido, Japan*

CHRISTOPHER A. LOWRY • *Department of Integrative Physiology, University of Colorado Boulder, Boulder, CO, USA; Department of Psychology and Neuroscience, University of Colorado Boulder, Boulder, CO, USA; Department of Physical Medicine and Rehabilitation, University of Colorado Anschutz Medical Campus, Aurora, CO, USA; Veterans Health Administration, Rocky Mountain Mental Illness Research Education and Clinical Center (MIRECC), Rocky Mountain Regional Veterans Affairs Medical Center (RMRVAMC), Aurora, CO, USA; Military and Veteran Microbiome: Consortium for Research and Education (MVM-CoRE), Aurora, CO, USA*

CHIARA LUCIFORA • *Dipartimento di Scienze Cognitive, Psicologiche, Pedagogiche e degli studi culturali, Università di Messina, Messina, Italy*

STEVEN F. MAIER • *Center for Neuroscience, University of Colorado Boulder, Boulder, CO, USA; Department of Psychology and Neuroscience, University of Colorado Boulder, Boulder, CO, USA*

IGNACIO JAVIER MARIN-BLASCO • *Institut de Neurociències, Universitat Autònoma de Barcelona, Barcelona, Spain*

GABRIELLA MARTINO • *Department of Clinical and Experimental Medicine, University of Messina, Messina, Italy*

HILARY A. MARUSAK • *Department of Psychiatry and Behavioral Neurosciences, School of Medicine, Wayne State University, Detroit, MI, USA*

HIROKAZU MATSUZAKI • *Laboratory of Pharmacology, Faculty of Pharmaceutical Sciences, Josai University, Sakado, Japan*

MOHAMMED R. MILAD • *Department of Psychiatry, NYU Grossman School of Medicine, New York, NY, USA; Nathan Kline Institute for Psychiatric Research, Orangeburg, NY, USA*

DYLAN B. MILLER • *Department of Psychiatry, NYU Grossman School of Medicine, New York, NY, USA*

PATRICIA MOLINA • *Institut de Neurociències, Universitat Autònoma de Barcelona, Barcelona, Spain; Unitat de Fisiologia Animal (Departament de Biologia Cel·lular, Fisiologia i Immunologia), Facultat de Biociències, Universitat Autònoma de Barcelona, Barcelona, Spain*

MICHAEL A. NITSCHE • *Department of Psychology and Neurosciences, Leibniz Research Centre for Working Environment and Human Factors, Dortmund, Germany; Department of Neurology, University Medical Hospital Bergmannsheil, Bochum, Germany*

SETH D. NORRHOLM • *Neuroscience Center for Anxiety, Stress, and Trauma (NeuroCAST), Department of Psychiatry and Behavioral Neurosciences, Wayne State University School of Medicine, Detroit, MI, USA*

SCOTT P. ORR • *Psychiatry Department, Massachusetts General Hospital and Harvard Medical School, Charlestown, MA, USA*

EDWARD F. PACE-SCHOTT • *Department of Psychiatry, Massachusetts General Hospital, Charlestown, MA, USA; Athinoula A. Martinos Center for Biomedical Imaging, Charlestown, MA, USA; Department of Psychiatry, Harvard Medical School, Charlestown, MA, USA*

SHIVANI PANDEY • *National Center for PTSD Women's Health Sciences Division, VA Boston Healthcare System, Boston, MA, USA*

MARIANA E. PEREIRA • *Donders Institute for Brain, Cognition and Behaviour, Radboud University Medical Center, Nijmegen, the Netherlands*

LAURA PEREZ-CABALLERO • *Institut de Neurociències, Universitat Autònoma de Barcelona, Barcelona, Spain; Departament de Psicobiologia i de Metodologia de les Ciències de la Salut, Universitat Autònoma de Barcelona, Barcelona, Spain*

CRAIG PETERS • *Department of Pharmacy Practice, Eugene Applebaum College of Pharmacy and Health Sciences, Wayne State University, Detroit, MI, USA*

SUZANNE L. PINELES • *National Center for PTSD Women's Health Sciences Division, VA Boston Healthcare System, Boston, MA, USA; Department of Psychiatry, Boston University Chobanian & Avedisian School of Medicine, Boston, MA, USA*

GRAZIANO PINNA • *Psychiatric Institute, University of Illinois, Chicago, IL, USA; UI Center on Depression and Resilience (UICDR), University of Illinois at Chicago, Chicago, IL, USA; Center for Alcohol Research in Epigenetics (CARE), Department of Psychiatry, University of Illinois at Chicago, Chicago, IL, USA*

JACOB RABER • *Department of Behavioral Neuroscience, Oregon Health & Science University, Portland, OR, USA; Departments of Neurology, Psychiatry, and Radiation Medicine and Division of Neuroscience, ONPRC, Oregon Health & Science University, Portland, OR, USA*

CHRISTINE A. RABINAK • *Department of Pharmacy Practice, Eugene Applebaum College of Pharmacy and Health Sciences, Wayne State University, Detroit, MI, USA*

MADELEINE M. RASSABY • *Department of Psychiatry, NYU Grossman School of Medicine, New York, NY, USA*

MOHAMMAD A. SALEHINEJAD • *Department of Psychology and Neurosciences, Leibniz Research Centre for Working Environment and Human Factors, Dortmund, Germany*

LUCA SPIRO SANTOVITO • *Psychiatric Institute, University of Illinois, Chicago, IL, USA*

HIROKI SHIKANAI • *Department of Pharmacology, Faculty of Pharmaceutical Sciences, Health Science University of Hokkaido, Hokkaido, Japan; Advanced Research Promotion Center, Health Science University of Hokkaido, Hokkaido, Japan*

TSUGUMI SHINDO • *Department of Pharmacology, Faculty of Pharmaceutical Sciences, Health Science University of Hokkaido, Hokkaido, Japan*

RACHEL SHOR • *National Center for PTSD Women's Health Sciences Division, VA Boston Healthcare System, Boston, MA, USA; Department of Psychiatry, Boston University Chobanian & Avedisian School of Medicine, Boston, MA, USA*

SARA STANCIU • *Emotional Brain Institute, The Nathan S. Kline Institute for Psychiatric Research, New York, NY, USA; Child and Adolescent Psychiatry, New York University Langone Medical Center, New York, NY, USA; Bordeaux Neurocampus, University of Bordeaux, Bordeaux, France*

CRISTINA A. J. STERN • *Departamento de Farmacologia, Universidade Federal do Paraná, Curitiba, PR, Brazil*

JOHN D. STERRETT • *Department of Integrative Physiology, University of Colorado Boulder, Boulder, CO, USA; BioFrontiers Institute, University of Colorado Boulder, Boulder, CO, USA*

DEBORAH SUCHECKI • *Department of Psychobiology, Escola Paulista de Medicina, Universidade Federal de São Paulo, São Paulo, Brazil*

REGINA M. SULLIVAN • *Emotional Brain Institute, The Nathan S. Kline Institute for Psychiatric Research, New York, NY, USA; Child and Adolescent Psychiatry, New York University Langone Medical Center, New York, NY, USA*

EILEEN RUTH SAMSON TORRES • *Department of Behavioral Neuroscience, Oregon Health & Science University, Portland, OR, USA*

ERIC RAUL VELASCO • *Institut de Neurociències, Universitat Autònoma de Barcelona, Barcelona, Spain*

CARMELO M. VICARIO • *Dipartimento di Scienze Cognitive, Psicologiche, Pedagogiche e degli studi culturali, Università di Messina, Messina, Italy*

ZHENFU WEN • *Department of Psychiatry, NYU Grossman School of Medicine, New York, NY, USA*

CRISTIAN A. ZAMBRANO • *Department of Integrative Physiology, University of Colorado Boulder, Boulder, CO, USA; Center for Neuroscience, University of Colorado Boulder, Boulder, CO, USA*

ROSEANNA M. ZANCA • *Emotional Brain Institute, The Nathan S. Kline Institute for Psychiatric Research, New York, NY, USA; Child and Adolescent Psychiatry, New York University Langone Medical Center, New York, NY, USA*

Chapter 1

Translational Model of Infant PTSD Induction: Methods for Infant Fear Conditioning

Roseanna M. Zanca, Sara Stanciu, Islam Ahmed, Christopher K. Cain, and Regina M. Sullivan

Abstract

Across species such as humans and rodents, pathological fear expression occurs in PTSD, and early life trauma has been shown to cause PTSD but also lays the foundation for later-life vulnerability to PTSD. Fear conditioning in rodents has been an important source for better understanding the neurobehavioral features of PTSD, yet technical difficulties required to adapt fear conditioning to very young pups and the age-specific challenges of working with a vulnerable population have presented major roadblocks for some labs. Here we present infant rodent fear conditioning methods for very young rats to better understand both the age-specific expression of PTSD and its use as a tool in understanding how infant trauma experiences capture specific neural events that can initiate the pathway to pathology. These methods include an experimental framework that incorporates the infant's developmental environment of warmth and maternal presence. We also present detailed infant handling procedures that enable consistent, replicable learning in these young infants. Finally, we introduce an infant maltreatment method termed the *Scarcity-Adversity model of low bedding*, which induces harsh maternal treatment of pups, and mimics one of the more robust predictors of PTSD in humans. The incorporation of fear conditioning following typical and atypical maternal care can provide a comprehensive animal model to explore age-specific characteristics of PTSD and define a causal link between infant trauma and later-life PTSD.

Key words Fear conditioning, Threat, Fear, PTSD, Infant, Trauma bonding, Amygdala, Social Buffering, Attachment figure

1 Introduction

Posttraumatic stress disorder (PTSD) is typically defined by trauma exposure and persistent dysfunctional emotional processing, such as intrusive thoughts, flashbacks, nightmares, and unregulated emotions, as described by DSM-5 Diagnostic Criteria for PTSD and ICD-1 [1–10] (see other chapters in this book: Marusak et al.,

---

Authors Christopher K. Cain and Regina M. Sullivan are co-last authors for this chapter.

Graziano Pinna (ed.), *Translational Methods for PTSD Research*, Neuromethods, vol. 198, https://doi.org/10.1007/978-1-0716-3218-5_1,
© The Author(s), under exclusive license to Springer Science+Business Media, LLC, part of Springer Nature 2023

Chap. 3; Suchecki et al., Chap. 11). Recent research has highlighted early life trauma, especially when inflicted by a parent, as a critical experience that can initiate PTSD in children and create a vulnerability for PTSD later in life [11, 12]. However, understanding the mechanistic link between early life trauma and PTSD at different stages of life, specifically, focusing on potential early life markers of later-life PTSD, has been challenging and is beginning to be addressed in animal models (see Suchecki et al., Chap. 11 in this book; [13]).

PTSD symptoms in adults are characterized by a mismatch between the defense behavior expressed and the actual level of threat [14]. Symptoms are also associated with neural deficits, most prominently within the hippocampus, amygdala, and prefrontal circuitry, although brain changes can be ubiquitous [15, 16]. Many of these adult symptoms and neural signatures are similar to those found in teens and older children, defined by abnormal fear and dysregulation of emotions, at least compared to typically developing children [17–23]. However, for the youngest children, it can be challenging to identify PTSD, as symptoms are superimposed on the backdrop of developmentally transitioning fear and emotional regulatory systems [11, 23–29]. These systems also exhibit vast individual behavioral variability across a nonlinear developmental trajectory, even for typically developing populations of children [30]. Adding to the challenge, limited linguistic and cognitive abilities make the expression of a clear trauma message difficult in young children. Perhaps the largest barrier to identifying and treating PTSD in very young children is that the major impact of trauma is typically delayed until peri-adolescence. Considerable progress has been made in recent decades, but major challenges remain, especially related to understanding the dynamic neural patterns mediating PTSD, as the brain expands during maturation and neural networks reorganize to accommodate behavioral transitions [31–37].

PTSD is unique among anxiety disorders in that it is defined by dysfunctional emotional memory processes related to specific traumatic events. One prominent model for studying PTSD-related emotional memory is classical or Pavlovian fear conditioning, where subjects learn that specific environmental stimuli predict harm. Fear conditioning provides a robust, standardized method for creating threat memories, measuring fear responses and assessing emotion regulation. Adding to its translational value, fear conditioning is effective across species and is sensitive to stressors that produce excessive or poorly regulated fear responses. Fear conditioning also occurs over most of development. While one of the most famous fear conditioning studies in children was the little Albert study, it appears unlikely to be a valid example of fear learning [38]. Nonetheless, children readily learn fear conditioning with reliable demonstration modeled in children as young as 4 years old [39–50]. Overall, these studies have shown that childhood fear learning is associated with the amygdala, and suggests a consistent neural basis for fear across development.

Due to imaging limitations in humans, fear learning neural circuits have been predominately uncovered by animal research in mechanistic causal experiments. Human imaging techniques have shown remarkable network convergence across a myriad of species, including humans, nonhuman primates, and rodents [51–54]. This data gathered on humans, when interpreted through the lens of animal research's heightened experimental control and techniques to probe the brain networks for causal mechanisms, provides critical information to better understand age-specific neurobehavioral changes underlying PTSD. Yet, while understanding early life PTSD is clearly an important mental health issue, there is a dearth of research in infant rodent fear conditioning, especially young infants, because of technical challenges that impede applying adult equipment and procedures to studies of infants. The purpose of this chapter is to outline these limitations and define age-specific fear conditioning methods for infant rats.

2 Basic Overview of Unique Learning Characteristics of the Three Learning Epochs

While extensive work remains to define the ontogeny of infant fear learning, considerable progress has been made on describing the development of fear learning in infant rats and excellent reviews already exist, without a focus, however, on methodology unique to infant learning [30, 55–59]. Thus, to anchor the importance of conducting age-appropriate fear conditioning, we categorize the developing rat pup into three age ranges based on neurobehavioral transitions in fear learning and motoric skills to navigate the shock grid, as illustrated in Fig. 1. The first two age epochs, termed the Sensitive Period and the Transitional Sensitive Period, require a custom-made conditioning setup and shock delivery to the foot or tail, while the oldest Post-Sensitive Period age range can use standard fear conditioning apparatus typically used for mice.

2.1 Sensitive Period for Attachment (Postnatal Days (PN)0-9)

Throughout the course of this age range, pups are primarily confined to the nest before the emergence of crawling. Pups have functional olfactory and somatosensory systems, but not vision and audition. The fear conditioning paradigm (odor-0.5 mA) does not support fear learning, rather it supports approach learning to the odor [60–62] suggesting pups do not differentially categorize aversive and appetitive events [63]. In fact, the odor-shock learning produces a learned odor that can replace the natural maternal odor. Specifically, similar to maternal odor, this shock-induced learned preference odor supports social behavior towards the mother and nipple attachment and appears indistinguishable

Neurobehavioral transitions in fear learning and age-appropriate fear conditioning

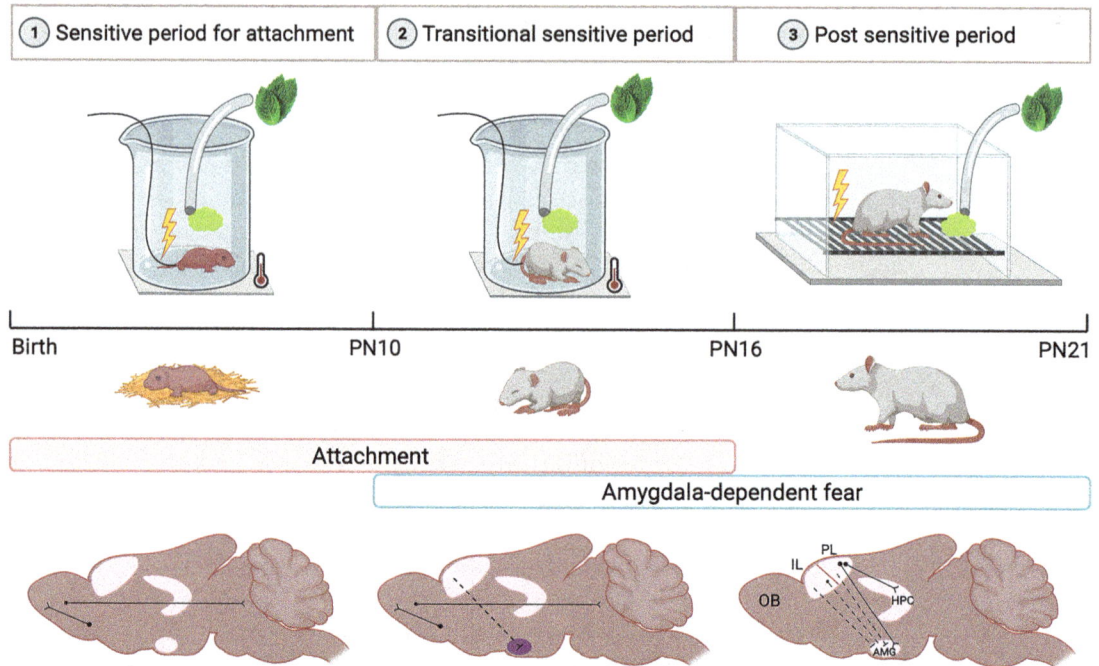

Fig. 1 The three age ranges with unique fear learning, behavioral transitions and, fear conditioning methods. During the Sensitive Period (birth-PN9) the odor-shock (0.5 mA to tail) fear conditioning paradigm does not support fear learning, rather it supports approach learning to odor. The amygdala is not engaged, rather the same neural network used within the nest for attachment learning is engaged, involving the olfactory bulb and piriform (olfactory) cortex. The Transitional Sensitive Period (PN10-PN15) is characterized by functional emergence of amygdala-dependent fear conditioning, using the basolateral (BLA), which is suppressed if the mother is present. PN10 is also the emergence of crawling in pups [10] as they begin to take brief excursions outside the nest, suggesting that within the nest, learning is suppressed by the mother, while outside the nest, a more adult-like learning system emerges [88, 126]. During these first two developmental periods, pup conditioning typically relies on olfaction since the auditory and visual systems are functionally emerging as the Transitional Sensitive Period ends. Pups during this stage of infancy cannot thermoregulate effectively without the mother, thus, the Plexiglass conditioning containers are heated to enable pups to maintain a thermoneutral body temperature (surgical water heating pads). Additionally, since pups are motorically immature and cannot stand on a grid floor typically used for fear conditioning, pups are outfitted with a tail or foot electrode to ensure the ventrum is not the shock target as interoceptive and exteroceptive pain is processed differently in pups [81]. The last developmental period, termed the Post-Sensitive Period (PN16-PN21) is characterized by the termination of the mother's ability to suppress fear learning. It is also characterized by the functional inclusion of the hippocampus and the prefrontal cortex into the fear network. Standard mouse cages can be used at this age, although ensure that the animal can support itself during the shock administration

from the natural maternal odor or odor learning using milk or tactile stimulation (mimics maternal licking) [61, 64, 65]. Importantly, if fear conditioning (pups alone) is initiated during the Sensitive Period (producing a preferred odor) and conditioning is continued into the Transitional Sensitive Period, pups continue to express an odor preference, not amygdala-dependent fear: although as discussed below, this repeated pain-associated preference learning produces later life pathology [66–68].

During this Sensitive Period, pups do not engage the amygdala to support this odor preference induced by fear conditioning [60] and instead engage a neural network used within the nest for attachment learning, engaging the olfactory bulb and piriform cortex with norepinephrine (NE) as sufficient and necessary to support this learning [69], although serotonin can enhance this learning [70]. While young pups show distinct responses to shock, milk, or stroking, they fail to distinguish between the reward value as the Unconditioned Stimulus (UCS) [63]. Together, these data support our working hypothesis that the mechanism for this robust attachment learning occurs across diverse UCSs, which is supported at the level of the locus coeruleus (LC), and similar norepinephrine (NE) release across UCS's [71–73]. Within the olfactory bulb, NE is required for pup odor learning [74–80].

2.2 Sensitive Period Non-amygdala Aversion Learning

However, non-amygdala aversion learning can be induced, although it requires malaise [81], induced by either LiCl or high shock (above 1.0 ma) [8, 82]. The amygdala becomes involved in this malaise aversion learning closer to weaning [83, 84].

2.3 Sensitive Period Precocial Amygdala-Dependent Aversion Learning

There are at least a few situations where amygdala-dependent fear learning has been demonstrated in this age range. First, injecting typically reared pups with exogenous corticosterone (CORT) either systemically or amygdala microinjections can induce fear learning in pups as young as PN5, and the infant amygdala is dependent upon CORT for plasticity to support learning [55, 85, 86]. Second, if pups are reared with a mother repeatedly engaging in rough maternal care, the amygdala will prematurely become involved in fear learning, likely through amygdala precocial maturation and bring about a premature end to the stress hyporesponsive period (SHRP) [86–89]. Third, in typically reared rats, a fearful mother releasing a fear pheromone can engage her pups' amygdala and support fear learning in pups as young as PN5 [90, 91].

2.4 Transitional Sensitive Period PN10-15

The next phase of pup learning begins with odor-0.5 mA fear conditioning paradigm supporting amygdala-dependent fear learning, as evidenced by avoidance of the conditioned stimulus (CS) and suppression of movement during a CS presentation in a confined environment [60]. Importantly, one age-specific feature of this learning epoch is that the presence of the mother blocks this amygdala-dependent fear learning through the blockade of amygdala plasticity [92, 93]. This process of social presence blocking pup fear learning is sometimes called "social buffering," although it is more correctly referred to as "social blockade" since the mother blocks pups' shock-induced stress response (corticosterone hormone) which, in turn, prevents fear conditioning. Since the maternal odor is learned, a novel odor (i.e., peppermint) can be conditioned to become a new maternal odor, which will also block this amygdala-dependent learning [67, 94, 95]. This "social blockade" of fear learning by maternal presence is not seen in pups that were reared by a mother repeatedly treating pups harshly, which is discussed below.

Amygdala-dependent fear conditioning at PN10 coincides with the emergence of crawling and pups' ability to make brief excursions outside of the nest [10]. We have suggested that without the protections of the mother outside of the nest, evolutionary pressure and the emergence of freezing within fear conditioning may have increased chances of survival.

It should also be noted that the broader circuit supporting fear learning in the epoch differs from the adult circuit, at least in part. For example, the hippocampus is not functionally integrated into the fear learning network, meaning that fear of the odor (cue learning) is not gated by hippocampal context until around weaning [90], although pups at the end of this age range seem to be able to learn context but retain it for a few minutes following conditioning until around weaning [96, 97]. In other words, in adults with both an amygdala and hippocampus engaged by fear conditioning, fear expression is enhanced if the context is the same for conditioning and testing, and attenuated in a novel testing environment. Pups at this age do not have the fear of the odor gated by context. Interestingly, this non-gated fear expression is a characteristic of PTSD [98].

2.5 Post-Sensitive Period PN16-23 (Weaning Age)

Pup fear learning is similar to adult learning in that pups exhibit fear learning that is not blocked by maternal presence. Similar to adults, social buffering can attenuate fear conditioning [99]. However, major learning differences remain, including functional inclusion of the prefrontal cortex (PFC) and adult-like extinction (as opposed to forgetting) [100–102] as this learning epoch begins and functional inclusion of the hippocampus and long-term memory of context learning as the epoch ends (context learning can be retained for minutes following acquisition in pups as this epoch begins) [96, 97, 103]. Standard mouse cages can be used at this age, although with the assurance that when the animal is shocked, the legs will be able to support the pup. Some younger pups in this age range cannot support themselves as they are being shocked and receive a shock to the ventrum, rather than the foot (see [81]. However, the fear circuit is still not fully mature and continues to change through adolescence [56, 104].

3 Fear Conditioning Basics

Fear conditioning is a simple paradigm that is readily applied to a myriad of species and across development. It involves temporal pairings of a neutral CS (e.g., sound) and an aversive unconditioned stimulus (US; e.g., electric shock). After conditioning, presentations of the CS elicit defensive responses to cope with the expected US. The neural circuits, cellular processes, and molecular mechanisms of fear conditioning are well-defined after decades of intense research in animals. It has also proven to be a valuable tool for understanding PTSD across species due to phylogenetically preserved circuits and behavior [105]. It should be noted that while any neutral sensory stimulus can be a CS, pup learning typically relies on olfaction, since vision and audition are late developing. Importantly, Rudy revealed that while the auditory and visual sensory systems become functional towards the end of the previous learning epoch, there was a delay (lasting a few days) before pups were capable of using these sensory cues in learning [106, 107]. Thus, auditory and visual cue fear conditioning emerges during this last learning epoch.

4 Pup-Specific Preparations for Fear Conditioning

4.1 Managing a Stress-Free Colony

Fear conditioning in young pups requires a healthy colony and non-stressed mothers that can withstand experimenter disruption. Indeed, working with a vulnerable population requires special considerations since stressors that have little impact on typical adult rodent colonies can directly impact breeding success, prenatal stress, and postnatal maternal care, all of which can alter neurobehavioral development of pups and decreases the probability of seeing robust, replicable learning in pups. While the colony considerations listed here are no different than those that should be applied to any rodent animal colony, we outline specific areas to check on a daily basis. (1) Animals are not housed with any other animals and are a satellite colony in our lab to permit oversight. Breeding takes place in this room, although animals are occasionally shipped in, as needed for breeding. If this is not possible, careful surveillance of all activities within your colony will help ensure a non-stressed colony. (2) A consistent animal caretaker that conducts cage cleaning at the same time of day, which in our colony is done biweekly. (3) Cleaning is done quietly, and animals are in the new cage back on the animal rack within a few minutes. Part of the nest is transferred to the new cage to ease the transition. Cages of pregnant females during the last days of pregnancy and those 2–3 days postpartum are not disturbed by cage cleaning, although some dirty bedding can be removed and replaced with clean bedding. (4) A nesting hutch measured at approximately 8 inches (20 cm) in diameter is helpful, we use a semicircle of PVC pipe approved by the veterinarian. (5) The animal caretaker and all lab staff have extensive training beyond proper handling of rats. For example, rats are never transferred by the tail, instead they are picked up by two hands and left to rest on the arm during transfer or other perfunctory needs. All lab staff should also possess the skills to know signs of rat mental and physical health. Overall, the animals' responses to the animal caretaker or lab staff entering the animal room should not be hypervigilance. (6) A no-cleaning card is placed on the cage so that it is not disturbed for a day before the experiment. (7) Mothers are bred 5–6 times with a 2-week recovery period between weaning of her pups and breeding.

4.2 Preparing Pups for an Experiment Is Done Within 5 Min from Nest to Conditioning

Removing pups from the cage for use in an experiment also requires special handling: (1) Before removing a cage from the rack, the cage is given a gentle tap and we wait until the mother is off her litter before the cage is moved. This prevents the mother from frantically trampling her pups, a stressful situation that warrants not using pups in an experiment for that day. A cart is placed next to the rack with a clean cage ready to accommodate the mother while gathering pups. (2) The cage is then placed on a cart, the top removed and placed over the clean cage, and the mother moved to a new cage. The pups are gently and rapidly removed with a scooping action to ensure the pups do not experience rough handling by the experimenter. (3) Once the pups are removed, the mother is placed back in the cage and the cage is returned to the rack within 1 min. (4) A second group of pups are not taken from the cage until the next day, or if needed, the litter is left undisturbed for at least 5 h following the last disturbance to the litter. This ensures that the mother and pups have recovered from being disturbed by the experimenter and responding at baseline.

Next, the pups are weighted, marked in three places (tail, ear, back) with a permanent Sharpy marker to identify pups for at least 1 day (a number on pups' back is easy for experimental identification but is sometimes removed by the mother, the tail and ear are typically left intact), the tail wiped with alcohol to prepare for the shock electrode attachment and urination stimulated to ensure the electrode is not wet during the experiment. Urination is stimulated by using a warm, wet sable hair paintbrush or kimwipe, using an action that mimics maternal licking.

Then, pups are outfitted with a custom-made electrode that is positioned on the base of the tail. The electrode consists of two pieces of 20-inch-long gray thread, one insulated by a small amount of micropore tape surrounding the flexible cloth wire and the other left exposed to ensure it is light weight (electrodes are separated by 0.5 mm just above the exposed wire, which is maintained with crazy glue).

Following an application of alcohol to the tail, the electrode is attached to the pup's tail using either Collodion (surgical glue) or is securely taped to the tail with micropore tape wrapped around the pup's tail. The pup is held still during the electrode attachment by swaddling it in a synthetic fur (tip of the nose to permit breathing and the base of tail exposed to permit electrode attachment), which is secured by Velcro. The pups will typically fall asleep. This process takes considerable practice but can be done in less than 30 s by a skilled technician, with close to 100% success rate of pups receiving a shock.

Pups are immediately placed in the age-appropriate fear conditioning apparatus within individual containers for young pups or the mouse conditioning apparatus for older pups. For the Sensitive Period and Transitional Sensitive Period pups, the conditioning apparatus temperature is held at a constant temperature to enable pups to maintain a thermoneutral body. In our lab we do this by placing the Plexiglas containers used for conditioning on a surgical heating pad, which has a consistent temperature (store-bought heating pads have inconsistent heating both temporally and spatially, leaving pups overheated or cold). Ambient and surface temperature of the Plexiglas tub is monitored with each CS-UCS trial using both surface and ambient temperature probes to ensure our temperature does not fluctuate by more than 0.1–0.2 °C during the experiment.

5 Age-Specific Fear Conditioning Methods

5.1 Young Pup Conditioning Procedure

We use a custom-built olfactory fear conditioning apparatus from Noldus Ethovision (see Fig. 2). Shock boxes, olfactometer, and cameras for recording behavior are controlled by the Noldus USB-IO box. The USB-IO box then allows for control of the odor-shock system via computer software, Ethovision XT. In the software, a custom template can be created for automatic odor-shock delivery during conditioning, and whole-body activity from each pup is also recorded live. Pups PN15 and younger are individually placed in a 600 mL clear beaker and cloth wire electrodes are attached to the base of the tail and then attached to the shock boxes via alligator clips. Pups are given a 10-min habituation period prior to conditioning, to ensure they have recovered from experimental handling. In pups younger than PN13, failure to habituate during this time typically indicates there has been rough handling by the mother or experimenter (see section above). Behavior is recorded during habituation and conditioning (30 s pre-CS, during 30 s CS, and 5 s shock for shock response).

After habituation, odor-shock conditioning begins with three conditioning groups. (1) Paired odor-shock – The paired pups condition receives inter-trial intervals (ITI) of 4 min. Consisting

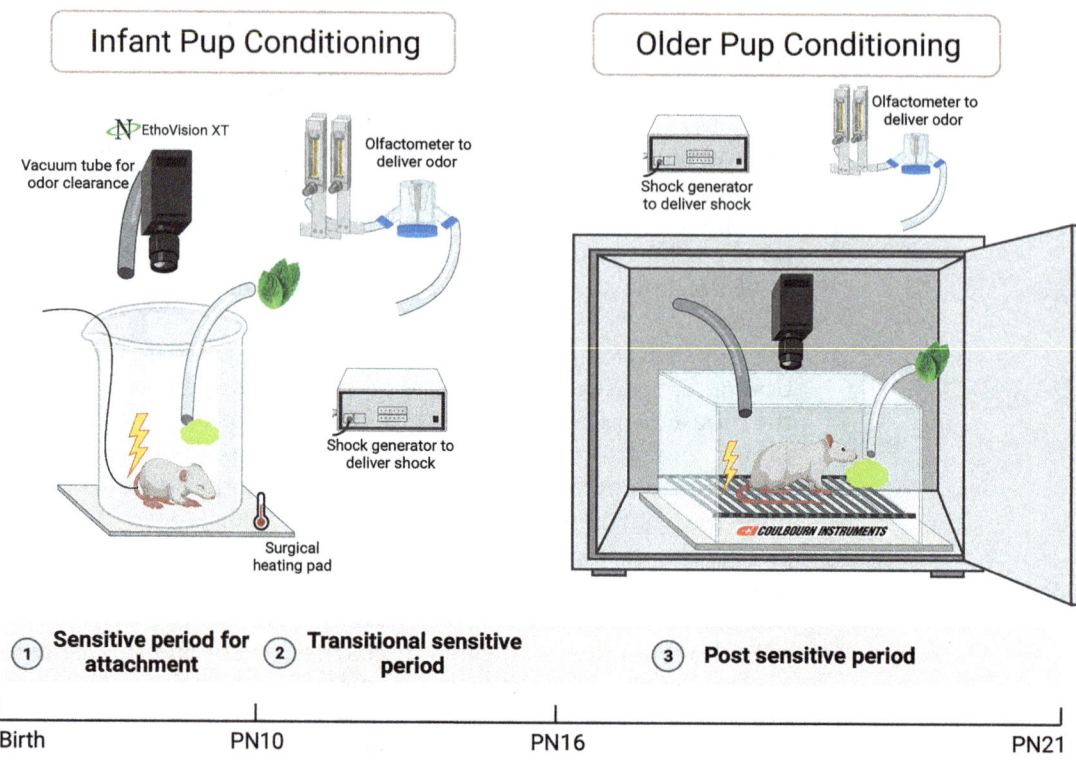

Fig. 2 Pups PN15 and younger are individually placed in a 600 mL clear beaker and cloth wire electrodes are attached to the base of the tail and then attached to the shock boxes. We use a custom build olfactory fear conditioning apparatus from Noldus Ethovision XT. Shock boxes, olfactometer, and cameras for recording behavior are controlled by the Noldus USB-IO box, which allows for control of the odor-shock system via computer software, Noldus Ethovision XT. Pups PN16 and older are generally mature enough to motorically stand on the shock grid, even after being shocked and can be conditioned using a mouse operant chamber system (Coulbourn Instruments). Because pups of this age see and hear, our mouse operant chamber systems are enclosed in a noise attenuating chamber. Shock boxes, olfactometer, and cameras for recording behavior are controlled by Actimetrics FreezeFrame software system and records whole body immobility. A vacuum tube is used to evacuate the odor from the beaker/cage for both conditioning apparatuses

of 30-s peppermint odor (CS) with 1 s 0.5 mA shock (US) delivered during the last 1 s of the odor. (2) Unpaired odor-shock conditioning – this group receives the 4-min ITI of peppermint odor at the same schedule as the paired animals, with a shock delivered between two odor presentations (2 min after the shock). (3) Odor only – this group receives the same 4-min ITI of peppermint odor without a shock. After conditioning has ended, animals are removed from the conditioning apparatus, the wire electrodes gently removed, and pups are placed back in their home cage while ensuring that the mother is not disturbed and does not trample her pups. All stimulus presentations and video recordings are delivered using the Noldus Ethovision system (https://www.noldus.com/ethovision-xt). This system also analyzes body activity levels and some individual behaviors, although the videos can also be hand scored using Boris to construct an ethogram (https://boris.readthedocs.io/en/latest/).

5.2 Setting Up Odor-Shock Conditioning Program in Noldus Ethovision XT

In the Noldus Ethovision XT software click "create new default experiment" to create a new program. Then the following steps are performed: (1) *Experiment settings:* In the experiment settings tab, make sure the cameras are being read by clicking the "eye" icon. Then go to arenas option and chose the number of arenas your experiment needs. This will be based on the number of subjects in your experiment (e.g. six arenas equal six subjects). (2) *Trial control settings:* Here, the automated control output for shock and odor delivery is written. This is a custom program created by Noldus Ethovision XT and is created based on the requests of the experimenter and their experiment. (3) *Trial List Settings:* Adjust the amount of trials based on your experiment here and then fill in the corresponding information for each subject. (4) *Arena Settings:* The number of arenas is already created in experimental settings (see above) and can be viewed here in box form. To change to the circular form required for the odor-shock conditioning protocol, simply click "Grab Background Image" once you have set up all beakers with the odor tubes (see pup conditioning protocol above). Then, choose "Arena 1" in the text box and click the circle option at the top and draw the circle over the beaker. Move the arrow that says, "Arena 1" from the box to the newly drawn circle (this will automatically change the arena to the circle) and then right click and delete the box. Then, click "Arena 2" and do the same for the second beaker. Continue this process for the rest of the beakers. Once done, choose "calibrate" and measure the length and width of the actual beakers with a ruler and write the mm measurement for each arena setting (this allows the Noldus Ethovision XT software to detect the real scale of the beakers). (5) *Trial Control Hardware Settings:* Within the arena section, click on "Trial Control Hardware" to ensure the Noldus IO-USB box is indeed connected to the software and detecting the shock and odor. Test that the odor and shockers are working by clicking on each shock box listed in the program and clicking "Test." Do the same for odor. Shock boxes can be changed and accommodated to the experiment by clicking on the drop menu where the shock box number is located and choosing a different shock box. (6) *Detection Settings:* This will be done after the pups are placed into their designated beakers and the wire electrodes are attached to the alligator clips (see pup odor-shock conditioning above). Once done, set the settings so the yellow portion on the screen is covering the pup. (7) *Acquisition settings:* This is where the trials start. Press the green button to begin the trial.

5.3 Older Pup Conditioning Procedure

The older Post-Sensitive Period pups show cue fear conditioning similar to adults, where the presence of an important social partner can reduce fear conditioning. Pups in the age range are generally mature enough to motorically stand on the shock grid (Coulbourn Instruments Precision Animal Shocker), even after being shocked and can be conditioned using a mouse operant chamber system, such as the system provided by Coulbourn Instruments (Habitest Operant Cage model). Since pups of this age are able to see and hear, our mouse operant chamber systems are enclosed in a noise attenuating chamber (Coulbourn Isolation Cubicle) with an LED light placed under the chamber. Shock boxes, olfactometer, and cameras for recording behavior are controlled by Actimetrics FreezeFrame software system and the whole body immobility is recorded. Pups are placed in a mouse operant conditioning chamber containing a shock grid floor and have an adaptation period of 10 min to recover from experimental handling. Next, odor-shock conditioning begins with three conditioning groups, using the same parameters used for younger pups, although the shock is 0.6 mA scrambled. (1) Paired odor-shock – 4 min ITI, with a 30-s peppermint odor (CS) and a 1 s 0.6 mA shock (US) overlapping with the last second of the odor. (2) Unpaired odor-shock conditioning – 4 min ITI with a 30-s peppermint odor and a shock delivered 2 min later. (3) Odor only – 4 min ITI of peppermint odor without a shock.

6 Maternal Presence During Conditioning

Since young pups are almost always with the mother, the natural environment is with the mother, which makes it a useful tool for assessing age-specific behaviors. *We use two procedures for maternal presence, which produce similar blockade of fear learning in younger pups and attenuation of fear learning in older pups: (1) an anesthetized mother placed where pups can smell the mother and/or interact during the entire conditioning procedure and (2) an awake mother placed in an olfactometer for delivery of the maternal odor in a more temporally controlled fashion. Since pups depend upon the maternal odor for social behavior and social buffering, the odor appears sufficient.*

6.1 Anesthetized Mother

To anesthetize a mother, a standard adult dose of pentobarbital or ketamine:zylene mixture can be used. However, if contact with the mother is permitted, it is best to use Urethane since it blocks the mother's milk ejection and eliminates the impact of milk delivery during fear conditioning. Once anesthetized, the mother is placed in the conditioning apparatus: younger pups can have the mother behind a screen or given direct access to the mother if the electrode is attached to pups' tail or foot. Older pups conditioned in standard conditioning chambers can use the mother to avoid the shock, and for this reason the mother is placed outside the shock box near the grid floor where the maternal odor enters the chamber. Pups' view of the mother can be blocked by covering the Plexiglas conditioning chamber.

6.2 Maternal Odor Delivered via an Olfactometer

This procedure does not differ across ages and conditioning equipment. The mother is placed into a large glass container on a screen elevated about two in above the floor of the container. The 50 LPM air flow (produced by an aquarium pump, air tank, or via an air compressor) enters the mother's chamber split between two tubes that are placed under the mother (under the screen) to ensure that maternal odor saturates the air before leaving the chamber via two ports at the top of the opposite side of the chamber. The air saturated with maternal odor leaving the chamber can be directly delivered to the pup, either under the shock grid floor or under a wire mesh floor for pups with a tail shock.

7 Testing Using Either a Cue Test or Y-Maze

There are a few different behavioral tests used to assess learning in rat pups, including a standard cue test (i.e. freezing, immobility) and a Y-maze with the CS odor vs. a familiar odor (relative avoidance/approach of the odor previously paired with shock). Both tests are given approximately 24 h after conditioning to assess long-term memory. Pups are taken from the nest in groups of 3–4 pups, tested within 15 min, and then returned to the nest. More pups are taken from the same litter after a 5 h delay, as the litter recovers from handling.

7.1 Cue-Test

This is a standard cue test used in adults, where the CS-only is presented during the test. During the Sensitive and Transitional Sensitive Period, pups show immobility rather than typical adult freezing. Younger pups are tested within the beakers used for conditioning in the same conditioning space, while the Post-Sensitive Period pups are tested in 2000 mL glass beakers and placed in an attenuating chamber (Med Associates, Inc) in another room [108, 109].

7.2 Y-Maze Test

An age-appropriate size is used for the Y-maze apparatus (PN9–13: 10, 8.5, 8 cm; PN17: 19, 10, 10 cm) and two arms (PN9–13: 24, 8.5, 8; PN17: 29, 10, 9.5 cm) with wire mesh on the floor of the maze (the textured floor is associated with greater movement). The CS odor is placed at the end of one ally, and familiar clean wood shavings (same brand as used in the cages) are used in the other arm, which are prepared before pups are taken from the nest (typically a few minutes for the odor to disperse). The alley in which the odor is placed is counterbalanced across testing days.

For individual pup testing, the pup is given three or five trials depending on the experiment. The pup is removed from the holding cage outside the testing room and gently and rapidly placed in the start box at the beginning of the two alleys of the Y-maze and the video recording started. To avoid any preference bias, the experimenter: (1) places the pup into the start box using both hands to equate touch across the body and (2) between trials, the direction in which the pup is placed in the start box is alternated (towards or away from the experimenter). At 2 s, the door is lifted giving the pups access to both alleys. The pup is given 30 s to choose an arm, with the distance down the alley to be considered a choice age-dependent: PN0 to PN11, the pup's nose and both front paws should be past the start of the arm. For pups aged PN12-18 the pup's nose, front paws, and half of its body should be past the start of the arm, and in older pups and adults, the entire body needs to pass the line at the start of the alley.

8 Methods to Induce a PTSD Phenotype That Can Be Uncovered by Fear Conditioning

Since one of the more robust predictors of childhood PTSD and vulnerability to later life PTSD is early life trauma, especially within a social context [9, 11, 12, 110], we present two models of adversity that engage the infant attachment system and mimic the outcome of maltreatment in children, illustrated in Fig. 3: *Scarcity-Adversity model of low bedding*, where pups receive repeated harsh maternal care for 5 days (i.e. maltreatment model) [66, 111–114] and a more controlled procedure of *Repeated fear conditioning* where paired odor-shock pups do not learn fear, but learn a new maternal odor, and later-life pathology that overlaps with maltreatment results [57, 68, 94, 115].

8.1 Scarcity-Adversity Model of Low Bedding (LB)

This manipulation can be started at any age, although beginning with at least 2 days during the Sensitive Period of Attachment, results in robust later life pathology. We typically begin this treatment at PN5 or PN8 and continue for 5 full days. Moreover, this LB procedure can be reduced to 90 min each day for 5 days, with similar outcome as the continuous 5-day procedure [111, 116–118]. A few variations of this LB model are currently used and provide varying levels of stress: our model does not produce any changes in weight gain in infancy or after, while other models are likely more stressful and produce some infant malnutrition (not atypical in many maltreatment cases in children) but weight is regained in later life [119]. All models are useful and important for better understanding early life adversity and are compared in a comprehensive review [118].

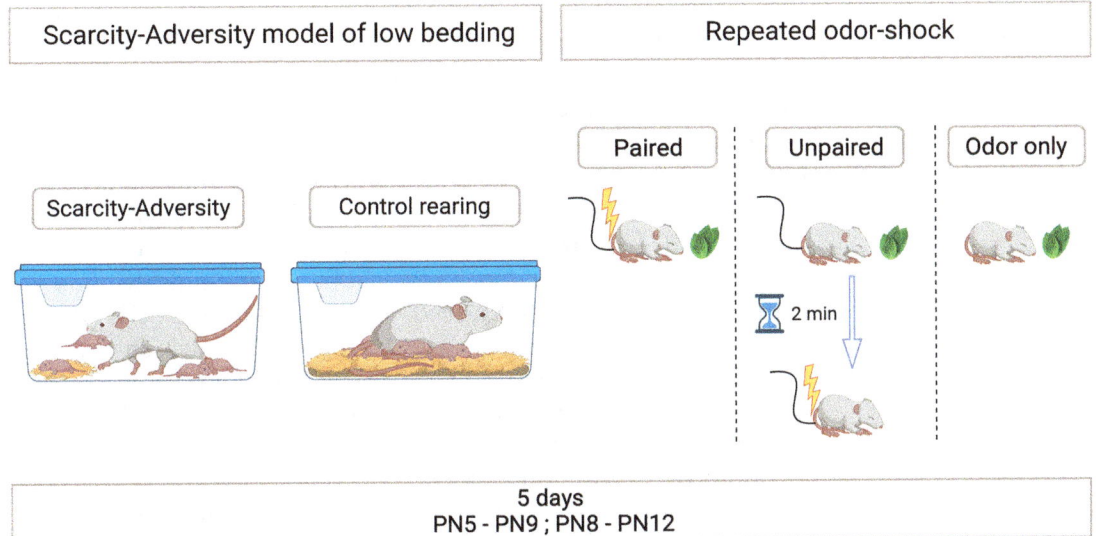

Fig. 3 Methods to induce a PTSD phenotype that can be uncovered by fear conditioning. The Scarcity-Adversity model consists in housing the mother and pups for 5 days in a limited bedding environment which decreases the mother's time with the pups and some rough handling. A calm mother who has access to enough bedding material for her pups is used as a control (control rearing). The control and the LB cages are housed on the same rack on the same shelf in the animal room and handled similarly (cleaned in the same way and at the same time). The Repeated fear conditioning protocol also takes place for 5 days, but care is taken to attach the tail-electrode delivering the shock in a new location every day

Pups rely on the maternal odor to identify their mother, and the maternal odor is diet-dependent. Therefore, pups do not distinguish between their own mother and similarly aged postpartum females, enabling the mother exhibiting harsh care to be their own mother or an alien mother. Although a similarly aged postpartum female is ideal [120].

The dam and her pups are housed in a standard cage with limited nesting/bedding material (100 mL or equal to 0.5″ layer over half the cage floor). This limited bedding environment increases nest building and some rough handling although never produces any physical evidence of maltreatment such as wounds and bruises. While pups show clear responses to the maltreatment, including vocalization and slight movement away from the mother, during periods of typical care, the adversity-reared pups are similar to control-reared pups. Indeed, similarly to infant-trauma-induced pathology in humans, the pathological effect of maltreatment generally emerges later in life. Our Scarcity-Adversity model of low bedding differs from other low bedding models in a few key features, which may underlie divergence in weight gain across procedures and potentially stress level differences: we replace soiled bedding with clean bedding (usually every 1–2 days) that provides the litter with typical olfactory environment and we do not include the wire mesh floor that is typically painful on the mother rat's paws

[121, 122]. At present, across labs, a distinction between these procedures has blurred [118, 123, 124]. A diversity of rodent infant trauma procedures are critical for modeling the diverse adversity experiences of children.

Typically, a control (amble bedding) and the LB cages are housed on the same rack on the same shelf in the animal room and handled similarly. The LB cages are cleaned as needed to ensure mother and pups remain clean: approximately twice during the continuous 5-day procedure the soiled bedding is quickly removed with a gloved hand and replaced with clean bedding. The same is done to the control cage. At times, an LB mother will remove food from the food hopper to use in the nest. When this occurs, the food is removed and we ensure only large food pellets are available. If it continues, the mother is not used.

Both the control and LB cage are videotaped and analyzed by Noldus Ethovision XT software for general movement in the cage (2 areas, nest and non-nest) and videos are hand scored using free Boris software for specific maternal and pup behaviors. Behaviors monitored include time pups and mother spend together, maternal nest building, mother's and pups' time in nest, maternal handling of pups (transporting pups, stepping on, dragging or jumping on pups, licking pups), and pups' ultrasonic vocalization.

8.2 Repeated Fear Conditioning

Procedures are the same as described above for odor-shock conditioning, although care is taken to attach the tail shock in a new location each day. The outcome of the different conditioning groups differs greatly. Paired pups show similar outcome to the Scarcity-Adversity LB pups, presumably because this conditioning is supporting the learning of a new maternal odor and engaging the attachment network [66, 112]. On the other hand, unpaired pups show increased anxiety behavior in adolescence [125].

Please see two other methods not presented here that can be used to induce infant trauma: (1) shocking pups in the mother's presence vs. shock alone to highlight the social context of trauma as targeting the amygdala [111, 116] and (2) a mother rat becoming fearful repeatedly when with her pups [90], and (3) a procedure in older infant using repeated shock [13].

9 Conclusion

Due to ethical, technical, and logistical reasons, experiments assessing the neurobiology of PTSD and assessment of the causal role of early-life trauma on PTSD in children and in later-life are rare. To address these issues, we turn to animal research to define mechanisms of expression of PTSD across the life span using age-appropriate methods and data interpretation. Here we have presented an infant rodent model that is strongly tied to the major methodology to assess PTSD in older children and adults, fear conditioning. Importantly, our age-specific methods of fear conditioning also capitalize on the age-specific role of the parent in gating trauma (or not as occurs in maltreatment), as we consider infant trauma and PTSD.

The strength of an animal model of infant PTSD and fear conditioning is that the neural network is well documented in adults and brain areas critical in fear conditioning are known to slowly develop in infant rats and children. Thus, while animal research cannot model all aspects of human PTSD, the strength of this developmental approach enables us to capitalize on the relatively known brain development of the infant rat to identify causal mechanisms in brain circuits, including the amygdala, hippocampus, and mPFC as critical for brain processing.

Acknowledgments

This work was supported by NIH R37HD083217 (RMS) and NIH MH114931 (CKC).

References

1. Wang X et al (2020) Cortical volume abnormalities in posttraumatic stress disorder: an ENIGMA-psychiatric genomics consortium PTSD workgroup mega-analysis. Mol Psychiatry
2. Hancock L, Bryant RA (2020) Posttraumatic stress, stressor controllability, and avoidance. Behav Res Ther 128:103591
3. Koenen KC et al (2017) Posttraumatic stress disorder in the World Mental Health Surveys. Psychol Med 47(13):2260–2274
4. McLaughlin KA et al (2015) Subthreshold posttraumatic stress disorder in the world health organization world mental health surveys. Biol Psychiatry 77(4):375–384
5. Sareen J (2014) Posttraumatic stress disorder in adults: impact, comorbidity, risk factors, and treatment. Can J Psychiatr 59(9): 460–467
6. LeardMann CA et al (2021) Comparison of posttraumatic stress disorder checklist instruments from diagnostic and statistical manual of mental disorders, fourth edition vs fifth edition in a large cohort of US military service members and veterans. JAMA Netw Open 4(4):e218072
7. Shalev A, Liberzon I, Marmar C (2017) Posttraumatic stress disorder. N Engl J Med 376(25):2459–2469
8. Smotherman WP, Robinson SR (1985) The rat fetus in its environment: behavioral adjustments to novel, familiar, aversive, and conditioned stimuli presented in utero. Behav Neurosci 99(3):521–530
9. Rameckers SA et al (2021) The impact of childhood maltreatment on the severity of childhood-related posttraumatic stress disorder in adults. Child Abuse Negl 120:105208
10. Bolles RC, Woods PJ (1964) The ontogeny of behavior in the albino rat. Anim Behav 12: 427–441
11. Hoeboer C et al (2021) The effect of parental emotional abuse on the severity and treatment of PTSD symptoms in children and adolescents. Child Abuse Negl 111:104775
12. Bohus M et al (2020) Dialectical behavior therapy for posttraumatic stress disorder (DBT-PTSD) compared with cognitive processing therapy (CPT) in complex presentations of PTSD in women survivors of childhood abuse: a randomized clinical trial. JAMA Psychiatry 77(12):1235–1245
13. Poulos AM et al (2014) Amnesia for early life stress does not preclude the adult development of posttraumatic stress disorder symptoms in rats. Biol Psychiatry 76(4):306–314
14. Hoffman AN et al (2022) Anxiety, fear, panic: an approach to assessing the defensive

behavior system across the predatory imminence continuum. Learn Behav 50:339

15. Kang JI et al (2020) Effect of combat exposure and posttraumatic stress disorder on telomere length and amygdala volume. Biol Psychiatry Cogn Neurosci Neuroimaging 5(7):678–687

16. Ousdal OT et al (2020) The association of PTSD symptom severity with amygdala nuclei volumes in traumatized youths. Transl Psychiatry 10(1):288

17. Hilberdink CE et al (2021) Dysregulated functional brain connectivity in response to acute social-evaluative stress in adolescents with PTSD symptoms. Eur J Psychotraumatol 12(1):1880727

18. Cisler JM, Herringa RJ (2021) Posttraumatic stress disorder and the developing adolescent brain. Biol Psychiatry 89(2):144–151

19. Ross MC, Cisler JM (2020) Altered large-scale functional brain organization in posttraumatic stress disorder: a comprehensive review of univariate and network-level neurocircuitry models of PTSD. Neuroimage Clin 27:102319

20. Fitzgerald JM et al (2020) Multi-voxel pattern analysis of amygdala functional connectivity at rest predicts variability in posttraumatic stress severity. Brain Behav 10(8):e01707

21. King AP et al (2016) Dopamine receptor gene DRD4 7-repeat allele X maternal sensitivity interaction on child externalizing behavior problems: independent replication of effects at 18 months. PLoS One 11(8):e0160473

22. Sicorello M et al (2021) Differential effects of early adversity and posttraumatic stress disorder on amygdala reactivity: the role of developmental timing. Biol Psychiatry Cogn Neurosci Neuroimaging 6(11):1044–1051

23. Killion BE, Weyandt LL (2020) Brain structure in childhood maltreatment-related PTSD across the lifespan: a systematic review. Appl Neuropsychol Child 9(1):68–82

24. Scheeringa MS et al (2003) New findings on alternative criteria for PTSD in preschool children. J Am Acad Child Adolesc Psychiatry 42(5):561–570

25. Bennett RS et al (2021) A systematic review of controlled-trials for PTSD in maltreated children and adolescents. Child Maltreat 26(3):325–343

26. Creech SK, Misca G (2017) Parenting with PTSD: a review of research on the influence of PTSD on parent-child functioning in military and veteran families. Front Psychol 8:1101

27. Moner N et al (2021) Assessment of PTSD and posttraumatic symptomatology in very young children: a systematic review. J Child Adolesc Psychiatr Nurs

28. De Young AC, Landolt MA (2018) PTSD in children below the age of 6 years. Curr Psychiatry Rep 20(11):97

29. Yang J et al (2021) Using deep learning to classify pediatric posttraumatic stress disorder at the individual level. BMC Psychiatry 21(1):535

30. Shechner T et al (2014) Fear conditioning and extinction across development: evidence from human studies and animal models. Biol Psychol 100:1–12

31. Richardson H et al (2018) Development of the social brain from age three to twelve years. Nat Commun 9(1):1027

32. Silvers JA et al (2017) vlPFC-vmPFC-amygdala interactions underlie age-related differences in cognitive regulation of emotion. Cereb Cortex 27(7):3502–3514

33. Casey BJ et al (2019) Development of the emotional brain. Neurosci Lett 693:29–34

34. Somerville LH et al (2018) The lifespan human connectome project in development: a large-scale study of brain connectivity development in 5–21 year olds. NeuroImage 183:456–468

35. Uddin LQ (2021) Cognitive and behavioural flexibility: neural mechanisms and clinical considerations. Nat Rev Neurosci 22(3):167–179

36. Gordon A et al (2021) Long-term maturation of human cortical organoids matches key early postnatal transitions. Nat Neurosci 24(3):331–342

37. Russell JD et al (2021) Pediatric PTSD is characterized by age- and sex-related abnormalities in structural connectivity. Neuropsychopharmacology 46(12):2217–2223

38. Powell RA, Schmaltz RM (2021) Did Little Albert actually acquire a conditioned fear of furry animals? What the film evidence tells us. Hist Psychol 24(2):164–181

39. Treanor M, Rosenberg BM, Craske MG (2021) Pavlovian learning processes in pediatric anxiety disorders: a critical review. Biol Psychiatry 89(7):690–696

40. Tottenham N et al (2019) Parental presence switches avoidance to attraction learning in children. Nat Hum Behav 3(10):1070–1077

41. McLaughlin KA et al (2016) Maltreatment exposure, brain structure, and fear conditioning in children and adolescents. Neuropsychopharmacology 41(8):1956–1964

42. Lambert HK, McLaughlin KA (2019) Impaired hippocampus-dependent associative learning as a mechanism underlying PTSD: a meta-analysis. Neurosci Biobehav Rev 107:729–749
43. Marusak HA et al (2019) Pediatric cancer, posttraumatic stress and fear-related neural circuitry. Int J Hematol Oncol 8(2):IJH17
44. Garrett A et al (2019) Longitudinal changes in brain function associated with symptom improvement in youth with PTSD. J Psychiatr Res 114:161–169
45. Reinhard J et al (2021) Fear conditioning and stimulus generalization in association with age in children and adolescents. Eur Child Adolesc Psychiatry 31:1581
46. Jovanovic T et al (2020) Impact of ADCYAP1R1 genotype on longitudinal fear conditioning in children: interaction with trauma and sex. Neuropsychopharmacology 45(10):1603–1608
47. Newall C et al (2017) The relative effectiveness of extinction and counter-conditioning in diminishing children's fear. Behav Res Ther 95:42–49
48. Dvir M et al (2019) Fear conditioning and extinction in anxious and non-anxious youth: a meta-analysis. Behav Res Ther 120:103431
49. Chen FR, Raine A, Gao Y (2021) Reduced electrodermal fear conditioning and child callous-unemotional traits. Res Child Adolesc Psychopathol 49(4):459–469
50. Dubi K et al (2008) Maternal modeling and the acquisition of fear and avoidance in toddlers: influence of stimulus preparedness and child temperament. J Abnorm Child Psychol 36(4):499–512
51. VanElzakker MB et al (2014) From Pavlov to PTSD: the extinction of conditioned fear in rodents, humans, and anxiety disorders. Neurobiol Learn Mem 113:3–18
52. LeDoux JE, Pine DS (2016) Using neuroscience to help understand fear and anxiety: a two-system framework. Am J Psychiatry 173(11):1083–1093
53. Goswami S et al (2013) Animal models of post-traumatic stress disorder: face validity. Front Neurosci 7:89
54. Phelps EA, LeDoux JE (2005) Contributions of the amygdala to emotion processing: from animal models to human behavior. Neuron 48(2):175–187
55. Sullivan RM, Holman PJ (2010) Transitions in sensitive period attachment learning in infancy: the role of corticosterone. Neurosci Biobehav Rev 34(6):835–844
56. Callaghan B et al (2019) Using a developmental ecology framework to align fear neurobiology across species. Annu Rev Clin Psychol 15:345–369
57. Roth TL, Sullivan RM (2005) Memory of early maltreatment: neonatal behavioral and neural correlates of maternal maltreatment within the context of classical conditioning. Biol Psychiatry 57(8):823–831
58. Tallot L, Doyere V, Sullivan RM (2016) Developmental emergence of fear/threat learning: neurobiology, associations and timing. Genes Brain Behav 15(1):144–154
59. Hunt PS et al (2007) Synapses, circuits, and the ontogeny of learning. Dev Psychobiol 49(7):649–663
60. Sullivan RM et al (2000) Good memories of bad events in infancy. Nature 407(6800):38–39
61. Sullivan RM et al (1990) Modified behavioral and olfactory bulb responses to maternal odors in preweanling rats. Brain Res Dev Brain Res 53(2):243–247
62. Raineki C, Moriceau S, Sullivan RM (2010) Developing a neurobehavioral animal model of infant attachment to an abusive caregiver. Biol Psychiatry 67(12):1137–1145
63. Camp LL, Rudy JW (1988) Changes in the categorization of appetitive and aversive events during postnatal development of the rat. Dev Psychobiol 21(1):25–42
64. Sullivan RM et al (1994) Bilateral 6-OHDA lesions of the locus coeruleus impair associative olfactory learning in newborn rats. Brain Res 643(1–2):306–309
65. Sullivan RM, Wilson DA (1994) The locus coeruleus, norepinephrine, and memory in newborns. Brain Res Bull 35(5–6):467–472
66. Raineki C et al (2015) Paradoxical neurobehavioral rescue by memories of early-life abuse: the safety signal value of odors learned during abusive attachment. Neuropsychopharmacology 40(4):906–914
67. Raineki C, Rincón Cortés M, Sullivan RM (2010) Infant social behavior dysfunction predicts adolescent depressive-like behaviors, in prep
68. Rincón Cortés M, Barr GA, Sullivan RM (2013) Infant attachment associated with pain produces later-life depressive-like behavior that is rescued by maternal odor and intra-amygdala serotonin manipulation. Society for Neuroscience Abstracts, 425.33
69. Sullivan RM et al (2000) Association of an odor with activation of olfactory bulb noradrenergic beta-receptors or locus coeruleus stimulation is sufficient to produce learned

approach responses to that odor in neonatal rats. Behav Neurosci 114(5):957–962
70. McLean JH, Harley CW (2004) Olfactory learning in the rat pup: a model that may permit visualization of a mammalian memory trace. Neuroreport 15(11):1691–1697
71. Nakamura S, Sakaguchi T (1990) Development and plasticity of the locus coeruleus: a review of recent physiological and pharmacological experimentation. Prog Neurobiol 34(6):505–526
72. Nakamura S, Kimura F, Sakaguchi T (1987) Postnatal development of electrical activity in the locus ceruleus. J Neurophysiol 58(3): 510–524
73. Kimura F, Nakamura S (1985) Locus coeruleus neurons in the neonatal rat: electrical activity and responses to sensory stimulation. Brain Res 355(2):301–305
74. Sullivan RM, Wilson DA, Leon M (1989) Norepinephrine and learning-induced plasticity in infant rat olfactory system. J Neurosci 9(11):3998–4006
75. Sullivan RM (2001) Unique characteristics of neonatal classical conditioning: the role of the amygdala and locus Coeruleus. Integr Physiol Behav Sci 36(4):293–307
76. Yuan Q et al (2002) Optical imaging of odor preference memory in the rat olfactory bulb. J Neurophysiol 87(6):3156–3159
77. Langdon PE, Harley CW, McLean JH (1997) Increased beta adrenoceptor activation overcomes conditioned olfactory learning deficits induced by serotonin depletion. Brain Res Dev Brain Res 102(2):291–293
78. McLean JH, Shipley MT (1991) Postnatal development of the noradrenergic projection from locus coeruleus to the olfactory bulb in the rat. J Comp Neurol 304(3):467–477
79. Yuan Q (2009) Theta bursts in the olfactory nerve paired with beta-adrenoceptor activation induce calcium elevation in mitral cells: a mechanism for odor preference learning in the neonate rat. Learn Mem 16(11):676–681
80. Shakhawat AM, Harley CW, Yuan Q (2012) Olfactory bulb alpha2-adrenoceptor activation promotes rat pup odor-preference learning via a cAMP-independent mechanism. Learn Mem 19(11):499–502
81. Haroutunian V, Campbell BA (1979) Emergence of interoceptive and exteroceptive control of behavior in rats. Science 205(4409): 927–929
82. Miller JS, Molina JC, Spear NE (1990) Ontogenetic differences in the expression of odor-aversion learning in 4- and 8-day-old rats. Dev Psychobiol 23(4):319–330
83. Shionoya K et al (2006) Development switch in neural circuitry underlying odor-malaise learning. Learn Mem 13(6):801–808
84. Raineki C et al (2009) Ontogeny of odor-LiCl vs. odor-shock learning: similar behaviors but divergent ages of functional amygdala emergence. Learn Mem 16(2):114–121
85. Ehrlich DE et al (2013) Postnatal maturation of GABAergic transmission in the rat basolateral amygdala. J Neurophysiol 110(4): 926–941
86. Thompson JV, Sullivan RM, Wilson DA (2008) Developmental emergence of fear learning corresponds with changes in amygdala synaptic plasticity. Brain Res 1200:58–65
87. Junod A et al (2019) Development of threat expression following infant maltreatment: infant and adult enhancement but adolescent attenuation. Front Behav Neurosci 13(130)
88. Moriceau S et al (2004) Corticosterone controls the developmental emergence of fear and amygdala function to predator odors in infant rat pups. Int J Dev Neurosci 22(5–6): 415–422
89. Santiago AN et al (2018) Early life trauma increases threat response of peri-weaning rats, reduction of axo-somatic synapses formed by parvalbumin cells and perineuronal net in the basolateral nucleus of amygdala. J Comp Neurol 526(16):2647–2664
90. Debiec J, Sullivan RM (2014) Intergenerational transmission of emotional trauma through amygdala-dependent mother-to-infant transfer of specific fear. Proc Natl Acad Sci U S A 111(33):12222–12227
91. Debiec J, Olsson A (2017) Social fear learning: from animal models to human function. Trends Cogn Sci 21(7):546–555
92. Moriceau S et al (2006) Dual circuitry for odor-shock conditioning during infancy: corticosterone switches between fear and attraction via amygdala. J Neurosci 26(25): 6737–6748
93. Moriceau S, Sullivan RM (2006) Maternal presence serves as a switch between learning fear and attraction in infancy. Nat Neurosci 9(8):1004–1006
94. Sevelinges Y et al (2007) Enduring effects of infant memories: infant odor-shock conditioning attenuates amygdala activity and adult fear conditioning. Biol Psychiatry 62(10):1070–1079
95. Moriceau S, Roth TL, Sullivan RM (2010) Rodent model of infant attachment learning and stress. Dev Psychobiol 52(7):651–660

96. Stanton ME et al (2021) Mechanisms of context conditioning in the developing rat. Neurobiol Learn Mem 179:107388
97. Robinson-Drummer PA et al (2018) Age and experience dependent changes in Egr-1 expression during the ontogeny of the context preexposure facilitation effect (CPFE). Neurobiol Learn Mem 150:1–12
98. Alexandra Kredlow M et al (2022) Prefrontal cortex, amygdala, and threat processing: implications for PTSD. Neuropsychopharmacology 47(1):247–259
99. Robinson-Drummer PA et al (2019) Infant trauma alters social buffering of threat learning: emerging role of prefrontal cortex in preadolescence. Front Behav Neurosci 13:132
100. Li S, Kim JH, Richardson R (2012) Differential involvement of the medial prefrontal cortex in the expression of learned fear across development. Behav Neurosci 126(2):217–225
101. Kim JH, Hamlin AS, Richardson R (2009) Fear extinction across development: the involvement of the medial prefrontal cortex as assessed by temporary inactivation and immunohistochemistry. J Neurosci 29(35):10802–10808
102. Kim JH, Richardson R (2007) A developmental dissociation of context and GABA effects on extinguished fear in rats. Behav Neurosci 121(1):131–139
103. Raineki C et al (2010) Functional emergence of the hippocampus in context fear learning in infant rats. Hippocampus 20(9):1037–1046
104. Meyer HC, Lee FS (2019) Translating developmental neuroscience to understand risk for psychiatric disorders. Am J Psychiatry 176(3):179–185
105. Verbitsky A, Dopfel D, Zhang N (2020) Rodent models of post-traumatic stress disorder: behavioral assessment. Transl Psychiatry 10(1):132
106. Rudy JW (1993) Contextual conditioning and auditory cue conditioning dissociate during development. Behav Neurosci 107(5):887–891
107. Moye TB, Rudy JW (1985) Ontogenesis of learning: VI. Learned and unlearned responses to visual stimulation in the infant hooded rat. Dev Psychobiol 18(5):395–409
108. Barnet RC, Hunt PS (2006) The expression of fear-potentiated startle during development: integration of learning and response systems. Behav Neurosci 120(4):861–872
109. Yap CS, Stapinski L, Richardson R (2005) Behavioral expression of learned fear: updating of early memories. Behav Neurosci 119(6):1467–1476
110. Crow TM et al (2021) The roles of attachment and emotion dysregulation in the association between childhood maltreatment and PTSD in an inner-city sample. Child Abuse Negl 118:105139
111. Opendak M et al (2021) Bidirectional control of infant rat social behavior via dopaminergic innervation of the basolateral amygdala. Neuron 109:4018
112. Raineki C et al (2012) Effects of early-life abuse differ across development: infant social behavior deficits are followed by adolescent depressive-like behaviors mediated by the amygdala. J Neurosci 32(22):7758–7765
113. Perry RE et al (2019) Developing a neurobehavioral animal model of poverty: drawing cross-species connections between environments of scarcity-adversity, parenting quality, and infant outcome. Dev Psychopathol 31(2):399–418
114. Roth T, Sullivan R (2005) Memory of early maltreatment: neonatal behavioral and neural correlates of maternal maltreatment within the context of classical conditioning. Biol Psychiatry 57(8):823–831
115. Rincón Cortés RCM, Sullivan RM (2010) Infant social behavior dysfunction predicts adolescent depressive-like behaviors. Society for Neuroscience, San Diego
116. Raineki C et al (2019) During infant maltreatment, stress targets hippocampus, but stress with mother present targets amygdala and social behavior. Proc Natl Acad Sci U S A 116(45):22821–22832
117. Keller SM, Doherty TS, Roth TL (2019) Pharmacological manipulation of DNA methylation normalizes maternal behavior, DNA methylation, and gene expression in dams with a history of maltreatment. Sci Rep 9(1):10253
118. Walker CD et al (2017) Chronic early life stress induced by limited bedding and nesting (LBN) material in rodents: critical considerations of methodology, outcomes and translational potential. Stress:1–28
119. Baram TZ et al (2012) Fragmentation and unpredictability of early-life experience in mental disorders. Am J Psychiatry 169(9):907–915
120. Al Ain S et al (2015) Newborns prefer the odor of milk and nipples from females matched in lactation age: comparison of two mouse strains. Physiol Behav 147:122–130
121. Avishai-Eliner S et al (2002) Stressed-out, or in (utero)? Trends Neurosci 25(10):518–524

122. Gilles EE, Schultz L, Baram TZ (1996) Abnormal corticosterone regulation in an immature rat model of continuous chronic stress. Pediatr Neurol 15(2):114–119
123. Brenhouse HC, Bath KG (2019) Bundling the haystack to find the needle: challenges and opportunities in modeling risk and resilience following early life stress. Front Neuroendocrinol 54:100768
124. Guadagno A et al (2020) It is all in the right amygdala: increased synaptic plasticity and perineuronal nets in male, but not female, juvenile rat pups after exposure to early-life stress. J Neurosci 40(43):8276–8291
125. Sarro EC, Sullivan RM, Barr G (2014) Unpredictable neonatal stress enhances adult anxiety and alters amygdala gene expression related to serotonin and GABA. Neuroscience 258:147–161
126. Moriceau S, Sullivan RM (2004) Unique neural circuitry for neonatal olfactory learning. J Neurosci 24(5):1182–1189

Chapter 2

Periadolescent Social Isolation Effects on Extinction of Conditioned Fear

Katherine Drummond and Jee Hyun Kim

Abstract

Not all who are trauma-exposed develop posttraumatic stress disorder (PTSD). An important feature in its emergence is adverse early-life experiences, which should be represented in preclinical models of PTSD. Here we present post-weaning social isolation as a rodent model of early-life adversity that leads to impaired extinction of conditioned fear, which is associated with PTSD. This model has reasonable face validity, as early social experiences play a critical role in risks of PTSD. We describe and discuss the various nuances in the different methodologies currently employed, and suggest ways to maximize reliability and reproducibility. A main focus is to avoid creating a sensory deprivation condition that leads to schizophrenic rather than PTSD symptoms. The goal of this protocol is to provide the reader with guidelines on how to study early-life stress with results that are translatable to the human condition.

Key words Social isolation, Stress, Early-life adversity, Development, Extinction, Conditioned fear, Rat, Mice, Adolescence

1 Introduction: Early-Life Adversity and Posttraumatic Stress Disorder

Posttraumatic stress disorder (PTSD) is a chronic disorder that develops in some individuals after traumatic experience/s, characterized by a series of psychological dysfunctions including hyperarousal, fear, and intrusion of the traumatic memory even when the individual is in an objectively safe situation [1]. Preclinical modeling of PTSD aims to ultimately improve understanding of biobehavioral etiology of PTSD to identify potential targets for novel pharmacotherapies, and screen drugs for their use as PTSD treatment in humans. While there currently is no animal model that encapsulates all criteria of PTSD, there are animal models that do capture different features and symptoms of PTSD that provide important insights to achieve such translational aims.

Only a minority of individuals exposed to a traumatic event develop PTSD [2, 3], and an important feature in the development

PTSD is adverse early-life experiences. For example, a large retrospective cohort study of 56,082 participants from UK Biobank showed that childhood maltreatment is most strongly associated with PTSD compared to any other mental disorders [4]. When 1,277,546 Danish adolescents aged ~16 never diagnosed with PTSD were followed for >10 years in a longitudinal study, increasing levels of adversities were significantly more likely to result in PTSD [5]. Although other covariates including genetic predisposition contribute towards PTSD etiology [6, 7], childhood/adolescent adversity is a consistent risk variable that should be represented in preclinical models of PTSD.

Post-weaning social isolation is a rodent model of early-life adversity involving social deprivation. In respect to PTSD, this model has reasonable face validity, as early social experiences play a critical role in risks of PTSD [8]. For example, having interpersonal stress significantly increased PTSD diagnosis rate in refugee minors, even after accounting for age, gender, and residence status [9]. Conversely, social support and capital can alleviate posttraumatic stress symptoms in disaster or war-exposed adolescents and young adults [10, 11]. Social isolation is particularly relevant in contemporary psychiatry, considering that it is proposed as the reason for increased PTSD symptoms during the COVID-19 pandemic, especially in youths [12, 13]. Isolation can increase the propensity of developing PTSD-like behavioral and biological responses such as anxiety, hypervigilance, and dysfunction of the prefrontal cortex [14, 15]. Importantly, post-weaning social isolation impairs extinction of conditioned fear in rats [16], which is an emerging fear dysregulation phenotype of people with PTSD [17–20]. In typical fear extinction, a fear-eliciting stimulus is repeatedly presented in the absence of any aversive outcomes, ultimately reducing the fear response. Extinction is evolutionarily conserved between humans and rodents [21], and extinction studies in humans and rodents have driven our understanding of exposure therapies to treat PTSD [17, 22, 23]. Social isolation induced extinction deficits may explain why of all PTSD subtypes, persistent/recurrent PTSD subtype is the most highly associated with increasing number of childhood adversities [24], with extinction failures potentially leading to the relapse of PTSD.

The isolated environment is generally self-explanatory, with the subjects being maintained singly with no *physical* contact with other members of the same species. For the aim of observing fear dysregulation without causing any sensory deficits, it is crucial that rodents can hear and smell other animals. That is, isolation-rearing with sensory deprivation in housing such as in individually ventilated cages may cause schizophrenia-like phenotype [25]. Stimulation of any form is usually kept to a minimum with only routine maintenance to limit extra-cage excitation. The social environment consists of a standard laboratory cage containing two to six animals of the same sex, which resembles the standard colony situation in many research facilities. Although the cage size and number of animals may vary, the subjects, much like in the isolated condition, are given minimum extra-cage stimulation.

2 Materials

2.1 Animals

We use outbred Charles River Sprague–Dawley rats (ArcCrl:CD (SD)IGS, Animal Resources Centre, Australia) that are bred locally to ensure a controlled environment from birth. Breeding for the provision of experimental subjects involves adult males and nulliparous females paired together for 5–7 days, followed by the removal of males from the mating housing.

It is possible to purchase order rodents to a rodent facility at weaning age, typically around post-natal day 21 (P21). However, given the known stress of shipping and potential early weaning from the supplier across a variety of strains [26, 27], it is recommended that rats are bred and born onsite. It should also be noted that the majority of isolation experiments to date have used Sprague Dawley, Long Evans, or Wistar rat strains, but there have been limited attempts to examine strain differences (see [28]).

2.2 Housing Facility and Cages

Isolated and socially housed rats should be held together in a local ethics committee approved holding room to ensure similar sensory experience (i.e., shared visual, auditory, and olfactory cues). The holding room is typically maintained with luminance on a 12:12 light–dark cycle with lighting on 07:00–19:00. The temperature and humidity is typically regulated at 22–25 °C and 55–60%, respectively. Holding room temperature should be maintained at the higher end of the recommended range (e.g. 24–25 °C), as is typically recommended for pregnant, lactating rats and pups up to 4 weeks of age [29]. If different housing conditions cannot be housed together in the same holding room, temperature should be consistent across isolated vs. grouped housing, as there is no consensus on the effects of isolation on body temperature. It is possible that isolated rats may experience lower body temperature during sleep, especially at younger ages when they have not gained enough body fat [30]; however, that is not consistently observed [31]. In general, male and female rats are housed in separate rooms [32], in which case male and female holding rooms should be as similar as possible.

We use grouped or isolated environments within identical open-top wire lidded opaque plastic cages, with a solid bottom that is lined with aspen bedding, a paper towel and tunnel in the same holding room. The cage size, thickness of bedding, amount of paper towels and number of tunnels should be consistent between isolated and grouped housing. For different studies, we have used

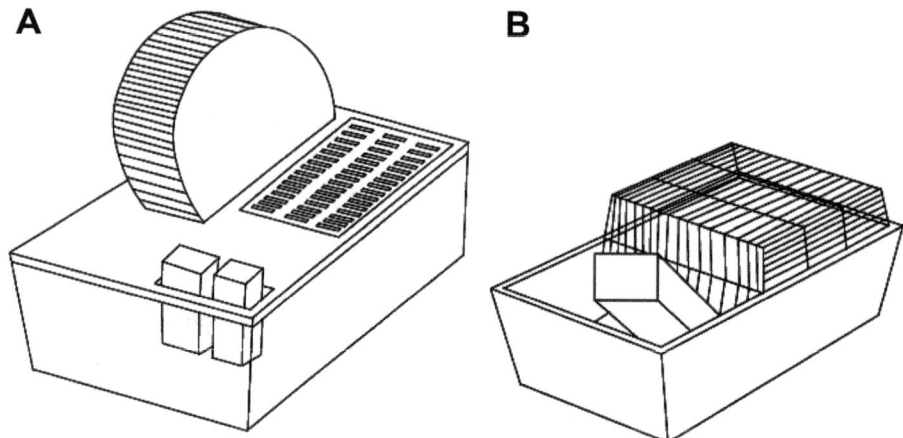

Fig. 1 (**a**) An open-top opaque plastic cage (50.8 × 40.64 × 20.96 cm, length × width × height) that contained a 35.56 cm diameter running wheel (Lafayette Instrument Company, Indiana, USA), which is locked for isolation-only groups [16]. (**b**) A standard open "high top" wire lidded opaque plastic cage (45 × 30 × 26 cm, length × width × height)

open-top cages fitting this description (Fig. 1) with grouped vs. isolation using identical cages with only rat number per cage as a variable (1 for isolated and 2–3 for grouped). While it is common to modify the size of the cage environment between singly and grouped housed rodents, sizes should be kept identical whenever possible. The size of the cage will depend on the chosen peer group size (or vice versa), as overcrowding stress should be avoided [33]. Only when the cages are kept identical between isolated and grouped housing we can infer that behavioral or brain differences are due to isolation rather than other aspects of the housing environment.

The types of cage utilized in post-weaning isolation studies vary across laboratories (For examples see: [34, 35]). Open-top housing is the most widely used cage environment. This cage allows for olfactory and auditory contact with other rodents. Typically, sidewalls are solid and opaque, so rodents cannot make visual contact with each other. They also allow rodents to dangle from the lid, providing a degree of exercise or physical stimulation, as well as visual contact. We strongly recommend similar open-top wire lidded solid bottom plastic cages to study isolation-induced PTSD symptoms compared to open-top barren, hanging, or sound-proof cages that have been used more historically. Specifically, hanging cages (see description in [36]) or specially designed cages with floors made of wire bars or mesh with an underhanging tray for the collection of urine and feces are not recommended. They likely are mildly stressful housing conditions, typically with limited bedding and nesting materials and restrictive size, likely reducing the opportunity for physical activity and resulting in locomotor and sensory dysfunction [36, 37]. We also do not

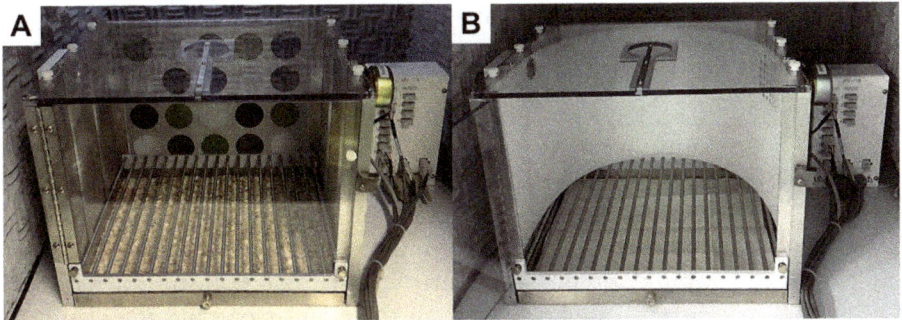

Fig. 2 Fear conditioning chambers. Contexts differed in terms of lighting, odor, wall shape/pattern, and bedding. (**a**) Context A had stainless steel side walls and a flat opaque rear wall with round stickers. A tray of aspen bedding was placed beneath the flooring. A white house light was always on. Chambers were cleaned with a eucalyptus-scented agent. (**b**) Context B had a curved wall made of white Perspex®. A tray lined with two paper towels was placed beneath the flooring. A white stimulus light was always on. Chambers were cleaned with 80% ethanol

recommend individually ventilated cages, which severely limit sensory cues, especially for the isolated rodents. Such cages would be targeting sensory deprivation over and above social isolation. Accordingly, we also do not recommend sound-proofing cages (e.g., [38]) or increasing the distance or spacing between cages that limit auditory, olfactory, and visual contact with other rodents (e.g., [39]).

2.3 Conditioned Fear Behavioral Apparatus

We use VideoFreeze® System (Med Associates, VT, USA) chambers ($31.8 \times 25.4 \times 26.7$ cm) for test acquisition and extinction of conditioned fear. Each chamber is enclosed in a sound attenuating box with a near infrared (NIR) light source. A high frequency speaker is located on the right wall of the chamber to deliver discrete auditory stimuli. The chamber floor comprises of stainless steel rods (19 rods, 0.5 cm in diameter and spaced 1.6 cm apart) that can deliver electric shocks. VideoFreeze® System is used to program the delivery of the auditory and shock stimuli, and to record all freezing. Two distinct contexts differing in visual, olfactory, and tactile experience located in separate rooms (Fig. 2) are typically used to differentiate conditioning context- vs. cue-induced fear behaviors.

3 Methods

3.1 Pre- and Post-weaning

It is ideal that pre-weaning environment are accounted for in experimental design or analysis, as they may confound social isolation effects. Daily inspection of female breeders for the birth of litters was conducted at 09:00–10:00 a.m. from 3 weeks since mating. Day of birth of experimental subjects is assigned as P0.

At P2, litters are sexed and culled to 12 pups [6±1 of each sex (50:50 sex ratio)] and weighed in our studies. The litter size should be determined primarily on local ethics regulations based on cage size. Pups remain co-housed with their dam and siblings from birth until weaning (P20-22). Animal husbandry standards typically require a cage change every 7–10 days, we thus recommend timing cage changes to ~P2 (standardizing litter size), ~P12, ~P21 (weaning) to minimize disruption. All cages should be provided with abundant bedding and nesting material and to reduce stress of cage change, old nesting material or partial cage change could be considered.

Experiment begins at weaning (range: P20-22) with rehousing of all cages. This is to include the purported sensitive period for social isolation (weaning to 60) for anxiety-related behaviors [40–42]. All rats are weighed at weaning to inform treatment allocation (i.e., groups should be weight-matched). We typically exclude rats that weigh more or less than 15% from their littermate average. Exclusions based on this criterion are very rare (<1 rat every 10 litters from our experience), which may be due to the standardization of litter size at P2.

Social housing at weaning typically groups rodents of the same sex, putatively to avoid unknown effects of direct sexual interactions during maturation [43]. We recommend group housing rodents together from different litters (i.e., unfamiliar peers rather than siblings or pre-weaning littermates) to ensure each group is counterbalanced for pre-weaning background [44]. Experiments on fear and anxiety-related behaviors also commonly group rats from different litters [45, 46]. Experimental design and planned statistical analyses should be carefully considered in this process.

3.2 Treatment Allocation

Treatment allocation is semi-random to ensure, at the time of weaning, that experimental subjects are comparable between conditions from the standpoint of their weights and apparent health. Isolation vs. grouped housing should account for distributions of body weights among treatment groups. Importantly, treatment allocation should ensure that rats from the same litter (dam) are evenly divided across rearing conditions to avoid litter effects. Ideally, no more than 1 rat per litter should be allocated per experimental group [44].

3.3 Handling, Weighing, and Cage Cleaning

Isolated and grouped rodents should be minimally handled. In general, rodents should be handled and weighed no more than once a week, given reports that more frequent handling nullifies social isolation effects [36, 47]. Therefore, a once-weekly handling session (~1 min) should coincide with weighing, cage cleaning/changing, and tail marking with an indelible pen. In our experience, such weekly handling is sufficient to minimize acute effects of handling on fear behavioral testing. We recommend two experimenters to be involved in handling to facilitate blinding (see Subheading 3.6 Blinding). With cleaning, some studies intentionally reduce the effect of environmental novelty, for instance, by providing handful of old bedding, or only partly replacing cage components (e.g. retaining cage lid for the duration of experimentation) [48]. We replace the cages with new clean cages and observed consistent effects of social isolation [16].

A minimal handling procedure [47] typically lasts 1 min, during which the rat is removed from its cage by its body and placed on the lap of the experimenter or on the top of a table. The back is rubbed by the gloved fingers of the experimenter and then the rat is returned to its cage. This handling procedure differs from other procedures, as it occurs during postweaning development, rather than consecutively in the days prior to behavioural testing (e.g. four, once-daily, handling sessions), which may result in differing effects between rearing conditions [49].

3.4 Isolation Duration

Standard isolation protocols vary isolation length from 2 to 8 weeks [50]. We use 4 weeks isolation duration because it has been shown to produce consistent changes in anxiety-related behaviors [40], and we are interested in behavioral testing primarily during adolescence [16, 51, 52]. It is likely that isolation may need to be continuous and chronic (>2 weeks) to robustly change behaviors (see discussions in [14, 15]). For this reason, and to avoid any acute effects of re-socialization, we recommend keeping the isolated rats isolated during behavioral testing, unless observing behavior after chronic re-socialization is the focus of the study.

3.5 Re-socialization

Some studies may require re-socialization following social isolation. We re-socialize isolated rats into the same group housed conditioned as described above. Notably, it is critical that rats that were always in grouped condition are re-housed into different groups when isolated rats are re-socialized so that all rats similarly undergo disrupted social hierarchy (as has been done previously [53, 54]).

We typically keep re-socialization duration at 4 weeks to match 4 weeks of preadolescent social isolation, and still observed persistent deficits in extinction of conditioned fear (Drummond et al. unpublished observations). Other studies have reported 2 weeks [53], 3 weeks [55], or longer durations [56–58]. Repeated re-socialization may be stressful due to reformations in social hierarchy and is not recommended. There is evidence that re-socialization with previously socially housed rats impacts brain and behavior differently to re-socialization with isolated rats [59]. A more commonly reported methods suggest that isolate-reared rodents should be housed together, which may reduce social defeat that may occur disproportionately high in previously isolated rodents [30, 59, 60].

3.6 Blinding and Counterbalancing

Blinding promotes comparable handling and treatment of the animals by experimenters. As rats are often tested when housed in different environments, it is critical that an experimenter assists the experimenter blinded to housing condition run the behavioral tests. Order of testing and chamber allocation should be counterbalanced between groups to produce the most reliable and replicable data.

3.7 Vaginal Lavage

Consider changing laboratory coats between any male and female rodent handling. If experimenters are lavaging female rodents, we suggest the following:

(i) at least one lavage should occur the day prior to behavioral testing.

(ii) all male subjects should be handled identically to control the handling of females related to lavaging.

(iii) During behavioral testing days, lavaging should occur after fear behavior but not within 1 h after testing, as the potential stress of the procedure may interfere with memory consolidation.

3.8 Standard Conditioned Fear Extinction Protocol

3.8.1 Habituation and Light Cycle

Rodents are typically brought into the behavioral testing area well before the commencement of testing. The timing varies between laboratories, especially with anecdotal evidence suggesting that rats require less time than mice. Within the same laboratory, each cohort should use the same habituation duration. We typically use 10–20 min.

We have pilot data that conditioned fear tested in the light phase produces much more consistent results (Zbukvic et al. unpublished observations), hence we recommend testing during the light phase. All cohorts should aim to begin testing at similar times of the day to minimize circadian rhythm effects.

3.8.2 Fear Conditioning, Extinction, and Test

Extinction of conditioned fear is one of the most extensively studied and replicated paradigm in the world and the present protocol cannot capture the entire range and justifications for all the differences reported in the literature, hence we describe here our typical protocol.

For acquisition of conditioned fear, rodents are placed in a novel chamber and after a 2 min baseline period were presented with six 10 s tone cue (5000 Hz, 80 dB) that co-terminates with a 1 s foot-shock (0.6–1.0 mA). Inter-trial interval (ITI) ranges from 85 to 135 s with a mean of 110 s. We have used 3–6 pairings depending on the age and sex of the rodents.

The next day, extinction involves rodents being placed in a novel context and after a 2 min baseline period, 10 s tone cue (5000 Hz, 80 dB) is presented 30–90 times with 10 s ITI. Extinction memory is then tested the next day with either the same protocol as the extinction protocol (often with reduced number of tone presentations), or with a blocked presentation of the tone cue (120 s) following a 2 min baseline period. Extinction protocol is used as a test when high amount of spontaneous recovery is expected at test (e.g., assessing adolescent or aging rodents [16, 61, 62]). The blocked presentation is often used when the within-session extinction the previous day was particularly effective (i.e., very low fear behavior shown at the end of extinction).

3.8.3 Behavioral Measure of Fear

Freezing is calculated via automated near-infrared video tracking provided by VideoFreeze® System (Med Associates) with a motion threshold 50 from 30 frames (1 s). These parameters are validated within our laboratory to most closely match the manual scoring by observers blind to any treatment conditions. We validate for every new study, something we recommend in other laboratories so that scoring sleeping as freezing can be avoided.

4 Conclusions

The aforementioned housing conditions are critical for researchers to consider to socially deprive the rodents while keeping most of the sensory experience intact. While the lack of stimulation arising from direct contact with conspecifics would be consistent with what isolated humans may experience, having a severely reduced movement and sensory experience of any kind that may be caused by hanging or individually ventilated cages would not represent an average isolated human experience.

Schrijver et al. (2002) shows that systematic variation of both social and inanimate background might help distinguish between robust results that are generalizable vs. idiosyncratic across different environmental conditions. Considering that sensory and social deprivation during development may indeed induce schizophrenia-relevant symptoms [25], translational psychiatry research may indeed benefit from systematic variation of social and/or inanimate stimulation to study the most appropriate psychiatric illnesses in animal models.

The potential effects of behavioral testing during isolation is worth discussing. Lukkes et al. (2009) reported previous isolation elevates freezing to a tone when rodents are no longer in isolation. This effect is not seen if the rodents remain in isolation [16, 32]. Together, the findings suggest that a current state of isolation is not driving the changes observed in conditioned fear behavior. In any case, re-socialization can lead to social defeat stress, hence we recommend chronic re-socialization before any behaviors are tested.

A limitation of the present protocol is that female adolescents appear to be more resilient against isolation effects on extinction of conditioned fear [16]. Interestingly, only in isolation females did exercise impair extinction recall [16], highlighting that isolation does increase vulnerability in females to PTSD-related behaviors. Effects of running have not been observed in group-housed females [16, 63]. There is speculation that unconstrained wheel running could be stress-provoking in female rats [64], supporting the notion that exercise in isolation summated to cause stress-induced impaired extinction.

Lastly, genetic differences are likely to affect susceptibility to isolation effects (gene-environment interactions). With the increasing investigation on the nature and permanence of isolation effects, future genetic predisposition studies may be well placed to identification of subsection of population more susceptible to these stress effects, which may further the translational value of animal models to understand PTSD in humans.

References

1. American Psychiatric Association, American Psychiatric Association, DSM-5 Task Force (2013) Diagnostic and statistical manual of mental disorders: DSM-5, 5th edn. American Psychiatric Association, Washington, DC
2. Chung MC, AlQarni N, AlMazrouei M, Al Muhairi S, Shakra M, Mitchell B, Al Mazrouei S, Al Hashimi S (2021) Posttraumatic stress disorder and psychiatric co-morbidity among Syrian refugees: the role of trauma exposure, trauma centrality, self-efficacy and emotional suppression. J Ment Health 30(6):681–689
3. Wiseman C, Croft J, Zammit S (2021) Examining the relationship between early childhood temperament, trauma, and post-traumatic stress disorder. J Psychiatr Res 144:427–433
4. Macpherson JM, Gray SR, Ip P, McCallum M, Hanlon P, Welsh P, Chan KL, Mair FS, Celis-Morales C, Minnis H et al (2021) Child maltreatment and incident mental disorders in middle and older ages: a retrospective UK Biobank cohort study. Lancet Reg Health Eur 11: 100224
5. El-Khoury F, Rieckmann A, Bengtsson J, Melchior M, Rod NH (2021) Childhood adversity trajectories and PTSD in young adulthood: a nationwide Danish register-based cohort study of more than one million individuals. J Psychiatr Res 136:274–280
6. Maihofer AX, Choi KW, Coleman JRI, Daskalakis NP, Denckla CA, Ketema E, Morey RA, Polimanti R, Ratanatharathorn A, Torres K et al (2022) Enhancing Discovery of Genetic Variants for Posttraumatic Stress Disorder Through Integration of Quantitative Phenotypes and Trauma Exposure Information. Biol Psychiatry 91(7):626–636
7. Duncan LE, Ratanatharathorn A, Aiello AE, Almli LM, Amstadter AB, Ashley-Koch AE, Baker DG, Beckham JC, Bierut LJ, Bisson J et al (2018) Largest GWAS of PTSD (N=20 070) yields genetic overlap with schizophrenia and sex differences in heritability. Mol Psychiatry 23(3):666–673
8. Koenen KC, Moffitt TE, Poulton R, Martin J, Caspi A (2007) Early childhood factors associated with the development of post-traumatic stress disorder: results from a longitudinal birth cohort. Psychol Med 37(2):181–192
9. Veeser J, Barkmann C, Schumacher L, Zindler A, Schon G, Barthel D (2023) Post-traumatic stress disorder in refugee minors in an outpatient care center: prevalence and associated factors. Eur Child Adolesc Psychiatry 32(3):419–426
10. Yuan G, Shi W, Lowe S, Chang K, Jackson T, Hall BJ (2021) Associations between posttraumatic stress symptoms, perceived social support and psychological distress among disaster-exposed Chinese young adults: a three-wave longitudinal mediation model. J Psychiatr Res 137:491–497
11. Duren R, Yalcin O (2021) Social capital and mental health problems among Syrian refugee adolescents: the mediating roles of perceived

social support and post-traumatic symptoms. Int J Soc Psychiatry 67:243–250
12. Mensi MM, Capone L, Rogantini C, Orlandi M, Ballante E, Borgatti R (2021) COVID-19-related psychiatric impact on Italian adolescent population: a cross-sectional cohort study. J Community Psychol 49(5): 1457–1469
13. Rohde C, Jefsen OH, Norremark B, Danielsen AA, Ostergaard SD (2020) Psychiatric symptoms related to the COVID-19 pandemic. Acta Neuropsychiatr 32(5):274–276
14. Lukkes JL, Watt MJ, Lowry CA, Forster GL (2009) Consequences of post-weaning social isolation on anxiety behavior and related neural circuits in rodents. Front Behav Neurosci 3:18
15. Fone KC, Porkess MV (2008) Behavioural and neurochemical effects of post-weaning social isolation in rodents-relevance to developmental neuropsychiatric disorders. Neurosci Biobehav Rev 32(6):1087–1102
16. Drummond KD, Waring ML, Faulkner GJ, Blewitt ME, Perry CJ, Kim JH (2021) Hippocampal neurogenesis mediates sex-specific effects of social isolation and exercise on fear extinction in adolescence. Neurobiol Stress 15: 100367
17. Alexandra Kredlow M, Fenster RJ, Laurent ES, Ressler KJ, Phelps EA (2022) Prefrontal cortex, amygdala, and threat processing: implications for PTSD. Neuropsychopharmacology 47(1):247–259
18. Rougemont-Bücking A, Linnman C, Zeffiro TA, Zeidan MA, Lebron-Milad K, Rodriguez-Romaguera J, Rauch SL, Pitman RK, Milad MR (2011) Altered processing of contextual information during fear extinction in PTSD: an fMRI study. CNS Neurosci Ther 17(4): 227–236
19. Milad MR, Orr SP, Lasko NB, Chang Y, Rauch SL, Pitman RK (2008) Presence and acquired origin of reduced recall for fear extinction in PTSD: results of a twin study. J Psychiatr Res 42(7):515–520
20. Orr SP, Milad MR, Metzger LJ, Lasko NB, Gilbertson MW, Pitman RK (2006) Effects of beta blockade, PTSD diagnosis, and explicit threat on the extinction and retention of an aversively conditioned response. Biol Psychol 73(3):262–271
21. Sevenster D, Visser RM, D'Hooge R (2018) A translational perspective on neural circuits of fear extinction: current promises and challenges. Neurobiol Learn Mem 155:113–126
22. Ganella DE, Kim JH (2014) Developmental rodent models of fear and anxiety: from neurobiology to pharmacology. Br J Pharmacol 171(20):4556–4574
23. Maren S, Phan KL, Liberzon I (2013) The contextual brain: implications for fear conditioning, extinction and psychopathology. Nat Rev Neurosci 14(6):417–428
24. Mota N, Bolton SL, Enns MW, Afifi TO, El-Gabalawy R, Sommer JL, Pietrzak RH, Stein MB, Asmundson GJG, Sareen J (2021) Course and predictors of posttraumatic stress disorder in the Canadian Armed Forces: a nationally representative, 16-year follow-up study: Cours et predicteurs du trouble de stress post-traumatique dans les Forces armees canadiennes: une etude de suivi de 16 ans nationalement representative. Can J Psychiatr 66(11): 982–995
25. Jones CA, Watson DJ, Fone KC (2011) Animal models of schizophrenia. Br J Pharmacol 164(4):1162–1194
26. Mogi K, Nagasawa M, Kikusui T (2011) Developmental consequences and biological significance of mother-infant bonding. Prog Neuro-Psychopharmacol Biol Psychiatry 35(5):1232–1241
27. Kikusui T, Mori Y (2009) Behavioural and neurochemical consequences of early weaning in rodents. J Neuroendocrinol 21(4):427–431
28. Gentsch C, Lichtsteiner M, Feer H (1981) Locomotor activity, defecation score and corticosterone levels during an openfield exposure: a comparison among individually and group-housed rats, and genetically selected rat lines. Physiol Behav 27(1):183–186
29. National Health and Medical Research Council (2013) Australian code for the care and use of animals for scientific purposes. National Health and Medical Research Council, Canberra
30. Whitaker LR, Degoulet M, Morikawa H (2013) Social deprivation enhances VTA synaptic plasticity and drug-induced contextual learning. Neuron 77(2):335–345
31. Carnevali L, Mastorci F, Graiani G, Razzoli M, Trombini M, Pico-Alfonso MA, Arban R, Grippo AJ, Quaini F, Sgoifo A (2012) Social defeat and isolation induce clear signs of a depression-like state, but modest cardiac alterations in wild-type rats. Physiol Behav 106(2): 142–150
32. Weiss IC, Pryce CR, Jongen-Rêlo AL, Nanz-Bahr NI, Feldon J (2004) Effect of social isolation on stress-related behavioural and neuroendocrine state in the rat. Behav Brain Res 152(2):279–295
33. Baranyi J, Bakos N, Haller J (2005) Social instability in female rats: the relationship

34. Rosenzweig MR, Bennett EL, Diamond MC (1972) Cerebral effects of differential experience in hypophysectomized rats. J Comp Physiol Psychol 79(1):56

35. Menich SR, Baron A (1984) Social housing of rats: life-span effects on reaction time, exploration, weight, and longevity. Exp Aging Res 10(2):95–100

36. Holson RR, Scallet AC, Ali SF, Turner BB (1991) "Isolation stress" revisited: isolation-rearing effects depend on animal care methods. Physiol Behav 49:1107–1118

37. Heidbreder CA, Weiss IC, Domeney AM, Pryce C, Homberg J, Hedou G, Feldon J, Moran MC, Nelson P (2000) Behavioral, neurochemical and endocrinological characterization of the early social isolation syndrome. Neuroscience 100(4):749–768

38. Greco AM, Gambardella P, Sticchi R, D'Aponte D, Di Renzo G, De Franciscis P (1989) Effects of individual housing on circadian rhythms of adult rats. Physiol Behav 45(2):363–366

39. Sherif F, Oreland L (1996) Effect of the GABA-transaminase inhibitor vigabatrin on exploratory behaviour in socially isolated rats. Behav Brain Res 72:135–140

40. Burke AR, McCormick CM, Pellis SM, Lukkes JL (2017) Impact of adolescent social experiences on behavior and neural circuits implicated in mental illnesses. Neurosci Biobehav Rev 76(Pt B):280–300

41. Arakawa H (2007) Ontogenetic interaction between social relationships and defensive burying behavior in the rat. Physiol Behav 90(5):751–759

42. Robbins TW, Jones GH, Wilkinson LS (1996) Behavioural and neurochemical effects of early social deprivation in the rat. J Psychopharmacol 10(1):39–47

43. Rivest RW (1991) Sexual maturation in female rats: hereditary, developmental and environmental aspects. Experientia 47(10): 1027–1038

44. Denenberg VH (1984) Some statistical and experimental considerations in the use of the analysis-of-variance procedure. Am J Phys Regul Integr Comp Phys 246:R403–R408

45. Lukkes JL, Burke AR, Zelin NS, Hale MW, Lowry CA (2012) Post-weaning social isolation attenuates c-Fos expression in GABAergic interneurons in the basolateral amygdala of adult female rats. Physiol Behav 107(5): 719–725

46. Yusufishaq S, Rosenkranz JA (2013) Post-weaning social isolation impairs observational fear conditioning. Behav Brain Res 242:142–149

47. Reboucas RC, Schmidek WR (1997) Handling and isolation in three strains of rats affect open field, exploration, hoarding and predation. Physiol Behav 62(5):1159–1164

48. Schrijver NC, Bahr NI, Weiss IC, Wurbel H (2002) Dissociable effects of isolation rearing and environmental enrichment on exploration, spatial learning and HPA activity in adult rats. Pharmacol Biochem Behav 73(1):209–224

49. Pritchard LM, Van Kempen TA, Zimmerberg B (2013) Behavioral effects of repeated handling differ in rats reared in social isolation and environmental enrichment. Neurosci Lett 536:47–51

50. Walker DM, Cunningham AM, Gregory JK, Nestler EJ (2019) Long-term behavioral effects of post-weaning social isolation in males and females. Front Behav Neurosci 13: 66

51. Cullity ER, Guérin AA, Madsen HB, Perry CJ, Kim JH (2021) Insular cortex dopamine 1 and 2 receptors in methamphetamine conditioned place preference and aversion: age and sex differences. Neuroanatomy Behav 3:e24

52. Cullity ER, Guerin AA, Perry CJ, Kim JH (2021) Examining sex differences in conditioned place preference or aversion to methamphetamine in adolescent and adult mice. Front Pharmacol 12:770614

53. Lukkes JL, Mokin MV, Scholl JL, Forster GL (2009) Adult rats exposed to early-life social isolation exhibit increased anxiety and conditioned fear behavior, and altered hormonal stress responses. Horm Behav 55(1): 248–256

54. Lukkes J, Vuong S, Scholl J, Oliver H, Forster G (2009) Corticotropin-releasing factor receptor antagonism within the dorsal raphe nucleus reduces social anxiety-like behavior after early-life social isolation. J Neurosci 29(32): 9955–9960

55. Tulogdi A, Toth M, Barsvari B, Biro L, Mikics E, Haller J (2014) Effects of resocialization on post-weaning social isolation-induced abnormal aggression and social deficits in rats. Dev Psychobiol 56(1):49–57

56. Arakawa H (2003) The effects of isolation rearing on open-field behavior in male rats depends on developmental stages. Dev Psychobiol 43(1):11–19

57. Weintraub A, Singaravelu J, Bhatnagar S (2010) Enduring and sex-specific effects of

adolescent social isolation in rats on adult stress reactivity. Brain Res 1343:83–92

58. Chen W, An D, Xu H, Cheng X, Wang S, Yu W, Yu D, Zhao D, Sun Y, Deng W et al (2016) Effects of social isolation and re-socialization on cognition and ADAR1 (p110) expression in mice. PeerJ 4:e2306

59. Makinodan M, Ikawa D, Yamamuro K, Yamashita Y, Toritsuka M, Kimoto S, Yamauchi T, Okumura K, Komori T, Fukami SI et al (2017) Effects of the mode of re-socialization after juvenile social isolation on medial prefrontal cortex myelination and function. Sci Rep 7(1):5481

60. Liu J, Dietz K, DeLoyht JM, Pedre X, Kelkar D, Kaur J, Vialou V, Lobo MK, Dietz DM, Nestler EJ et al (2012) Impaired adult myelination in the prefrontal cortex of socially isolated mice. Nat Neurosci 15(12): 1621–1623

61. Short AK, Bui V, Zbukvic IC, Hannan AJ, Pang TY, Kim JH (2022) Sex-dependent effects of chronic exercise on cognitive flexibility but not hippocampal Bdnf in aging mice. Neuronal Sig 6(1):NS20210053

62. Perry CJ, Ganella DE, Nguyen LD, Du X, Drummond KD, Whittle S, Pang TY, Kim JH (2020) Assessment of conditioned fear extinction in male and female adolescent rats. Psychoneuroendocrinology 116:104670

63. Pietropaolo S, Feldon J, Alleva E, Cirulli F, Yee BK (2006) The role of voluntary exercise in enriched rearing: a behavioral analysis. Behav Neurosci 120(4):787–803

64. James MH, Campbell EJ, Walker FR, Smith DW, Richardson HN, Hodgson DM, Dayas CV (2014) Exercise reverses the effects of early life stress on orexin cell reactivity in male but not female rats. Front Behav Neurosci 8(18):244

Chapter 3

Using Virtual Reality to Study Fear and Extinction in Children and Adolescents

Hilary A. Marusak, Craig Peters, and Christine A. Rabinak

Abstract

Fear learning and extinction paradigms have proven to be robust translational tools to study the development, maintenance, and treatment of fear-based disorders, including anxiety and posttraumatic stress disorder (PTSD). However, these disorders frequently begin during childhood and adolescence, which requires a developmental approach and the use of age-appropriate tools to study fear extinction in developmental populations. Here, we detail a novel virtual reality paradigm we developed and validated to study fear learning and extinction in children and adolescents. Necessary equipment and software, considerations, and guidelines are also provided.

Key words Fear conditioning, Anxiety disorders, Posttraumatic stress disorder, Children, Adolescents, Fear extinction, Conditioned stimulus

1 Introduction

Fear-based disorders, including anxiety and posttraumatic stress disorder (PTSD), are characterized by a failure to appropriately inhibit or extinguish fear [1]. Fear extinction has been extensively described in humans and animal models [2] using Pavlovian fear extinction paradigms, which involve pairing an innocuous cue (conditioned stimulus; CS) with an aversive outcome (unconditioned stimulus, US). After repeated CS-US pairings, the CS begins to elicit conditioned fear responses. After repeated presentations of the CS in the absence of the US, however, a new memory is formed that competes with the fear memory trace (i.e., extinction [3]). Convergent research by our group and others in adults with anxiety and PTSD and animal models of anxiety [4–8] points to deficits in the ability to *recall* the extinction memory, causing inappropriate and persistent fear expression. These deficits appear to normalize following treatment [9]. Given the evolutionary conservation across species, fear extinction paradigms have emerged as a robust behavioral assay used in translational studies of the pathophysiology of fear-based disorders. In addition, the main evidence-based treatment for fear-based disorders in adults and in youth (i.e., cognitive behavioral therapy [10–12]) relies on principles of extinction learning. Thus, fear extinction paradigms have led to the development of novel treatment approaches for fear-based disorders [13].

Most of the research works on fear conditioning have focused on adults and have thus led to the development of therapies that target mature neural circuitry. However, anxiety and PTSD frequently emerge during adolescence [14–16], which requires a developmental approach to our understanding of fear-based disorders and more effective treatments and preventive interventions. Further, because existing therapies and medications were developed for adults, a comparative lack of knowledge about the development of fear extinction may limit successful treatment outcomes in youth [17]. Indeed, almost 50% of youth do not reach full recovery following current treatments [18]. As compared to adult-onset psychopathology, adolescent-onset psychopathology is associated with poorer outcomes (e.g., greater suicide risk, poorer treatment response) and increased risk of developing another psychiatric disorder later in life (e.g., depression) [19–22]. Therefore, there is a need for more research and age-appropriate tools to study fear extinction and risk of fear-based disorders in developmental populations.

This chapter describes a novel virtual reality paradigm we developed to study fear learning and extinction in children and adolescents [23]. This protocol occurs over the course of two consecutive days, and was adapted from a well-vetted contextual Pavlovian fear extinction paradigm widely used in adults [4, 24]. Our adapted paradigm was designed to be (1) more life-like and engaging for children, (2) age-appropriate, and (3) capture a variety of indicators of conditioned fear that are not commonly measured in laboratory-based paradigms (e.g., behavior). In particular, our virtual reality paradigm utilizes virtual environments and human avatars as stimuli, to model real-world events that children are likely to experience and are linked to psychopathology (e.g., interpersonal violence, abuse) [25, 26]. For the US, we selected an aversive white noise burst as a robust but well-tolerated stimulus, to minimize dropout in young populations. Dependent measures of conditioned fear include subjective, physiological, and behavioral measures, which reflect different aspects of fear and may map onto different symptom dimensions (e.g., hyperarousal, avoidance). Our paradigm is available upon request in various formats, including in 3D for use in VR and 2D for a computer/laptop or in the MRI scan environment, at www.wsuthinklab.com. Our group has also developed analogous 2D and 3D VR paradigms for adults, including healthy adults and adults with anxiety or PTSD. These paradigms are available upon request at www.tnp2lab.org.

2 Equipment, Materials, and Setup

2.1 Participants

We have utilized this fear conditioning and extinction protocol in children as young as 6 years of age. This includes healthy children, children with histories of trauma exposure, and children with high anxiety and/or PTSD symptoms.

2.2 Equipment and Software

An equipment and materials list is provided in Table 1. To administer the paradigm in the VR environment, a VR head-mounted display (e.g., Vive Cosmos) is required. The experimental computer should also be a high-performance rendering PC with a VR-capable graphics card, such as a gaming PC, though a less-intensive 2D version can be run on a regular PC or in the MR scan environment. If physiology is a desired dependent measure, equipment for recording physiology will be needed, along with a recording computer and software (e.g., BIOPAC's MP160 system and AcqKnowledge software). Development software is required to build your own task, and we use Vizard (WorldViz, Inc.), a python-based VR development platform for researchers. The experiment can be exported from Vizard as an executable file to be run on any PC.

2.3 Practice Session

All children are acclimated to the VR environment with a practice session. First, participants are seated in a chair and fitted with the VR headset. Ensure that the participant can view the full environment and hear the audio clearly. This practice session occurs in a separate virtual context from the experimental contexts. In our VR paradigm, the practice session occurs in a virtual art gallery (see Fig. 1a). Audio-recorded instructions accompany the practice session, and participants are acclimated to noises of the virtual environment (e.g., people talking quietly in the art gallery). During the practice session, participants learn to move around the virtual environment with the VR controllers, joystick, mouse, keyboard, or other response device (e.g., MR-compatible button box). They also practice submitting the subjective ratings that they complete during the experimental phases. If physiological responses are also being recorded, this is a good opportunity to have children practice keeping their arm still. For example, in our VR paradigm, participants move around the environment and submit responses using their dominant hand, and physiology (e.g., skin conductance responses [SCR]) is recorded from their non-dominant hand. Therefore, the non-dominant hand remains as still as possible. This practice session can be repeated as many times as needed, to ensure understanding.

2.4 Stimuli and Timing

We selected avatars as the CSs to model the interpersonal nature of fear-inducing events that children commonly experience and are linked to the development of fear-based disorders [25, 26]. During

Table 1
Equipment and materials list

Name	Description	Example	Needed to run in VR	Needed to collect physiology (e.g., SCR)	Needed to run in MRI scanner
VR head-mounted display	To display stimuli in 3D to participant	Vive Cosmos (HTC Corp)	Yes	No	No
High performance PC	To run the VR task	G5 Gaming Desktop (Dell Inc.)	Yes	No	Yes
Physiology recording equipment, software, and supplies	To collect physiological measures of conditioned fear	MP160 Data Acquisition System (BIOPAC Inc) with relevant amplifiers (e.g, EDA electrodermal activity amplifier), parts[b] (e.g., leads), software (e.g., AcqKnowledge) and supplies (e.g., Disp. RT Dry Electrodes, Gel101)	No	Yes	No[a]
Parallel port to USB adapter	To capture event codes from the VR task PC to the physiology recording system	Cortech Solutions SD-MS-TCPUA	No	Yes	No
Physiology recording PC	To record physiological measures of conditioned fear	Latitude 5420 (Dell Inc.)	No	Yes	No
Response device	To capture behavior and ratings	Outside of scanner: joystick, controller, keyboard Inside of scanner: MR-compatible response device (e.g., Lumina 2 × 2 Response Pad)	Yes	No	Yes
MR-compatible audio system	To deliver the US in the MRI scanner	Sensimetrics Model S14 system with MR-compatible earbuds (e.g., Comply Foam Ear Canal Tips)	No	No	Yes
MR-compatible project and video system	To display the task in the MRI scanner	Silent Vision™ (Avotec Inc.)	No	No	Yes

[a]MR-compatible versions of equipment and supplies are required if physiology will be collected in the MRI scan environment
[b]Parallel port to USB adapter may be needed for the data acquisition equipment (e.g., BIOPAC system) to receive event codes from the task PC (e.g., Cortech Solutions SD-MS-TCPUA)

Fig. 1 VR task stimuli. (**a**) training context, (**b**) test contexts, (**c**) virtual avatars, serving as conditioned stimuli (CS)

the paradigm (Fig. 2a), an aversive white noise burst (US) is paired with an avatar (CS), and participants are instructed that they "may or may not hear a loud noise" before every phase. First, participants undergo fear conditioning in a novel context (CXT+), which becomes associated as a "danger" context because of the presence of the US. Ten minutes later, fear extinction learning is performed in a "safe" context in which the CS that was previously paired with the US (CS+) is presented in the absence of the US (CXT−). For example, in our paradigm, the CXT+ may be assigned to a plaza or a stadium (see Fig. 1b). Twenty-four hours later, participants return to the lab for a second visit. During the second visit, participants complete a test of extinction recall in the CXT−, followed by a test of fear renewal in the CXT+ 10 min later. This second visit is optional, but is of particular interest to our group given that impaired extinction recall is linked to the pathophysiology of fear-based disorders [4–8]. It is also possible to manipulate the interval between extinction learning and recall to test extinction memory retention over longer periods, e.g., 1 week [27]. Contexts are matched on size and layout but comprised of different colors, features, textures, sounds, and backgrounds. Contexts are randomly assigned to CXT+ or CXT− across participants. At the start of the experiment, three avatars are randomly assigned to CS− (never paired with the US), CS + E (extinguished), or CS + U (unextinguished). CSs are adult male, but were designed to vary in hair color/style, clothing, and race to match the demographic makeup of our population (see Fig. 1c). Trial order is pseudorandomized for each phase, such that no more than two of the same trial types occur in a row. Each trial is followed by a 4–9 s

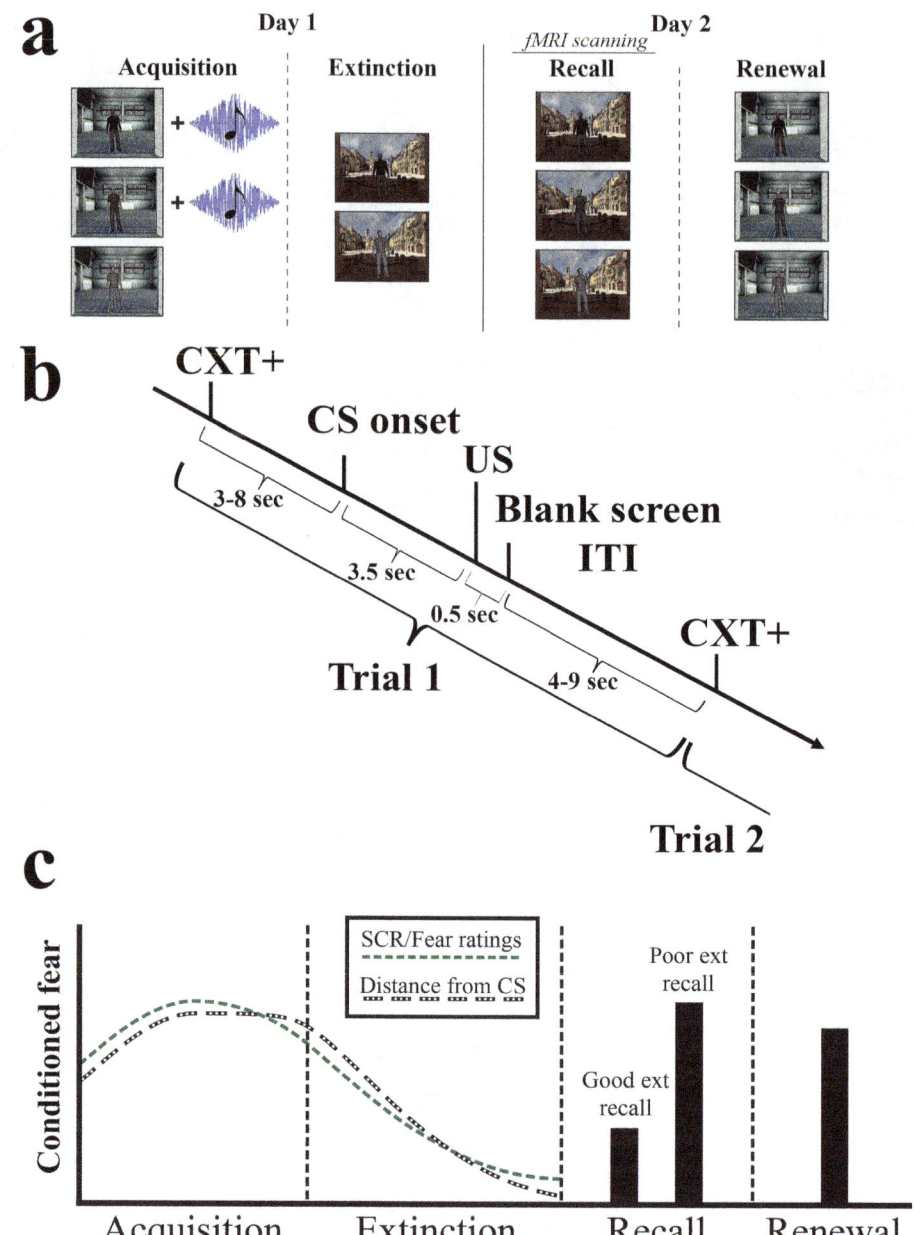

Fig. 2 Experiment overview. (**a**) Experimental phases, (**b**) trial timing, (**c**) hypothesized outcomes. Panel **a**: A white noise burst served as the US, and was paired with two out of three CSs (male avatars). Of note, fMRI scanning can occur during any phase. Panel **b**: For reinforced (CS+ trials), the US co-terminated with the CS. For trials that are not reinforced (CS− trials), the CS is presented for 4 s. Panel **c**: Measures of conditioned (e.g., SCRs, fear and US expectancy ratings, distance from the CS) are expected to increase over the course of fear conditioning and decrease over the course of extinction learning. During extinction recall (24 h later), high conditioned fear may indicate poor extinction recall. High conditioned fear during the renewal phase indicates the return of fear due to change in context from the safety context. CXT+, context. CS, conditioned stimulus; ITI, inter-trial interval; US, unconditioned stimulus; SCR, skin conductance responses

Table 2
Task timing

Phase	Approximate length (min)	CS− presentations	CS + E presentations	CS + U presentations	Total trials
Fear conditioning	7.3	8	8 (6 reinforced)	8 (6 reinforced)	24
Extinction learning	6.8	8	8		16
Extinction recall	7.3	8	8	8	24
Fear renewal	7.3	8	8	8	24

Numbers indicate the number of presentations/trials. Reinforced indicates trials that co-terminate with presentation of the US. *CS−* conditioned stimulus never paired with the US, *CS + E* conditioned stimulus that is paired with the US and subsequently extinguished, *CS + U* conditioned stimulus that is paired with the US and not extinguished

inter-trial interval (see Fig. 2b). This longer inter-trial interval is suggested to maintain ability to distinguish SCRs or BOLD responses (if during fMRI) between trials. For CS+ conditioning trials, a 95 dB 0.5 s white noise burst is delivered for the last 0.5 s of the trial (75% reinforcement), coinciding with CS/CXT offset. The US was selected based on prior fear conditioning studies in anxious youth [28] and partial reinforcement (75%) was selected to reduce potential habituation to the US and enable assessment of SCRs or fMRI response to the CS+ without the confound of US presentation during fear learning [29]. However, the first CS+ trial in the conditioning phase is always reinforced. Trial numbers are provided in Table 2 and trial timing is provided in Fig. 2b. Using this protocol, we observed that children were able to successfully learn and extinguish fear within-session [23]. The number of trials may be increased to ensure that all subjects fully extinguish fear within-session, for example. These considerations should be weighed carefully based on age of the participants and overall burden of timing. The entire protocol lasts ~28.7 min.

2.5 Setup

Time-of-day should be matched across participants to avoid known circadian fluctuations in fear conditioning and extinction [30, 31]. Other potential considerations may include stress, stress hormones, and gonadal hormones, particularly estradiol, which have all been shown to modulate fear extinction [32]. Experimental parameters should be counterbalanced across participants, including (1) trial order (e.g., psuedorandomized), (2) assignment of the context to the CXT+, and (3) designations of the avatars at CS−, CS + E, and CS + U. Participants should also be monitored throughout for task adherence, excess motion, and distress. Only one participant in our study discontinued fear conditioning after hearing the white noise burst, but were able to continue by completing the task in 2D on the computer screen rather than in the VR environment. Children should be closely monitored for distress and may discontinue at any time. Discontinuation rates should be monitored, and may indicate an adjustment to the intensity of the US, for example.

2.6 Dependent Measures

Our paradigm captures subjective, physiological, and behavioral measures of conditioned fear, which may capture unique aspects of fear [29, 33]. In our validation study in 6-12-year-old children [23], we observed areas of both convergence and divergence across these conditioned fear measures, which is similar to prior studies in adults. Hypothesized outcomes are outlined in Fig. 2c.

2.6.1 Subjective Ratings

In our paradigm, subjective ratings were captured in the beginning, middle, and end of each phase. Depending on the goals of the study, it may also be beneficial to capture ratings during each trial [34]; however, this significantly increases task duration. We captured subjective ratings with two questions that addressed fear and US expectancy, respectively: (1) "How scary is this?" (1–5 where 1 = "Not scary at all," 5 = "Very scary") and (2) "Do you think you will hear a loud sound with this"? (1–5 where 1 = "definitely no," 3 = "unsure," 5 = "definitely yes"). Participants were shown the CS in the CXT when asked to perform ratings, and were shown all CSs that were presented in the phase. Ratings for each trial are recorded in an experimental log file. At the end of each phase, we also ask participants whether they had a strategy in the "game," and if so, what the strategy was (open-ended). We also ask them why they thought they heard the loud sound sometimes but not others (open-ended).

2.6.2 Physiology

Physiological aspects of conditioned fear were captured by measuring SCRs, a noninvasive indicator of autonomic nervous system activity, specifically sympathetic arousal [35]. SCRs are the most commonly used indices of fear responding in humans [29]. Additional physiological measures may be collected such as ECG/EKG, fear potentiated-startle reflex (EMG eyeblink), HR, pupilar response, fMRI, and EEG/ERP. However, the latter may not be practical with concurrent use of a VR head-mounted display. We have successfully implemented this task in the MRI scanner and measured BOLD fMRI responses concurrently [36, 37]. For SCRs, we use two Ag/AgCl electrodes attached to the first and second digits of the nondominant hand. We use an MP160 BIOPAC amplifier and AcqKnowledge software (BIOPAC Systems, Inc.) to acquire the SCR trace (continuously sampled at 1000 Hz) and to capture SCRs using event-related electrodermal response analysis. SCRs are defined as maximum of the SCR trace within a 0.5–4.5 s window following CS onset (only for trials without the US), and accounting for a baseline (2 s window before CS onset). Visual inspection of the SCR trace is also suggested. SCRs are commonly transformed (e.g., square-root transformation) and values less than 0.02 μS are typically considered to be "non-responses" [38].

2.6.3 Behavior

Our VR paradigm captures participants' left-to-right (x) and forward-to-backward (z) movement in the virtual environment, on a trial-wise basis. During the experimental phases, movement is constricted to the bounds of the hallway and participants are not permitted to move beyond the CS or go backwards from the start position. Movement is captured at two time points during each trial: (1) "exploratory" movement in the environment, prior to the CS appearing, and (2) "CS" movement, which tracks approach/avoidant behaviors after CS onset. Of note, our paradigm returns the participant to the "start" position at the end of the hallway just prior to CS onset, to standardize start distance for "CS" movement. In our previous validation study, we found that children approach the CS+ more over the course of extinction learning, suggesting successful extinction. That is, they kept more distance (z) from the avatar at the beginning of extinction learning as compared to the end, measured during the "CS" movement period. With our paradigm, it is also possible to track individual movements on a trial-wise basis, which provides a robust parallel to behavioral studies in rodent models that track behaviors and movements (e.g., freezing, avoidance [39]). These data are available in experimental log files. We imported these log files into MATLAB software to create an animation of the paths taken over the course of the experiment.

3 Data Analysis

We have used repeated-measures ANOVAs to assess the effects of CS-type (CS+, CS−) and time (first half/first trial, second half/second trial) on dependent variables, for each phase. Additional factors may include group (e.g., PTSD, control). Fear and expectancy ratings should generally increase over the course of fear conditioning, and decrease over the course of extinction learning. SCRs and distance kept from the CS should also decrease over the course of extinction learning. We found that, overall, pre-adolescent children (i.e., ages 6–11 years) have poor extinction recall, indicated by elevated SCRs, avoidant behaviors, and subjective ratings at the start of extinction recall as compared to the end of extinction learning [37]. For fMRI analyses, we suggest creating contrasts for each CS type (> implicit baseline) to examine overall responses, and for CS + E vs. CS + U. The latter contrast isolates responses that are specific to extinction recall and independent from recall of conditioning (see [24, 37, 40]). Early (first half), late (second half), and overall trials can be modeled for each CS type. FMRI analyses can implement two complementary approaches: (1) a hypothesis-driven region of interest (ROI)-based analysis; and (2) exploratory whole-brain voxel-wise analysis. ROIs related to extinction recall are included but not limited to the amygdala, ventromedial prefrontal cortex/ventral anterior cingulate cortex, and the hippocampus. These ROIs can be defined by anatomical atlases [41] and/or functional activation peaks from meta-analyses [42, 43]. Additional details can be found in prior work [24, 37, 40].

4 Troubleshooting

Poor or altered SCRs: Low SCR response or many non-responses may indicate poor electrode placement or the presence of artifact. Indeed, SCRs are modulated by temperature, time-of-day, stress, sleep, caffeine, neuropsychiatric disease, medications, and may differ by age and race/ethnicity [44]. There are also unique effects of recording SCRs within an MRI environment, including gradient noise from the scanner and the psychological impact of a strange, loud environment. For a full discussion of SCRs, see [38]. Task timing may also contribute, and we refer the reader elsewhere for more discussion of these consideration [29].

Poor discrimination between cues or contexts: Particularly during fear conditioning, we and others have observed poor ability to discriminate between cues (e.g., CS+ vs. CS−) in children, which may be relevant to fear generalization and/or risk of anxiety [37, 45]. However, if increased discriminability is desired, many aspects of the cues or contexts can be manipulated, e.g., clothing, hair, accessories, height, sounds, colors, textures, and lighting. Discriminability between stimuli may be tested prior to the start of the experiment.

5 Concluding Remarks

Fear learning and extinction paradigms have proven useful for understanding neurobiological mechanisms involved in the pathogenesis, maintenance, and treatment of fear-based disorders. However, the majority of this work has been done in adults or in animal models, despite evidence that most fear-based disorders can be traced back to childhood or adolescence [14, 15]. In this chapter, we describe a novel VR fear learning and extinction protocol that we developed for pediatric and adolescent populations. Using these approaches, one can test for neurodevelopmental mechanisms leading to the emergence of fear-based disorders and examine the impact of risk factors (e.g., family history, childhood trauma exposure) on these processes. One may also use these approaches to test the effectiveness of preventive interventions to stem the etiology of fear-based disorders in at-risk populations, or to evaluate the impact of clinical treatments for youth with already-established symptoms and/or a diagnosis.

Acknowledgements

Thank you to the children and families who generously shared their time to participate in our research studies. Thank you also to staff of the TNP2 and THINK labs, particularly Allesandra Iadipaolo, Farrah Elrahal, and Austin Morales. Dr. Marusak is supported by NIH grants K01MH119241 and R21HD105882, and Dr. Rabinak is supported by R33MH11935.

References

1. Jovanovic T, Norrholm SD, Blanding NQ et al (2010) Impaired fear inhibition is a biomarker of PTSD but not depression. Depress Anxiety 27:244–251
2. Raber J, Arzy S, Bertolus JB et al (2019) Current understanding of fear learning and memory in humans and animal models and the value of a linguistic approach for analyzing fear learning and memory in humans. Neurosci Biobehav Rev 105:136
3. Quirk GJ, Mueller D (2008) Neural mechanisms of extinction learning and retrieval. Neuropsychopharmacology 33:56–72
4. Milad MR, Pitman RK, Ellis CB et al (2009) Neurobiological basis of failure to recall extinction memory in posttraumatic stress disorder. Biol Psychiatry 66:1075–1082
5. Knox D, George SA, Fitzpatrick CJ et al (2012) Single prolonged stress disrupts retention of extinguished fear in rats. Learn Mem 19:43–49
6. Garfinkel SN, Abelson JL, King AP et al (2014) Impaired contextual modulation of memories in PTSD: an fMRI and psychophysiological study of extinction retention and fear renewal. J Neurosci 34:13435–13443
7. Duits P, Cath DC, Lissek S et al (2015) Updated meta-analysis of classical fear conditioning in the anxiety disorders. Depress Anxiety 32:239
8. Marin MF, Zsido RG, Song H et al (2017) Skin conductance responses and neural activations during fear conditioning and extinction recall across anxiety disorders. JAMA Psychiatry 74:622
9. Helpman L, Marin MF, Papini S et al (2016) Neural changes in extinction recall following prolonged exposure treatment for PTSD: a longitudinal fMRI study. NeuroImage Clin 12:715–723
10. Read KL, Puleo CM, Wei C et al (2013) Cognitive–behavioral treatment for pediatric anxiety disorders BT – Pediatric anxiety disorders: A clinical guide. In: Vasa RA, Roy AK, editors. New York, NY: Springer New York, 269–287
11. Cohen JA, Mannarino AP (2015) Trauma-focused cognitive behavior therapy for traumatized children and families. Child Adolesc Psychiatr Clin N Am 24:557–570
12. Villabø MA, Compton SN (2019) Cognitive behavioral therapy. In: Pediatric anxiety disorders. Academic, London
13. Rabinak CA, Angstadt M, Sripada CS et al (2013) Cannabinoid facilitation of fear extinction memory recall in humans. Neuropharmacology 64:396–402
14. Paus T, Keshavan M, Giedd JN (2008) Why do many psychiatric disorders emerge during adolescence? Nat Rev Neurosci 9:947–957
15. Kessler RC, Berglund P, Demler O et al (2005) Lifetime prevalence and age-of-onset distributions of DSM-IV disorders in the National Comorbidity Survey Replication. Arch Gen Psychiatry 62:593

16. Essau CA, Conradt J, Petermann F (2000) Frequency, comorbidity, and psychosocial impairment of anxiety disorders in German adolescents. J Anxiety Disord 14:263–279
17. Liberman LC, Lipp OV, Spence SH et al (2006) Evidence for retarded extinction of aversive learning in anxious children. Behav Res Ther 44:1491–1502
18. Warwick H, Reardon T, Cooper P et al (2017) Complete recovery from anxiety disorders following cognitive behavior therapy in children and adolescents: a meta-analysis. Clin Psychol Rev 52:77
19. Woodward LJ, Fergusson DM (2001) Life course outcomes of young people with anxiety disorders in adolescence. J Am Acad Child Adolesc Psychiatry 40:1086–1093
20. Weissman MM, Wolk S, Wickramaratne P et al (1999) Children with prepubertal-onset major depressive disorder and anxiety grown up. Arch Gen Psychiatry 56:794–801
21. Pine DS, Cohen P, Gurley D et al (1998) The risk for early-adulthood anxiety and depressive disorders in adolescents with anxiety and depressive disorders. Arch Gen Psychiatry 55:56
22. Buckner JD, Schmidt NB, Lang AR et al (2008) Specificity of social anxiety disorder as a risk factor for alcohol and cannabis dependence. J Psychiatr Res 42:230
23. Marusak HA, Peters CA, Hehr A et al (2017) A novel paradigm to study interpersonal threat-related learning and extinction in children using virtual reality. Sci Rep 7:16840
24. Milad MR, Wright CI, Orr SP et al (2007) Recall of fear extinction in humans activates the ventromedial prefrontal cortex and hippocampus in concert. Biol Psychiatry 62:446–454
25. Kessler RC, McLaughlin KA, Green JG et al (2010) Childhood adversities and adult psychopathology in the WHO world mental health surveys. Br J Psychiatry 197:378–385
26. Finkelhor D, Turner H a, Shattuck A et al (2013) Violence, abuse, and crime exposure in a national sample of children and youth. JAMA Pediatr 167:614–621
27. Hammoud MZ, Peters C, Hatfield JRB et al (2019) Influence of Δ9-tetrahydrocannabinol on long-term neural correlates of threat extinction memory retention in humans. Neuropsychopharmacology 44:1769–1777
28. Shechner T, Britton JC, Ronkin EG et al (2015) Fear conditioning and extinction in anxious and nonanxious youth and adults: examining a novel developmentally appropriate fear-conditioning task. Depress Anxiety 32:277–288
29. Lonsdorf TB, Menz MM, Andreatta M et al (2017) Don't fear 'fear conditioning': methodological considerations for the design and analysis of studies on human fear acquisition, extinction, and return of fear. Neurosci Biobehav Rev 77:247
30. Pace-Schott EF, Spencer RMC, Vijayakumar S et al (2013) Extinction of conditioned fear is better learned and recalled in the morning than in the evening. J Psychiatr Res 47:1776–1784
31. Albrecht A, Stork O (2017) Circadian rhythms in fear conditioning: an overview of behavioral, brain system, and molecular interactions. Neural Plast
32. Stockhorst U, Antov MI (2016) Modulation of fear extinction by stress, stress hormones and estradiol: a review. Front Behav Neurosci 9:359
33. Beckers T, Krypotos A-M, Boddez Y et al (2013) What's wrong with fear conditioning? Biol Psychol 92:90–96
34. Prenoveau JM, Craske MG, Liao B et al (2013) Human fear conditioning and extinction: timing is everything…or is it? Biol Psychol 92:59
35. Laine CM, Spitler KM, Mosher CP et al (2009) Behavioral triggers of skin conductance responses and their neural correlates in the primate amygdala. J Neurophysiol 101:1749
36. Marusak HA, Hehr A, Bhogal A et al (2021) Alterations in fear extinction neural circuitry and fear-related behavior linked to trauma exposure in children. Behav Brain Res 398:112958
37. Marusak HA, Peters C, Hehr A et al (2018) Poor between-session recall of extinction learning and hippocampal activation and connectivity in children. Neurobiol Learn Mem 156:86
38. Boucsein W, Fowles DC, Grimnes S et al (2012) Publication recommendations for electrodermal measurements. Psychophysiology 49:1017
39. Moita MAP, Rosis S, Zhou Y et al (2004) Putting fear in its place: remapping of hippocampal place cells during fear conditioning. J Neurosci 24:7015
40. Rabinak CA, Angstadt M, Lyons M et al (2014) Cannabinoid modulation of prefrontal-limbic activation during fear extinction learning and recall in humans. Neurobiol Learn Mem 113:125–134
41. Tzourio-Mazoyer N, Landeau B, Papathanassiou D et al (2002) Automated anatomical labeling of activations in SPM using a

macroscopic anatomical parcellation of the MNI MRI single-subject brain. NeuroImage 15:273–289

42. Suarez-Jimenez B, Albajes-Eizagirre A, Lazarov A et al (2019) Neural signatures of conditioning, extinction learning, and extinction recall in posttraumatic stress disorder: a meta-analysis of functional magnetic resonance imaging studies. Psychol Med 50(9):1442–1451

43. Fullana MA, Albajes-Eizagirre A, Soriano-Mas C et al (2018) Fear extinction in the human brain: a meta-analysis of fMRI studies in healthy participants. Neurosci Biobehav Rev 88:16–25

44. Alexandra Kredlow M, Pineles SL, Inslicht SS et al (2017) Assessment of skin conductance in African American and Non–African American participants in studies of conditioned fear. Psychophysiology 54:1741

45. Jovanovic T, Nylocks KM, Gamwell KL et al (2014) Development of fear acquisition and extinction in children: effects of age and anxiety. Neurobiol Learn Mem 113:135–142

Chapter 4

Assessing the Role of Sleep in the Regulation of Emotion in PTSD

Ihori Kobayashi, Mariana E. Pereira, Kilana D. Jenkins, Fred L. Johnson III, and Edward F. Pace-Schott

Abstract

Sleep disturbances such as trouble falling and staying asleep and recurrent trauma nightmares are common symptoms of posttraumatic stress disorder (PTSD). Although initially considered only to be a symptom of PTSD; evidence that sleep disturbance plays a critical role in the development and persistence of this disorder is accruing. In the present chapter, key literature on sleep in PTSD, methodological factors that need to be considered when planning sleep studies with PTSD populations, and psychophysiological variables closely associated with sleep in PTSD are reviewed. This is followed by a description of both traditional and novel methods for assessing sleep and sleep-associated phenomena (e.g., traditional and quantitative polysomnography, actigraphy, subjective sleep and dream measures, fear conditioning and extinction, and brain imaging during sleep) as well as a discussion of the unique methodological challenges associated with studying sleep in PTSD.

Key words PTSD, Sleep, Nightmare, Polysomnography, Actigraphy, Fear conditioning and extinction, Heart rate variability, Cortisol, DLMO, Brain imaging

1 Introduction

Difficulty initiating and maintaining sleep and recurrent nightmares about trauma are common symptoms and two of the diagnostic criteria of posttraumatic stress disorder (PTSD) [1–3]. Specifically, 50–91% people with PTSD experience difficulty initiating and/or maintaining nighttime sleep, and 50–70% experience trauma-related nightmares [3–9]. A recent meta-analysis of polysomnographic (PSG) studies revealed that individuals with PTSD have reduced total sleep time (TST), sleep efficiency (SE), and slow wave sleep (SWS) and increased wake after sleep onset (WASO) compared with those without PTSD [10].

Given converging evidence from prospective studies examining associations between sleep and PTSD symptoms, it has been hypothesized that sleep impairments are not only symptoms of PTSD but also contributors to the development and persistence of the disorder [11]. In prospective studies, subjective sleep disturbance prior to trauma and insomnia and nightmares soon after trauma predicted development of PTSD [12–15]. Likewise, fragmentation of rapid-eye-movement (REM) sleep and increased autonomic arousal during REM sleep following trauma exposure were predictive of subsequent PTSD [16, 17]. In treatment studies, improvement in sleep-related symptoms during cognitive behavioral interventions for PTSD was associated with reduction of other PTSD symptoms [18–20]. Reduction of autonomic arousal during REM sleep was observed in PTSD patients who responded to cognitive behavioral therapy [21]. Because sleep influences memory processes critical for regulating emotion, including fear conditioning, extinction of conditioned fear, safety learning, and habituation, it has been hypothesized that this may be a mechanism by which sleep contributes to the course of PTSD [11, 22, 23]. Given the potential roles of sleep in PTSD, there is a significant need for research to elucidate the possible cause–effect relationships between sleep, sleep disturbance, and the emotional, cognitive, and neural pathophysiological processes that underlie PTSD and to apply this information to the development of effective intervention strategies.

1.1 Overview of Sleep Assessment Methods Used in PTSD Research

Both subjective and objective sleep measures are used in PTSD research. Subjective measures include sleep and dream questionnaires and diaries. Questionnaires are typically used to retrospectively assess habitual sleep patterns and quality in the recent past (e.g., past 2 weeks) whereas diaries are completed daily often in the morning upon awakening. Both questionnaires and diaries are low-cost, easy-to-use, and able to assess sleep quality and patterns based on the participant's perceptions; however, accuracy of estimations of some sleep parameters, such as sleep onset latency (SOL) and WASO, is relatively poor compared with objective measures (*see* Subheadings 2.3 and 2.4 for more information about subjective sleep measures).

Standard PSG and actigraphy are the most commonly used objective sleep measures in PTSD research. PSG is the gold standard for sleep assessment. It consists of various electrophysiological sensors [e.g., electroencephalogram (EEG), electromyogram (EMG), and electrooculogram (EOG)] for identifying sleep stages and movement events as well as additional sensors for respiratory event monitoring (e.g., pulse oximetry and respiratory belts). PSG is costly and requires well-trained technicians. Sleep laboratory environments, in which PSG recordings often take place, do not capture participants' naturalistic sleep in their home environments. To mitigate this potential problem, ambulatory PSG has been used, but the number of studies is limited (*see* Subheading 2.1 for more information about PSG). Sleep actigraphy, a wristwatch-shaped device usually worn on the wrist, is capable of long-term (e.g., 4+ weeks) monitoring of sleep in participants' regular sleep environments. Actigraphy uses an accelerometer device to estimate sleep and wake state for each epoch (typically in 30–120 s) based on detected limb movement activity. Actigraphy is an ambulatory assessment option that has been gaining popularity in PTSD research due to its ease of use, long-term recording capability, and lower cost compared to PSG. Actigraphy, however, is not able to differentiate sleep stages as can be done using PSG (see Subheading 2.2 for more information about actigraphy).

In addition to the conventional sleep measures described above, more recently, methodologies such as high-density EEG (hd-EEG) and magnetoencephalography (MEG) have been used in sleep research. Hd-EEG includes a greater number of EEG electrodes (up to 256 channels) than standard PSG, which greatly improves spatial resolution. Hd-EEG has recently been used to search for novel sleep-related biomarkers of PTSD. Relatively high spatial resolution of hd-EEG enables researchers to examine aspects of sleep EEG that are difficult to capture using standard PSG, such as topographical distributions of certain characteristics of EEG signals and sources of specific EEG activities (current source localization). For example, a hd-EEG study found that veterans with PTSD had lower delta power (1–4 Hz) in the centro-parietal regions during non-REM sleep and elevated high-frequency power in the antero-frontal regions during both REM and non-REM sleep compared with veterans without PTSD [24]. MEG is a noninvasive neurophysiological recording method that detects magnetic signals induced by electrical activity in the brain. MEG has high spatial and temporal resolution allowing measurement of changes in neural activities at specific brain regions and also at the network level [25]. MEG is also able to perform source localization with better spatial precision than EEG because magnetic signals are not distorted by varying conductivities of different tissues [26]. A major disadvantage of MEG is the cost needed for purchasing and maintaining the instrument, which is considerably greater than EEG. Use of MEG in PTSD research has increased in the past decade; however, this technique has not yet been applied to research on sleep and PTSD [25].

In addition to insomnia and recurrent nightmares related to the trauma, individuals with PTSD, compared to those without PTSD, are more likely to report having been told or having suspected themselves that they acted out dreams; however, dream enactment and other complex behaviors associated with PTSD have rarely been observed in sleep laboratory settings [27, 28]. A laboratory video-PSG study with 394 veterans found that 64% and 74% of veterans with PTSD and veterans with both PTSD and traumatic brain injury (TBI), respectively, reported history of dream enactment whereas only 27% of veterans without either PTSD or TBI reported dream enactment [28]. Despite the higher rate of dream enactment among veterans with PTSD, dream enactment behaviors were not captured on video [28]. Nevertheless, a complex syndrome involving disruptive nocturnal behaviors, dream enactment, and sympathetic arousal, usually resulting from severe military trauma, has been proposed as a distinct sleep disorder, trauma-associated sleep disorder [29–31]. To further characterize dream enactment and other complex nocturnal behaviors associated with PTSD, video recording in conjunction with PSG or other sleep measures is needed at home as well as in the laboratory.

1.2 Factors That Need to Be Considered When Planning Sleep Studies in PTSD Research

1.2.1 Sleep Environment

People with PTSD self-report sleep disturbances at home more consistently and with greater magnitude compared with laboratory PSG findings [32–35]. An increased sense of safety in the sleep laboratory, where sleep technicians are watching over the participants all night, has been hypothesized to account for this discrepancy [35]. Consistent with this hypothesis, sexual assault survivors with PTSD report that they perceived the sleep laboratory safer, quieter, and more comfortable compared with their homes [33]. Studies of urban-residing minorities have shown that neighborhood disorder, such as noises, crime, and disruptive activities, as well as perceived problems within an individual's building are associated with greater sleep disturbance [36, 37]. These findings underscore the importance of considering potential impacts of the sleep environment when determining sleep-recording methods (e.g., laboratory vs. home/ambulatory).

1.2.2 First-Night Effect

On the first night of sleep recording in the laboratory, participants often experience worse sleep, such as disrupted sleep initiation and maintenance, increased lighter sleep, and reduced REM sleep, than on subsequent nights [38]. This phenomenon is called the first-night effect. Some studies have suggested a differential influence of the first-night effect on people with and without PTSD. In those studies, individuals with PTSD had no changes in sleep from the first to second night or better sleep on the first night, whereas participants without PTSD exhibited worse sleep on the first night [24, 32, 39]. To reduce the impact of the first-night effect, PSG study protocols often include an adaptation night before collecting baseline data.

1.2.3 Sleep Apnea and Other Sleep Disorders

People with PTSD often experience symptoms of sleep disorders other than insomnia and trauma-related nightmares. One of the most common sleep disorders comorbid with PTSD is sleep apnea. The prevalence of sleep apnea in people with PTSD varies widely by study (0–90%) [40], but a meta-analysis estimated that 75% of people with PTSD have at least mild obstructive sleep apnea (OSA) [41] – a greater prevalence than that found in the general population: 9–38% in men and 13–33% in women [42]. Sleep apnea is typically accompanied by sleep disturbances similar to those evident in individuals with PTSD, including sleep fragmentation and decreased deep (slow wave) sleep [43–45]. In addition, people with PTSD sometimes experience symptoms of parasomnias, including dream enactment, REM without atonia, and sleep paralysis [28, 29, 46, 47]. Therefore, it is necessary to test for, and account for, comorbid sleep disorders when assessing the sleep of PTSD patients in both clinical and research settings.

1.2.4 Traumatic Brain Injury (TBI) and Physical Injury

Both PTSD and sleep disturbances often co-occur with TBI. Rates of comorbid PTSD in individuals with TBI are estimated at 11–19% in civilians and 37–48% in military/veteran populations [48–50]. Approximately half of individuals with TBI experience some form of sleep disturbance such as insomnia, hypersomnia, and OSA [51]. Specific sleep alterations in TBI include reduced SE and TST and increased WASO [52]. Despite frequent comorbidity, few studies have been conducted to try to distinguish the relative contributions of TBI vs. PTSD on the sleep disturbance of individuals diagnosed with both disorders [28]. PTSD and sleep disturbances can also develop after physical injury. In a U.S. national sample of patients hospitalized following a physical (non-TBI) injury, approximately 23% had probable PTSD 1 year after injury [53]. In a sample of orthopedic trauma patients presented at a trauma clinic, 86% reported poor sleep quality [54]. Specific sleep disturbances often reported by physical injury patients are difficulty initiating sleep, frequent awakenings, and reduced TST [54, 55]. Pain, discomfort, and mental health problems are risk factors for sleep disturbance in physical injury patients [54–56]. Given the overlap among sleep disturbances associated with TBI, other physical injury, and PTSD, interpretations of sleep study results in participants with any combination of these frequently comorbid conditions should be performed with caution.

1.2.5 Duration of PTSD, Time Elapsed Since Trauma, and Age of Participants

These three factors often overlap in studies of people with current PTSD [34] and have impacts on sleep, especially REM sleep. Mellman and colleagues [57] reported positive correlations of duration of PTSD with percentage of REM sleep and average length of REM segments as well as a negative correlation with REM latency in young adult (age 18 and 35 years) African Americans with current PTSD. Both Kobayashi et al. [34] and Zhang et al.'s [10] meta-analyses revealed effects of age on differences in REM sleep amount between people with and without PTSD. Specifically, Kobayashi et al. found relatively increased REM sleep in PTSD compared with control only in studies with mean age of participants above 42.4 years, and Zhang et al. found decreased REM sleep in PTSD only in studies with mean age below 30 years. In the studies included in Kobayashi et al.'s meta-analysis, the mean age of participants and duration of time since trauma exposure were positively associated. These findings suggest that aspects of REM sleep in PTSD may change over time, which could partially explain discrepant findings among PSG studies.

1.2.6 Overnight Sleep vs. Daytime Nap

Both overnight sleep and daytime nap studies have been used to assess the relationships of sleep with emotional memories and psychological interventions (*see* Subheading 2.5.1 for more information about methods for examining associations of sleep with fear conditioning and extinction memory). Overnight sleep studies require more resources (e.g., time, funding, and an overnight sleep facility) than nap paradigms. Although the assumption that the effects of sleep obtained during the nighttime and during the day are qualitatively equivalent is parsimonious, this has not actually been demonstrated empirically. Sleep architectures, duration of sleep, and timing in the circadian cycle differ between overnight sleep and daytime naps. Durations of sleep opportunity in nap studies typically allow for only one non-REM/REM sleep cycle (if that, since the cycle is typically 90–120 min in duration), and it is not unusual for daytime sleep periods to be devoid of REM sleep [58, 59]. For overnight sleep, it has been hypothesized that SWS, which dominates in the first half of sleep period, and REM sleep, which dominates in the latter half of sleep, sequentially contribute to memory consolidation [60]. Despite the differences between nighttime sleep and daytime naps, meta-analyses did not reveal significant differences in the effects of nighttime vs. daytime sleep on recall of emotional stimuli or fear extinction memory [61, 62].

1.3 Assessment and Experimental Methods Often Used in Sleep Research in PTSD

1.3.1 Emotional Memory and Fear-Based Behaviors

It has been hypothesized that sleep contributes to the development and maintenance of PTSD symptoms through its impacts on emotional memory and fear-based behaviors. Fear conditioning and extinction (*see* Subheading 2.5.1 for more information about methods for examining associations of sleep with fear conditioning and extinction memory) and script-driven imagery are methods often used to examine emotional memory processes and fear-based behaviors that are hypothesized to be associated with sleep disturbances in those with PTSD. Studies employing fear conditioning and/or extinction paradigms have suggested that post-learning sleep, especially REM sleep, facilitates retention of differential responses to safety versus danger cues learned during conditioning as well as retention of extinction memories [59, 63–65]. Studies in this area have mainly been conducted on healthy volunteers, although a few have included participants with conditions that are known to impact sleep [66, 67]. Straus and colleagues [66] examined associations of overnight REM sleep parameters with fear conditioning and extinction in veterans with PTSD and found that a higher percentage of REM sleep after fear extinction was associated with more rapid recall of the safety stimulus during subsequent extinction recall. Bottary and colleagues [67] reported that in people with insomnia, greater percentage of REM sleep and longer REM sleep segment length predicted worse extinction memory recall. Conversely, in good sleepers, longer REM segment length was predictive of better extinction memory recall.

The presence of recurrent, distressing, intrusive memory of a traumatic event is both a symptom of PTSD and a predictor of PTSD severity [68, 69]. Analogue trauma paradigms have been used to determine the effects of sleep in the development of intrusive memories in healthy volunteers. In analogue trauma paradigms, participants are exposed to stimuli depicting trauma (e.g., traumatic film clips) before overnight sleep or daytime naps. Findings from studies have been mixed. Some have found that, after watching trauma film clips, sleep, compared with wakefulness, reduced intrusive memories of analogue trauma [70, 71]. In addition, some aspects of post-exposure REM sleep (e.g., lower REM density, higher REM theta activity) predicted fewer intrusive memories [70, 72]. However, findings from some studies suggested an increased frequency of intrusive memories following post-exposure sleep [73] while others have revealed no significant effects of sleep [74] [for a review, see [75]].

Script-driven imagery of trauma has long been used to probe psychophysiological reactivity to trauma-related stimuli in individuals with PTSD [76, 77]. In this approach, prerecorded short (e.g., 30 sec – 3 min) individualized scripts developed by researchers based on participants' own traumatic experiences are presented after which the participants are instructed to imagine the traumatic experience as vividly as possible. During this imagery task, self-reported emotional reactions, various psychophysiological responses (e.g., heart rate, skin conductance, and EMG), and neural activations are recorded. Individuals with PTSD often exhibit greater emotional and psychophysiological reactivity to script-driven imagery tasks than do trauma-exposed individuals without PTSD [76, 78, 79], and PTSD treatment can reduce psychophysiological reactivity to this task [78, 80, 81]. Script-driven imagery has been used in a limited number of sleep studies. In one such study, it was found that female physical or sexual assault survivors who reported higher anxiety following script-driven imagery of trauma exhibited relatively poor sleep quality if they had been diagnosed with PTSD. However, the direction of the association was reversed in women without PTSD [82]. Rhudy and colleagues [83, 84] examined response to script-driven imagery of participants' recurrent nightmares in their studies of trauma-exposed individuals. They found that script-driven imagery of nightmare induced psychophysiological responses regardless of PTSD diagnostic status and that cognitive behavioral therapy for nightmare reduced psychological reactivity to the imagery task.

Much of what is currently known about the role of sleep in the regulation of emotional responsivity is based on findings from studies on healthy volunteers with no known psychopathologies. Since a defining characteristic of stress-related disorders like PTSD is emotional dysregulation, the extent to which findings from these studies generalize to patient populations is questionable. Accordingly, more studies on individuals with PTSD are needed.

1.3.2 Experimental Simulation of Exposure and Exposure Therapy

Based on findings indicating that sleep, especially REM sleep, facilitates retention of extinction memory in healthy individuals, researchers have now initiated studies to examine whether sleep following exposure to feared stimuli enhances effects of the exposure intervention in individuals with anxiety disorders or PTSD. Early studies, conducted with individuals suffering from fear of spiders, used simulated exposure techniques (e.g., virtual reality, video of spiders) [85, 86]. These studies revealed that both post-exposure daytime naps [86] and overnight sleep [85] enhanced the therapeutic effects of the simulated exposure compared to wakefulness. Recent studies have similarly utilized exposure therapy for more complex conditions, such as social anxiety and PTSD. Pace-Schott and colleagues [58] examined the effects of a nap versus wakefulness following two of five exposure sessions in a brief exposure-based group therapy for social anxiety. Post-exposure naps reduced psychophysiological responses to a social-stress challenge from pre- to posttreatment; however, naps did not enhance the effects of exposure therapy on ratings of social anxiety disorder symptoms. Kobayashi and colleagues [87] tested the hypothesis that improving sleep with suvorexant (a dual orexin receptor antagonist approved for treating insomnia) following written narrative exposure therapy sessions would enhance its therapeutic effects of exposure in individuals with PTSD. They found that suvorexant, compared with placebo, administered following two evening written narrative exposure sessions enhanced subjective between-session habituation, but not PTSD symptom reductions. Both studies suggested potential benefits of sleep following exposure therapy sessions for anxiety disorders and PTSD. However, further studies are needed to determine whether, and the extent to which, sleep enhancement constitutes an efficacious adjunctive treatment strategy.

1.3.3 Indices of Autonomic Nervous System (ANS)

Measures of autonomic activity in sleep are increasingly being used to explore the bases of the emotional regulatory functions of sleep in trauma-exposed individuals [17, 88–91]. In PTSD research, ANS activity has been indexed by various measures, including heart rate, heart rate variability (HRV) (*see* Subheading 2.5.2 for more information about measures of HRV), galvanic skin response, and urinary or plasma concentrations of catecholamines [21, 90, 92, 93]. Studies of heart rate and HRV showed that individuals with PTSD, compared to those without PTSD, had greater ANS arousal [i.e., higher sympathetic nervous system (SNS) activity and/or lower parasympathetic nervous system (PNS) activity] during resting wakefulness as well as during stress [94]. Higher SNS activity, as indicated by higher concentrations of urinary and plasma norepinephrine, has also been shown in individuals with PTSD [95]. Converging lines of evidence indicate that people with PTSD have elevated ANS arousal during sleep, and a blunted daytime-to-nighttime reduction of ANS arousal compared with control participants with and without trauma exposure [88, 90, 91, 96–101]. In addition, HRV during REM sleep has been associated with the development of PTSD symptoms and with treatment outcomes: Mellman and colleagues [17] assessed HRV during sleep within a month after serious injury and found that patients who developed significant PTSD symptoms at 2 months post-injury had higher autonomic arousal, as indexed by HRV, during REM sleep compared with those who did not subsequently experience significant PTSD symptoms. Nishith and colleagues [21] reported that autonomic arousal during REM sleep, as indexed by HRV, was reduced from pre- to posttreatment in PTSD patients who responded to cognitive behavioral therapy for PTSD. In contrast, autonomic arousal increased in those who withdrew from treatment. These findings suggest that, in addition to its potential as a biomarker of stress-related pathology, ANS activity during REM sleep may impact the physiological mechanisms that underlie the development and maintenance of PTSD.

1.3.4 Assessments of Hypothalamic-Pituitary-Adrenal (HPA) Axis Function

A large literature exists examining cortisol baseline levels, diurnal variations, and the cortisol awakening response (CAR) in PTSD [for a review, see [102]; for more information about methods to obtain a diurnal profile of cortisol, please *see* Subheading 2.5.3]. In many studies, but not all, lower basal salivary and 24-h urinary cortisol levels have been found in individuals with PTSD compared with those without this disorder [103–105]. In addition, in adult trauma survivors, lower urinary or blood cortisol concentrations immediately after trauma exposure were found to be predictive of a trend toward more severe subsequent PTSD symptoms [106]. Yehuda [102] hypothesized that altered HPA function is a preexisting vulnerability factor that influences psychophysiological responsivity at the time of trauma exposure and that low cortisol levels lead to failure to modulate SNS response to trauma and trauma reminders, which in turn results in abnormal consolidation of trauma memory and perpetuation of PTSD symptoms.

It is well documented that cortisol has a circadian rhythm that is influenced by sleep. Cortisol levels peak shortly after morning awakening and decline over the course of the day until early morning when they start increasing again [107, 108]. Findings of studies examining CAR and PTSD have been mixed. Some studies did not find differences in CAR between people with and without PTSD [101, 108, 109] whereas others report reduced CAR in individuals with PTSD compared with those without [107, 110]. A limited number of studies with small samples have examined cortisol levels during sleep. Yehuda and colleagues [108] reported that plasma cortisol levels were lower in veterans with PTSD than in healthy controls between 4:30 am and 5:00 am due to a prolonged nadir. Van Liempt and colleagues [101] reported that people with PTSD trended toward lower plasma cortisol levels during the first half of the sleep period (vs. controls both with and without trauma exposure) and that lower cortisol levels were associated with greater amounts of SWS.

1.3.5 Measures of Chronotype and Circadian Rhythm

Circadian rhythms are physiological and behavioral changes with a 24-h periodicity. Such rhythms are controlled by intracellular clocks consisting of interlocking positive and negative feedback cycles for the transcription and translation of the so-called clock genes that reliably produce specific outputs at specific times over a 24-h period [111, 112]. Such cycles are tightly synchronized among the cells of the suprachiasmatic nucleus of the hypothalamus, which serves as a master clock controlling similar 24-h molecular oscillators in cells elsewhere in the central nervous system and throughout the periphery [113]. Although in healthy individuals circadian rhythms are entrained to ambient environmental and social "zeitgebers," the time at which sleep is initiated widely varies with only the extremes considered to be actual circadian sleep disorders such as delayed sleep phase syndrome. Just as there are individual differences in the amount of sleep individuals need each night [114], there are also individual differences in "chronotype" with those who prefer to stay up late and "sleep in late" in the morning called "evening types" or "owls" and those who prefer to retire earlier in the evening and arise earlier in the morning called "morning types" or "larks." Chronotype is at least partly determined by genetics [for review and discussion of chronotypes, see [115]]. Evening chronotype has been associated with increased neuroticism [116] and various psychiatric disorders, including depression, substance abuse, and anxiety disorders, as well as sleep disturbances (e.g., insomnia, nightmares) [117]. Research to determine the relationship between PTSD and chronotype has been limited, but some studies have been conducted. In military veterans, evening chronotype was associated with more severe lifetime PTSD symptoms and more frequent and intense nightmares [118]. In a small study of military personnel with PTSD ($n = 7$), individuals with PTSD had blunted nocturnal dim light melatonin onset (DLMO, an index of circadian rhythm timing; please *see* Subheading 2.5.4 for methods for obtaining DLMO) compared with healthy controls [119, 120].

1.3.6 Brain Imaging

Neuroimaging methods are an essential tool to bridge animal models of PTSD and human research identifying across species shared regions of fear and extinction circuitry. However, recent meta-analyses [121, 122] and network-based analyses [123] have shown that experimental fear conditioning in humans recruit much broader regions of the forebrain and brainstem. Both task-based and resting-state functional imaging studies continue to enhance our knowledge of the neuropathophysiological correlates of PTSD. Structural neuroimaging, including diffusion tensor imaging (fractional anisotropy), provides information on the structure and integrity of both gray and white matter.

Structural magnetic resonance imaging (MRI) studies have consistently revealed reduced hippocampal volume in individuals diagnosed with PTSD, although this is a common finding among individuals exposed to trauma regardless of whether PTSD or other stress-related disorders have been diagnosed [124] [For an in-depth meta-analysis, please see [125, 126]]. This remains controversial as other studies of monozygotic twins discordant for PTSD suggest that smaller hippocampi may constitute a risk factor for rather than a result of trauma [127]. Also associated with PTSD are reduced prefrontal cortex (PFC) volume [128] and decreased neuronal density in the anterior cingulate cortex (ACC) [129]. In addition, it has been found that amygdala activation was prominent in anxiety and PTSD patient populations compared to controls [130]. Accordingly, findings from these and other studies suggest that abnormalities in circuits involving the ventromedial PFC (vmPFC), amygdala, and hippocampus underlie PTSD [131]. For more detailed information on neuroimaging findings in PTSD, please refer to [127, 132, 133].

Human neuroimaging research has shown that areas activated during REM sleep, including the amygdala, dorsal ACC (dACC), insular cortex, vmPFC, and hippocampus, overlap with areas commonly activated during fear conditioning (dACC, amygdala) and extinction recall (vmPFC, hippocampus) [11]. A meta-analysis of functional MRI studies revealed differences in the neural activity pattern associated with fear conditioning and extinction in individuals with PTSD vs. trauma-exposed controls [134]. Individuals with PTSD exhibited greater activation in medial PFC (mPFC) and in an anterior hippocampus-amygdala area during conditioning, in the anterior hippocampus-amygdala area during extinction learning, and in the anterior hippocampus-amygdala area and medial prefrontal regions during extinction recall [134]. Seo and colleagues [89] examined the influence of nighttime sleep on neuronal activation patterns during fear conditioning and extinction learning and recall in trauma-exposed individuals with and without PTSD. They found that among participants with PTSD, better sleep quality (e.g., lower nightmare frequency, shorter SOL, lower ANS arousal during REM sleep) was associated with greater activation of emotion-regulatory-related regions (e.g., middle frontal cortex, premotor cortex, dorsal medial PFC). This suggests a testable hypothesis that improving the sleep of PTSD patients will result in improved emotion regulation.

To date, use of functional brain imaging techniques to assess neural activity during sleep (*see* Subheading 2.5.5 for more information about brain imaging methods during sleep recording) has been extremely limited in PTSD research. Germain and colleagues [135] examined neural activity during REM sleep and wakefulness in a small sample of combat-exposed veterans with and without PTSD ($n = 6$ each group) using a fluorodeoxyglucose (FDG) positron emission tomography (PET) method. During both wakefulness and REM sleep, greater levels of metabolic activity (i.e., neuronal FDG uptake) were found in veterans with PTSD compared to those without PTSD. Differences were evident in regions implicated in arousal regulation, fear response, and reward processing – including the ventrolateral and ventromedial frontal cortex, ACC, thalamus, right amygdala, basal ganglia, raphe nuclei, and locus coeruleus.

2 Methods

2.1 Polysomnography (PSG)

This section describes PSG methodologies often used in PTSD research with adult participants. Specific procedures vary depending on the PSG system and experimental approaches that are being employed (e.g., lab-based vs. ambulatory recording, sleep parameters of interests). Researchers should select the most appropriate methods to achieve their study goals.

2.1.1 Selection of a PSG System

Features of PSG systems that should be carefully considered include: battery life (if the system is not powered by standard AC outlets), memory capacity, number of channels, the types of sensors included (e.g., EMG, EOG, pulse oximetry), sampling rate, and self-application capability. The importance of other features will vary based on the research questions being addressed, but the ability to monitor sleep remotely in real time and the ability to capture synchronized video recordings (e.g., to diagnose REM behavior disorder) are generally desirable when studying PTSD patients. The American Academy of Sleep Medicine (AASM) Scoring Manual recommends a sampling rate of at least 200 Hz [136]. If quantitative EEG (qEEG) analysis is performed, the sampling rate must be at least twice as high as the highest frequency of interest (*see* **Note 1** in Subheading 2.1.8).

2.1.2 Ambulatory PSG

One important consideration when planning a PSG study on a participant with PTSD is whether to perform it in the sleep laboratory or in the participant's home using an ambulatory system. The ability to capture sleep in the participant's home sleep environment is a major "Pro" for ambulatory monitoring systems, in part, because the PTSD patient's sense of security can impact his/her sleep quality and quantity (for more information about the impacts of the sleep environment on the PTSD patient's sleep, see Subheading 1.2.1). Likewise, if the goal is to capture relatively rare events such as highly veridical trauma nightmares, then the odds are increased by utilizing a home monitoring system over multiple consecutive nights (multiple-night lab-based studies can be prohibitively expensive). An additional advantage of home studies is the reduced staffing costs – especially if the home study is not being monitored remotely in real time (although these savings are offset somewhat by the purchase price of the ambulatory PSG system).

On the Con side, ambulatory PSG significantly reduces the degree of control by the investigator such as the ability to detect and repair equipment failure (e.g., detached electrodes) in real time. It also limits the ability to conduct research that requires sleep stage manipulations (e.g., experimental SWS fragmentation), instrumental awakenings (e.g., for sleep-stage specific dream reports) and some sleep-dependent memory protocols (e.g., targeted memory reactivation). In addition, advanced electrophysiology methods such as high-density EEG montages (which facilitate source localization) or invasive physiological measures can usually only be carried out in laboratories. Moreover, for many of these systems the electrodes must be applied by a technician either in the participant's home or in the laboratory. However, ambulatory PSG systems that enable self-application of the electrodes are being developed.

2.1.3 Preparation

Prior to the night of the first sleep recording, researchers should inform participants about how to prepare for the visit and what to expect during the sleep recording, including possible first-night effects, the duration of their stay in the lab, availability of meals, things to pack for the stay, and any medications and substances to avoid prior to and during the visit (*see* **Notes 2** and **3** in Subheading 2.1.8).

Preparing all equipment and supplies before the arrival of participants will help obtain high-quality PSG signals. Equipment and supplies for PSG generally include: PSG amplifier unit, computer, gold cup surface electrodes, electrocardiogram (ECG)/EMG snap electrode cables and electrodes, pulse oximetry, respiratory belts, nasal cannula, thermistor sensor, skin prep gel, conductive paste, cotton tipped applicators, small gauze squares (e.g., 2″ × 2″), surgical tape, scissors, tape measure, skin marker, exam gloves, comb, and hairclips (*see* **Notes 4** and **5** in Subheading 2.1.8).

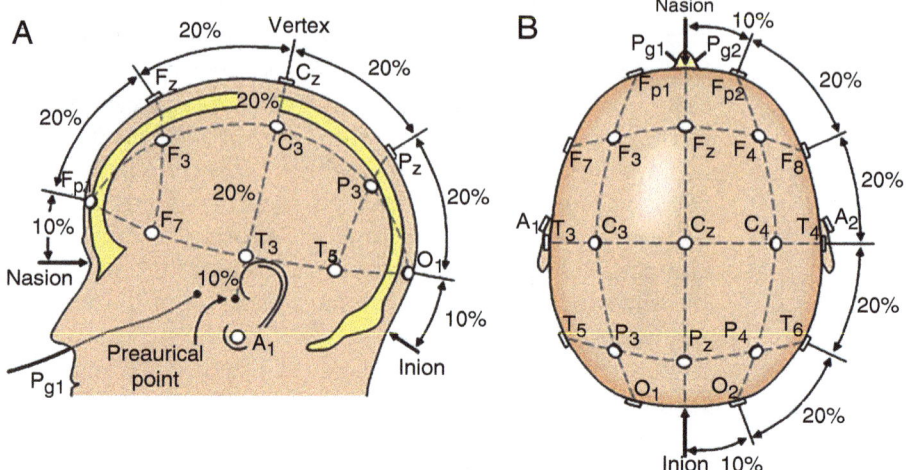

Fig. 1 The International 10–20 System seen from (**a**) left and (**b**) above the head. *A* ear lobe, *C* central, *F* frontal, *FP* frontal polar, *O* occipital, *P* parietal, *Pg* nasopharyngeal, *T* temporal. (Reproduced from Foldvary-Schaefer and Grigg-Damberger 2012 [235] with permission from Elsevier)

2.1.4 Placement of Sensors

For standard PSG, the International 10–20 System (*see* Fig. 1) is used for the placement of EEG electrodes. Using a tape measure, locate and mark electrode placement sites. A montage for standard PSG includes frontal (F3 and F4), central (C3 and C4), and occipital (O1 and O2) electrodes referenced to contralateral mastoids (M1 and M2; behind the ears) for EEG, E1 (1 cm lateral and 1 cm below to the left outer canthus) and E2 (1 cm lateral and 1 cm above to the right outer canthus), EOG electrodes both referenced to M2, two chin EMG electrodes (2 cm below the mandible and 2 cm on either side of the midline) referenced to an EMG electrode placed 1 cm above the mandible on the midline, and two ECG leads (below right clavicle and left fifth intercostal space) (*see* **Note 6** in Subheading 2.1.8). In addition, ground/reference electrodes are placed at the frontal pole (FP) and the midcentral (Cz) site. Recommended specific reference sites may vary with the system being used, especially for ambulatory systems. For monitoring of respiratory events and leg movements, additional sensors, including nasal pressure sensor, oronasal thermistor, chest and abdominal respiratory belts, pulse oximetry, and EMG electrodes on the anterior tibialis muscles of both legs are added [136]. Screening for OSA and periodic limb movement disorder is often performed on the first (acclimation) night before the baseline PSG night.

Ensure that appropriate digital filters are applied. For visual staging, apply a filter ranging from 0.3 Hz to 35 Hz for EEG and EOG and from 10 Hz to 100 Hz for EMG [136]. For qEEG analyses, higher upper cut-off frequency ("low-pass") filter may be used for processing of higher frequency bands.

Prepare electrode placement sites by scrubbing each site using a cotton tipped applicator with a small amount of skin prep gel. If needed, use a comb and hairclips to expose the scalp at the attachment sites. Remove the gel from the skin surface with clean gauze. Scoop conductive paste with a gold electrode cup and attach the cup on a site, with paste side down, by applying moderate pressure. Ensure that all wires of electrodes point toward the participant's back. To secure the cup on skin with hair, cover the cup with the prepared small gauze square, with paste side down. On skin without hair, use pieces of surgical tape to secure the cup. Technicians should ensure that all leads are plugged in to correct slots on the amplifier unit. Bundle EEG, EOG, and EMG wires using surgical tape like a loose low ponytail.

Check impedances of electrodes and repeat the skin preparation and electrode placement procedures for the electrodes with high impedances. The AASM Scoring Manual recommends 5 KΩ or less [136]. Perform calibrations to ensure that signals respond to various physiological changes, such as opening and closing eyes, looking at various directions, and holding their breath.

2.1.5 Initiation of Sleep Recording

To protect the integrity of electrode placements, assist the participant with getting into bed in the sleeping room without pulling any wires. For ambulatory monitoring at home, if the participant is required to initiate recording, provide detailed instructions and a reminder to start recording before their expected bedtime. Phone calls or text messaging by staff can help ensure data capture.

2.1.6 Scoring

Both visual staging and qEEG analysis approaches have been used in PTSD research. The AASM criteria [136] and Rechtschaffen and Kales (R&K) criteria [137] with the epoch lengths of 30 or 20 s are often used in visual staging of PSG records in PTSD research. R&K criteria were the first standardized sleep staging rules published in 1968 [138] and describe scoring rules for seven stages (wake, stage 1, 2, 3, and 4, stage REM, and movement time). The AASM developed a new scoring guideline by modifying R&K criteria in 2007 [139]. The AASM scoring manual describes criteria for wake (Stage W), three non-REM sleep stages (N1, N2, and N3) and REM sleep (Stage R) as well as rules for scoring arousals, respiratory events, cardiac events, and leg movements. Stages 3 and 4 defined in the R&K standard were combined into N3 in the AASM manual [140].

Sleep parameters generally derived from visual sleep staging include TST, SOL, REM latency, WASO, SE, time in each stage, percent of TST in each stage, and REM density (the number of rapid eye movements per minute during REM sleep). In PTSD research, indices of REM sleep fragmentation, such as average REM segment length [67, 141] and the number of arousals occurring during REM sleep [142] are sometimes obtained as REM sleep abnormalities, particularly, fragmented REM sleep, has been implicated in pathogenesis and pathophysiology of PTSD [141].

For qEEG analysis, power spectral density (PSD: $\mu V^2/Hz$) for each frequency bin is commonly estimated using Fast Fourier transform (FFT). FFT can be performed using commercially available signal processing software or custom-made programs. The EEG signal is divided into consecutive or overlapping segments. Researchers need to select an appropriate segment length considering the underlying signal stationarity, frequency resolution, and alignment with the visual sleep staging epoch length [143, 144]. To reduce effects of segment edges, a tapered window function [e.g., a Hanning window, see [145]] is applied to each segment. Artifacts need to be removed before performing FFT by rejecting segments containing artifacts or cancelling or correcting the artifacts [146]. The artifact processing can be performed manually or using artifact detection and removal algorithms. The average band power (μV^2) for different frequency bands [e.g., delta (0–4 Hz), theta (4–8 Hz), alpha (8–13 Hz), beta (13–30 Hz), actual frequency ranges vary slightly by study] is then computed by approximating an integral of PSD over a frequency band. Relative power is often computed by dividing the power of a frequency band by the total power in the whole spectrum. Along with PSD of standard frequency bands, some studies estimate power in the slow oscillation (e.g., <0.75 Hz), spindle-frequency (sigma; 9–16 Hz) and gamma frequency bands as well as performing automated spindle density and amplitude analyses [24, 147–149]. The average band power obtained for each segment is often further averaged over a period of interest, such as a visual staging epoch and a bout of a certain sleep stage.

2.1.7 High-Density EEG (hd-EEG)

Hd-EEG provides higher spatial resolution than standard EEG by recording signals from a greater number of electrodes (up to 256 channels) [150]. To place EEG, EMG, and EOG electrodes, a net or a cap with embedded pre-wired electrodes is applied to the head. Data are topographically analyzed to examine spatial distributions of different EEG frequency power bands. With the greater and denser coverage of the head surface, differences in distributions of certain power band magnitudes across the scalp can be compared between groups (e.g., participants with a disorder vs. control) and between experimental conditions. The high spatial and temporal resolutions of hd-EEG facilitates localization of the origin and examination of propagation of specific EEG signals (e.g., sleep spindles, slow waves) or spectral power bands using source localization software [e.g., [151]].

2.1.8 Notes: PSG

Note 1: The feasibility of performing qEEG analyses on data collected from battery operated systems is limited because these systems tend to restrict sampling rate in order to maximize battery life.

Note 2: Some hair styles (e.g., a sewn-in wig) do not allow EEG sensors to be attached directly to the scalp. Before scheduling the first PSG visit, researchers, if applicable, should discuss with participants the possibility of changing their hair style for the PSG recording night.

Note 3: Participants may request study staff of a specific gender to perform PSG sensor application and/or overnight monitoring in the lab, preferences that may be associated with their trauma history. If possible, a research team should include both male and female staff trained to perform these tasks to improve participants' sense of safety and comfort.

Note 4: To avoid inducing allergic reactions, select supplies and equipment made of hypoallergic materials.

Note 5: To make the process of electrode placement smoother, prepare small pieces (2–3 inches) of surgical tapes and spread a small amount of conductive paste on one side of each small gauze square.

Note 6: If the participant is uncomfortable with the researcher placing an ECG electrode on the fifth intercostal space, the researcher may let the participant place the electrode for themselves or use an alternative location below the left clavicle.

2.2 Actigraphy

Wrist actigraphy can be used to objectively measure sleep and wakefulness over long periods (multiple weeks) in the participants' home environment. As mentioned above, this is a great advantage in studying sleep of people with PTSD whose sleep may be influenced by perceived safety of their sleep environments. This section focuses on the use of actigraphy to assess the timing and duration of sleep in adult participants with PTSD.

2.2.1 Selection of an Actigraphy System

Before choosing an actigraphy system, researchers should review validation study findings of the specific device being considered. Actigraphy systems designed to measure daytime physical activity are usually not validated for measuring sleep [152]. A meta-analysis of studies examining agreement between research-grade actigraphy and PSG revealed that actigraphy is able to detect sleep (.83 average sensitivity) better than wakefulness (.51 average specificity) and more accurately estimates sleep parameters in healthy adults compared to adults with physical or psychiatric conditions [153]. In healthy adults, actigraphy significantly underestimated SOL compared with PSG (which is the "gold standard"), but discrepancies between actigraphy and PSG were not significant for TST, SE, or

WASO. In adults with some health conditions, actigraphy significantly overestimated TST and SE and underestimated SOL [153]. A meta-analysis of people with insomnia found that FDA-approved actigraphy devices significantly and consistently underestimated SOL compared with PSG [154]. Therefore, data obtained using actigraphy should be interpreted (and possibly caveated) based on the most up-to-date information on validity of the actigraphy device.

Both research-grade actigraphy and consumer-grade wearable activity trackers have been used in sleep research. Consumer wearable activity trackers cost less than research-grade actigraphy and have the capability to connect to a mobile application allowing users to monitor their nightly sleep. Despite these advantages and widespread use of consumer-grade wearable devices, it should be noted that the validity of the sleep/wake data produced by many of these devices has not always been adequately assessed, and for those that have been tested, the results have varied considerably [155–157].

Additional considerations when selecting an actigraphy system include battery life, battery type (e.g., rechargeable), memory capacity, event marker and light sensor availability, capability to administer rating scales (e.g., pain, fatigue), and especially whether or not they are waterproof and shock resistant for extended, continuous use.

2.2.2 Preparation

Before preparing an actigraphy unit for recording, ensure that data from the previous recording session were downloaded and saved and that the battery is fully charged or replaced with a new battery (*see* **Note 1** in Subheading 2.2.5). Although specific procedures to prepare an actigraphy device for recording vary by system, an actigraphy device is normally connected to a computer and initialized using the system-specific software. During the initialization, some systems allow users to set a data collection start time and select a data collection mode (e.g., zero crossing) and epoch length (e.g., 30 s, 1 min). Select the mode and epoch length used in the validation studies of the actigraphy system unless the study goals preclude the use of these settings. In addition, carefully review the maximum recording duration limits in terms of both battery life and data storage, the latter being specific to the epoch length selected. The shorter the epoch length is, the shorter the maximum recording duration becomes.

2.2.3 Initiation of Recording

Actigraphy units should be thoroughly sanitized between participants. Actigraphy is usually worn on the nondominant wrist like a wristwatch (*see* **Notes 2** and **3** in Subheading 2.2.5). Ensure that the wristband is not wrapped too tightly or too loosely around the wrist. Providing an instruction sheet and explaining each instruction to participants will help ensure that high quality data is obtained. The instruction sheet should include information regarding the following: the importance of continuously wearing the actigraphy device; whether the device is waterproof or water-resistant and, if not waterproof, situations in which it should be removed (e.g., swimming/submerging in the water); activities should be avoided with actigraphy (e.g., contact sports); the importance of keeping the sleep diary throughout the actigraphy monitoring period; how to complete the sleep diary; and when the event marker button should be pressed. Participants are often instructed to press the event marker button when they start trying to fall asleep (i.e., at lights-out), wake up in the middle of sleep, wake up for the last time in the morning, get out of bed for the day, and try to take a nap. Participants should also be instructed not to press the event marker button when they realize that they forgot to press the button since event marks placed at later time points are sometimes confusing for researchers when scoring the data. Sleep diaries that are used in conjunction with wrist actigraphy often include items that can help the researcher interpret event marker data, such as times when the actigraphy device was removed from and placed back on the wrist (see more detailed methodological considerations for sleep diaries in the Subheading 2.3).

The duration of actigraphy monitoring can vary depending on the research needs. However, 7–14 days, including at least one weekend is typically adequate for determining a participant's typical sleep/wake patterns [155].

2.2.4 Scoring

Connect the actigraphy device to a computer and download the data. Review the data presented in a graph (*see* Fig. 2 for a sample graph of actigraphy data) and ensure that the data do not have substantial gaps (e.g., missing data due to extended off-wrist periods), low battery power, or other device malfunctions. Trim the periods at the beginning and end of the recording when the device was not on the wrist. Referring to the off-wrist periods reported on the sleep diary, manually mark the periods during which the actigraphy device was removed from the wrist.

If the actigraphy software allows users to select a scoring algorithm or sensitivity level, select the algorithm and sensitivity level that were used when the system was validated. Scoring algorithms are able to determine whether the participant was asleep or awake for each epoch. Typically, algorithms for sleep scoring are applied to portions of the recording period identified as in-bed periods (i.e., periods when the participant was trying to sleep [158]; also called a "Down" or "Rest" interval) to avoid overestimation of sleep time. Although some actigraphy software has a function to automatically identify in-bed periods, researchers should manually set in-bed periods for high quality sleep scoring (*see* **Note 4** in Subheading 2.2.5) using information from event markers and diaries. Studies

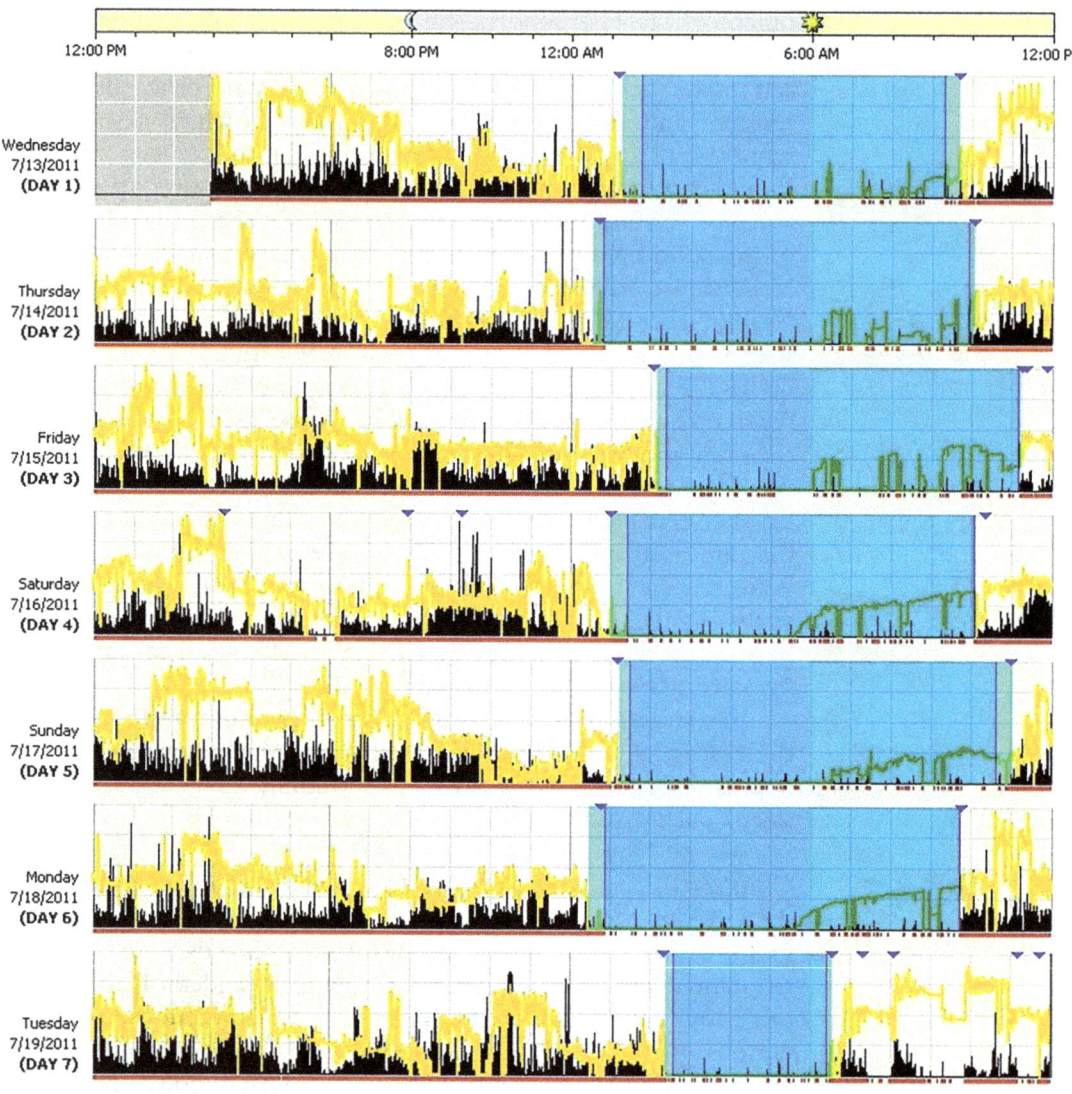

Fig. 2 An example graph presentation of actigraphy data. Note that monitoring days are centered at midnight. The black vertical lines indicate activity levels, and the yellow lines indicate white light intensity. The light blue areas indicate the in-bed intervals which can be manually set by researchers or automatically set by the software. The darker blue areas indicate sleep intervals within the in-bed intervals that are automatically set by the software. The red dots/lines below the activity indicate the epochs scored as wake by the scoring algorithm. Blue inverted triangles show the times when the event marker button was pressed

have shown better validity and reliability for manually specified in-bed periods compared to automatically determined periods [159, 160]. Although researchers have published protocols for manually identifying the beginning and end of in-bed intervals for primary sleep periods, there are no widely accepted standardized protocols. The published protocols, in general, instruct that multiple sources of information be utilized, including the amount of activity captured by actigraphy, the event marker, the sleep diary, and the light sensor, to determine in-bed intervals [155, 158–160]. In published protocols, the beginning of an in-bed period is characterized by sharp reductions of activity and light intensity that occur around the time the participant begins trying to go to sleep as indicated by the event marker and sleep diary followed by very low activity and light intensity that continue for at least several minutes (e.g., 2.5–5 min) (*see* **Note 5** in Subheading 2.2.5). The end of an in-bed interval is generally characterized by increased activity and light intensity that continue for at least several minutes. If naps are also of interest to the research, in-bed intervals for naps can also be manually specified using similar methods. However, for naps, in-bed intervals are marked only when the sleep diary or event marker indicates that the participant tried to nap because low activity and light intensity can also indicate quiet wakefulness rather than naps [158].

The actigraphy software provides estimates of various sleep/wake parameters using mathematical algorithms. The sleep variables generated by actigraph software typically include times of sleep onset and offset, time in bed, TST, SOL, WASO, and SE. Reporting and interpretation of SOL require caution [155] as estimation of SOL requires inputs from the sleep diary and/or event marker which is often inaccurate or unavailable and therefore has relatively limited validity compared to the other estimated variables [153, 154].

2.2.5 Notes: Actigraphy

Note 1: Investigators should estimate battery life (based on the date printed on the battery packaging) conservatively as this duration may change as the device ages. Investigators can periodically exchange a unit for a fully charged one if recording for extended periods. Individual devices used should be documented using serial numbers.

Note 2: Researchers should know what materials are used on the surface of the actigraphy unit and make sure participants are not allergic to any of the materials.

Note 3: For participants concerned about skin irritations, actigraphy can be placed over cloth such as a thin wristband unless the unit includes a sensor that requires direct contact with skin. Prior to determining whether this is a viable option, researchers should ensure that placing the unit over cloth does not result in epochs being recorded as off-wrist [155].

Note 4: Downloading and editing data as soon as the actigraphy device is retrieved from the participant is a best practice for obtaining high quality data. If there are any difficulties in identifying off-wrist periods or in-bed periods, ask participants for clarification before the participant leaves the lab.

Note 5: Some individuals with PTSD habitually leave a light or TV on when they sleep to avoid the dark. Therefore, the light intensity may not be a reliable indicator of the beginning of in-bed periods. This is something that could be asked and recorded along with other demographic study data and/or queried via sleep diary on a daily basis.

2.3 Sleep Questionnaires and Diaries

Sleep questionnaires and diaries are widely used in PTSD research as low-cost and easy-to-use methods to gather information about habitual sleep and subjective sleep quality. However, sleep parameters derived from subjective sleep measures, especially sleep questionnaires, are often not consistent with those obtained using PSG or actigraphy [161–163]. Both people with and without PTSD tend to overestimate sleep onset latency and underestimate WASO compared to objective sleep measures [161, 163] – limitations should be kept in mind when using questionnaires and diaries in sleep studies.

2.3.1 Common Sleep Questionnaires

Sleep questionnaires designed to assess sleep quality and symptoms of common sleep disturbance, such as insomnia, are often used in PTSD research. The following are brief descriptions of some of these questionnaires most often administered to adult participants:

The Pittsburgh Sleep Quality Index (PSQI) [164] is used to assess habitual sleep patterns, quality, and sleep-related behaviors during the past month. It consists of questions about habitual bedtime and rise time, SOL, TST, and rating scales to assess the frequency of different sleep problems, sleep quality, and sleep-related behaviors. Seven component scores (the range of each component score: 0–3) (e.g., subjective sleep quality, sleep disturbances, and daytime dysfunction) are obtained, and a global score (0–21) is computed by summing all component scores. A global score of >5 indicates clinically significant poor sleep quality.

The Insomnia Severity Index (ISI) [165] consists of seven 5-point (0–4) scales assessing severity of insomnia symptoms and their impacts over the past 2 weeks. A total score of ≥10 is able to detect insomnia cases with 86.1% sensitivity and 87.7% specificity in the general population [166].

The Epworth Sleepiness Scale (ESS) [167] is used to assess an individual's general level of daytime sleepiness "in recent times." It consists of eight 4-point scales [0 (*would never doze*) – 3 (*high chance to dozing*)] measuring tendency to doze off or fall asleep in different situations (e.g., sitting and reading, in a car as a passenger, talking to someone). Although different cutoff scores have been proposed to identify excessive daytime sleepiness, a total score of >10 is widely used and able to detect excessive daytime sleepiness with 93.5% sensitivity and 100% specificity [168].

The STOP Questionnaire [169] is a brief screener for OSA consisting of four yes/no questions: *S*nore loudly, *T*ired during daytime, *O*bserved stopped breathing during sleep, and high blood *P*ressure. Individuals who answer "Yes" to ≥2 of these questions are categorized as being at high risk of OSA. The STOP Questionnaire is also used in combination with additional clinical information: body mass index (*B*MI), *a*ge, *n*eck circumference, and *g*ender, and this alternative scoring method is called the STOP-BANG scoring model. BMI > 35 kg/m^2, age > 50 years, neck circumference > 40 cm, and being male gender are categorized as a "Yes" answer. If the participant answered "Yes" to ≥3 of the STOP-BANG questions, that individual is considered as high risk for OSA. The STOP-BANG scoring model demonstrated greater sensitivity in detecting OSA (apnea hypopnea index >5) than the STOP questionnaire alone (sensitivity of 65.6% vs. 83.6% in preoperative adult patients) [169].

The Dysfunctional Beliefs About Sleep questionnaire (DBAS) [165] was developed to assess maladaptive cognitions and attitudes regarding sleep and insomnia that are hypothesized to contribute to the perpetuation of sleep disturbances. The original DBAS [165] consists of 30 statements for which participants/patients indicate their level of agreement on a 100 mm visual analogue scale with the left end labeled "Strongly disagree" and the right end labeled "Strongly agree." The statements include, for example: "I need 8 h of sleep to feel refreshed and function well during the day" and "When I have trouble falling asleep or getting back to sleep after nighttime awakening, I should stay in bed and try harder." Responses can be analyzed both qualitatively and quantitatively. Qualitative application is particularly useful in clinical settings where potentially sleep-disruptive cognitions are initially assessed and addressed in treatment. For quantitative use, average of scores for the entire scale or subscales are computed. Abbreviated versions with 10 and 16 selected items (DBAS-10, DBAS-16) have been developed [170, 171]. A comparison of psychometric properties of these versions revealed that, overall, the DBAS-16 had better reliability and validity than the 30- and 10-item versions [172].

The Ford Insomnia Response to Stress Test (FIRST) [173] is a recently developed, brief scale in which the respondent estimates the degree to which nine specific daytime events would impact the likelihood of their having difficulty falling asleep on a 4-level scale [1 (*Not likely*) – 4 (*Very likely*)]. Individuals at high risk for clinically significant insomnia characteristically show elevated scores on this scale.

2.3.2 Questionnaires Designed to Assess Trauma/PTSD-Related Sleep Disturbances

In addition to commonly used sleep questionnaires, PTSD studies often include questionnaires specifically designed to capture trauma/PTSD-related sleep disturbances that are not assessed in common sleep/insomnia questionnaires. Questionnaires designed specifically to assess trauma/PTSD-related sleep disturbances are a fairly recent development, and use of these instruments is still limited, except for the PSQI Addendum for PTSD [27], the first published questionnaire focusing on PTSD-related sleep disturbances. The following are brief descriptions of trauma/PTSD-related sleep questionnaires.

The Pittsburgh Sleep Quality Index Addendum for PTSD (PSQI-A) [27] is a rating scale designed to assess disruptive nocturnal behaviors that are often experienced by people with PTSD but are not measured in the original PSQI (*see* Subheading 2.3.1 for the description of the PSQI). The PSQI-A consists of seven 4-point [0 (Not during the past month) – 3 (three or more times a week)] rating scales assessing the frequency of seven disruptive nocturnal behaviors (e.g., hot flashes, memories, or nightmares of a traumatic experience) in the past month and three clinical/informative items (e.g., anxiety and anger associated with the memories/nightmares). A global score, summing the seven rating scale scores, of ≥4 was able to identify women with PTSD with 94% sensitivity and 82% specificity [27] and male military veterans with PTSD with 71% sensitivity and 82% specificity [174].

The Fear of Sleep Inventory (FOSI) [175] is a self-report measure developed to assess fear of sleep and nighttime vigilance levels that are hypothesized to contribute to the persistence of sleep disturbances following trauma exposure [176]. The FOSI consists of 23 statements about thoughts and behaviors suggesting a fear of sleep or nighttime vigilance. Examples of these statements are: "I was fearful of letting my guard down while sleeping" and "I slept with something or someone in bed with me to help me feel safe." Participants indicate the frequency of each thought or behavior in the past month on a 5-point [0 (*Not at all*) – 4 (*Nearly every night*)] scale. A FOSI-Short Form (FOSI-SF) with 13 items has been developed. The total scores of both FOSI and FOSI-SF have good convergent validity with PTSD and insomnia severity [176, 177].

2.3.3 Administration, Scoring, and Interpretation of Sleep Questionnaires

Sleep questionnaires are administered using a paper-and-pencil or electronic format. Electronic questionnaires often include data quality control functions, such as allowing researchers to set a valid response range and save time entering data and checking accuracy. However, regardless of questionnaire format, researchers should review participants' responses immediately after they are turned in and, if applicable, ask participants to clarify any possibly invalid or ambiguous answers (*see* **Note 1** in Subheading 2.3.6).

Although sleep questionnaires have some advantages over objective measures, such as lower cost and burden on participants especially compared to PSG, results should be interpreted with caution as various biases inherent in retrospective self-report measures may undermine the validity of data. Biases relevant to sleep questionnaires include inability to accurately perceive or recall sleep experiences over the past few weeks, tendency to recall nights with worse symptoms than other nights, and tendency to recall sleep that meets an expectation of disturbed sleep [163].

2.3.4 Sleep Diaries

A sleep diary is a subjective measure used to prospectively monitor daily sleep-related symptoms and behaviors that can affect sleep (e.g., caffeine intake). Unlike sleep questionnaires, which require participants to recall sleep over an extended time, sleep diaries are completed daily usually in the morning immediately after awakening; therefore, impacts of some of the biases associated with sleep questionnaires (*see* Subheading 2.3.4) are mitigated. In fact, studies have shown that sleep indices (e.g., TST, SOL) measured by PSG and actigraphy were significantly correlated with respective parameters measured by sleep diaries, but not with sleep questionnaires, in individuals both with and without PTSD [161, 178].

The Consensus Sleep Diary (CSD) was developed in 2012 by a panel of sleep experts in collaboration with potential users to standardize data collection procedures [179]. Prior to the development of the CSD, no standardized or widely accepted sleep diaries existed, which made it difficult to interpret and integrate data across studies. There are two versions: the Core CSD and the Expanded CSD. Both are formatted to collect 7 days of data and include standardized instructions. The Core CSD consists of nine items with a focus on the most critical parameters, including the time getting into bed, the time trying to go to sleep, SOL, number of awakenings, total duration of awakenings, time of final awakening, the time of getting out of bed for the day, a sleep quality rating, and open-ended comments. Participants are instructed to complete the diary every morning within 1 h of getting out of bed. The Expanded CSD has two versions, for morning and evening, and includes optional items in addition to the core items. The optional items are included to gather information about the time between the final awakening and the time getting out of bed, estimation of TST, napping, consumption of alcohol and caffeinated drinks, and use of sleep medication. A validation study of the CSD with a community sample revealed that the CSD accurately measured sleep timing as measured by a single-channel EEG recording and actigraphy, but overestimated TST and SE [180].

2.3.5 Administration and Scoring of Sleep Diaries

Sleep diaries can be administered using a paper-and-pencil or electronic format. The online submission capability of digital diaries is useful for ensuring timely completion of diary entries. Regardless of the format, a daily reminder soon after participants' habitual rise-time and additional reminders, if needed, can be provided to help ensure compliance. Researchers should go over all of the diary instructions with participants prior to starting the sleep monitoring period and review their responses as soon as receiving them (*see* **Note 2** in Subheading 2.3.6). It has been recommended that sleep diaries be administered over a minimum of 2 consecutive weeks to help ensure that habitual (as opposed to transitory) sleep characteristics are captured [181].

Common sleep parameters derived from sleep diaries are subjective TIB, TST, SOL WASO, and SE. The CSD does not specify the method used to compute these parameters. As a result, different definitions of the denominator for computing SE have been used and suggested (e.g., denominator defined as duration between the time the participant tried to go to sleep and the time of getting out of bed for the day; duration between the time the participant got into bed and the time of getting out of bed for the day) [180, 182, 183]. Researchers should choose the most appropriate formula to achieve their study goals and consistently apply the formula to all participants in the study.

2.3.6 Notes: Sleep Questionnaires and Diaries

Note 1: Mix-ups of AM and PM in reporting bedtime and risetime are one of the most common errors that affect data accuracy significantly. Ask participants for clarification if they report an unusual bedtime or risetime. It's often useful to specify that the date they enter on the form should be that of the evening before the sleep period, even if that sleep period was initiated after midnight.

Note 2: Participants often have difficulty in differentiating the time they got into bed (Item 1 in the CSD) and the time they tried to go to sleep (Item 2) and understanding the time from which SOL (Item 3) should be counted. Using example situations to explain each item can help participants understand the instructions. When reviewing diary entries, watch out for common errors participants make when they do not understand definitions of these items. With correct understanding of the instructions for these items, the time they began trying to sleep is never before the time they got into bed, and the reported SOL is always the difference between the time they began trying to sleep and the time they think they went to sleep. If researchers detect signs of misunderstanding (e.g., Item 2 time is before Item 1 time; reported SOL is time between Item 1 and Item 2 times), check the participant's understanding of the instructions and correct the errors together.

2.4 Nightmare Measures

Repetitive nightmares that resemble an experienced trauma have been characterized as being among the highly specific "hallmark" symptoms of PTSD [135, 184, 185]. Although, in the past, nightmares have been explicitly associated with bad dreams that awaken the sleeper, it is now believed that bad dreams and nightmares are expressions of the same phenomenon, differing in their intensity [e.g., [186]]. This section outlines nightmare assessment approaches that have been used in PTSD research.

2.4.1 Nightmare Questionnaires and Diaries

There are a variety of nightmare questionnaires. They have generally not been validated to the extent that sleep quality questionnaires have, and their administration procedures and interpretation is less standardized. Among retrospective measures, the simplest is the single item, dichotomous (yes/no) nightmare question on the Clinician Administered PTSD Scale for DSM-5 [187], the PTSD checklist for DSM-5 [188], the Life Event Checklist for DSM-5 [189], the Acute Stress Disorder Interview [190], or the Impact of Event Scale [191]. Examples of questionnaires designed to delve deeper into the nightmare experience by assessing the characteristics of nightmares include the Disturbing Dream and Nightmare Severity Index [192] and the Nightmare Distress Questionnaire [193]. In addition, a sleep diary that provides longitudinal data on nightmares is useful to determine frequency (recall rate) or content variables related to daytime symptoms.

PTSD studies, along with a standard sleep diary, often include a nightmare questionnaire or diary that queries the extent to which recalled nightmares are exactly like (replicative), similar to (mixed), possibly related to, or unrelated to an actual traumatic experience [e.g., [194, 195]]. Some dream diaries and questionnaires in PTSD studies also include Likert-type scales on which individuals rate different aspects of their nightmares including specific emotions (e.g., fearfulness), vividness, intensity, similarity to an actual prior traumatic experience, and physical reactions (e.g., racing heart) [13, 195, 196]. Detailed content analyses are enabled by including in the sleep diary an open-ended question on dream/nightmare content [197].

2.4.2 Dream Reporting during PSG or in Conjunction with Other Sleep Monitoring

Oral or written reporting of dream content during PSG, actigraphy, or diary-only sleep monitoring has been used in PTSD research. Although the number of PSG studies have been limited, studies utilizing ambulatory systems have proved useful [198]. To obtain reports, participants are instructed to press a device's event marker button whenever awoken by a nightmare. Upon waking, they then write or, preferably, audio record the content of the nightmare using a time-stamped recorder [198, 199]. Reports of dreams not causing nocturnal awakening should be collected in the morning immediately upon waking and before other activities make recall difficult. Same-night written or audiotaped reports (including morning reports if there was no awakening) minimize memory recall failures and biases associated with retrospective surveys. Moreover, PSG recordings capture the psychophysiological characteristics of the sleep state immediately prior to any awakenings from nightmares. Notably, trauma-associated nightmares can arise from either REM or non-REM sleep [198, 200].

2.5 Methods to Examine Associations Between Sleep and Relevant Psychophysiological Phenomena

2.5.1 Fear Conditioning and Extinction

This section focuses on approaches to examining associations between sleep and fear conditioning and extinction learning and memory. For example, a commonly employed 2-day experimental fear conditioning and extinction protocol developed by Mohammed Milad [201] and designed for use in the MRI environment has been used extensively in sleep-related studies [89, 202–205]. In Session 1, during Fear Conditioning, partial reinforcement (63%) with a mild electric shock produces conditioned skin conductance responses (SCR) to two differently colored lamps (CS+), but not a third color (CS-) in a virtual "conditioning context." Immediately afterward, during Extinction Learning, one CS+ (CS+E) but not the other (CS+U) is extinguished by un-reinforced presentations within a virtual "extinction context." After a delay (the durations of which, along with other experimental manipulations, vary across studies), Extinction Recall is tested in the extinction context. Contextual Fear Renewal and/or reinstatement can also be tested in the conditioning or a novel context. For detailed methods of fear conditioning and extinction, including advantages and disadvantages of measuring fear-related arousal using SCR, fear-potentiated startle (FPS) with facial electromyography, and/or subjective (shock expectancy) measures, please refer to Chap. 5 of this book.

Table 1 presents examples of study designs used to examine associations between overnight sleep and fear conditioning and extinction learning and memory in healthy adult volunteers. In studies examining associations of sleep characteristics (e.g., amount of certain sleep stages) with fear conditioning and extinction learning and memory, sleep is often recorded using PSG without any direct experimental manipulations of sleep. In addition, observational approaches are used in studies examining the effects of fear conditioning learning on sleep (Protocols 3 & 4 in Table 1). Sleep manipulations, such as total sleep deprivation, REM sleep deprivation, and sleep restriction have been used to examine effects on extinction learning and memory of sleep versus wakefulness or the effects of specific sleep stages or sleep timing (e.g., REM sleep, first or latter half of sleep). However, approaches that manipulate sleep often confound the effects of sleep with the consequences of sleep deprivation (e.g. stress, fatigue, and reduced vigilance) during the subsequent learning or recall tasks [64, 206]. To reduce these confounding effects, Menz and colleagues [64, 207] (Protocol 8 in Table 1) included a recovery sleep night. Comparisons of overnight sleep with daytime wakefulness instead of overnight sleep deprivation [202] (Protocol 6 in Table 1) eliminate the effects of sleep deprivation; however, this approach introduces a circadian

Table 1
Examples of study designs to examine relationships between overnight sleep and fear conditioning and extinction learning and memory

#						References
Impacts of sleep on FC learning						
	1–4 night(s)	Day 1				
1	Sleep manipulation or observation	FC				[236–238]
Impacts of sleep on FC memory recall						
	Night 0	Day 1	Night 1	Day 2		
2		FC	Sleep manipulation	Recall		[206]
Impact of FC on post-FC sleep and impacts of sleep on FC recall						
3	Baseline sleep	FC	Sleep observation	Recall		[63]
Impacts of FC on post-FC sleep and impacts of sleep on fear extinction learning						
	Night 0	Day 1	Night 1	Day 2		
4	Habituation/screening sleep	FC	Sleep observation	Ext.		[239]
Impacts of sleep on fear extinction recall						
	Day 1	Night 1	Day 2	Night 2	Day 3	
5	FC	Habituation/screening sleep	Ext.	Sleep manipulation	Recall	[240]
6	[Daytime wake group] FC & Ext (AM) – Daytime wake (12 h) – Recall (PM) [Evening sleep group] FC & Ext (PM)	Normal Sleep (12 h)	Recall (AM)			[202]
Impacts of sleep on fear extinction learning and recall						
7	FC	Sleep manipulation	Ext	Sleep manipulation	Recall	[241]
Impacts of sleep on FC recall and extinction recall						
8	FC only CS (CS+ N vs. CS- N) FC & Ext. CS (CS+E vs. CS- E)	Sleep manipulation	Normal wake activity	Recovery sleep	Recall	[64, 207]

Note. *FC* fear conditioning, *Ext.* fear extinction, *CS+ E* conditioned stimulus that was subject to extinction, *CS+ N* conditioned stimulus that was not subject to extinction, *CS- E* control stimulus that was presented during conditioning and extinction; *CS- N* = control stimulus that was presented during conditioning only rhythm confound – a significant concern since Pace-Schott and colleagues [203] have demonstrated that fear extinction is better learned and generalized in the morning than in the evening.

Daytime napping paradigms have also been used in studies examining associations of sleep with fear conditioning and extinction memory. Experimental manipulations (e.g., daytime napping vs. remaining awake) and the characteristics of a nap (e.g., amounts of various sleep stages) have been associated with fear conditioning and extinction learning and recall [59, 208, 209]. Napping paradigms require less time than overnight sleep studies, and confounding effects of sleep deprivation and circadian rhythms can be minimized. However, naps lack the characteristic architecture of nocturnal sleep, and key sleep stages (e.g. REM, SWS) may be absent during naps.

2.5.2 Nocturnal Heart Rate Variability (HRV)

Many studies examining autonomic activity during sleep have focused on the strength of PNS outflow as indexed from ECG data by the root mean square of the successive differences in the R-R interval (RMSSD) or by the absolute power of the high frequency band (0.15–0.40 Hz) of variations in the R-R interval (HF-power). These two highly correlated variables reflect modulation of heart rate by the vagus nerve, the main pathway of PNS outflow to the periphery from brainstem nuclei. Although the low frequency (LF) to HF power ratio (LF/HF; a.k.a., sympathovagal balance) has been used in the past as a measure of relative sympathetic or parasympathetic dominance, its use has become controversial due to the inaccuracy of the LF band as a measure of sympathetic tone [210–212]. Generally, at least five continuous minutes or more of ECG with a high sampling rate (minimum of 500 Hz) is recommended, especially for measuring HF-power [210–212]. Nonetheless, these criteria are frequently not met, especially when computing RMSSD from PSG (which is typically recorded with a sampling rate of ~250 Hz). Commercial analytic systems, such as Kubios software (Kubios Oy, Kuopio, Finland) provide artifact rejection (e.g., ectopic beats, premature ventricular contractions) and compute both time and frequency-domain HRV measures from ECG data. These data are saved in European Data Format (EDF) that is common to PSG systems, as well as other data formats. Typically, HF-power is logarithmically transformed to normalize data distribution prior to analysis.

2.5.3 Cortisol

It is generally recommended that measurement of salivary cortisol be carried out using unstimulated, whole saliva collected by the passive drool technique [213] using materials provided by commercial laboratories such as Salimetrics, LLC (Carlsbad, CA). Once obtained, saliva samples should be stored in airtight vials, stored at −20 °C and analyzed within approximately 0.5 year (to avoid slow evaporative changes). Unbound cortisol is analyzed using enzyme immunoassay (EIA) with commercially available kits or by commercial laboratories. To control for inter-individual variability, a diurnal profile of cortisol levels can be obtained to compute an area-under-curve (AUC) baseline [214] and a provide values with which concentration values obtained at other times may be normalized. To obtain this profile, five to six samples are typically taken, for example: immediately upon waking, 30 min after waking (in order to capture the cortisol awakening response or CAR), at noon or just before lunch (whichever is first), at 5 PM or just before dinner (whichever is first) and at 10:00 PM. Tested individuals should be instructed not to eat, smoke, drink caffeinated beverages or exercise, for 1 h prior to providing saliva samples.

2.5.4 Chronotype and Circadian Markers

Chronotype is a circadian trait classification of individuals' time-of-day preference (e.g., morningness and eveningness). It can be assessed subjectively using the Morningness/Eveningness Questionnaire (MEQ) [215] as well as a growing number of briefer questionnaires such as the Composite Scale of Morningness [216, 217] and the Munich Chronotype Questionnaire [218]. Subjective measures of chronotype map well onto objective measures of the output of the circadian clock such as described below.

The dim-light melatonin onset assay or DLMO has become widely used in circadian and sleep laboratories [219]. Procedures very similar to those used in the laboratory have now been adapted for ambulatory assessment, thereby greatly extending opportunities for its use [220]. In the ambulatory assay, on the sample collection evening, the participant is directed to collect a total of 8 saliva samples: 1 saliva sample each hour beginning 7 h before their usual bedtime and ending 1 h after their usual bedtime. They are provided nightlights, bulb adapters, dim bulbs, and blackout goggles which are worn throughout the procedure. They are also instructed not to eat anything within 30 min of a sample and to rinse their mouth with water after eating. Pre-labeled wide-opening polystyrene tubes are provided to the participant, who is instructed to fill each tube at least half way with saliva (which ensure a volume of at least 2 mL). As each sample is collected, participants document the tube number and clock time on a log and store each sample in their home freezer. Frozen saliva samples and logs are subsequently turned in to the researcher during the next visit to the laboratory. Samples can then be sent to a commercial laboratory such as Solidphase, Inc. (Portland, ME) for assay. Such laboratories often use the Bühlmann Direct Saliva Melatonin radioimmunoassay kit (Buhlmann Diagnostics Corp., Amherst, NH). This assay is based on the Kennaway G280 anti-melatonin antibody [221]. Melatonin phase is determined from the DLMO as the time at which the saliva levels reach 3 pg/mL [222]. This threshold accounts for the lower melatonin levels in saliva compared to plasma [223, 224]. Linear interpolation between adjacent samples is used if necessary to determine DLMO. A far less costly, albeit less accurate, correlate of DLMO is the mid-sleep point on "free nights" (those without imposed bedtime or rise time such as weekend or vacation nights) [225, 226].

2.5.5 Structural and Functional Brain Imaging

The most popular neuroimaging techniques applied in sleep research are PET, MRI, single-photon emission computed tomography (SPECT), and near-infrared spectroscopy (NIRS). PET and SPECT imaging use radioactive tracers to quantify and visualize metabolic changes in the brain, for example, blood flow and local chemical composition. Structural MRI exploits the decay rate property of proton spins to distinguish different brain tissues. After disturbing spins of hydrogen atoms by applying a radiofrequency pulse at the resonant frequency, spins return to baseline and realign with the magnetic field (generally 1.5 or 3T in human studies) at time rates determined by regional chemical composition. There are several MRI sequences dedicated to defining tissues in the brain. For instance, T1-weighted imaging provides superior gray and white matter contrast, and T2-weighted sequences usually contrast cerebrospinal fluid and brain tissues. In functional MRI (fMRI), the blood-oxygen-level-dependent (BOLD) signal, measures hemodynamic changes as a proxy for neuronal activity. NIRS uses an infrared light signal, which can penetrate skin, scalp, and dura to reach the cortex, after which sensors ("optodes") measure the backscattered signal. In NIRS, differences in absorption of near infrared light by oxygenated and deoxygenated hemoglobin are indirect measures of neural activity. PET imaging has better image sensitivity and resolution than SPECT, whereas the latter is less costly and uses lower doses of radioactive tracer. Since fMRI can measure brain hemodynamics without the need for radioactive tracers, this technique has become the most widely used research neuroimaging modality allowing repeated measurements in populations with contraindications for use of radioactive tracers such as newborns. In addition, fMRI has a better spatiotemporal resolution and lower costs compared to PET and SPECT. NIRS has several advantages over the other neuroimaging methods as it is motion tolerant, quiet, portable and relatively inexpensive, all features of great utility for sleep studies. However, compared to other modalities, NIRS has poorer spatial resolution, a lower signal-to-noise ratio and is unable to visualize deep brain structures with images restricted to lateral cortices. For studies of the brain during sleep, each neuroimaging method (or combination of methods, e.g., high density EEG and PET) provides capabilities, advantages, and disadvantages (*see* Table 2).

Sleep neuroimaging imposes challenges summarized in three main areas: (i) *hardware challenges*: If, for example, PSG needs to be

Table 2
Comparison between neuroimaging methods

Modality	Image principle	Resolution	Time	Information	Cost
PET	Different radioactive tracers are used depending on the image purpose. Gamma-rays are detected by a camera to form multi-dimensional images	≈3 mm	sec-min	Physiological, molecular	High
SPECT	Gamma-emitting radioactive tracers are detected by a gamma camera acquiring multiple 2D images	≈7 mm	min	Physiological, molecular	High
NIRS	The absorbed ratio of infrared light is measured as this ratio changes depending on oxygenated hemoglobin concentration	≈1 cm	sec-min	Physiological, molecular	Medium
fMRI	BOLD signal results from detected changes in magnetic field on oxygenated brain blood flow	≈3 mm	sec-min	Physiological, molecular	High
sMRI	Strong magnetic field forces spins to align, MR sensors detect spins decay as the energy released and the time decay depends on local tissue chemical composition	≈1 mm	sec-min	Anatomical	High

Note. *PET* positron emission tomography, *SPECT* single-photon emission computed tomography, *NIRS* near-infrared spectroscopy, *fMRI* functional magnetic resonance imaging, *sMRI* structural magnetic resonance imaging, *BOLD* blood-oxygen-level-dependent, *MR* magnetic resonance

recorded in an MR scanner, then specially designed equipment for this purpose will need to be purchased. However, standard PSG systems can be used in a PET scanner; (ii) *experimental setup challenges*: for both MRI and PET, head movement must be restricted (which can reduce comfort and interfere with sleep onset). MRI scans are very noisy. For NIRS, the arrangement of EEG electrodes and optical channels is a primary concern, whereas for PET and SPECT, discomfort at the site of radiotracer injection is a concern; and (iii) *data analysis challenges*: simultaneous EEG-fMRI requires the removal of unsystematic artifacts from the EEG data. Additional EEG, EOG, and EMG electrodes (required for PSG recordings) compound the challenges, such as using auxiliary electrodes and channels that can, in turn, lead to hardware constraints and signal distortions due to reference electrodes and cardio-ballistic artifacts – problems can be addressed by using EEG systems designed for use in the MR environment. If such EEG system is not used, one way to overcome such problems may be by duplicating frontal and temporal channels that are then re-referenced to contralateral mastoids to record EOG and EMG signals. In addition, to visualize eye movements and muscle atonia, band-pass filters need to be adjusted in a manner that is not consistent with the AASM scoring manual [136]. Especially if the research question concerns phasic and tonic REM sleep, observing EMG activity drops may be challenging. It is also important to mention the correct placement of the EEG amplifier and its cables inside the scanner room, following the manufacturer's recommendations, to avoid signal loss and artifacts due to possible cable movements, especially when the amplifier is placed inside the scanner bore.

As mentioned above, the scanner noise is a well-known methodological constraint of sleep imaging studies. Studies have shown that REM sleep is suppressed by acoustic noise, explaining the much lesser REM compared to non-REM sleep seen in neuroimaging studies [227]. Recent developments in MRI promise silent sequences that can prevent sleep disturbances due to loud scanner noise [228]. Adaptation nights prior to the experimental session can help reduce drop-out rates as participants can habituate to the scanner equipment, movement restrictions, and habituate to the scanner environment. Sleep loss has also been used to exacerbate sleepiness and facilitate subsequent sleep in the scanner. However, prior sleep loss can potentially alter sleep architecture (e.g., increase slow wave activity), which may impact the generalizability of the findings [229]. Moreover, MRI-compatible noise-canceling headphones are commercially available and can help mitigate scanner-noise problems. They are also useful for administering acoustic stimulation in the MR environment.

Movement restrictions can be a significant challenge in sleep neuroimaging studies. Participants cannot move their head, much less change positions, which produce imaging artifacts in MRI and PET studies. In addition, for PET studies, the participant will also have an immobilized arm with an indwelling catheter for radiotracer injection. These restrictions accompanied by lengthy recordings account for the high drop-out rates in sleep neuroimaging studies. However, technological innovations in the area of portable MRI that may reduce these challenges are on the horizon [230, 231].

Since sleep neuroimaging studies must use simultaneous PSG recordings to perform accurate sleep scoring, another major challenge is the removal of non-systematic artifacts from the data – a particular challenge for MRI studies. As private companies mainly design their own artifact removal software, modification of such to adapt to any individual or sequence-based artifact removal algorithm is often impractical. Alternative methods of scoring sleep, for example, based on electrocardiogram or respiratory signals instead of a full PSG set up (EEG, EOG, and EMG) may eventually provide a workable solution [232] that increases subject comfort and reduces data collection complexity.

This brief overview of imaging methods summarizes the main points to consider when planning sleep neuroimaging studies. For more detailed recommendations on the experimental design of each method, we refer the reader to Moehlman et al. [233] and Shah et al. [234]. Combining neuroimaging methods with electrophysiology recordings has opened new research avenues and helped researchers cross-correlate animal and human studies. Neuroimaging provides information necessary to uncover underlying sleep neuronal mechanisms from a whole-phenomenon perspective by focusing on different sleep stages as well as stage-specific events such as sleep spindles, K-complexes, slow waves, and rapid eye movements. PTSD research can benefit from sleep studies from several perspectives. For instance, investigating the structural and functional neural correlates of nightmares may provide new insights for developing novel treatments. For example, a better understanding of the neural mechanisms underlying lucid dreaming may reveal ways in which PTSD patients might gain control of and overcome nightmares.

The limitations imposed by the complexity of sleep neuroimaging designs should be carefully considered when planning imaging of sleep in PTSD. Nonetheless, the feasibility of sleep neuroimaging studies is already a reality that will enable great discoveries in all areas related to sleep and sleep disturbances.

3 Conclusion

Sleep studies in PTSD utilize various methods of assessing sleep and relevant psychophysiological processes. Traditional assessment approaches such as PSG, sleep diaries, and HRV have been widely used in sleep studies in PTSD research. However, to date, the use of cutting-edge brain imaging methods to explore the relationship between sleep and PTSD has been extremely limited. Individuals with PTSD have been included in only a small number of studies that examine the role of sleep in emotion regulation processes, such as fear conditioning and extinction, and exposure therapy. To facilitate the translation of research findings into clinical practice, more studies with clinical samples are needed.

Acknowledgements

This chapter was written while Ihori Kobayashi held an NRC Research Associateship award at Performance Assessment and Chemical Evaluation (PACE) Laboratory, Behavioral Biology Branch, Center for Military Psychiatry and Neuroscience, Walter Reed Army Institute of Research.

Disclaimer The contents, opinions, and assertions contained herein are the private views of the authors and are not to be considered as official or reflecting the view of Walter Reed Army Institute of Research, the Department of the Army, or the Department of Defense.

References

1. Ohayon MM, Shapiro CM (2000) Sleep disturbances and psychiatric disorders associated with posttraumatic stress disorder in the general population. Compr Psychiatry 41(6):469–478
2. American Psychiatric Association (2013) Diagnostic and statistical manual of mental disorders, 5th edn. American Psychiatric Association, Washington, DC
3. Milanak ME, Zuromski KL, Cero I, Wilkerson AK, Resnick HS, Kilpatrick DG (2019) Traumatic event exposure, posttraumatic stress disorder, and sleep disturbances in a national sample of US adults. J Trauma Stress 32(1):14–22
4. Wittmann L (2007) PTSD: posttraumatic sleep disorder? Sleep Hypn 9(1):1–5
5. Leskin GA, Woodward SH, Young HE, Sheikh JI (2002) Effects of comorbid diagnoses on sleep disturbance in PTSD. J Psychiatr Res 36(6):449–452
6. Creamer JL, Brock MS, Matsangas P, Motamedi V, Mysliwiec V (2018) Nightmares in United States military personnel with sleep disturbances. J Clin Sleep Med 14(3):419–426
7. Miller MW, Wolf EJ, Kilpatrick D, Resnick H, Marx BP, Holowka DW et al (2013) The prevalence and latent structure of proposed DSM-5 posttraumatic stress disorder symptoms in US national and veteran samples. Psychol Trauma Theory Res Pract Policy 5(6):501
8. Grossman ES, Hoffman YS, Shrira A, Kedar M, Ben-Ezra M, Dinnayi M et al (2019) Preliminary evidence linking complex-PTSD to insomnia in a sample of Yazidi genocide survivors. Psychiatry Res 271:161–166
9. Neylan TC, Marmar CR, Metzler TJ, Weiss DS, Zatzick DF, Delucchi KL et al (1998) Sleep disturbances in the Vietnam generation: findings from a nationally representative sample of male Vietnam veterans. Am J Psychiatry 155(7):929–933
10. Zhang Y, Ren R, Sanford LD, Yang L, Zhou J, Zhang J et al (2019) Sleep in posttraumatic stress disorder: a systematic review and meta-analysis of polysomnographic findings. Sleep Med Rev 48:101210
11. Pace-Schott EF, Germain A, Milad MR (2015) Sleep and REM sleep disturbance in the pathophysiology of PTSD: the role of extinction memory. Biol Mood Anxiety Disord 5(1):1–19
12. Kobayashi I, Sledjeski EM, Spoonster E, Fallon WF, Delahanty DL (2008) Effects of early nightmares on the development of sleep disturbances in motor vehicle accident victims. J Trauma Stress 21(6):548–555
13. Mellman TA, David D, Bustamante V, Torres J, Fins A (2001) Dreams in the acute aftermath of trauma and their relationship to PTSD. J Trauma Stress 14(1):241–247
14. Bryant RA, Creamer M, O'Donnell M, Silove D, McFarlane AC (2010) Sleep disturbance immediately prior to trauma predicts subsequent psychiatric disorder. Sleep 33(1):69–74
15. Gehrman P, Seelig A, Jacobson I, Boyko E, Hooper T, Gackstetter G et al (2012) Predeployment sleep duration and insomnia symptoms as risk factors for new-onset mental health disorders following military deployment. Sleep 36(7):1009–1018
16. Mellman TA, Bustamante V, Fins AI, Pigeon WR, Nolan B (2002) REM sleep and the early development of posttraumatic stress disorder. Am J Psychiatry 159(10):1696–1701
17. Mellman TA, Knorr BR, Pigeon WR, Leiter JC, Akay M (2004) Heart rate variability during sleep and the early development of posttraumatic stress disorder. Biol Psychiatry 55(9):953–956
18. Lommen MJJ, Grey N, Clark DM, Wild J, Stott R, Ehlers A (2015) Sleep and treatment outcome in posttraumatic stress disorder: results from an effectiveness study. Depress Anxiety 33:575–583
19. Zalta AK, Pinkerton LM, Valdespino-Hayden Z, Smith DL, Burgess HJ, Held P

20. et al (2020) Examining insomnia during intensive treatment for veterans with posttraumatic stress disorder: does it improve and does it predict treatment outcomes? J Trauma Stress 33(4):521–527
21. López CM, Lancaster CL, Gros DF, Acierno R (2017) Residual sleep problems predict reduced response to prolonged exposure among veterans with PTSD. J Psychopathol Behav Assess 39(4):755–763
22. Nishith P, Duntley SP, Domitrovich PP, Uhles ML, Cook BJ, Stein PK (2003) Effect of cognitive behavioral therapy on heart rate variability during REM sleep in female rape victims with PTSD. J Trauma Stress 16(3):247–250
23. Germain A, Buysse DJ, Nofzinger E (2008) Sleep-specific mechanisms underlying posttraumatic stress disorder: integrative review and neurobiological hypotheses. Sleep Med Rev 12(3):185–195
24. Straus LD, Drummond S, Risbrough VB, Norman SB (2017) Sleep disruption, safety learning, and fear extinction in humans: implications for posttraumatic stress disorder. In: Behavioral neurobiology of PTSD. Springer, pp 193–205
25. Wang C, Ramakrishnan S, Laxminarayan S, Dovzhenok A, Cashmere JD, Germain A et al (2020) An attempt to identify reproducible high-density EEG markers of PTSD during sleep. Sleep 43(1):zsz207
26. Wilson TW, Heinrichs-Graham E, Proskovec AL, McDermott TJ (2016) Neuroimaging with magnetoencephalography: a dynamic view of brain pathophysiology. Transl Res 175:17–36
27. Baillet S (2017) Magnetoencephalography for brain electrophysiology and imaging. Nat Neurosci 20(3):327–339
28. Germain A, Hall M, Krakow B, Shear M, Buysse D (2005) A brief sleep scale for posttraumatic stress disorder: Pittsburgh sleep quality index addendum for PTSD. Anxiety Disord 19:233–244
29. Elliott JE, Opel RA, Pleshakov D, Rachakonda T, Chau AQ, Weymann KB et al (2020) Posttraumatic stress disorder increases the odds of REM sleep behavior disorder and other parasomnias in Veterans with and without comorbid traumatic brain injury. Sleep 43(3):zsz237
30. Mysliwiec V, Brock MS, Creamer JL, O'Reilly BM, Germain A, Roth BJ (2018) Trauma associated sleep disorder: a parasomnia induced by trauma. Sleep Med Rev 37:94–104
31. Barone DA (2020) Dream enactment behavior—a real nightmare: a review of posttraumatic stress disorder, REM sleep behavior disorder, and trauma-associated sleep disorder. J Clin Sleep Med 16(11):1943–1948
32. Brock MS, Powell TA, Creamer JL, Moore BA, Mysliwiec V (2019) Trauma associated sleep disorder: clinical developments 5 years after discovery. Curr Psychiatry Rep 21(9):1–11
33. Hurwitz TD, Mahowald MW, Kuskowski M, Engdahl BE (1998) Polysomnographic sleep is not clinically impaired in Vietnam combat veterans with chronic posttraumatic stress disorder. Biol Psychiatry 44(10):1066–1073
34. Lipinska G, Thomas KG (2017) Better sleep in a strange bed? Sleep quality in South African women with posttraumatic stress disorder. Front Psychol 8:1555
35. Kobayashi I, Boarts JM, Delahanty DL (2007) Polysomnographically measured sleep abnormalities in PTSD: a meta-analytic review. Psychophysiology 44(4):660–669
36. Woodward SH, Bliwise DL, Friedman MJ, Gusman DF (1996) Subjective versus objective sleep in Vietnam combat veterans hospitalized for PTSD. J Trauma Stress 9(1):137–143
37. Chambers EC, Pichardo MS, Rosenbaum E (2016) Sleep and the housing and neighborhood environment of urban Latino adults living in low-income housing: the AHOME study. Behav Sleep Med 14(2):169–184
38. Hall Brown T, Mellman TA (2014) The influence of PTSD, sleep fears, and neighborhood stress on insomnia and short sleep duration in urban, young adult, African Americans. Behav Sleep Med 12(3):198–206
39. Ding L, Chen B, Dai Y, Li Y (2022) A meta-analysis of the first-night effect in healthy individuals for the full age spectrum. Sleep Med 89:159–165
40. Herbst E, Metzler TJ, Lenoci M, McCaslin SE, Inslicht S, Marmar CR et al (2010) Adaptation effects to sleep studies in participants with and without chronic posttraumatic stress disorder. Psychophysiology 47(6):1127–1133
41. Krakow BJ, Ulibarri VA, Moore BA, McIver ND (2015) Posttraumatic stress disorder and sleep-disordered breathing: a review of comorbidity research. Sleep Med Rev 24:37–45
42. Zhang Y, Weed JG, Ren R, Tang X, Zhang W (2017) Prevalence of obstructive sleep apnea in patients with posttraumatic stress disorder and its impact on adherence to continuous

42. Senaratna CV, Perret JL, Lodge CJ, Lowe AJ, Campbell BE, Matheson MC et al (2017) Prevalence of obstructive sleep apnea in the general population: a systematic review. Sleep Med Rev 34:70–81

43. Bardwell WA, Moore P, Ancoli-Israel S, Dimsdale JE (2000) Does obstructive sleep apnea confound sleep architecture findings in subjects with depressive symptoms? Biol Psychiatry 48(10):1001–1009

44. Kimoff RJ (1996) Sleep fragmentation in obstructive sleep apnea. Sleep 19(suppl_9):S61–S66

45. Breslau N, Roth T, Burduvali E, Kapke A, Schults L, Roehrs T (2005) Sleep in lifetime posttraumatic stress disorder: a community-based polysomnographic study. Correct Arch Gen Psychiatry 62(2):172

46. Abrams MP, Mulligan AD, Carleton RN, Asmundson GJ (2008) Prevalence and correlates of sleep paralysis in adults reporting childhood sexual abuse. J Anxiety Disord 22(8):1535–1541

47. Mellman TA, Aigbogun N, Graves RE, Lawson WB, Alim TN (2008) Sleep paralysis and trauma, psychiatric symptoms and disorders in an adult African American population attending primary medical care. Depress Anxiety 25(5):435–440

48. Van Praag DL, Cnossen MC, Polinder S, Wilson L, Maas AI (2019) Post-traumatic stress disorder after civilian traumatic brain injury: a systematic review and meta-analysis of prevalence rates. J Neurotrauma 36(23):3220–3232

49. Loignon A, Ouellet MC, Belleville G (2020) A systematic review and meta-analysis on PTSD following TBI among military/veteran and civilian populations. J Head Trauma Rehabil 35(1):E21–E35

50. Iljazi A, Ashina H, Al-Khazali HM, Lipton RB, Ashina M, Schytz HW et al (2020) Post-traumatic stress disorder after traumatic brain injury—a systematic review and meta-analysis. Neurol Sci 41(10)

51. Mathias J, Alvaro P (2012) Prevalence of sleep disturbances, disorders, and problems following traumatic brain injury: a meta-analysis. Sleep Med 13(7):898–905

52. Grima N, Ponsford J, Rajaratnam SM, Mansfield D, Pase MP (2016) Sleep disturbances in traumatic brain injury: a meta-analysis. J Clin Sleep Med 12(3):419–428

53. Zatzick DF, Rivara FP, Nathens AB, Jurkovich GJ, Wang J, Fan M et al (2007) A nationwide US study of post-traumatic stress after hospitalization for physical injury. Psychol Med 37(10):1469–1480

54. Swann MC, Batty M, Hu G, Mitchell T, Box H, Starr A (2018) Sleep disturbance in orthopaedic trauma patients. J Orthop Trauma 32(10):500–504

55. Raymond I, Ancoli-Israel S, Choinière M (2004) Sleep disturbances, pain and analgesia in adults hospitalized for burn injuries. Sleep Med 5(6):551–559

56. Budh CN, Hultling C, Lundeberg T (2005) Quality of sleep in individuals with spinal cord injury: a comparison between patients with and without pain. Spinal Cord 43(2):85–95

57. Mellman TA, Kobayashi I, Lavela J, Wilson B, Hall Brown TS (2014) A relationship between REM sleep measures and the duration of post-traumatic stress disorder in a Young adult urban minority population. Sleep 37:1321–1326

58. Pace-Schott EF, Bottary RM, Kim S, Rosencrans PL, Vijayakumar S, Orr SP et al (2018) Effects of post-exposure naps on exposure therapy for social anxiety. Psychiatry Res 270:523–530

59. Spoormaker VI, Sturm A, Andrade KC, Schröter MS, Goya-Maldonado R, Holsboer F et al (2010) The neural correlates and temporal sequence of the relationship between shock exposure, disturbed sleep and impaired consolidation of fear extinction. J Psychiatr Res 44(16):1121–1128

60. Stickgold R, Walker MP (2013) Sleep-dependent memory triage: evolving generalization through selective processing. Nat Neurosci 16(2):139–145

61. Schenker MT, Ney LJ, Miller LN, Felmingham KL, Nicholas CL, Jordan AS (2021) Sleep and fear conditioning, extinction learning and extinction recall: a systematic review and meta-analysis of polysomnographic findings. Sleep Med Rev 59:101501

62. Schäfer SK, Wirth BE, Staginnus M, Becker N, Michael T, Sopp MR (2020) Sleep's impact on emotional recognition memory: a meta-analysis of whole-night, nap, and REM sleep effects. Sleep Med Rev 51:101280

63. Marshall AJ, Acheson DT, Risbrough VB, Straus LD, Drummond SP (2014) Fear conditioning, safety learning, and sleep in humans. J Neurosci 34(35):11754–11760

64. Menz MM, Rihm JS, Salari N, Born J, Kalisch R, Pape HC et al (2013) The role of sleep and sleep deprivation in consolidating fear memories. NeuroImage 75:87–96

65. Pace-Schott EF, Tracy LE, Rubin Z, Mollica AG, Ellenbogen JM, Bianchi MT et al (2014) Interactions of time of day and sleep with between-session habituation and extinction memory in young adult males. Exp Brain Res 232:1–16
66. Straus LD, Norman SB, Risbrough VB, Acheson DT, Drummond SP (2018) REM sleep and safety signal learning in posttraumatic stress disorder: a preliminary study in military veterans. Neurobiol Stress 9:22–28
67. Bottary R, Seo J, Daffre C, Gazecki S, Moore KN, Kopotiyenko K et al (2020) Fear extinction memory is negatively associated with REM sleep in insomnia disorder. Sleep 43(7):zsaa007
68. Michael T, Ehlers A, Halligan SL, Clark D (2005) Unwanted memories of assault: what intrusion characteristics are associated with PTSD? Behav Res Ther 43(5):613–628
69. Porcheret K, Iyadurai L, Bonsall MB, Goodwin GM, Beer SA, Darwent M et al (2020) Sleep and intrusive memories immediately after a traumatic event in emergency department patients. Sleep 43(8):zsaa033
70. Kleim B, Wysokowsky J, Schmid N, Seifritz E, Rasch B (2016) Effects of sleep after experimental trauma on intrusive emotional memories. Sleep 39(12):2125–2132
71. Zeng S, Lau EYY, Li SX, Hu X (2021) Sleep differentially impacts involuntary intrusions and voluntary recognitions of lab-analogue traumatic memories. J Sleep Res 30(3): e13208
72. Sopp MR, Brueckner AH, Schäfer SK, Lass-Hennemann J, Michael T (2019) REM theta activity predicts re-experiencing symptoms after exposure to a traumatic film. Sleep Med 54:142–152
73. Porcheret K, Holmes EA, Goodwin GM, Foster RG, Wulff K (2015) Psychological effect of an analogue traumatic event reduced by sleep deprivation. Sleep 38(7):1017–1025
74. Porcheret K, van Heugten–van der Kloet D, Goodwin GM, Foster RG, Wulff K, Holmes EA (2019) Investigation of the impact of total sleep deprivation at home on the number of intrusive memories to an analogue trauma. Transl Psychiatry 9(1):1–13
75. Davidson P, Pace-Schott E (2021) Go to bed and you might feel better in the morning—the effect of sleep on affective tone and intrusiveness of emotional memories. Curr Sleep Med Rep 7(2):31–46
76. Pitman RK, Orr SP, Forgue DF, de Jong JB, Claiborn JM (1987) Psychophysiologic assessment of posttraumatic stress disorder imagery in Vietnam combat veterans. Arch Gen Psychiatry 44(11):970–975
77. Pitman RK, Orr SP, Forgue DF, Altman B, de Jong JB, Herz LR (1990) Psychophysiologic responses to combat imagery of Vietnam veterans with posttraumatic stress disorder versus other anxiety disorders. J Abnorm Psychol 99(1):49–54
78. Lindauer RT, van Meijel EP, Jalink M, Olff M, Carlier IV, Gersons BP (2006) Heart rate responsivity to script-driven imagery in posttraumatic stress disorder: specificity of response and effects of psychotherapy. Psychosom Med 68(1):33–40
79. Barkay G, Freedman N, Lester H, Louzoun Y, Sapoznikov D, Luckenbaugh D et al (2012) Brain activation and heart rate during script-driven traumatic imagery in PTSD: preliminary findings. Psychiatry Res Neuroimaging 204(2–3):155–160
80. Wangelin BC, Tuerk PW (2015) Taking the pulse of prolonged exposure therapy: physiological reactivity to trauma imagery as an objective measure of treatment response. Depress Anxiety 32(12):927–934
81. Shalev AY, Orr SP, Pitman RK (1992) Psychophysiologic response during script-driven imagery as an outcome measure in posttraumatic stress disorder. J Clin Psychiatry 53(9): 324–326
82. Babson KA, Badour CL, Feldner MT, Bunaciu L (2012) The relationship of sleep quality and PTSD to anxious reactivity from idiographic traumatic event script-driven imagery. J Trauma Stress 25(5):503–510
83. Rhudy JL, Davis JL, Williams AE, McCabe KM, Bartley EJ, Byrd PM et al (2010) Cognitive-behavioral treatment for chronic nightmares in trauma-exposed persons: assessing physiological reactions to nightmare-related fear. J Clin Psychol 66(4):365–382
84. Rhudy JL, Davis JL, Williams AE, McCabe KM, Byrd PM (2008) Physiological–emotional reactivity to nightmare-related imagery in trauma-exposed persons with chronic nightmares. Behav Sleep Med 6(3):158–177
85. Pace-Schott EF, Verga PW, Bennett TS, Spencer R (2012) Sleep promotes consolidation and generalization of extinction learning in simulated exposure therapy for spider fear. J Psychiatr Res 46(8):1036–1044
86. Kleim B, Wilhelm F, Temp L, Margraf J, Wiederhold B, Rasch B (2014) Sleep enhances exposure therapy. Psychol Med 44(7):1511–1519
87. Kobayashi I, Mellman TA, Cannon A, Brown I, Boadi L, Howell MK et al (2022)

Blocking the orexin system following therapeutic exposure promoted between session habituation, but not PTSD symptom reduction. J Psychiatr Res 145:222–229

88. Kobayashi I, Lavela J, Mellman TA (2014) Nocturnal autonomic balance and sleep in PTSD and resilience. J Trauma Stress 27(6): 712–716

89. Seo J, Oliver KI, Daffre C, Moore KN, Gazecki S, Lasko NB et al (2021) Associations of sleep measures with neural activations accompanying fear conditioning and extinction learning and memory in trauma-exposed individuals. Sleep 45:zsab261

90. Kobayashi I, Lavela J, Bell K, Mellman TA (2016) The impact of posttraumatic stress disorder versus resilience on nocturnal autonomic nervous system activity as functions of sleep stage and time of sleep. Physiol Behav 164:11–18

91. Ulmer CS, Hall MH, Dennis PA, Beckham JC, Germain A (2018) Posttraumatic stress disorder diagnosis is associated with reduced parasympathetic activity during sleep in US veterans and military service members of the Iraq and Afghanistan wars. Sleep 41(12): zsy174

92. Gramlich MA, Smolenski DJ, Norr AM, Rothbaum BO, Rizzo AA, Andrasik F et al (2021) Psychophysiology during exposure to trauma memories: comparative effects of virtual reality and imaginal exposure for posttraumatic stress disorder. Depress Anxiety 38(6):626–638

93. Young EA, Breslau N (2004) Cortisol and catecholamines in posttraumatic stress disorder: an epidemiologic community study. Arch Gen Psychiatry 61(4):394–401

94. Schneider M, Schwerdtfeger A (2020) Autonomic dysfunction in posttraumatic stress disorder indexed by heart rate variability: a meta-analysis. Psychol Med 50(12):1937–1948

95. Pan X, Kaminga AC, Wen SW, Liu A (2018) Catecholamines in post-traumatic stress disorder: a systematic review and meta-analysis. Front Mol Neurosci 11:450

96. Yehuda R, Siever LJ, Teicher MH, Levengood RA, Gerber DK, Schmeidler J et al (1998) Plasma norepinephrine and 3-methoxy-4-hydroxyphenylglycol concentrations and severity of depression in combat posttraumatic stress disorder and major depressive disorder. Biol Psychiatry 44(1):56–63

97. Mellman T, Kumar A, Kulick-Bell R, Kumar M, Nolan B (1995) Nocturnal/daytime urine noradrenergic measures and sleep in combat-related PTSD. Biol Psychiatry 38: 174–179

98. Agorastos A, Boel JA, Heppner PS, Hager T, Moeller-Bertram T, Haji U et al (2013) Diminished vagal activity and blunted diurnal variation of heart rate dynamics in posttraumatic stress disorder. Stress 16(3):300–310

99. Woodward SH, Arsenault NJ, Voelker K, Nguyen T, Lynch J, Skultety K et al (2009) Autonomic activation during sleep in post-traumatic stress disorder and panic: a mattress Actigraphic study. Biol Psychiatry 66(1): 41–46

100. Bertram F, Jamison AL, Slightam C, Kim S, Roth HL, Roth WT (2014) Autonomic arousal during Actigraphically estimated waking and sleep in male veterans with PTSD. J Trauma Stress 27(5):610–617

101. van Liempt S, Arends J, Cluitmans PJM, Westenberg HGM, Kahn RS, Vermetten E (2013) Sympathetic activity and hypothalamo-pituitary–adrenal axis activity during sleep in post-traumatic stress disorder: a study assessing polysomnography with simultaneous blood sampling. Psychoneuroendocrinology 38(1):155–165

102. Yehuda R (2009) Status of glucocorticoid alterations in post-traumatic stress disorder. Ann N Y Acad Sci 1179(1):56–69

103. Pan X, Wang Z, Wu X, Wen SW, Liu A (2018) Salivary cortisol in post-traumatic stress disorder: a systematic review and meta-analysis. BMC Psychiatry 18(1):1–10

104. Pan X, Kaminga AC, Wen SW, Wang Z, Wu X, Liu A (2020) The 24-hour urinary cortisol in post-traumatic stress disorder: a meta-analysis. PLoS One 15(1):e0227560

105. Schumacher S, Niemeyer H, Engel S, Cwik JC, Laufer S, Klusmann H et al (2019) HPA axis regulation in posttraumatic stress disorder: a meta-analysis focusing on potential moderators. Neurosci Biobehav Rev 100: 35–57

106. Morris MC, Hellman N, Abelson JL, Rao U (2016) Cortisol, heart rate, and blood pressure as early markers of PTSD risk: a systematic review and meta-analysis. Clin Psychol Rev 49:79–91

107. Lauc G, Zvonar K, Vukšić-Mihaljević Z, Flögel M (2004) Post-awakening changes in salivary cortisol in veterans with and without PTSD. Stress Health J Int Soc Investig Stress 20(2):99–102

108. Yehuda R, Teicher MH, Trestman RL, Levengood RA, Siever LJ (1996) Cortisol regulation in posttraumatic stress disorder and

108. major depression: a chronobiological analysis. Biol Psychiatry 40:79–88
109. Eckart C, Engler H, Riether C, Kolassa S, Elbert T, Kolassa I (2009) No PTSD-related differences in diurnal cortisol profiles of genocide survivors. Psychoneuroendocrinology 34(4):523–531
110. Wahbeh H, Oken BS (2013) Salivary cortisol lower in posttraumatic stress disorder. J Trauma Stress 26(2):241–248
111. Takahashi JS (2015) Molecular components of the circadian clock in mammals. Diabetes Obes Metab 17:6–11
112. Takahashi JS (2017) Transcriptional architecture of the mammalian circadian clock. Nat Rev Genet 18(3):164–179
113. Dibner C, Schibler U, Albrecht U (2010) The mammalian circadian timing system: organization and coordination of central and peripheral clocks. Annu Rev Physiol 72:517–549
114. Kitamura S, Katayose Y, Nakazaki K, Motomura Y, Oba K, Katsunuma R et al (2016) Estimating individual optimal sleep duration and potential sleep debt. Sci Rep 6(1):1–9
115. Gentry NW, Ashbrook LH, Fu Y, Ptáček LJ (2021) Human circadian variations. J Clin Invest 131(16):e148282
116. Adan A, Archer SN, Hidalgo MP, Di Milia L, Natale V, Randler C (2012) Circadian typology: a comprehensive review. Chronobiol Int 29(9):1153–1175
117. Kivelä L, Papadopoulos MR, Antypa N (2018) Chronotype and psychiatric disorders. Curr Sleep Med Rep 4(2):94–103
118. Hasler BP, Insana SP, James JA, Germain A (2013) Evening-type military veterans report worse lifetime posttraumatic stress symptoms and greater brainstem activity across wakefulness and REM sleep. Biol Psychol 94(2):255–262
119. Paul MA, Love RJ, Jetly R, Richardson JD, Lanius RA, Miller JC et al (2019) Blunted nocturnal salivary melatonin secretion profiles in military-related posttraumatic stress disorder. Front Psych 10:882
120. Love R, Rhind S, Rakesh J, Richardson D, Lanius R, MacDonald M et al (2018) S253. Sleep disturbances, disruption of circadian rhythm and loss of daily melatonin secretion in CAF military personnel suffering from post-traumatic stress disorder. Biol Psychiatry 83(9):S446–S447
121. Fullana MA, Albajes-Eizagirre A, Soriano-Mas C, Vervliet B, Cardoner N, Benet O et al (2018) Fear extinction in the human brain: a meta-analysis of fMRI studies in healthy participants. Neurosci Biobehav Rev 88:16–25
122. Fullana M, Harrison B, Soriano-Mas C, Vervliet B, Cardoner N, Ávila-Parcet A et al (2016) Neural signatures of human fear conditioning: an updated and extended meta-analysis of fMRI studies. Mol Psychiatry 21(4):500–508
123. Wen Z, Seo J, Pace-Schott EF, Milad MR (2022) Abnormal dynamic functional connectivity during fear extinction learning in PTSD and anxiety disorders. Mol Psychiatry 27:1–9
124. Kim EJ, Pellman B, Kim JJ (2015) Stress effects on the hippocampus: a critical review. Learn Mem 22(9):411–416
125. Kitayama N, Vaccarino V, Kutner M, Weiss P, Bremner JD (2005) Magnetic resonance imaging (MRI) measurement of hippocampal volume in posttraumatic stress disorder: a meta-analysis. J Affect Disord 88(1):79–86
126. Karl A, Schaefer M, Malta LS, Dörfel D, Rohleder N, Werner A (2006) A meta-analysis of structural brain abnormalities in PTSD. Neurosci Biobehav Rev 30(7):1004–1031
127. Pitman RK, Rasmusson AM, Koenen KC, Shin LM, Orr SP, Gilbertson MW et al (2012) Biological studies of post-traumatic stress disorder. Nat Rev Neurosci 13(11):769–787
128. Carrion VG, Weems CF, Richert K, Hoffman BC, Reiss AL (2010) Decreased prefrontal cortical volume associated with increased bedtime cortisol in traumatized youth. Biol Psychiatry 68(5):491–493
129. Kitayama N, Quinn S, Bremner JD (2006) Smaller volume of anterior cingulate cortex in abuse-related posttraumatic stress disorder. J Affect Disord 90(2–3):171–174
130. Etkin A, Wager TD (2007) Functional neuroimaging of anxiety: a meta-analysis of emotional processing in PTSD, social anxiety disorder, and specific phobia. Am J Psychiatry 164(10):1476–1488
131. Hayes JP, Hayes SM, Mikedis AM (2012) Quantitative meta-analysis of neural activity in posttraumatic stress disorder. Biol Mood Anxiety Disord 2(1):1–13
132. Spoormaker VI (2018) PTSD, arousal, and disrupted (REM) sleep. In: Sleep and combat-related post traumatic stress disorder. Springer, pp 227–232
133. Shalev A, Liberzon I, Marmar C (2017) Posttraumatic stress disorder. N Engl J Med 376(25):2459–2469

134. Suarez-Jimenez B, Albajes-Eizagirre A, Lazarov A, Zhu X, Harrison BJ, Radua J et al (2020) Neural signatures of conditioning, extinction learning, and extinction recall in posttraumatic stress disorder: a meta-analysis of functional magnetic resonance imaging studies. Psychol Med 50(9): 1442–1451

135. Germain A, James J, Insana S, Herringa RJ, Mammen O, Price J et al (2013) A window into the invisible wound of war: functional neuroimaging of REM sleep in returning combat veterans with PTSD. Psychiatry Res Neuroimaging 211(2):176–179

136. The American Academy of Sleep Medicine (2020) The AASM manual for the scoring of sleep and associated events: rules, terminology and technical specifications. Version 2.6. American Academy of Sleep Medicine, Darien

137. Rechtschaffen A, Kales A (eds) (1968) A manual for standardized terminology, techniques and scoring system for sleep stages in human subjects. U.S. National Institute of Neurological Diseases and Blindness, Neurological Information Network, Washington, DC

138. Novelli L, Ferri R, Bruni O (2010) Sleep classification according to AASM and Rechtschaffen and Kales: effects on sleep scoring parameters of children and adolescents. J Sleep Res 19(1p2):238–247

139. Iber C, Ancoli-Israel S, Chesson A, Quan ST (2007) The AASM manual for the scoring of sleep and associated events. Rules, terminology and technical specifications. American Academy of Sleep Medicine, Westchester, IL

140. Moser D, Anderer P, Gruber G, Parapatics S, Loretz E, Boeck M et al (2009) Sleep classification according to AASM and Rechtschaffen & Kales: effects on sleep scoring parameters. Sleep 32(2):139–149

141. Mellman TA, Pigeon WR, Nowell PD, Nolan B (2007) Relationships between REM sleep findings and PTSD symptoms during the early aftermath of trauma. J Trauma Stress 20(5):893–901

142. Lipinska G, Thomas KG (2019) The interaction of REM fragmentation and night-time arousal modulates sleep-dependent emotional memory consolidation. Front Psychol 10: 1766

143. Motamedi-Fakhr S, Moshrefi-Torbati M, Hill M, Hill CM, White PR (2014) Signal processing techniques applied to human sleep EEG signals—a review. Biomed Signal Process Control 10:21–33

144. Achermann P (2009) EEG analysis applied to sleep. Epileptologie 26:28–33

145. Rusterholz T, Achermann P, Dürr R, Koenig T, Tarokh L (2017) Global field synchronization in gamma range of the sleep EEG tracks sleep depth: artifact introduced by a rectangular analysis window. J Neurosci Methods 284:21–26

146. Islam MK, Rastegarnia A, Yang Z (2016) Methods for artifact detection and removal from scalp EEG: a review. Neurophysiologie Clinique/Clin Neurophysiol 46(4):287–305

147. Denis D, Bottary R, Cunningham TJ, Zeng S, Daffre C, Oliver KL et al (2021) Sleep power spectral density and spindles in PTSD and their relationship to symptom severity. Front Psych 12:766647

148. Wang C, Laxminarayan S, Ramakrishnan S, Dovzhenok A, Cashmere JD, Germain A et al (2020) Increased oscillatory frequency of sleep spindles in combat-exposed veteran men with post-traumatic stress disorder. Sleep 43(10):zsaa064

149. Moon SY, Choi YB, Jung HK, Lee YI, Choi SH (2018) Increased frontal gamma and Posterior Delta powers as potential neurophysiological correlates differentiating posttraumatic stress disorder from anxiety disorders. Psychiatry Investig 15(11): 1087–1093

150. Pisarenco I, Caporro M, Prosperetti C, Manconi M (2014) High-density electroencephalography as an innovative tool to explore sleep physiology and sleep related disorders. Int J Psychophysiol 92(1):8–15

151. Michel CM, Brunet D (2019) EEG source imaging: a practical review of the analysis steps. Front Neurol 10:325

152. Full KM, Kerr J, Grandner MA, Malhotra A, Moran K, Godoble S et al (2018) Validation of a physical activity accelerometer device worn on the hip and wrist against polysomnography. Sleep Health 4(2):209–216

153. Conley S, Knies A, Batten J, Ash G, Miner B, Hwang Y et al (2019) Agreement between actigraphic and polysomnographic measures of sleep in adults with and without chronic conditions: a systematic review and meta-analysis. Sleep Med Rev 46:151–160

154. Smith MT, McCrae CS, Cheung J, Martin JL, Harrod CG, Heald JL et al (2018) Use of actigraphy for the evaluation of sleep disorders and circadian rhythm sleep-wake disorders: an American Academy of sleep medicine systematic review, meta-analysis, and GRADE assessment. J Clin Sleep Med 14(7): 1209–1230

155. Ancoli-Israel S, Martin JL, Blackwell T, Buenaver L, Liu L, Meltzer LJ et al (2015)

The SBSM guide to actigraphy monitoring: clinical and research applications. Behav Sleep Med 13(sup1):S4–S38

156. Depner CM, Cheng PC, Devine JK, Khosla S, De Zambotti M, Robillard R et al (2020) Wearable technologies for developing sleep and circadian biomarkers: a summary of workshop discussions. Sleep 43(2):zsz254

157. Chinoy ED, Cuellar JA, Huwa KE, Jameson JT, Watson CH, Bessman SC et al (2021) Performance of seven consumer sleep-tracking devices compared with polysomnography. Sleep 44(5):zsaa291

158. Patel SR, Weng J, Rueschman M, Dudley KA, Loredo JS, Mossavar-Rahmani Y et al (2015) Reproducibility of a standardized actigraphy scoring algorithm for sleep in a US Hispanic/Latino population. Sleep 38(9):1497–1503

159. Boyne K, Sherry DD, Gallagher PR, Olsen M, Brooks LJ (2013) Accuracy of computer algorithms and the human eye in scoring actigraphy. Sleep Breath 17(1):411–417

160. Chow CM, Wong SN, Shin M, Maddox RG, Feilds KL, Paxton K et al (2016) Defining the rest interval associated with the main sleep period in actigraph scoring. Nat Sci Sleep 8:321–328

161. Kobayashi I, Huntley E, Lavela J, Mellman TA (2012) Subjectively and objectively measured sleep with and without posttraumatic stress disorder and trauma exposure. Sleep 35(7):957–965

162. Ghadami MR, Khaledi-Paveh B, Nasouri M, Khazaie H (2015) PTSD-related paradoxical insomnia: an actigraphic study among veterans with chronic PTSD. J Inj Violence Res 7(2):54–58

163. Slightam C, Petrowski K, Jamison AL, Keller M, Bertram F, Kim S et al (2018) Assessing sleep quality using self-report and actigraphy in PTSD. J Sleep Res 27(3): e12632

164. Buysse DJ, Reynolds CF, Monk TH, Berman SR, Kupfer DJ (1989) The Pittsburgh sleep quality index: a new instrument for psychiatric practice and research. Psychiatry Res 28(2): 193–213

165. Morin CM (1993) Insomnia: psychological assessment and management. Guilford Press

166. Morin CM, Belleville G, Bélanger L, Ivers H (2011) The insomnia severity index: psychometric indicators to detect insomnia cases and evaluate treatment response. Sleep 34(5): 601–608

167. Johns MW (1991) A new method for measuring daytime sleepiness: the Epworth sleepiness scale. Sleep 14(6):540–545

168. Rosenthal LD, Dolan DC (2008) The Epworth sleepiness scale in the identification of obstructive sleep apnea. J Nerv Ment Dis 196(5):429–431

169. Chung F, Yegneswaran B, Liao P, Chung SA, Vairavanathan S, Islam S et al (2008) STOP questionnaire: a tool to screen patients for obstructive sleep apnea. J Am Soc Anesthesiol 108(5):812–821

170. Morin CM, Vallieres A, Ivers H (2007) Dysfunctional beliefs and attitudes about sleep (DBAS): validation of a brief version (DBAS-16). Sleep 30(11):1547–1554

171. Espie CA, Inglis SJ, Harvey L, Tessier S (2000) Insomniacs' attributions: psychometric properties of the dysfunctional beliefs and attitudes about sleep scale and the sleep disturbance questionnaire. J Psychosom Res 48(2):141–148

172. Chung KF, Ho FY, Yeung WF (2016) Psychometric comparison of the full and abbreviated versions of the dysfunctional beliefs and attitudes about sleep scale. J Clin Sleep Med 12(6):821–828

173. Drake C, Richardson G, Roehrs T, Scofield H, Roth T (2004) Vulnerability to stress-related sleep disturbance and hyperarousal. Sleep 27(2):285–291

174. Insana SP, Hall M, Buysse DJ, Germain A (2013) Validation of the Pittsburgh sleep quality index addendum for posttraumatic stress disorder (PSQI-A) in US male military veterans. J Trauma Stress 26(2):192–200

175. Zayfert C, DeViva J, Pigeon W, Goodson J (2006) Fear of sleep and nighttime vigilance in trauma-related insomnia: a preliminary report on the fear of sleep inventory. The International Society for Traumatic Stress Studies, November 4–7, Hollywood, CA

176. Pruiksma KE, Taylor DJ, Ruggero C, Boals A, Davis JL, Cranston C et al (2014) A psychometric study of the fear of sleep inventory-short form (FoSI-SF). J Clin Sleep Med 10(5):551–558

177. Huntley ED, Hall Brown TS, Kobayashi I, Mellman TA (2014) Validation of the Fear of Sleep Inventory (FOSI) in an urban young adult African American sample. J Trauma Stress 27:103–107

178. Calhoun PS, Wiley M, Dennis MF, Means MK, Edinger JD, Beckham JC (2007) Objective evidence of sleep disturbance in women with posttraumatic stress disorder. J Trauma Stress 20(6):1009–1018

179. Carney CE, Buysse DJ, Ancoli-Israel S, Edinger JD, Krystal AD, Lichstein KL et al (2012) The consensus sleep diary: standardizing

180. Dietch J, Taylor D (2021) Evaluation of the Consensus Sleep Diary in a community sample: comparison with single-channel electroencephalography, actigraphy, and retrospective questionnaire. J Clin Sleep Med 17(7):1389–1399
181. Wohlgemuth WK, Edinger JD, Fins AI, Sullivan RJ (1999) How many nights are enough? The short-term stability of sleep parameters in elderly insomniacs and normal sleepers. Psychophysiology 36(2):233–244
182. Reed DL, Sacco WP (2016) Measuring sleep efficiency: what should the denominator be? J Clin Sleep Med 12(2):263–266
183. Maich KH, Lachowski AM, Carney CE (2018) Psychometric properties of the consensus sleep diary in those with insomnia disorder. Behav Sleep Med 16(2):117–134
184. Ross RJ, Ball WA, Sullivan KA, Caroff SN (1989) Sleep disturbance as the hallmark of posttraumatic stress disorder. Am J Psychiatry 146(6):697–707
185. Phelps AJ, Forbes D, Creamer M (2008) Understanding posttraumatic nightmares: an empirical and conceptual review. Clin Psychol Rev 28(2):338–355
186. Robert G, Zadra A (2014) Thematic and content analysis of idiopathic nightmares and bad dreams. Sleep 37(2):409–417
187. Weathers FW, Blake DD, Schnurr PP, Kaloupek DG, Marx BP, Keane TM (2013) The clinician-administered PTSD scale for DSM-5 (CAPS-5). National Center for PTSD
188. The PTSD Checklist for DSM-5 (PCL-5) (2013) Scale available from the National Center for PTSD [Internet]. Available from: www.ptsd.va.gov
189. Weathers FW, Blake DD, Schnurr PP, Kaloupek DG, Marx BP, Keane TM (2013) The life events checklist for DSM-5 (LEC-5). Instrument available from the National Center for PTSD at www.ptsd.va.gov
190. Bryant RA, Harvey AG, Dang ST, Sackville T (1998) Assessing acute stress disorder: psychometric properties of a structured clinical interview. Psychol Assess 10(3):215–220
191. Weiss DS (2004) The impact of event scale-revised. In: Wilson JP, Keane TM (eds) Assessing psychological trauma and PTSD, 2nd edn. Guilford Press, New York, pp 168–189
192. Krakow B (2006) Nightmare complaints in treatment-seeking patients in clinical sleep medicine settings: diagnostic and treatment implications. Sleep 29(10):1313–1319
193. Belicki K (1992) The relationship of nightmare frequency to nightmare suffering with implications for treatment and research. Dreaming 2(3):143–148
194. Mäder T, Oliver KI, Daffre C, Kim S, Orr SP, Lasko NB et al (2021) Autonomic activity, posttraumatic and nontraumatic nightmares, and PTSD after trauma exposure. Psychol Med 53:1–10
195. Cranston CC, Miller KE, Davis JL, Rhudy JL (2017) Preliminary validation of a brief measure of the frequency and severity of nightmares: the trauma-related nightmare survey. J Trauma Dissociation 18(1):88–99
196. Forbes D, Phelps A, McHugh T (2001) Treatment of combat-related nightmares using imagery rehearsal: a pilot study. J Trauma Stress 14(2):433–442
197. Pigeon W, Carr M, Mellman T (2021) Dream content associated with the development of PTSD. Int J Dream Res 14:136–140
198. Phelps AJ, Kanaan RA, Worsnop C, Redston S, Ralph N, Forbes D (2018) An ambulatory polysomnography study of the post-traumatic nightmares of post-traumatic stress disorder. Sleep 41(1):zsx188
199. Woodward SH, Arsenault NJ, Murray C, Bliwise DL (2000) Laboratory sleep correlates of nightmare complaint in PTSD inpatients. Biol Psychiatry 48(11):1081–1087
200. Woodward SH, Arsenault NJ, Michel GE, Santerre CS, Groves WK, Stewart WK (2000) Polysomnographic characteristics of trauma-related nightmares. Sleep 23(S2):A356–A357
201. Milad MR, Wright CI, Orr SP, Pitman RK, Quirk GJ, Rauch SL (2007) Recall of fear extinction in humans activates the ventromedial prefrontal cortex and hippocampus in concert. Biol Psychiatry 62(5):446–454
202. Pace-Schott EF, Milad MR, Orr SP, Rauch SL, Stickgold R, Pitman RK (2009) Sleep promotes generalization of extinction of conditioned fear. Sleep 32(1):19–26
203. Pace-Schott EF, Spencer RMC, Vijayakumar S, Ahmed NAK, Verga PW, Orr SP et al (2013) Extinction of conditioned fear is better learned and recalled in the morning than in the evening. J Psychiatr Res 47(11):1776–1784
204. Seo J, Moore KN, Gazecki S, Bottary RM, Milad MR, Song H et al (2018) Delayed fear extinction in individuals with insomnia disorder. Sleep 41(8):zsy095
205. Seo J, Pace-Schott EF, Milad MR, Song H, Germain A (2021) Partial and total sleep deprivation interferes with neural correlates of

consolidation of fear extinction memory. Biol Psychiatry Cogn Neurosci Neuroimaging 6(3):299–309

206. Zenses A, Lenaert B, Peigneux P, Beckers T, Boddez Y (2020) Sleep deprivation increases threat beliefs in human fear conditioning. J Sleep Res 29(3):e12873

207. Menz MM, Rihm JS, Buchel C (2016) REM sleep is causal to successful consolidation of dangerous and safety stimuli and reduces return of fear after extinction. J Neurosci 36(7):2148–2160

208. Davidson P, Carlsson I, Jönsson P, Johansson M (2018) A more generalized fear response after a daytime nap. Neurobiol Learn Mem 151:18–27

209. Sturm A, Czisch M, Spoormaker VI (2013) Effects of unconditioned stimulus intensity and fear extinction on subsequent sleep architecture in an afternoon nap. J Sleep Res 22(6):648–655

210. Laborde S, Mosley E, Thayer JF (2017) Heart rate variability and cardiac vagal tone in psychophysiological research–recommendations for experiment planning, data analysis, and data reporting. Front Psychol 8:213

211. Liddell BJ, Kemp AH, Steel Z, Nickerson A, Bryant RA, Tam N et al (2016) Heart rate variability and the relationship between trauma exposure age, and psychopathology in a post-conflict setting. BMC Psychiatry 16(1):1–9

212. Shaffer F, Ginsberg JP (2017) An overview of heart rate variability metrics and norms. Front Public Health 5:258

213. Adam EK, Kumari M (2009) Assessing salivary cortisol in large-scale, epidemiological research. Psychoneuroendocrinology 34(10): 1423–1436

214. Pruessner JC, Kirschbaum C, Meinlschmid G, Hellhammer DH (2003) Two formulas for computation of the area under the curve represent measures of total hormone concentration versus time-dependent change. Psychoneuroendocrinology 28(7):916–931

215. Horne JA, Östberg O (1976) A self-assessment questionnaire to determine morningness-eveningness in human circadian rhythms. Int J Chronobiol 4(2):97–110

216. Randler C, Jankowski KS (2014) Evidence for the validity of the composite scale of morningness based on students from Germany and Poland–relationship with sleep–wake and social schedules. Biol Rhythm Res 45(4): 653–659

217. Smith CS, Reilly C, Midkiff K (1989) Evaluation of three circadian rhythm questionnaires with suggestions for an improved measure of morningness. J Appl Psychol 74(5):728

218. Zavada A, Gordijn MC, Beersma DG, Daan S, Roenneberg T (2005) Comparison of the Munich Chronotype Questionnaire with the Horne-Östberg's morningness-eveningness score. Chronobiol Int 22(2):267–278

219. Pandi-Perumal SR, Smits M, Spence W, Srinivasan V, Cardinali DP, Lowe AD et al (2007) Dim light melatonin onset (DLMO): a tool for the analysis of circadian phase in human sleep and chronobiological disorders. Prog Neuro-Psychopharmacol Biol Psychiatry 31(1):1–11

220. Pullman RE, Roepke SE, Duffy JF (2012) Laboratory validation of an in-home method for assessing circadian phase using dim light melatonin onset (DLMO). Sleep Med 13(6): 703–706

221. Kennaway D, Frith R, Phillipou G, Matthews C, Seamark R (1977) A specific radioimmunoassay for melatonin in biological tissue and fluids and its validation by gas chromatography-mass spectrometry. Endocrinology 101(1):119–127

222. Danilenko KV, Verevkin EG, Antyufeev VS, Wirz-Justice A, Cajochen C (2014) The hockey-stick method to estimate evening dim light melatonin onset (DLMO) in humans. Chronobiol Int 31(3):349–355

223. Benloucif S, Burgess HJ, Klerman EB, Lewy AJ, Middleton B, Murphy PJ et al (2008) Measuring melatonin in humans. J Clin Sleep Med 4(1):66–69

224. Voultsios A, Kennaway DJ, Dawson D (1997) Salivary melatonin as a circadian phase marker: validation and comparison to plasma melatonin. J Biol Rhythm 12(5):457–466

225. Burgess HJ, Savic N, Sletten T, Roach G, Gilbert SS, Dawson D (2003) The relationship between the dim light melatonin onset and sleep on a regular schedule in young healthy adults. Behav Sleep Med 1(2): 102–114

226. Reiter AM, Sargent C, Roach GD (2020) Finding DLMO: estimating dim light melatonin onset from sleep markers derived from questionnaires, diaries and actigraphy. Chronobiol Int 37(9–10):1412–1424

227. Czisch M, Wehrle R (2010) Sleep. In: Mulert C, Lemieux L (eds) EEG-fMRI: physiological basis, technique, and applications. Springer, Berlin, pp 279–301

228. Schmitter S, Diesch E, Amann M, Kroll A, Moayer M, Schad L (2008) Silent echo-planar imaging for auditory FMRI. MAGMA 21(5): 317–325

229. Wang Y, Duan W, Lei X (2020) Impaired coupling of the Brain's default network during sleep deprivation: a resting-state EEG study. Nat Sci Sleep 12:937

230. Corea JR, Flynn AM, Lechêne B, Scott G, Reed GD, Shin PJ et al (2016) Screen-printed flexible MRI receive coils. Nat Commun 7(1):1–7

231. Cooley CZ, McDaniel PC, Stockmann JP, Srinivas SA, Cauley SF, Sliwiak M et al (2021) A portable scanner for magnetic resonance imaging of the brain. Nat Biomed Eng 5(3):229–239

232. Sun H, Ganglberger W, Panneerselvam E, Leone MJ, Quadri SA, Goparaju B et al (2020) Sleep staging from electrocardiography and respiration with deep learning. Sleep 43(7):zsz306

233. Moehlman TM, de Zwart JA, Chappel-Farley MG, Liu X, McClain IB, Chang C et al (2019) All-night functional magnetic resonance imaging sleep studies. J Neurosci Methods 316:83–98

234. Shah NJ, Oros-Peusquens A, Arrubla J, Zhang K, Warbrick T, Mauler J et al (2013) Advances in multimodal neuroimaging: hybrid MR–PET and MR–PET–EEG at 3 T and 9.4 T. J Magn Reson 229:101–115

235. Foldvary-Schaefer N, Grigg-Damberger MM (2012) Identifying interictal and ictal epileptic activity in polysomnograms. Sleep Med Clin 7(1):39–58

236. Feng P, Becker B, Feng T, Zheng Y (2018) Alter spontaneous activity in amygdala and vmPFC during fear consolidation following 24 h sleep deprivation. NeuroImage 172:461–469

237. Lerner I, Lupkin SM, Sinha N, Tsai A, Gluck MA (2017) Baseline levels of rapid eye movement sleep may protect against excessive activity in fear-related neural circuitry. J Neurosci 37(46):11233–11244

238. Peters AC, Blechert J, Sämann PG, Eidner I, Czisch M, Spoormaker VI (2014) One night of partial sleep deprivation affects habituation of hypothalamus and skin conductance responses. J Neurophysiol 112(6):1267–1276

239. Spoormaker V, Gvozdanovic G, Sämann P, Czisch M (2014) Ventromedial prefrontal cortex activity and rapid eye movement sleep are associated with subsequent fear expression in human subjects. Exp Brain Res 232(5):1547–1554

240. Spoormaker VI, Schröter MS, Andrade KC, Dresler M, Kiem SA, Goya-Maldonado R et al (2012) Effects of rapid eye movement sleep deprivation on fear extinction recall and prediction error signalling. Hum Brain Mapp 33:2362–2376

241. Straus LD, Acheson DT, Risbrough VB, Drummond SP (2017) Sleep deprivation disrupts recall of conditioned fear extinction. Biol Psychiatry Cogn Neurosci Neuroimaging 2(2):123–129

Chapter 5

Pavlovian Conditioning and Extinction Methods for Studying the Neurobiology of Fear Learning in PTSD

Dylan B. Miller, Madeleine M. Rassaby, Zhenfu Wen, and Mohammed R. Milad

Abstract

We describe a human threat conditioning and extinction paradigm used to study posttraumatic stress disorder as well as other psychiatric disorders. This paradigm was designed based on rodent threat conditioning paradigms for its application in translational neuroscience research. It is a passive viewing emotional learning and memory task composed of four phases. The first is the threat conditioning phase, followed by a threat extinction learning phase. Subsequently, memory for extinction learning is evaluated in extinction recall after a time delay (typically 24 hours after extinction learning). The last phase is the renewal of threat responding after extinction. Participants undergo this paradigm while in an fMRI scanner to identify brain correlates during all emotional learning phases. The duration of each experimental phase is approximately 13 minutes. Various time delays can be inserted between phases, depending on the aims of the experiment. Other experimental measures such as skin conductance responses and subjective measures could also be acquired during all experimental phases.

Key words Fear, Posttraumatic stress disorder, Anxiety disorders, Pavlovian conditioning, Conditioning paradigm, Fear extinction, Fear circuit, fMRI, Neurobiology, Skin conductance response

1 Introduction

Posttraumatic stress disorder (PTSD) is a debilitating disorder characterized by the continuous experiencing of intense and disturbing thoughts and feelings after the occurrence of a traumatic event. Individuals diagnosed with PTSD often experience strong negative reactions to stimuli that remind them of the trauma, causing them to chronically fear and frequently avoid situations in which triggers are present [1]. Abnormal fear processing is a core feature of PTSD. While it is common to experience a traumatic

Supplementary Information The online version contains supplementary material available at https://doi.org/10.1007/978-1-0716-3218-5_5.

event, and even to develop a conditioned fear response to something involved in this event, most people's fear is eventually extinguished by experiencing safe situations involving the originally feared stimulus. However, those with PTSD retain a persistent fear response despite repeated exposure to the stimulus with no adverse consequence. Specifically, individuals with PTSD demonstrate dysfunction in the recall of the fear extinction memory [2–5].

Paradigms designed to study threat conditioning and its extinction in humans have been in use for decades by experimental psychologists [6, 7]. The experimental designs were simple and mostly aimed to examine psychophysiological and subjective indices of associative learning and memory in humans. These earlier versions of the paradigms involved the pairing of simple stimuli, such as geometric shapes presented on a computer screen (e.g., blue square, red circle), with the delivery of a mild aversive cue (e.g., an electric shock or a loud noise; [7–9]. Extinction of threat conditioning followed, but the memory for extinction learning was rarely evaluated. These studies led to a wealth of information that guided the development of several associative learning and memory theories and models [10]. Moreover, clinicians and psychologists began to test threat conditioning and extinction in fear- and anxiety-based disorders such as PTSD. The use of these paradigms began to give insights into the mechanisms of the etiology/maintenance of some anxiety- and fear-based disorders, thereby highlighting their utility in examining the brain mechanisms that are associated with impaired emotional learning and memory in these disorders [11–13].

In the early 1990s neuroimaging tools became mainstream and were commonly used to study the human brain. In parallel, electrophysiological studies in rodents began to show exquisite details regarding anatomical specificity and unique temporal engagement of key brain nodes mediating threat conditioning and its extinction [14]. These synergistic advances dictated the need to develop threat conditioning and extinction paradigms that fit better within the functional magnetic resonance imaging (fMRI) environment. Also, the new paradigms needed to enable the dissection of different phases of learning within a session (early vs. late extinction) given that animal data began to show distinct neural activations associated with these phases. Motivated by all the knowledge gained from rodents and our experience in fMRI, we developed, tested, and validated the threat conditioning and extinction paradigm we describe herein [11, 15, 16]. Since its original validation and implementations, the paradigm has now been widely used across several laboratories and several modified versions of it have been published [4, 17–25]. Notably, the paradigm has been used to assess the abnormalities in fear-based disorders such as PTSD [11–13]. We provide rationale for key design features that distinguish this paradigm from others, and highlight variations for the ways in which the paradigm can be implemented and analyzed.

There are four primary phases in this paradigm: (1) threat conditioning, (2) threat extinction learning, (3) threat extinction memory retention, and (4) renewal of conditioned responses. Each phase has a total of 32 trials, lasting approximately 13 minutes. Each trial begins with the presentation of a picture of a room (context A) that contains a lamp that is seemingly turned off. Then, after 3 seconds, the lamp appears to turn on, revealing its color (either blue, red, or yellow). The light stays on for a duration of 6 seconds.

The primary application of this protocol is to examine the neural correlates of fear and threat conditioning, extinction learning, and extinction memory recall and fear renewal. This is important in the context of studying PTSD due to the dysfunction in fear extinction recall that is present in patients with this disorder [2–5]. This paradigm's primary design is to enable the acquisition of fMRI data utilizing blood-oxygen-level-dependent (BOLD) signals. Acquisition of psychophysiological data such as skin conductance responses (SCR)/electrodermal activity (EDA), and subjective reports can be added to the protocol. Each of the experimental phases is designed to last for a duration long enough to enable the study of temporal dynamics of learning-induced neural plasticity, based on substantial evidence from rodent studies. Thus, each experimental phase can be, and in fact should be, divided into blocks of 4 or 8 trials, depending on the temporal resolution intended to be studied based on the study aims. Data from animal studies show that averaging neural activity across an entire learning session (e.g., during the conditioning or extinction phases) can lead to the loss of critical information, such as the neural signal associated with early or late learning-induced neural activations (for example see [26–28]). Thus, it is imperative that the researcher does not analyze the results by averaging the fMRI data for a given cue across an entire learning phase.

1.1 Comparison with Other Methods

One of the major goals for the design of this paradigm is to examine the neural correlates of extinction memory in the fMRI scanner. As it is well known in the field, fMRI measures relative difference in activation between two different stimuli. Prior paradigms used two cues: a conditioned stimulus followed by a shock (CS+) and a conditioned stimulus not followed by a shock (CS−). To examine extinction memory, we would have to contrast the activations associated with the CS+ after extinction learning (a cue that is now safe because conditioned responses associated with it are now extinguished) with those induced by the CS− (a safe cue because it was never associated with an aversive cue). As such, during the memory recall test with the older two-cue protocol, we would be comparing a safe cue (CS+) with another safe cue (CS−) [7]. The concern is that this contrast might not reveal the best or strongest fMRI signal associated with extinction memory.

To resolve this issue, we introduced a third conditioned stimulus to the protocol. Thus, a unique feature of this protocol is the conditioning of two visual cues with the aversive stimulus, and then only extinguishing one of them, while having a third stimulus not being followed by the shock. Thus, during the extinction memory test, the experimenter will have data from three different cues: one conditioned and extinguished (we call this the extinguished CS+; CS+E), a conditioned cue that is associated with the shock but was not extinguished (we call this the unextinguished CS+, CS+U), and the conditioned cue that is never followed by a shock (the CS−). To test for neural correlates associated with the extinction memory, the optimal contrast would therefore be CS+E (a cue that had been conditioned and extinguished) vs. CS+U (a cue that had been conditioned only).

2 Materials

2.1 Subjects

- Human subjects: participants are required to complete a diagnostic interview to assess for the presence and primary diagnosis of PTSD, a drug screen, and pregnancy test (for female participants – as there may be unknown adverse effects of fMRI and stimulations to a fetus). Participants should be literate and have normal or corrected to normal vision and hearing.
    - CAUTION: Informed consent must be obtained from all subjects. The study protocol must be approved by the appropriate institutional review board and be in compliance with relevant laws and regulations.
- Depending on the goal of your study, it may be appropriate to recruit both healthy participants and participants who meet the criteria for PTSD to analyze how they behaviorally and neurobiologically differ in fear conditioning, extinction, recall, and renewal.

2.2 General Equipment

- Basic psychophysiological laboratory equipment for recording SCR/EDA: recording electrodes (e.g. Biopac EL507/EL509), isotonic recording gel (e.g. Biopac GEL101), lead cables.
- Stimulation equipment: stimulating electrodes (e.g. Biopac EL503/EL508), electrode gel (e.g. Biopac GEL100/GEL104), skin preparation gel (e.g. Biopac ELPREP), lead cables.
- Questionnaires and evaluation forms used to assess personality traits and affective states as well as the prevalence of characteristics associated with PTSD. These include, but are not

limited to: NEO-Five Factor Inventory (NEO-FFI), Anxiety Sensitivity Index (ASI), Beck Anxiety Inventory (BAI), Beck Depression Inventory (BDI-II), Edinburgh Handedness Survey, State-Trait Anxiety Index (STAI), Pittsburgh Sleep Quality Index (PSQI), Posttraumatic Cognitions Inventory (PTCI), Posttraumatic Diagnostic Scale (PDS-5), Quality of Life Enjoyment and Satisfaction Questionnaire (Q-LES-Q-SF), Mindful Attention Awareness Scale (MAAS), Childhood Trauma Questionnaire (CTQ), etc.

2.3 Experimental Room

- Stimulus presentation software (e.g., Presentation (Neurobehavioral Systems, http://www.neurobs.com/)) that is programmed to present visual stimuli and send information to a Biopac Modular Instruments System (see below).
- AcqKnowledge (Biopac Systems Inc.) that records both SCR/EDA measurements from the modular Biopac System and the timestamped event that triggers from the computer running the stimulus program.
- Two computers, one for visual stimulus presentations and one for data collection with AcqKnowledge. PC Requirements depend on software systems used in stimuli presentation and data acquisition. Presentation (Neurobehavioral Systems) requires a minimum 4 GB RAM (6 GB recommended) and Windows (32- or 64-bit) as an operating system. We use AcqKnowledge (Biopac) for data acquisition, which requires a minimum 4 GB RAM (8 GB recommended), and a 64-bit (Windows or Mac) operating system.

2.4 Biopac Systems Inc. Modular Instrument System

- A Biopac Modular Instruments System is used to control stimulation and communicate SCR/EDA data collection across multiple systems.
- The MP160 acts as the central hub of the modular system, collecting data from up to 16 channels and communicating with the machine running AcqKnowledge.
- A Biopac Electrodermal Activity Amplifier (EDA100C) records and transmits SCR/EDA measurements to the MP160 through EL507/509 electrodermal electrodes and LEAD110A electrode leads.
- A Biopac High Level Transducer Interface Module (HLT100C) digitizes information from 16 input and 2 output channels in order to report it to the AcqKnowledge software through the MP160 module.
- The Stimulator Module (STM100C) provides modifiable pulse and waveform stimulus outputs to use in stimulus-response experiments.

- The Stimulation Isolation Adaptor (STMISOC) controls the intensity of the shock (either in voltage or current, we recommend current (mA)) administered by the STM100C shocking module.
- INISOA/OUTISOA analog signal isolator and Current Feedback Monitor Cables (CBLCFMA) are recommended to be used with any voltage stimulator to monitor and record the actual current delivered to the participant by connecting to a channel in the HLT100C module.

3 Methods

3.1 Experimental Design, Components, and Structure

A key factor in the initial design of this protocol was to develop a system to concurrently collect psychophysiology data and fMRI BOLD signal data. The latency present in both BOLD and SCR reactions are compatible, such that the optimal timing of stimulus presentations in this protocol adequately accommodates data collection for both measures. Additionally, the systems required for collecting SCR data can be implemented in an MRI environment without complications. Fear-potentiated startle, for example, is difficult to implement in an MRI system because it inherently involves movement that deteriorates the quality of the imaging data. Similarly, the equipment used to measure eye blink or pupil dilation reactions may leave artifacts or obstructions in imaging data.

SCR is one of the most popular and oldest techniques used to record autonomic arousal [7, 29] and has been extensively used in emotional learning studies. It is a calculated measurement of the electric current that can be passed between two superficial electrodes, usually placed on the subject's hand. A single response is associated with a time-stamped event and compares a pre-event baseline to a short-term, post-event maximum skin conductance. There are a wide variety of measurements that can be used to detect and analyze a subject's response to stimuli. We elect to use SCR, and recommend its use, because it is easy to understand, safe and noninvasive for subjects, relatively affordable, and provides a consistent measure of anxiety and threat response. This protocol can easily be adapted to include other measures applicable for different hypotheses. These other measures include, but are not limited to, heart rate, eye blink, fear-potentiated startle, and pupil dilation. A concise summary of these techniques and others, as well as notes for their respective paradigm design, can be found in [7] or [29].

The paradigm should take place over the course of 2–3 consecutive days, depending on the goals of the study and whether the investigators choose to insert a break between conditioning and extinction. This protocol contains four primary phases, which

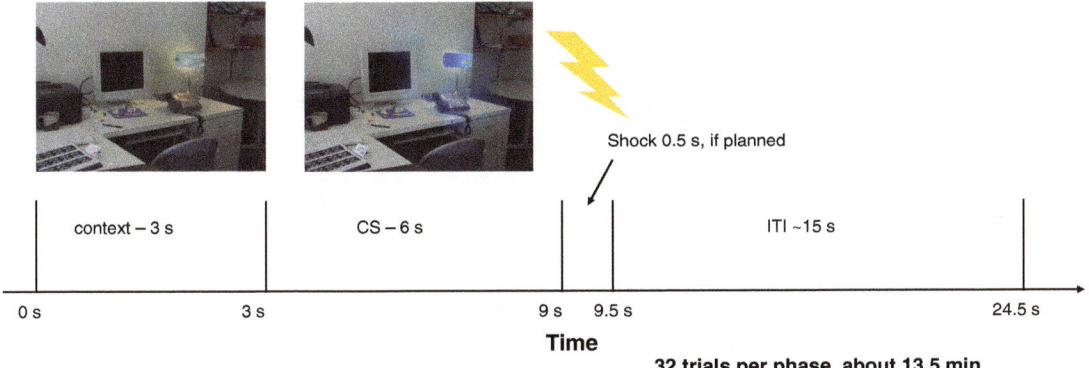

Fig. 1 Trial structure. The figure represents the structure of one trial of any phase of the fear conditioning and extinction paradigm. In each trial, the context (uncolored lamp) is displayed for 3 seconds, the CS (conditioned stimulus – colored lamp) is displayed for 6 seconds and followed by a 0.5 second shock. Finally, an inter-trial interval (ITI) is displayed for an average of 15 seconds, ranging from 12–18 seconds

correspond to different components of classical conditioning: threat conditioning, threat extinction, extinction recall, and renewal. In conditioning, the subject learns to associate a neutral stimulus, in this case an image of a colored lamp, with an unconditioned, naturally aversive stimulus, in this case a brief electric shock to the foot. The former is herein referred to as a conditioned stimulus (CS) and the latter is referred to as the unconditioned stimulus (US). This protocol utilizes three iterations of the CS across the whole experiment: a red lamp, a blue lamp, and a yellow lamp. These neutral stimuli are presented in a digital image that places the lamp in a broader context (CX), of which there are two options: an office desk or a bookshelf. In each instance, the subject is first shown the context with the lamp seemingly turned off, then after 3 seconds the lamp appears to turn on, revealing its color. Figure 1 illustrates the full structure of a single trial.

In conditioning, two lamp colors will be followed by a shock (CS+'s) while the remaining color is never followed by a shock (CS−). The CS+ presentations are followed by a shock with a 62.5% (5/8) reinforcement rate to provide ambiguity that slows the rate of both threat conditioning and extinction. Conditioning contains 32 trials and is divided into two halves, where only one CS + iteration is presented in each half, pseudo-randomly interspersed with an equivalent number of CS− presentations. Consequently, there are 8 presentations of each CS+ color and 16 total presentations of CS−.

In extinction, one of the previously conditioned light colors and the CS− color from conditioning are presented without any shocks administered. Sixteen trials of each of the two CS's are presented in a pseudo-random order. The CS+ presented in extinction is herein referred to as CS+E or the extinguished CS+ while the

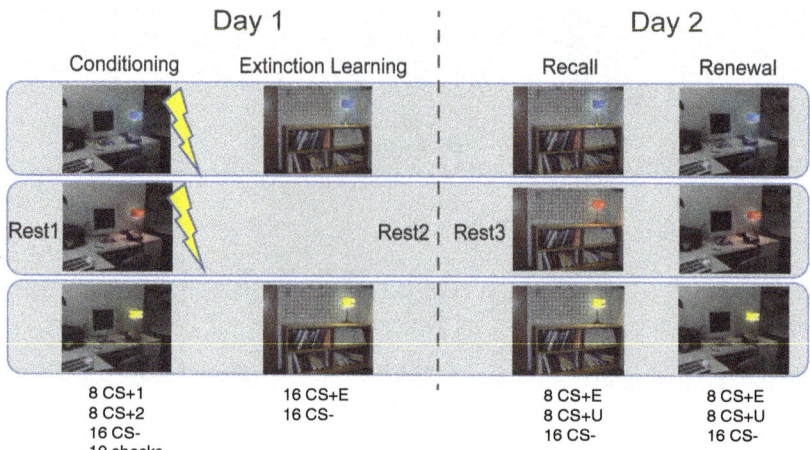

Fig. 2 Fear conditioning and extinction paradigm. The figure provides a visual representation of the full two-day paradigm, including conditioning, extinction learning, extinction recall, and renewal. This could also occur over 3 days, adding a rest in between conditioning and extinction learning, depending on the goal of the research. As can be seen below, conditioning and renewal are in the same context, while extinction learning and recall are in a different context (represented using different rooms)

remaining CS+ from conditioning is referred to as CS+U or the unextinguished CS+. Both recall and renewal are structured in the same way as conditioning; they are divided into two halves, where only one of the CS+'s are paired with the CS− in each half. There are no shocks in the recall and renewal phases. A visual representation of these four phases is presented in Fig. 2.

The retrieval of a threat extinction or threat conditioning memory is highly dependent on the contextual information associated with those memories [15, 30, 31]. Consequently, the context associated with the CS is modulated across the experiment in order to test concepts such as threat generalization and renewal. Conditioning is conducted using one context (CX+) while both extinction and recall occur in a separate context (CX-) in this protocol. Renewal is a return to the context associated with the threat conditioning memory and thus uses CX+ [15].

In order to account for order effects, we use eight variations of the paradigm, which are randomly assigned to each subject. These vary and counter-balance the colors assigned to each CS type, the context image used as CX+, which CS+ type appears first during conditioning, recall, and renewal, as well as the distribution of CS+ and CS− trials within each run. Once a subject is assigned a paradigm version (labeled CS1R to CS8R), this version will be used for all phases.

3.2 Procedure

3.2.1 Intake and Assessment

1. When recruiting subjects, develop a questionnaire that assesses if a subject likely meets the inclusion and exclusion criteria for your study. These criteria are safety and hypothesis driven. Interested volunteers are generally pre-screened using a phone interview that assesses the primary inclusion and exclusion criteria for the study. If a volunteer meets general criteria, they are scheduled for an in-person screening to further evaluate whether they meet criteria for a PTSD diagnosis, and to assess other mental health conditions and obtain physical health history.

2. Obtain informed consent from the potential subject, as required and in compliance with all relevant IRB and legal requirements.

3. Have a trained clinician administer clinical assessments to determine the presence and primary diagnosis of PTSD (e.g., Clinician Administered PTSD Scale (CAPS-5)), and other psychiatric evaluations using questionnaires, such as the Structured Clinical Interview (SCID-5) or Mini Internal Neuropsychiatric Interview (MINI). Interpretation and utilization of questionnaires is hypothesis driven, though we typically include the following: Clinician Administered PTSD Scale (CAPS-5), Structured Clinical Interview (SCID-5), Mini Internal Neuropsychiatric Interview (MINI), Life Events Checklist (LEC-5), Columbia-Suicide Severity Rating Scale (C-SSRS), and Clinical Global Impression Scale (CGI-S).

3.2.2 Subject Preparation

1. On day 1 of the experiment, subjects will undergo shock expectancy, habituation, conditioning, and extinction in an experimental room outside of the scanner. Once you and the subject are in the experimental room, explain to the subject what will be occurring that day, such as what images they will see and the different phases of the experiment. Remind the subject that a highly annoying but not painful electric stimulation (US) will be administered and that they will select the level of stimulation that they find highly annoying but not painful. They are also free to end the experiment at any time. The language used in describing tasks to subjects is very important; Supplemental Document 1 "Experimenter Script" contains a sample script that a technician can follow for explaining the shock calibration, habituation, and conditioning steps to the participant.

2. Make sure the subject washes their hands (but they should not use hand sanitizer or alcohol wipes), apply isotonic gel to the EDA electrodes, and attach recording electrodes to the palm of the non-dominant hand. Recording electrodes should be directly adjacent to each other. Clean the subject's foot (the

foot that is on the same side as the recording electrodes) with the exfoliating gel and attach the stimulation electrodes. Apply gel to electrodes before attaching. Verify that the SCR electrodes are functioning properly by opening a test recording in AcqKnowledge to see the live output from the electrodes. There should be a visible rise in SCR if the subject takes a deep breath.

3. The shock calibration Presentation shows the subject an '+' shape on a black background that is replaced by an 'x' shape, which signals a shock. The technician begins calibration with the shock set at a very low electric current and slowly and incrementally increases the shock at each repetition, asking the subject to rate the unpleasant nature of the shock. Once it appears that any additional increase in shock intensity would be painful to the subject, stop increasing the intensity of the shock and use that level for the remainder of the experiment. It may be useful to ask the subject to rate the shock from one to five, where five is painful, and aim to select a current that they feel is at a four on that scale. Inform the participant that this level of stimulation will be used for the entire experiment and will not be changed during the experiment (i.e., if they receive a stimulation it will be at this level).

4. Open and begin the habituation Presentation file. The habituation Presentation shows subjects the images that will be used as the CS's in the CX+ without any shocks. This phase is about 6 minutes long. Open and begin an AcqKnowledge graph and record SCR.

3.2.3 Conditioning

1. This can occur inside or outside an fMRI scanner.
2. Open the conditioning Presentation file.
   (a) CRITICAL: There are eight versions of the Presentation file for each phase; use the same version of the paradigm to which the subject has been assigned across all phases.
3. Before beginning the presentation, open a new recording file in AcqKnowledge, or whichever program you are using to collect SCR data, and name the file. A suggested file naming convention could be 1ABC001_PHASE, where the first number indicates the paradigm version, followed by a six-character subject ID, and finishing with an underscore and a three-letter abbreviation of the phase (e.g. cond, ext, rec, ren).
4. Start recording in AcqKnowledge before beginning the Presentation software and start Presentation.
5. When the Presentation file finishes, name the log file from the presentation software using the same naming convention as the AcqKnowledge file and save it.

6. Document any observed movement and have the subject complete a questionnaire that inquires about any patterns of shocks they observed.

3.2.4 Extinction

1. This can occur inside or outside an fMRI scanner.

2. Open the extinction Presentation file and a new recording in AcqKnowledge, following the same procedure as in the conditioning phase.

3. Document any observed movement and have the subject complete a questionnaire that inquires about any patterns of shocks they observed.

4. If using a 2- or 3-day paradigm, allow approximately a full day to pass between extinction and the next session.

3.2.5 Recall and Renewal

1. On day 2 of the experiment, subjects will undergo recall and renewal in an fMRI scanner, as recall is the phase with which brain activation abnormalities are associated with a PTSD diagnosis. Have the subject complete a questionnaire that documents what patterns they remember from the previous phases. This is used as a validation for their recorded responses during recall.

2. Attach shock electrodes to the subject's foot, as if they are to be shocked again. Apply isotonic gel to the EDA electrodes, and attach recording electrodes to the palm of the non-dominant hand.

3. Before beginning recall and renewal, run a few structural scans, such as T1-weighted and T2-weighted images for anatomical reference. Other image modalities such as resting-state fMRI, arterial spin labeling (ASL) can also be scanned based on the goal of the study.

4. Recall and renewal are performed consecutively in one session and can be reordered or counter-balanced depending on the goals of the study. They can be recorded into distinct AcqKnowledge files, in which case repeat steps 8–11 for each run, as with the other phases, or into one longer AcqKnowlege file. In the latter case, the two phases will need to be separated before processing or analysis.

5. Document any observed movement and have the subject complete a questionnaire that inquires about any patterns of shocks they observed.

4 Notes

4.1 Timing

This paradigm is typically conducted on two consecutive days, with habituation, conditioning, and extinction occurring on day 1, and recall and renewal occurring on day 2. Depending on the goal of your study, it may be appropriate to spread the study across three consecutive days, with a 1-day break between conditioning and extinction to administer an intervention to observe its effect on extinction learning on day 2. Conditioning, extinction, recall, and renewal are all approximately 13 minute long runs with 32 stimulus trials in each. Each trial includes an inter-trial interval (ITI), which pseudo-randomly varies between 12 and 18 seconds, with the average length being 15 seconds. After following the CX (uncolored lamp image) for 3 seconds, and then the CS (colored lamp image) either followed by a 0.5 second stimulation or no stimulation. In conditioning, all three colored lamps are presented pseudo-randomly, with only two of the colored lamps followed by a stimulation. In extinction, one of the colored lamps that was previously followed by a stimulation and the lamp that was not followed by a stimulation are pseudo-randomly presented. Recall and renewal are similar to conditioning but no images are followed by a stimulation. A total of 32 trials (ITI – CX – CS) are conducted for each phase of the experiment. This is the general experimental design of human Pavlovian classical fear conditioning and extinction paradigms, and it can be manipulated to best fit a specific study [7].

4.2 Analysis

Based on the collected SCR data, reaction values can be calculated for each trial by subtracting the square-root transformed mean skin conductance level observed during the last 2 seconds of context preceding CS onset from the maximal skin conductance level reached during CS presentation. While different analytic approaches to analyze SCR data have been used [32–37], recent studies did not show advantages of a specific method [38, 39]. It is out of the scope here to make recommendations on a specific method of SCR analyses. Subjects' data should be analyzed to verify that assumptions inherent to future statistical inferences are met. Each subject should show evidence of conditioning: reactions to both CS+'s in conditioning must be higher than to the CS− by the end of the run and the reactions to the CS− should fall over time. Similarly, subjects should show evidence of extinction: reactions to the CS+E should be lower at the end of extinction than at the beginning of the phase. Participants who showed no evidence of conditioning or who moved excessively during the experiment should be excluded from the analyses. The trials of the same CS type can be divided into blocks (e.g., 4 trials per block) to improve signal-to-noise ratio. Other SCR-based measures such as the extinction retention index can be calculated. Statistical analyses can be conducted on the SCR-based measures. For example, a repeated-measures Analysis of Variance (ANOVA) can be used to test the mean reaction across experimental phases in a group of participants, with the CS type and blocks as within-subject factors.

Data quality of the fMRI data should be checked first to exclude images with low quality (e.g., large motion artifacts, bad brain coverage) before analysis. The data quality assessment can be assisted by software like MRIQC [40]. The structural and functional images can then be preprocessed following established pipelines in neuroimaging analysis. The standard steps for fMRI preprocessing often include (1) Realignment and Slice-Timing Correction, which correct motion artifacts and timing errors in the functional images; (2) Coregistration and Normalization, which align the functional and structural images and transform them to a standardized space (e.g., MNI space); (3) Spatial Smoothing to suppress spatial noise and enhance the signal to noise ratio (SNR). The preprocessing steps can be done with a variety of software such as SPM, FSL, and AFNI. Other software that implement automated analysis pipelines, such as the fMRIPrep, can also be used for the analysis [41]. The preprocessed fMRI data are fed to first-level analysis based on the general linear model (GLM). The design matrix is constructed according to the study aims. Considering that the learning is intrinsically dynamic, we recommend to divide trials of the same type (e.g., CS+) into blocks of 4 or 8 trials, construct each block as a separate regressor in the design matrix. The regressors are constructed by convolving the box-car function of the stimuli with the hemodynamic response function. Other confounds such as scanner drifts, motion parameters, white matter/cerebral spinal fluid noise can be added into the GLM model. The statistical maps derived from the first-level analysis are used for second-level analysis to identify brain activations.

4.3 Anticipated Results

Before discussing the expected results on a macro or study scale, it is important to set baseline expectations for what an individual SCR should look like. The structure of a SCR is well understood and has been reported previously. Following a stimulus presentation, there should be between 1 and 3 seconds of lag before any reaction is seen, followed by another 1 to 3 seconds of rising SCR [29]. Consequently, the peak of a single reaction should occur between approximately 2 and 6 seconds following a stimulus presentation. This structure is a useful marker for troubleshooting stimulus timing or shock administration.

There are general patterns that are expected across all phases, as evidenced by previous research and investigations using this paradigm. During conditioning, there should be a strong downward trend in the SCR for CS− trials, such that there is a clear differentiation between CS− and the CS+ trials. Additionally, there should not be any pattern of difference between the two CS+ stimuli during this phase. During extinction, it is expected that SCR for CS− will remain low throughout, though it may start higher than it ended in conditioning. SCR for CS+E should decline significantly over the course of the run as the subject learns to now associate that stimulus with safety. In recall, if the extinction memory is expressed more than the conditioning memory (representing strong extinction recall), the SCR for CS+E will be lower than that of CS+U. However, if the conditioning or threat memory is expressed more than the extinction memory (such as may be present in patients with PTSD), SCR for CS+E will not be statistically distinguishable from that of CS+U. A succinct way to calculate and analyze CS+E against CS+U, particularly during recall, is to calculate the difference between those SCR values by trial for each subject. These values can then be treated and analyzed in the same way as other SCR values; calculate the mean for each trial by group and perform an ANOVA if desired. By using this approach, there is a more direct analysis on the differentiation between the CS+E and CS+U. In renewal, assuming that any renewal occurs, there should be a slightly higher SCR in the first few trials when compared to recall.

In PTSD patients, it would be expected for the SCR values during the presentation of the CS+E and CS+U to be similar, as they may exhibit poor extinction memory recall, and therefore display a fear response to all conditioned stimuli [4, 5, 11]. Additionally, we would expect PTSD patients to display greater dACC and amygdala activation, and less vmPFC and hippocampus activation during the presentation of the CS+E in extinction recall compared to healthy controls [2–5, 22, 42]. It is possible that these brain activations may look more similar to those of healthy controls with the successful implementation of a treatment.

5 Conclusions

This fear conditioning and extinction paradigm has been widely used by researchers to examine emotion and memory, and in recent years has been implemented to study fear-based disorders such as PTSD. Rodent-to-human translational research, in addition to a rapid growth in the use of functional neuroimaging, has allowed us to explore the underlying neurobiological mechanisms of PTSD [14, 28, 43]. The fear circuit involves the amygdala, hippocampus, ventromedial prefrontal cortex (vmPFC), and dorsal anterior cingulate cortex (dACC). The vmPFC is activated during extinction learning and during the retrieval and expression of an extinction memory, and the dACC is activated during the expression of a fear memory [16, 44–47].

Additionally, the magnitude of vmPFC and hippocampus activations during extinction recall positively correlates with the magnitude of the expression of the extinction memory [5, 16]. A number of functional imaging studies have demonstrated that patients with PTSD show greater amygdala activation during extinction learning, in addition to reduced hippocampus and vmPFC activation, but greater dACC activation during extinction recall [2, 4, 5, 22, 42, 43]. These abnormal activations might contribute to excessive fear responses (marked by higher amygdala and dACC activation), and a failure to inhibit this response (marked by decreased vmPFC activation; [3, 4, 22] in PTSD patients. Additionally, the hypoactivation of the hippocampus may point to a deficit in the ability to identify safe contexts, and difficulty with adaptive fear learning and its subsequent extinction [3, 4, 43].

Skin conductance response (SCR) is also a valuable measure in assessing abnormalities associated with PTSD [29]. Studies have shown that SCR, a physiological indicator of stress/fear response, is higher in patients diagnosed with PTSD compared to healthy controls during extinction recall [4, 11] and other studies have reported abnormal SCR in PTSD during fear conditioning and during extinction learning [4, 5, 12, 13, 48].

There are, however, some challenges with the implementation of SCR and its integration with neuroimaging data. One challenge has been that SCR is highly variable, and sometimes participants present an absence of reactivity to the conditioned stimuli in their skin conductance. This leads to an abundance of data exclusion due to a "lack of conditionability," which could result in a potential sample bias [49]. However, studies have found that these participants produce an interesting atypical pattern of neurological responses to the presentations of the CS+ versus CS−. Including and dividing the results of the participants who expressed lower and higher fear acquisition through SCR could provide a valuable insight into the differences in fear-encoding and fear-expression between these groups [50].

Another challenge has been found in the methods of analysis of neuroimaging data. Some researchers report a lack of consistent evidence for the activation of the amygdala during conditioning, and the vmPFC in extinction learning [51]. Part of this inconsistency is due to the fact that the amygdala signal is transient, and often appears at the very beginning of the CS+ presentation (around 15 ms) [26, 27, 52].

Averaging the amygdala and vmPFC signals across multiple trials and using a stringent statistical threshold could explain why these signals are often not detected [53]. We have had success in observing amygdala and vmPFC activation with this paradigm when using techniques such as breaking up trials into time multiple blocks, and contrasting activations between different stimuli (CS+ vs. CS−) and analyzing the differential responses, as have other researchers [4, 5, 14, 16, 28, 45, 54–56].

Integrating techniques such as neuroimaging and SCR into fear learning paradigms as a means to study PTSD has developed over time from an understanding that PTSD is marked by excessive fear and a disrupted ability to extinguish this fear. These techniques can be utilized to further investigate the behavioral and neurobiological abnormalities of PTSD and other anxiety disorders, and guide the development of potential interventions.

References

1. APA (2013) Trauma- and stressor-related disorders. In: Diagnostic and statistical manual of mental disorders, 5th edn. https://doi.org/10.1176/appi.books.9780890425596.dsm07
2. Shvil E, Rusch HL, Sullivan GM, Neria Y (2013) Neural, psychophysiological, and behavioral markers of fear processing in PTSD: a review of the literature. Curr Psychiatry Rep 15(5):358–358. https://doi.org/10.1007/s11920-013-0358-3
3. Hughes KC, Shin LM (2011) Functional neuroimaging studies of post-traumatic stress disorder. Expert Rev Neurother 11(2):275–285. https://doi.org/10.1586/ern.10.198
4. Garfinkel SN, Abelson JL, King AP, Sripada RK, Wang X, Gaines LM, Liberzon I (2014) Impaired contextual modulation of memories in PTSD: an fMRI and psychophysiological study of extinction retention and fear renewal. J Neurosci 34(40):13435–13443. https://doi.org/10.1523/jneurosci.4287-13.2014
5. Milad MR, Pitman RK, Ellis CB, Gold AL, Shin LM, Lasko NB, Zeidan MA, Handwerger K, Orr SP, Rauch SL (2009) Neurobiological basis of failure to recall extinction memory in posttraumatic stress disorder. Biol Psychiatry 66(12):1075–1082. https://doi.org/10.1016/j.biopsych.2009.06.026
6. Holt DJ, Coombs G, Zeidan MA, Goff DC, Milad MR (2012) Failure of neural responses to safety cues in schizophrenia. Arch Gen Psychiatry 69(9):893–903. https://doi.org/10.1001/archgenpsychiatry.2011.2310
7. Lonsdorf TB, Menz MM, Andreatta M, Fullana MA, Golkar A, Haaker J, Heitland I, Hermann A, Kuhn M, Kruse O, Meir Drexler S, Meulders A, Nees F, Pittig A, Richter J, Römer S, Shiban Y, Schmitz A, Straube B, Vervliet B, Wendt J, Baas JMP, Merz CJ (2017) Don't fear 'fear conditioning': methodological considerations for the design and analysis of studies on human fear acquisition, extinction, and return of fear. Neurosci Biobehav Rev 77:247–285. https://doi.org/10.1016/j.neubiorev.2017.02.026
8. LaBar KS, Gatenby JC, Gore JC, LeDoux JE, Phelps EA (1998) Human amygdala activation during conditioned fear acquisition and extinction: a mixed-trial fMRI study. Neuron 20(5):937–945. https://doi.org/10.1016/s0896-6273(00)80475-4
9. Büchel C, Dolan RJ, Armony JL, Friston KJ (1999) Amygdala-hippocampal involvement in human aversive trace conditioning revealed through event-related functional magnetic resonance imaging. J Neurosci 19(24):10869–10876. https://doi.org/10.1523/jneurosci.19-24-10869.1999
10. Phelps EA (2006) Emotion and cognition: insights from studies of the human amygdala. Annu Rev Psychol 57:27–53. https://doi.org/10.1146/annurev.psych.56.091103.070234
11. Milad MR, Orr SP, Lasko NB, Chang Y, Rauch SL, Pitman RK (2008) Presence and acquired origin of reduced recall for fear extinction in PTSD: results of a twin study. J Psychiatr Res 42(7):515–520. https://doi.org/10.1016/j.jpsychires.2008.01.017
12. Duits P, Cath DC, Lissek S, Hox JJ, Hamm AO, Engelhard IM, van den Hout MA, Baas JM (2015) Updated meta-analysis of classical fear conditioning in the anxiety disorders. Depress Anxiety 32(4):239–253. https://doi.org/10.1002/da.22353
13. Lissek S, Powers AS, McClure EB, Phelps EA, Woldehawariat G, Grillon C, Pine DS (2005) Classical fear conditioning in the anxiety disorders: a meta-analysis. Behav Res Ther 43(11):1391–1424. https://doi.org/10.1016/j.brat.2004.10.007
14. Milad MR, Quirk GJ (2012) Fear extinction as a model for translational neuroscience: ten years of progress. Annu Rev Psychol 63:129–151. https://doi.org/10.1146/annurev.psych.121208.131631
15. Milad MR, Orr SP, Pitman RK, Rauch SL (2005) Context modulation of memory for fear extinction in humans. Psychophysiology 42(4):456–464. https://doi.org/10.1111/j.1469-8986.2005.00302.x

16. Milad MR, Wright CI, Orr SP, Pitman RK, Quirk GJ, Rauch SL (2007) Recall of fear extinction in humans activates the ventromedial prefrontal cortex and hippocampus in concert. Biol Psychiatry 62(5):446–454. https://doi.org/10.1016/j.biopsych.2006.17.011

17. Hamacher-Dang TC, Merz CJ, Wolf OT (2015) Stress following extinction learning leads to a context-dependent return of fear. Psychophysiology 52(4):489–498. https://doi.org/10.1111/psyp.12384

18. Holt DJ, Lebron-Milad K, Milad MR, Rauch SL, Pitman RK, Orr SP, Cassidy BS, Walsh JP, Goff DC (2009) Extinction memory is impaired in schizophrenia. Biol Psychiatry 65(6):455–463. https://doi.org/10.1016/j.biopsych.2008.09.017

19. Lebron-Milad K, Abbs B, Milad MR, Linnman C, Rougemount-Bucking A, Zeidan MA, Holt DJ, Goldstein JM (2012) Sex differences in the neurobiology of fear conditioning and extinction: a preliminary fMRI study of shared sex differences with stress-arousal circuitry. Biol Mood Anxiety Disord 2:7. https://doi.org/10.1186/2045-5380-2-7

20. Linnman C, Zeidan MA, Furtak SC, Pitman RK, Quirk GJ, Milad MR (2012) Resting amygdala and medial prefrontal metabolism predicts functional activation of the fear extinction circuit. Am J Psychiatry 169(4):415–423. https://doi.org/10.1176/appi.ajp.2011.10121780

21. Pace-Schott EF, Milad MR, Orr SP, Rauch SL, Stickgold R, Pitman RK (2009) Sleep promotes generalization of extinction of conditioned fear. Sleep 32(1):19–26

22. Rougemont-Bücking A, Linnman C, Zeffiro TA, Zeidan MA, Lebron-Milad K, Rodriguez-Romaguera J, Rauch SL, Pitman RK, Milad MR (2011) Altered processing of contextual information during fear extinction in PTSD: an fMRI study. CNS Neurosci Ther 17(4):227–236. https://doi.org/10.1111/j.1755-5949.2010.00152.x

23. Shvil E, Sullivan GM, Schafer S, Markowitz JC, Campeas M, Wager TD, Milad MR, Neria Y (2014) Sex differences in extinction recall in posttraumatic stress disorder: a pilot fMRI study. Neurobiol Learn Mem 113:101–108. https://doi.org/10.1016/j.nlm.2014.02.003

24. Zeidan MA, Lebron-Milad K, Thompson-Hollands J, Im JJY, Dougherty DD, Holt DJ, Orr SP, Milad MR (2012) Test-retest reliability during fear acquisition and fear extinction in humans. CNS Neurosci Ther 18(4):313–317. https://doi.org/10.1111/j.1755-5949.2011.00238.x

25. van 't Wout M, Mariano TY, Garnaat SL, Reddy MK, Rasmussen SA, Greenberg BD (2016) Can transcranial direct current stimulation augment extinction of conditioned fear? Brain Stimul 9(4):529–536. https://doi.org/10.1016/j.brs.2016.03.004

26. Quirk GJ, Repa JC, LeDoux JE (1995) Fear conditioning enhances short-latency auditory responses of lateral amygdala neurons: parallel recordings in the freely behaving rat. Neuron 15(5):1029–1039. https://doi.org/10.1016/0896-6273(95)90092-6

27. Quirk GJ, Armony JL, LeDoux JE (1997) Fear conditioning enhances different temporal components of tone-evoked spike trains in auditory cortex and lateral amygdala. Neuron 19(3):613–624. https://doi.org/10.1016/S0896-6273(00)80375-X

28. Milad MR, Quirk GJ (2002) Neurons in medial prefrontal cortex signal memory for fear extinction. Nature 420(6911):70–74. https://doi.org/10.1038/nature01138

29. Dawson ME, Schell AM, Filion DL (2007) The electrodermal system. In: Cacioppo JT, Tassinary LG, Berntson GG (eds) Handbook of psychophysiology, 3rd edn. Cambridge University Press, New York, pp 159–181. https://doi.org/10.1017/CBO9780511546396.007

30. Harris JA, Jones ML, Bailey GK, Westbrook RF (2000) Contextual control over conditioned responding in an extinction paradigm. J Exp Psychol Anim Behav Process 26(2):174–185. https://doi.org/10.1037//0097-7403.26.2.174

31. Bouton ME (2000) A learning theory perspective on lapse, relapse, and the maintenance of behavior change. Health Psychol 19(1s):57–63. https://doi.org/10.1037/0278-6133.19.suppl1.57

32. Bach DR, Friston KJ, Dolan RJ (2013) An improved algorithm for model-based analysis of evoked skin conductance responses. Biol Psychol 94(3):490–497. https://doi.org/10.1016/j.biopsycho.2013.09.010

33. Bach DR, Flandin G, Friston KJ, Dolan RJ (2009) Time-series analysis for rapid event-related skin conductance responses. J Neurosci Methods 184(2):224–234. https://doi.org/10.1016/j.jneumeth.2009.08.005

34. Peri T, Ben-Shakhar G, Orr SP, Shalev AY (2000) Psychophysiologic assessment of aversive conditioning in posttraumatic stress disorder. Biol Psychiatry 47(6):512–519. https://doi.org/10.1016/s0006-3223(99)00144-4

35. Orr SP, Metzger LJ, Lasko NB, Macklin ML, Peri T, Pitman RK (2000) De novo conditioning in trauma-exposed individuals with and

without posttraumatic stress disorder. J Abnorm Psychol 109(2):290–298
36. Marin MF, Hammoud MZ, Klumpp H, Simon NM, Milad MR (2020) Multimodal categorical and dimensional approaches to understanding threat conditioning and its extinction in individuals with anxiety disorders. JAMA Psychiatry 77(6):618–627. https://doi.org/10.1001/jamapsychiatry.2019.4833
37. Marin M-F, Zsido RG, Song H, Lasko NB, Killgore WDS, Rauch SL, Simon NM, Milad MR (2017) Skin conductance responses and neural activations during fear conditioning and extinction recall across anxiety disorders. JAMA Psychiatry 74(6):622–631. https://doi.org/10.1001/jamapsychiatry.2017.0329
38. Homan P, Lin Q, Murrough JW, Soleimani L, Bach DR, Clem RL, Schiller D (2017) Prazosin during threat discrimination boosts memory of the safe stimulus. Learn Mem Cold Spring Harbor 24(11):597–601. https://doi.org/10.1101/lm.045898.117
39. Lonsdorf T, Kuhn M, Gerlicher A (2021) Navigating the manifold of skin conductance response quantification approaches. PsyArXiv. https://doi.org/10.31234/osf.io/9h2kd
40. Esteban O, Birman D, Schaer M, Koyejo OO, Poldrack RA, Gorgolewski KJ (2017) MRIQC: advancing the automatic prediction of image quality in MRI from unseen sites. PLoS One 12(9):e0184661. https://doi.org/10.1371/journal.pone.0184661
41. Esteban O, Markiewicz CJ, Blair RW, Moodie CA, Isik AI, Erramuzpe A, Kent JD, Goncalves M, DuPre E, Snyder M, Oya H, Ghosh SS, Wright J, Durnez J, Poldrack RA, Gorgolewski KJ (2019) fMRIPrep: a robust preprocessing pipeline for functional MRI. Nat Methods 16(1):111–116. https://doi.org/10.1038/s41592-018-0235-4
42. Liberzon I, Sripada CS (2008) The functional neuroanatomy of PTSD: a critical review. Prog Brain Res 167:151–169. https://doi.org/10.1016/s0079-6123(07)67011-3
43. Negreira AM, Abdallah CG (2019) A review of fMRI affective processing paradigms used in the neurobiological study of posttraumatic stress disorder. Chronic Stress 3:2470547019829035. https://doi.org/10.1177/2470547019829035
44. Milad MR, Rosenbaum BL, Simon NM (2014) Neuroscience of fear extinction: implications for assessment and treatment of fear-based and anxiety related disorders. Behav Res Ther 62:17–23. https://doi.org/10.1016/j.brat.2014.08.006
45. Phelps EA, Delgado MR, Nearing KI, LeDoux JE (2004) Extinction learning in humans: role of the amygdala and vmPFC. Neuron 43(6):897–905. https://doi.org/10.1016/j.neuron.2004.08.042
46. Fullana MA, Albajes-Eizagirre A, Soriano-Mas C, Vervliet B, Cardoner N, Benet O, Radua J, Harrison BJ (2018) Fear extinction in the human brain: a meta-analysis of fMRI studies in healthy participants. Neurosci Biobehav Rev 88:16–25. https://doi.org/10.1016/j.neubiorev.2018.03.002
47. Kalisch R, Korenfeld E, Stephan KE, Weiskopf N, Seymour B, Dolan RJ (2006) Context-dependent human extinction memory is mediated by a ventromedial prefrontal and hippocampal network. J Neurosci 26(37):9503–9511. https://doi.org/10.1523/jneurosci.2021-06.2006
48. Blechert J, Michael T, Vriends N, Margraf J, Wilhelm FH (2007) Fear conditioning in post-traumatic stress disorder: evidence for delayed extinction of autonomic, experiential, and behavioural responses. Behav Res Ther 45(9):2019–2033. https://doi.org/10.1016/j.brat.2007.02.012
49. Lonsdorf TB, Klingelhöfer-Jens M, Andreatta M, Beckers T, Chalkia A, Gerlicher A, Jentsch VL, Meir Drexler S, Mertens G, Richter J, Sjouwerman R, Wendt J, Merz CJ (2019) Navigating the garden of forking paths for data exclusions in fear conditioning research. elife 8. https://doi.org/10.7554/eLife.52465
50. Marin M-F, Barbey F, Rosenbaum BL, Hammoud MZ, Orr SP, Milad MR (2020) Absence of conditioned responding in humans: a bad measure or individual differences? Psychophysiology 57(1):e13350. https://doi.org/10.1111/psyp.13350
51. Fullana MA, Harrison BJ, Soriano-Mas C, Vervliet B, Cardoner N, Àvila-Parcet A, Radua J (2016) Neural signatures of human fear conditioning: an updated and extended meta-analysis of fMRI studies. Mol Psychiatry 21(4):500–508. https://doi.org/10.1038/mp.2015.88
52. Quirk GJ, Armony JL, Repa JC, Li XF, LeDoux JE (1996) Emotional memory: a search for sites of plasticity. Cold Spring Harb Symp Quant Biol 61:247–257
53. Fullana MA, Albajes-Eizagirre A, Soriano-Mas C, Vervliet B, Cardoner N, Benet O, Radua J, Harrison BJ (2019) Amygdala where art thou? Neurosci Biobehav Rev 102:430–431. https://doi.org/10.1016/j.neubiorev.2018.06.003

54. Milad MR, Rauch SL, Pitman RK, Quirk GJ (2006) Fear extinction in rats: implications for human brain imaging and anxiety disorders. Biol Psychol 73(1):61–71. https://doi.org/10.1016/j.biopsycho.2006.01.008

55. Schiller D, Levy I, Niv Y, LeDoux JE, Phelps EA (2008) From fear to safety and back: reversal of fear in the human brain. J Neurosci 28(45):11517–11525. https://doi.org/10.1523/jneurosci.2265-08.2008

56. Zhou F, Geng Y, Xin F, Li J, Feng P, Liu C, Zhao W, Feng T, Guastella AJ, Ebstein RP, Kendrick KM, Becker B (2019) Human extinction learning is accelerated by an angiotensin antagonist via ventromedial prefrontal cortex and its connections with basolateral amygdala. Biol Psychiatry 86(12):910–920. https://doi.org/10.1016/j.biopsych.2019.07.007

Chapter 6

Reconciling Translational Disparities Between Empirical Approaches to Better Understand PTSD

Seth D. Norrholm, Timothy J. Cilley Jr., and Tanja Jovanovic

Abstract

The translational study of human affective states, including fear, is dependent on the implementation of experimental methodologies that target neurobehavioral response systems that are highly conserved across species. For the study of posttraumatic stress disorder (PTSD), as an example, investigators should be able to capture phenomenological features that encompass physiological arousal, defensive and behavioral activation, distress, associative learning, and adaptation within those processes. The psychophysiological study of traumatized populations has incorporated indices sensitive to the previously mentioned features with most success occurring within applications that include classical, or Pavlovian, conditioning. In general, the most commonly employed psychophysiological paradigms, i.e. skin conductance response and fear-potentiated startle, have identified PTSD-associated exaggerated conditioned fear expression, impaired extinction learning, and an overgeneralization of fear responses. In the present review, we describe common methods of collecting psychophysiological data in humans and discuss the convergence and divergence of these physiological indices with an emphasis on leveraging the advantages provided by each and to inform future innovations.

Key words Psychophysiology, Fear, Trauma, Translational, Stress, Affective neuroscience

1 Introduction

1.1 Translational Methods to Study Fear-Related Psychophysiological Features of PTSD

Translational techniques for the study of posttraumatic stress disorder (PTSD) and its associated neurobehavioral features are frequently developed within the context of Pavlovian fear conditioning paradigms and activation of neural circuits that underlie fear- and stressor-related behaviors. Such paradigms build on the foundation of animal research, and take advantage of a wealth of studies that have established the neurocircuitry of fear using methods that can be applied in humans [1, 2]. In human research, fear conditioning, and its subsequent extinction, is usually studied with objective, physiological measures of arousal and somatic fear responses as well as subjective, psychological methods, including evaluative ratings of US-expectancy, perceived fear, or negative versus positive affect [3]. In short, these paradigms involve the repeated pairing of a previously neutral stimulus (e.g., geometric shape) with an aversive consequence (e.g., electric shock) such that the neutral stimulus assumes the fear-eliciting properties of a conditioned stimulus (CS+); a process thought to underlie the development and maintenance of clinically relevant associative memories in anxiety-related pathologies. The extinction of conditioned fear responses then occurs through the repeated presentation of the CS+ without a negative outcome over many trials in most mammalian-based paradigms. As has been noted in several examples in the available literature, the investigation of extinction learning allows for the observation of discordance, both behaviorally and physiologically, between two competing memory traces, the original CS–US association and the newly acquired CS–no US association [4–6]. This is a critical transition believed to differentiate normal fear "metabolism" from the symptomatology of fear and anxiety disorders [7].

The paradigms employed as described above most often capture responses that are related to an organism's autonomic nervous system arousal and orienting behaviors and can be implemented with relative ease and high reliability. The most commonly used psychophysiological measures are skin conductance response (SCR, also sometimes known as Galvanic skin response or GSR) and eyeblink startle; however, pupil dilation and heart rate have also been used recently [8–12]. SCR is highly reactive to both external (environment specific) and internal (participant specific) changes that occur within an experimental session and are expressed largely independent from valence (positive or negative emotionality [13, 14]). SCR typically emerges within 0.5–4 s after stimulus onset and peaks within approximately 5 s after initiation [15, 16]. Acoustic startle measures, including fear-potentiation of this innate reflex (fear-potentiated startle, FPS), allow investigation of general arousal as well as sensitivity to the valence of a particular stimulus [7, 17]. In contrast to SCR, acoustic startle responses occur within approximately 120 ms after noise probe presentation and tend to habituate at a slower rate. Subjective measures of stimulus valence [18] and US-expectancy [19] are often collected concurrently with either SCR, FPS, or both [20] and can be captured on a trial-by-trial basis. Although both SCR and FPS can be measured peripherally to quantify fear learning, the neural underpinnings may differ (see Fig. 1). Skin conductance is thought to be a measure of sympathetic nervous system activity, such that stressful experiences activate the sympathoadrenal system to release adrenaline [21], which in turn evokes sweat gland activity [22]. SCR are consistently reported to be higher in PTSD patients [23]. On the other hand, FPS reflects more direct amygdala modulation of the

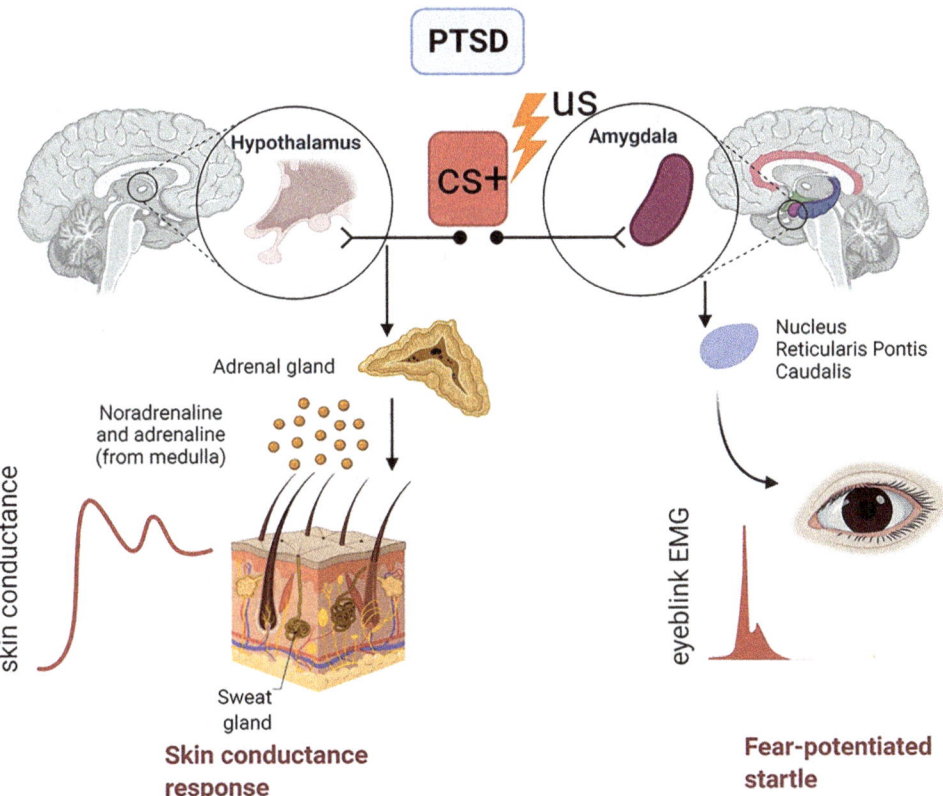

Fig. 1 Central nervous system, peripheral substrates, and primary output organs for conditioned fear expression using skin conductance and fear-potentiated startle measures. (Figure created in Biorender)

startle reflex. Animal studies have shown that fear conditioning increases amygdala activity which then influences the startle circuit via the nucleus reticularis pontis caudalis [24].

2 Materials

2.1 Participants

Human participants included in psychophysiological analyses of fear learning are most often recruited from an academic or clinical setting through physical and virtual advertisement, and those that are interested in a particular study are screened according to the objectives of that study. Because some studies are seeking individuals with a particular history (e.g., psychological trauma) or clinical diagnosis (e.g., PTSD) while others are recruiting psychiatrically healthy controls, and studies require that individuals are able to perceive the study stimuli, the screening process often includes an assessment of medical and psychiatric history, current living situation, medication history, an inventory of significant life events, and, in some cases, visual and auditory acuity.

To screen for the absence or presence of a psychological condition, investigators will most often employ the Structured Clinical Interview for DSM-IV/5 (SCID-IV [25] SCID-5 [26] or the Mini International Neuropsychiatric Interview (MINI) [27]. Assessment of trauma history and PTSD symptoms for studies focused on this area can be achieved through administration of the self-report PTSD Symptom Scale [28, 29], PTSD Checklist [30], and Childhood Trauma Questionnaire (if this is a developmental period of interest [31]) as well as semi-structured interviews including the Traumatic Events Interview [32] and the Clinician Administered PTSD Scale for DSM-5 [33]. Additional common intrinsic factors (i.e., hormonal, neuroendocrine) that are targeted for empirical study include menstrual cycle phase, estrogen/progesterone levels, or pregnancy status for women [34–36] and corticosteroids and their principle metabolites [37]. Participant age may be a factor to consider not only because of the potential contribution of hormonal fluctuations and age-associated changes but also as the incidence of hearing loss can increase with age as well and this can be exclusionary for acoustic startle-based testing. Common comorbidities that may be important to consider with traumatized populations include substance or alcohol use which can be assessed via self-report questionnaires (e.g., Alcohol Use Disorders Identification Test; AUDIT [38]) breathalyzer for alcohol, or urine drug screens (UDS). Given that recreational drugs such as cannabis [39, 40] and prescription medication like benzodiazepines [41] can influence psychophysiology, UDS testing may be required to exclude individuals with positive tests.

2.2 Equipment

Human fear conditioning tasks typically involve two sets of computer-based programs, one for the presentation of experimental stimuli and a second for the collection of psychophysiological data. Frequently used software packages for the presentation of fear conditioning cues, both auditory and visual, include SuperLab (Cedrus, San Pedro, CA, USA), E-prime (Psychology Software Tools, Pittsburgh, PA, USA), Psychlab (Precision Instruments, Cambridge, MA, USA), and Presentation (Neurobehavioral Systems, Berkeley, CA, USA) as examples. Conditioned stimuli (CS) are often geometric shapes or images presented on a computer screen [42–45], and unconditioned stimuli (US) are typically either an electric shock [46], loud noise or scream [47–50], airblast directed at the larynx [51], or a compound stimulus consisting of more than one of these [49]. Shock duration is brief, produced by a voltage stimulator (e.g., STM200 or Stimsola, BIOPAC, Goleta, CA, USA) and delivered through Ag/AgCl electrodes placed on the skin of the wrist, hypothenar muscle, or ankle. Shock intensity is usually titrated to a level specific to each participant and a subjective rating of "highly annoying but not painful" is typically targeted [15, 44, 52]. A common shock work-up procedure will increase electrical stimulation from zero by increments of 0.5 mA until the subject's reported level is reached. Airblasts to the larynx are also brief (250 ms) with an intensity of 140 p.s.i [43], controlled by a solenoid switch attached to a compressed air tank on one end and a small metal nozzle secured by customized tubing. Airblast intensity is not tailored on a case-by-case basis but may be adjusted based on the participant pool's age or vulnerabilities [53]. There are several considerations to evaluate when selecting the modality and qualitative features of a US including, but not limited to, translational applicability across species, contribution of organism sensitivity to the stimulus, potential subjective or appraisal-based connotations (humans only), and the degree of physiological and behavioral activation of stress response systems invoked [48].

2.3 Fear-Potentiated Startle

Recent applications of human fear-potentiated startle measures have used several systems including BIOPAC MP150/160 (Biopac Inc., Goleta, CA, USA) modules including the electromyogram (EMG) transducer amplifier to measure the eyeblink muscle contraction. This hardware apparatus is available with a wireless system of electrodes, leads, and receptacles (BioNomadix) or a fully wired setup. The Coulbourn (S75-01) system has also been used by several laboratories [54–56] and the FaceEMG Cap-MR has recently been put into use by the Center for Depression, Anxiety, and Stress at Harvard University. The Marine Resiliency Study at the University of California San Diego has used the SR-HLAB EMG system (San Diego Instruments, San Diego, CA, USA) to assess startle [57]. There are likely more systems being used; however, these are the ones most commonly reported in the literature.

2.4 Skin Conductance Response

SCR data collection has also been captured most frequently using a system such as the BIOPAC MP150/160 or Coulbourn (S71-23) and an associated GSR transducer amplifier. Data signal is captured from participants through the use of two Ag/AgCl electrodes filled with isotonic electrolyte gel that are placed on the hand or fingers.

2.5 US-Expectancy and Valence Ratings

Measures of a participant's expectancy of the US or subjective appraisal of a stimulus' emotional content or valence is often accomplished with an input device (e.g., keypad, joystick, or standard QWERTY keyboard) linked to the data acquisition apparatus such that the investigators can capture a trial-by-trial CS assessment from the participant [58]. Ratings of US-expectancy and stimulus valence can be obtained concurrently and often employ the use of an emotional self-assessment manakin [59, 60].

3 Methods

3.1 Fear Conditioning, Extinction, and Return of Fear

A typical mammalian fear conditioning paradigm will often consist of habituation, fear conditioning (also termed acquisition), extinction training, and extinction recall. During habituation, the research subject is placed in the experimental context and exposed to the CSs that will be paired with the US during the initial fear conditioning procedures. During fear conditioning or acquisition, the subject is repeatedly presented with pairings of the reinforced conditioned stimulus (CS+) with the aversive US. If presented on its own, the US will elicit a robust unconditioned response (UR) without any additional learning having occurred and this response will manifest itself as an increase in skin conductance and the acoustic startle response relative to each index's baseline. Once the CS+/US contingency is acquired, presentation of the CS+ alone will elicit an increased fear response (via startle, skin conductance, or expectancy/valence rating) without the US; this is now termed a conditioned response (CR). The inclusion of a nonreinforced CS− allows the investigators to observe increases in fear specific to the reinforced CS+.

Fear acquisition is often presented as a series of sequential "blocks" with each block consisting of randomly interleaved trials of each type: CS+, CS−, as well as noise probe alone (abbreviated NA) in startle studies. Several applications have used three blocks of conditioning with four trials of each type for a total of 12 trials each of the CS+, CS−, and NA. One factor that is often variable across fear conditioning studies is the reinforcement rate between the CS+ and the subsequent US. This rate is typically between 50% and 100% depending on study objectives as partial reinforcement during acquisition can delay the development of fear acquisition [20] as well as extinction learning [61]. Intertrial intervals are also randomized with most occurring between 9 and 30 s between each trial.

Extinction is the new learning process by which a previously reinforced CS+ is repeatedly presented without the US such that the subject learns that what was once aversive or dangerous is now safe. Successful extinction learning will result in a diminished capacity of the CS+ to elicit a CR and, ideally, a more prominent extinction memory. Following extinction learning, one can assess the degree to which a conditioned stimulus remains extinguished (termed extinction recall or retention) or to which the CR reemerges (termed the return of fear).

3.2 Fear-Potentiated Startle

Electromyographic activity of the *orbicularis oculi* muscle can be reliably used as a measure of both baseline and fear-potentiated startle by placing two 5 mm Ag/AgCl electrodes approximately 1 cm under the pupil and 1 cm below the lateral canthus

[45]. Impedance is typically measured prior to data collection and more conservative approaches aim to keep this value below 6 kiloohms. These data can be collected with the EMG module of the BIOPAC MP150/160 system, sampled at 1000 Hz, digitized at 16 bit A/D resolution, and amplified. Exported raw EMG data from a data acquisition package such as AcqKnowledge (BIOPAC, Goleta, CA, USA) can be filtered with low- and high-frequency cut-offs at 28 and 500 Hz, respectively (Blumenthal et al. 2005). Startle magnitude on a particular trial (expressed in microVolts) is defined as the peak amplitude of the *orbicularis oculi* contraction 20–200 ms after the presentation of an acoustic startle probe, which is typically a very brief (40 ms) burst of white noise set at a relatively high intensity (90–108 dB [A] SPL) delivered binaurally through headphones. Sound pressure and decibel level are calibrated via sound level meter (e.g., Model 9104, Exair, Cincinnati, OH, USA).

CS trials can last from 5 to 12 s, depending on the study. In one example of a CS+ trial, a shape CS is presented for a total of 6 s. An acoustic startle probe (40 ms, 108 dB [A] SPL) is then presented toward the end of the trial (at 5210 ms) and is followed 500 ms later by a 250-ms, 140 p.s.i. airblast to the larynx (US). This presentation sequence is based on seminal work by Davis and colleagues (1993) who showed that potentiation is greatest when startle is elicited in close temporal proximity to delivery of the US [62]. For this example, the CS− trial (or a nonreinforced CS+ trial during extinction) is also 6 s in duration with the startle probe presented toward the end of the trial with no subsequent presentation of the US. Fear-potentiated startle is expressed as the difference in mean startle magnitude on CS trials (either CS+ or CS−) and mean startle magnitude to the noise alone (sometimes referred to as intertrial interval, or ITI, startle) during each block of a session. Some groups have not included an acoustic startle probe on every CS trial as an effort to delay habituation of the reflex (e.g., 75% of CS trials include noise probe [15]) and this is consistent with the early rodent work on which contemporary human applications are based (e.g [63]).

3.3 Skin Conductance Response

SCR is a component of electrodermal activity that measures a change in skin conductance level elicited by a stimulus, and can be reliably assessed by placing two Ag/AgCl electrodes, prefilled with isotonic paste, on the hypothenar surface or medial phalanges of the non-dominant hand (especially if using an input device such as a keypad). When assessing FPS and SCR concurrently on a traditionally wired system such as the BIOPAC MP150, the SCR electrodes can also serve as an effective ground electrode for EMG. Electrodermal activity is typically 1 Hz low-pass filtered and 0.05 Hz high-pass filtered when capturing skin conductance. SCR (expressed in microSiemens) on a particular experimental trial is often defined as the average increase in response when comparing the 1 s-period prior to CS onset with the peak response in the 3–6 s period after the CS appears (the CS is usually presented for at least 6 s in applications such as the one described here).

3.4 US-Expectancy and Valence Ratings

The capture of US-expectancy and valence ratings can occur on a trial-by-trial basis [45], in between experimental sessions (i.e., between fear conditioning and extinction [44]), or at the conclusion of an experimental session through exit interview [64]. Participants are often provided with verbal instructions on the use of the input device used to collect ratings and the inclusion of verbal instructions and trial-by-trial ratings has been associated with more robust fear learning [58, 65–67]. For US-expectancy measures, participants routinely respond with one button if they expect the US on a particular trial (e.g., button marked '+'), a different button if they do not expect the US (e.g., button marked '−'), or a third button if they are unsure (e.g., button marked '0'). Similar applications have used a more continuous measure of expectancy like a visual analog scale (i.e., rating from 0 to 100 on certainty [20]) and the methods for obtaining US-expectancy can be easily adapted to capture valence or fear ratings on a particular trial.

3.5 Common Elements

3.5.1 Associations Between Measurements and Indices

It is not surprising that when conditioned stimuli contain a high degree of emotional salience or a particularly strong valence (e.g., images of violence), there is a strong association between startle and skin conductance recordings – both indices are increased with highly aversive stimuli. But clearly one measure is not required for the expression of the other, as potentiated startle responses to CS+ can persist for a period of time after skin conductance responses have habituated [64, 68]. As we will discuss in a later section, the strongest linkage between FPS and SCR appears when investigators employ trace (a time gap exists between CS offset and US onset) versus delay (CS–US co-terminate) conditioning. In short, investigators that have employed trace conditioning have seen a high correlation between the strength of SCR and FPS responses after conditioning, a requirement for declarative CS–US awareness, as well as recruitment of context encoding brain regions such as hippocampus [15, 64].

Recent work from Constantinou and colleagues (2021) has shown a positive correlation between participants reports of US-expectancy and SCR during both the fear conditioning and extinction phases of fear learning [42]. Further, this group demonstrated a positive association between US-expectancy during extinction training and ratings of negative affect during this phase. Notably, these associations were strongest at times during the acquisition and extinction phases where learning specific to the CS–US contingency (late acquisition, early and late extinction) has been shown to be the most robust [7, 42, 69, 70].

Table 1
Shows a summary of the three measures across different paradigms

	Arousal	Valence	Conditioned fear expression	CS+/CS− discrimination	Extinction learning (within session)	Extinction retention (between session)
Fear-potentiated startle	++[a]	+++	+++	+++	+++	++[b]
Skin conductance	++[a]	−−	+++	+++	−−	++[b]
US expectancy	++	+++	+++	+++	++	++[b]

+++ – supported by strong empirical evidence; ++ – supported by moderate empirical evidence
−− – not well supported by empirical evidence
[a]Both of these measures are subject to within-session habituation which must be accounted for when incorporating into experimental design and temporal parameters
[b]This test can be especially context-dependent based on parameters under which acquisition, extinction learning, and extinction recall occur (ABA, ABB, ABC contextual design)

Weike and colleagues (2007) discuss experimental situations in which there is weak association between physiological fear responses and declarative awareness of CS–US contingency [71]. This is primarily a concern in applications during which the participant sets the intensity level of the US such as the shock "work-up" procedure [46]. With this type of work-up, there is a risk of the participant setting the US intensity at a level insufficient to produce a "true" fear response. In other words, the participant learns that the CS predicts the US [72] but does not emotionally or physiologically show fear of the US. However, based on three decades of work across numerous groups, we can confidently report that all three measures (FPS, SCR, and US-expectancy) display greater responses to a reinforced CS+ as compared to a safe CS− [7]. Table 1 shows a summary of the three measures across different paradigms.

3.5.2 Relevance to Posttraumatic Stress Disorder (PTSD)

Using paradigms such as those described above has revealed that, in general, PTSD is associated with elevated fear responses [16, 54, 73, 74], impaired fear inhibition [43, 75], and deficient fear extinction as compared to psychiatric and healthy control groups [7, 55, 76–78]. It is important to note, however, that the available data is not unambiguous and there are reported differences based on the methodological approaches taken as well as the in the interpretation and analysis of psychophysiological indices. For example, while some groups have found a high degree of CS+/CS− discrimination with SCR in PTSD patients [44, 74] others have shown a blunted SCR discrimination in these groups [16, 54].

With respect to US-expectancy measures, it has often been the case that traumatized populations with PTSD symptoms will show a discrepancy between cognitive reports of danger versus safety as compared to physiological measures. In other words, reports have shown that while participants are able to correctly identify cues as safe (non-reinforced) or threatening (reinforced), conditioned fear responses continue to manifest themselves in the presence of safety cues [79, 80]. Thus, it remains possible that prefrontal cortical control over amygdala output continues to show impairment while higher order assessments are left intact. On the other hand, some studies that have used complex stimuli have shown that PTSD subjects also show higher expectancy of the US on CS− trials, suggesting that there may be more nuanced differences in some cognitive awareness of safety as well [80].

3.6 Paradigmatic Strengths

SCR methods lend themselves well to coupling with brain imaging techniques and procedures. The test electrodes provide little to no interference with the collection of brain scans and investigators can determine the level of activation in specific target brain regions and the relationship between brain region activation and activity of the sympathetic and parasympathetic divisions of the autonomic nervous system [16, 78, 81]. In addition, a growing body of work has demonstrated a strong association between physiological SCR and FPS discrimination between the reinforced CS+ and non-reinforced CS− and this same discrimination in US-expectancy measures during fear conditioning, or acquisition [42, 45, 71, 82, 83]. However, this tight linkage between physiological indices and psychological measures is not as clear during extinction training and extinction recall. Fear extinction studies using SCR as a primary psychophysiological outcome have shown great utility in illustrating between-session extinction effects, i.e. extinction retention [78, 81, 84, 85], while FPS extinction studies have shown a capacity to observe both within-session extinction and between-session extinction effects [58, 77, 86, 87].

There has been some recent work that has successfully integrated functional neuroimaging with fear-potentiated startle procedures [56]. In their concurrent examination of FPS and neural activity, Kuhn and colleagues identified valence-specific activation that recruits both the brainstem (e.g., nucleus reticularis pontis caudalis) and the specific divisions within the centromedial amygdala in a manner that is markedly different from that which follows general arousal. The latter study also represents an important confirmation of the highly conserved neuroanatomical substrates underlying mammalian acoustic startle, and by extension, fear learning.

3.7 Discrepancies/Limitations

The degree to which psychophysiological measures of fear are closely associated to one another or to which they represent common or disparate neurobiological underpinnings remains a topic of debate within the field [88]. A putative single process model in which there exists one complex "fear mechanism" was discussed in the early 2000s [89, 90] while a dual-process model in which associative and subjective fear learning share distinct underpinnings emerged during the same time frame [91, 92] and continues to be debated [42, 88, 93–95]. In addition, evidence has accumulated that both supports (SCR [83, 96, 97]; FPS [86]) and argues against (FPS [98–100]) close association between US-expectancy awareness and physiological responses. Finally, there are equally mixed results, drawn largely from the skin conductance literature, with respect to the association between subjective appraisal of positive or negative valence and physiological responding [82, 97, 101].

It should also be noted here that applications of both SCR and FPS measures must account for the significant habituation that can occur with the two indices especially if testing occurs across multiple sessions over several minutes. In general, SCR tends to habituate faster than FPS and this difference may be due to the presence of excitatory acoustic startle probes in the latter versus the former (which does not require an external stimulus to capture its activity). Further, it may be that the habituation differences between SCR and FPS account for the aforementioned sensitivity of SCR to between-session effects and FPS to within-session effects.

Consistent with the notion of distinct neural underpinnings for SCR and FPS, there have been reports in the literature that in addition to these indices not being tightly associated with one another there exists the possibility that the two show *opposite* effects [92, 102]. This is likely the result of participant-specific and methodological considerations including, but not limited to, contingency awareness, recruitment of limbic brain regions implicated in contextual processing (e.g., hippocampus), pharmacological manipulation (e.g., corticosteroids), and whether or not delay (CS–US co-terminate on a trial) or trace (time elapses between end of CS presentation and US delivery) conditioning is employed.

Early work by Hamm and Vaitl (1996) explored differential responding between SCR and FPS measures with a focus on the aversive nature of the US as well as contingency awareness [64]. The authors of the study found SCR discrimination between reinforced and nonreinforced CSs regardless of whether the US was aversive (electric shock) or non-aversive (simple vibration). However, using the same USs in different experiments with FPS, potentiated startle was only observed with an aversive US. Interestingly, the authors also found that FPS was present in participants regardless of awareness of the CS–US contingency, whereas differential SCR responding (CS+ vs. CS−) was only observed in those with declarative knowledge of the CS–US relationship. A study that examined contingency awareness and FPS to the CS+ and CS– separately found that FPS to the CS+ emerged very quickly, i.e. prior to declarative awareness, while FPS to the CS– decreased after awareness of the association between the CS and the absence of the US, suggesting that awareness may not be necessary for fear to be acquired, but may be necessary for safety learning [103]. In line with the two models discussed previously (single vs. dual processing of fear), there is also evidence that suggests declarative contingency awareness is requisite for the development of FPS expression and CS discrimination. Both Grillon (2002) and Purkis and Lipp (2001) observed FPS only in participants that were aware of the CS–US contingency [97, 104]; these findings support a single process mechanism for associative learning in human subjects [89, 90].

While recognizing the ambiguity in available evidence, Hamm and Weike (2005) presented a model in which FPS and SCR are mediated by two distinct fear learning processes: an amygdala-dependent, subcortical mechanism that governs reflexive fear responses (e.g., FPS) and is not dependent on contingency awareness and a second cortical- and hippocampal-dependent mechanism that encodes declarative learning/contingency awareness (and, as such, reflected in SCR). A large body of literature collected in the time since this model was presented suggests a more complex interaction of both reflexive and cognitive processes that may operate at different temporal windows during fear learning experiences such as those mediating defensive reflexes, arousal, and appraisal [15, 105].

Another important consideration when looking at psychophysiological measures of fear response is demographic variability. Ethnic and racial differences have been observed in startle and skin conductance. Baseline startle, i.e. response to the acoustic probe alone, has been found to be lower in Black American compared to White American samples [106]; however, startle modulation, such as FPS, may reduce these effects as a change from baseline may to some extent account for individual and race-related differences. SCR appears to be much more susceptible to such factors, with one study showing higher SCR in Hispanic males compared to non-Hispanic White males [107]. Of greater concern to the field, and to generalizing results across studies, is the observation that Black Americans are often excluded from fear conditioning studies due to low SCR levels [108]. In fact, classifying individuals as responders based on predetermined cutoffs, has disproportionately categorized Black Americans as SCR non-responders [108], resulting in underrepresentation of diverse samples in this research. Given the high rates of PTSD in Black American urban populations [109, 110], it is imperative that we better understand the underlying causes of these race-related differences. Some studies have suggested that these differences stem from biological factors, such as sweat chemistry (i.e. less chloride) [108] or skin tone (i.e. for methods that rely on LED light, such as wearables) [111], while others have noted that adverse life experiences in large part account for race-related differences in threat reactivity [112].

3.8 Recommendations/Future Directions

Taken together, the body of literature discussed in the current review suggests an approach in which the indices of fear most often applied to the study of PTSD fear-related phenomenology, whether they are psychophysiological or psychological/cognitive, be employed concurrently to allow investigators to capitalize on the individual strengths of each measure while overcoming the inherent weaknesses in others. An integrative approach such as this may facilitate the identification of PTSD-related biomarkers and targets for clinical intervention. As described in a recent study by Constantinou et al. (2021), an approach that includes multiple concurrent psychophysiological measures provides a greater workspace with regard to associations among indices within populations of interest (e.g., traumatized vs. non-traumatized), development of theoretical models of fear learning, as well as stronger analyses of treatment outcome measures, moderators and mediators, and individual as well as racial and ethnic differences. In addition, an integrative approach will also allow investigators in the field to better disentangle the heterogeneity present in the current body of literature as it relates to inhibitory learning and extinction deficits in fear-, trauma-, stressor-, and anxiety-related disorders. Lastly, the most often used psychophysiological measures are likely mediated by divergent neural, cognitive, affective, and temporal processes and, as such, it behooves the translational researcher to "cast a wide net" in experimental design and methodology.

4 Conclusions

It is becoming increasingly clear that many of the discrepancies between psychophysiological measures such as SCR and FPS are closely linked to experimental parameters including, but not limited to, stimulus selection, reinforcement rate, temporal relationships between acquisition and extinction learning, and participant awareness of the CS–US contingencies. As an example, there is considerable evidence that factors such as declarative knowledge or awareness is necessary for some types of conditioning (trace) but not others (delay) and that stimuli with an associated valence (negative or positive) require less declarative knowledge. What is clear to translational researchers at this point is that the underlying neurobiology mediating fear- and stressor-related behaviors, both adaptive and maladaptive, as well as the human conditions that are modeled preclinically recruit several physiological processes including, but not limited to, the sympathoadrenal system, the hypothalamic-pituitary-adrenal axis, limbic circuitry spanning the extended amygdala, and associative learning cortices. As such, the best practice continues to be the implementation of concurrent objective and subjective methods targeting the overlapping and divergent neural systems underlying the stressor-, fear-, and anxiety-related features of PTSD.

References

1. Bowers ME, Ressler KJ (2015) An overview of translationally informed treatments for posttraumatic stress disorder: animal models of Pavlovian fear conditioning to human clinical trials. Biol Psychiatry 78(5):E15–E27
2. Milad MR, Quirk GJ (2012) Fear extinction as a model for translational neuroscience: ten years of progress. Annu Rev Psychol 63(1): 129–151
3. VanElzakker MB et al (2014) From Pavlov to PTSD: the extinction of conditioned fear in rodents, humans, and anxiety disorders. Neurobiol Learn Mem 113:3–18
4. Milad MG, Quirk GJ (2002) Neurons in medial prefrontal cortex signal memory for fear extinction. Nature 420:70–74
5. Vervliet B, Craske MG, Hermans D (2013) Fear extinction and relapse: state of the art. Annu Rev Clin Psychol 9:215–248
6. Myers KM, Davis M (2002) Behavioral and neural analysis of extinction. Neuron 36(4): 567–584
7. Norrholm SD, Jovanovic T (2018) Fear processing, psychophysiology, and PTSD. Harv Rev Psychiatry 26(3):129–141
8. Bach DR, Melinscak F (2020) Psychophysiological modelling and the measurement of fear conditioning. Behav Res Ther 127: 103576
9. Staib M, Castegnetti G, Bach DR (2015) Optimising a model-based approach to inferring fear learning from skin conductance responses. J Neurosci Methods 255:131–138
10. Lonsdorf TB et al (2015) Sex differences in conditioned stimulus discrimination during context-dependent fear learning and its retrieval in humans: the role of biological sex, contraceptives and menstrual cycle phases. J Psychiatry Neurosci 40(6):368–375
11. Inslicht SS et al (2021) Randomized controlled experimental study of hydrocortisone and D-cycloserine effects on fear extinction in PTSD. Neuropsychopharmacology 47:1945
12. Stout DM et al (2018) Neural measures associated with configural threat acquisition. Neurobiol Learn Mem 150:99–106
13. Bradley MM et al (2008) The pupil as a measure of emotional arousal and autonomic activation. Psychophysiology 45(4):602–607
14. Hamm AO, Stark R (1993) Sensitization and aversive conditioning: effects on the startle reflex and electrodermal responding. Integr Physiol Behav Sci 28(2):171–176
15. Leuchs L, Schneider M, Spoormaker VI (2019) Measuring the conditioned response: a comparison of pupillometry, skin conductance, and startle electromyography. Psychophysiology 56(1):e13283
16. Glover EM et al (2011) Tools for translational neuroscience: PTSD is associated with heightened fear responses using acoustic startle but not skin conductance measures. Depress Anxiety 28(12):1058–1066
17. Blumenthal TD et al (2005) Committee report: guidelines for human startle eyeblink electromyographic studies. Psychophysiology 42:1–15
18. De Houwer J, Thomas S, Baeyens F (2001) Associative learning of likes and dislikes: a review of 25 years of research on human evaluative conditioning. Psychol Bull 127(6): 853–869
19. Cavanagh K, Davey GC (2001) The effect of mood and arousal on UCS expectancy biases. J Behav Ther Exp Psychiatry 32(1):29–49
20. Cornelisse S et al (2014) Delayed effects of cortisol enhance fear memory of trace conditioning. Psychoneuroendocrinology 40:257–268
21. Bierhaus A et al (2003) A mechanism converting psychosocial stress into mononuclear cell activation. Proc Natl Acad Sci U S A 100(4): 1920–1925
22. Tronstad C et al (2017) Detection of sympathoadrenal discharge by parameterisation of skin conductance and ECG measurement. Annu Int Conf IEEE Eng Med Biol Soc 2017:3997–4000
23. Liberzon I et al (1999) Neuroendocrine and psychophysiologic responses in PTSD: a symptom provocation study. Neuropsychopharmacology 21(1):40–50

24. Davis M, Lee Y (1997) Fear and anxiety: possible roles of the amygdala and bed nucleus of the stria terminalis. Cognit Emot 11:277–306
25. First MB et al (1996) Structured clinical interview for DSM-IV axis I disorders non-patient edition (SCID-I/NP, Version 2.0). Biometrics Research Department, New York State Psychiatric Institute, New York
26. First MB et al (2015) Structured clinical interview for DSM-5—research version. American Psychiatric Association, Arlington
27. Sheehan DV, Lecrubier Y, Sheehan KH (1998) The MINI-International Neuropsychiatric Interview (MINI): the development and validation of a structured diagnostic interview for DSM-IV and ICD-10. J Clin Psychiatry 1959:22–33
28. Falsetti S et al (1993) The modified PTSD symptom scale: a brief self-report measure of posttraumatic stress disorder. Behav Ther 16:161–162
29. Foa EB, Tolin DF (2000) Comparison of the PTSD symptom scale-interview version with the clinician administered PTSD scale. J Trauma Stress 13(2):181–191
30. Weathers FW et al (2013) The PTSD checklist for DSM-5 (PCL-5) – extended criterion a [Measurement instrument]. Available from: https://www.ptsd.va.gov/
31. Bernstein DP et al (2003) Development and validation of a brief screening version of the Childhood Trauma Questionnaire. Child Abuse Negl 27:169–190
32. Schwartz AC et al (2005) Posttraumatic stress disorder among African Americans in an inner city mental health clinic. Psychiatr Serv 56(2):212–215
33. Weathers FW, Bovin MJ, Lee DJ, Sloan DM, Schnurr PP, Kaloupek DG, Keane TM, Marx BP (2018) The Clinician-Administered PTSD Scale for DSM-5 (CAPS-5): Development and initial psychometric evaluation in military veterans. Psychol Assess 30(3):383–395. https://doi.org/10.1037/pas0000486
34. Glover EM et al (2012) Estrogen levels are associated with extinction deficits in women with posttraumatic stress disorder. Biol Psychiatry 72(1):19–24
35. Glover EM et al (2013) Inhibition of fear is differentially associated with cycling estrogen levels in women. J Psychiatry Neurosci 38(3):120129
36. Michopoulos V et al (2015) Psychophysiology and posttraumatic stress disorder symptom profile in pregnant African-American women with trauma exposure. Arch Womens Ment Health 18(4):639–648
37. Grillon C et al (2006) Cortisol and DHEA-S are associated with startle potentiation during aversive conditioning in humans. Psychopharmacology 186(3):434–441
38. Bohn MJ, Babor TF, Kranzler HR (1995) The Alcohol Use Disorders Identification Test (AUDIT): validation of a screening instrument for use in medical settings. J Stud Alcohol 56(4):423–432
39. Kedzior KK, Wehmann E, Martin-Iverson M (2016) Habituation of the startle reflex depends on attention in cannabis users. BMC Psychol 4(1):50
40. Fusar-Poli P et al (2009) Distinct effects of {delta}9-tetrahydrocannabinol and cannabidiol on neural activation during emotional processing. Arch Gen Psychiatry 66(1):95–105
41. Grillon C et al (2006) The benzodiazepine alprazolam dissociates contextual fear from cued fear in humans as assessed by fear-potentiated startle. Biol Psychiatry 60(7):760–766
42. Constantinou E et al (2021) Measuring fear: association among different measures of fear learning. J Behav Ther Exp Psychiatry 70:101618
43. Jovanovic T et al (2009) Posttraumatic stress disorder may be associated with impaired fear inhibition: relation to symptom severity. Psychiatry Res 167(1–2):151–160
44. Pohlchen D et al (2021) Examining differences in fear learning in patients with obsessive-compulsive disorder with pupillometry, startle electromyography and skin conductance responses. Front Psych 12:730742
45. Norrholm SD et al (2006) Conditioned fear extinction and reinstatement in a human fear-potentiated startle paradigm. Learn Mem 13(6):681–685
46. Grillon C, Ameli R (1998) Effects of threat of shock, shock electrode placement and darkness on startle. Int J Psychophysiol 28(3):223–231
47. Lambert E et al (2021) Fear conditioning in women with anorexia nervosa and healthy controls: a preliminary study. J Abnorm Psychol 130(5):490–497
48. Glenn CR, Lieberman L, Hajcak G (2012) Comparing electric shock and a fearful screaming face as unconditioned stimuli for fear learning. Int J Psychophysiol 86(3):214–219
49. Kredlow MA, Orr SP, Otto MW (2018) Who is studied in de novo fear conditioning paradigms? An examination of demographic and

stimulus characteristics predicting fear learning? Int J Psychophysiol 130:21–28
50. Hamm AO, Vaitl D, Lang PJ (1989) Fear conditioning, meaning, and belongingness: a selective association analysis. J Abnorm Psychol 98(4):395–406
51. Norrholm SD et al (2008) Timing of extinction relative to acquisition: a parametric analysis of fear extinction in humans. Behav Neurosci 122(5):1016–1030
52. Grillon C, Ameli R (1998) Effects of threat and safety signals on startle during anticipation of aversive shocks, sounds, or airblasts. J Psychophysiol 12:329–337
53. Jovanovic T et al (2020) Impact of ADCYAP1R1 genotype on longitudinal fear conditioning in children: interaction with trauma and sex. Neuropsychopharmacology 45(10):1603–1608
54. Orr SP et al (2000) De novo conditioning in trauma-exposed individuals with and without posttraumatic stress disorder. J Abnorm Psychol 109(2):290–298
55. Pole N et al (2009) Prospective prediction of posttraumatic stress disorder symptoms using fear potentiated auditory startle responses. Biol Psychiatry 65(3):235–240
56. Kuhn M et al (2020) The neurofunctional basis of affective startle modulation in humans: evidence from combined facial electromyography and functional magnetic resonance imaging. Biol Psychiatry 87(6):548–558
57. Glenn DE et al (2017) Fear learning alterations after traumatic brain injury and their role in development of posttraumatic stress symptoms. Depress Anxiety. Epub
58. Warren VT et al (2014) Human fear extinction and return of fear using reconsolidation update mechanisms: the contribution of on-line expectancy ratings. Neurobiol Learn Mem 113:165–173
59. de Haan MIC et al (2018) The influence of acoustic startle probes on fear learning in humans. Sci Rep 8(1):14552
60. Bradley MM, Lang PJ (1994) Measuring emotion: the self-assessment manikin and the semantic differential. J Behav Ther Exp Psychiatry 25(1):49–59
61. Leonard DW (1975) Partial reinforcement effects in classical aversive conditioning in rabbits and human beings. J Comp Physiol Psychol 88:596–608
62. Grillon C et al (1993) Measuring the time course of anticipatory anxiety using the fear-potentiated startle reflex. Psychophysiology 30(4):340–346
63. Davis M et al (1993) Fear-potentiated startle: a neural and pharmacological analysis. Behav Brain Res 58:175–198
64. Hamm AO, Vaitl D (1996) Affective learning: awareness and aversion. Psychophysiology 33:698–710
65. Kindt M, Soeter M, Vervliet B (2009) Beyond extinction: erasing human fear responses and preventing the return of fear. Nat Neurosci 12(3):256–258
66. Merz CJ et al (2012) Oral contraceptive usage alters the effects of cortisol on implicit fear learning. Horm Behav 62(4):531–538
67. Merz CJ et al (2012) Neuronal correlates of extinction learning are modulated by sex hormones. Soc Cogn Affect Neurosci 7(7):819–830
68. Bradley MM, Lang PJ, Cuthbert BN (1993) Emotion, novelty, and the startle reflex: habituation in humans. Behav Neurosci 107:970–980
69. Norrholm SD et al (2015) Fear load: the psychophysiological over-expression of fear as an intermediate phenotype associated with trauma reactions. Int J Psychophysiol 98(2 Pt 2):270–275
70. Norrholm SD et al (2013) Differential genetic and epigenetic regulation of catechol-O-methyltransferase is associated with impaired fear inhibition in posttraumatic stress disorder. Front Behav Neurosci 7:30
71. Weike AI, Schupp HT, Hamm AO (2007) Fear acquisition requires awareness in trace but not delay conditioning. Psychophysiology 44:170–180
72. Rescorla RA (1988) Pavlovian conditioning. It's not what you think it is. Am Psychol 43(3):151–160
73. Briscione MA, Jovanovic T, Norrholm SD (2014) Conditioned fear associated phenotypes as robust, translational indices of trauma-, stressor-, and anxiety-related behaviors. Front Psych 5:88
74. Young DA et al (2018) Association among anterior cingulate cortex volume, psychophysiological response, and PTSD diagnosis in a Veteran sample. Neurobiol Learn Mem 155:189–196
75. van Rooij SJH, Jovanovic T (2019) Impaired inhibition as an intermediate phenotype for PTSD risk and treatment response. Prog Neuro-Psychopharmacol Biol Psychiatry 89:435–445
76. Deslauriers J et al (2018) COMT val158met polymorphism links to altered fear conditioning and extinction are modulated by PTSD

and childhood trauma. Depress Anxiety 35(1):32–42
77. Acheson DT et al (2015) Conditioned fear and extinction learning performance and its association with psychiatric symptoms in active duty Marines. Psychoneuroendocrinology 51:495–505
78. Milad MR et al (2009) Neurobiological basis of failure to recall extinction memory in posttraumatic stress disorder. Biol Psychiatry 66(12):1075–1082
79. Jovanovic T et al (2013) Acute stress disorder versus chronic posttraumatic stress disorder: inhibition of fear as a function of time since trauma. Depress Anxiety 30(3):217–224
80. Rabinak CA et al (2017) Acquisition of CS-US contingencies during Pavlovian fear conditioning and extinction in social anxiety disorder and posttraumatic stress disorder. J Affect Disord 207:76–85
81. Garfinkel SN et al (2014) Impaired contextual modulation of memories in PTSD: an fMRI and psychophysiological study of extinction retention and fear renewal. J Neurosci 34(40):13435–13443
82. Dawson ME et al (2007) Under what conditions can human affective conditioning occur without contingency awareness? Test of the evaluative conditioning paradigm. Emotion 7(4):755–766
83. Lovibond PF (1992) Tonic and phasic electrodermal measures of human aversive conditioning with long duration stimuli. Psychophysiology 29(6):621–632
84. Milad MR et al (2006) Fear conditioning and extinction: influence of sex and menstrual cycle in healthy humans. Behav Neurosci 120(6):1196–1203
85. Milad MR et al (2007) Recall of fear extinction in humans activates the ventromedial prefrontal cortex and hippocampus in concert. Biol Psychiatry 62(5):446–454
86. Norrholm SD et al (2011) Versatility of fear-potentiated startle paradigms for assessing human conditioned fear extinction and return of fear. Front Behav Neurosci 5:77
87. Norrholm SD et al (2010) Fear extinction in traumatized civilians with posttraumatic stress disorder: relation to symptom severity. Biol Psychiatry (in press)
88. LeDoux JE, Brown R (2017) A higher-order theory of emotional consciousness. Proc Natl Acad Sci U S A 114(10):E2016–E2025
89. Lipp OV, Purkis HM (2005) No support for dual process accounts of human affective learning in simple Pavlovian conditioning. Cognit Emot 19:269–282
90. Lovibond PF, Shanks DR (2002) The role of awareness in Pavlovian conditioning: empirical evidence and theoretical implications. J Exp Psychol Anim Behav Process 28(1):3–26
91. Ohman A, Mineka S (2001) Fears, phobias, and preparedness: toward an evolved module of fear and fear learning. Psychol Rev 108:483–522
92. Hamm AO, Weike AI (2005) The neuropsychology of fear learning and fear regulation. Int J Psychophysiol 57(1):5–14
93. LeDoux JE, Pine DS (2016) Using neuroscience to help understand fear and anxiety: a two-system framework. Am J Psychiatry 173(11):1083–1093
94. LeDoux JE (2014) Coming to terms with fear. Proc Natl Acad Sci 111(8):2871–2878
95. Fanselow MS, Pennington ZT (2018) A return to the psychiatric dark ages with a two-system framework for fear. Behav Res Ther 100:24–29
96. Lovibond PF, Davis NR, O'Flaherty AS (2000) Protection from extinction in human fear conditioning. Behav Res Ther 38(10):967–983
97. Purkis HM, Lipp OV (2001) Does affective learning exist in the absence of contingency awareness? Learn Motiv 32(1):84–99
98. Sevenster D, Beckers T, Kindt M (2012) Instructed extinction differentially affects the emotional and cognitive expression of associative fear memory. Psychophysiology 49(10):1426–1435
99. Sevenster D, Beckers T, Kindt M (2014) Fear conditioning of SCR but not the startle reflex requires conscious discrimination of threat and safety. Front Behav Neurosci 8:32
100. Soeter M, Kindt M (2010) Dissociating response systems: erasing fear from memory. Neurobiol Learn Mem 94(1):30–41
101. Blechert J et al (2008) When two paradigms meet: does evaluative learning extinguish in differential fear conditioning? Learn Motiv 39(1):58–70
102. Soeter M, Kindt M (2011) Noradrenergic enhancement of associative fear memory in humans. Neurobiol Learn Mem 96:263–271
103. Jovanovic T et al (2006) Contingency awareness and fear inhibition in a human fear-potentiated startle paradigm. Behav Neurosci 120(5):995–1004
104. Grillon C (2002) Associative learning deficits increase symptoms of anxiety in humans. Biol Psychiatry 51(11):851–858

105. Leuchs L et al (2017) Neural correlates of pupil dilation during human fear learning. NeuroImage 147:186–197
106. Hasenkamp W et al (2008) Differences in startle reflex prepulse in European-American and African-Americans. Psychophysiology 45(5):876–882
107. Martinez KG et al (2014) Ethnic differences in physiological responses to fear conditioned stimuli. PLoS One 9(12):e114977
108. Kredlow AM et al (2017) Assessment of skin conductance in African American and Non-African American participants in studies of conditioned fear. Psychophysiology 54(11):1741–1754
109. Goldmann E et al (2011) Pervasive exposure to violence and posttraumatic stress disorder in a predominantly African American Urban Community: the Detroit Neighborhood Health Study. J Trauma Stress 24(6): 747–751
110. Gillespie CF et al (2009) Trauma exposure and stress-related disorders in inner city primary care patients. Gen Hosp Psychiatry (in press)
111. Nelson BW et al (2020) Guidelines for wrist-worn consumer wearable assessment of heart rate in biobehavioral research. NPJ Digit Med 3:90
112. Harnett NG et al (2019) Negative life experiences contribute to racial differences in the neural response to threat. NeuroImage 202: 116086

Chapter 7

An Integrative Model for Endophenotypes Relevant to Posttraumatic Stress Disorder (PTSD): Detailed Methodology for Inescapable Tail Shock Stress (IS) and Juvenile Social Exploration (JSE)

Nathan D. Andersen, John D. Sterrett, Gabriel W. Costanza-Chavez, Cristian A. Zambrano, Michael V. Baratta, Matthew G. Frank, Steven F. Maier, and Christopher A. Lowry

Abstract

Posttraumatic stress disorder (PTSD) is a trauma- and stressor-related disorder that is a source of significant societal and economic costs. Although it is not possible to fully model human psychiatric disorders using animal models, physiological responses to trauma and stressors, including hypothalamic-pituitary-adrenal (HPA) axis responses, autonomic nervous system responses, and immune responses, are highly conserved across mammalian species. Each of these physiological response systems, in turn, has been implicated in determining risk of development of PTSD symptoms, or contributing to PTSD severity, in humans, suggesting that understanding mechanisms underlying these responses may lead to novel therapeutic strategies for the prevention or treatment of PTSD. Furthermore, individual variability in physiological responses to trauma and stressors is thought to be an important determinant of stress vulnerability or stress resilience; therefore, understanding mechanisms underlying individual variability in physiological responses to trauma or stressor exposure has promise to increase our understanding of mechanisms underlying vulnerability to development of PTSD and persistence of PTSD symptoms. Here we describe a model of inescapable stress exposure in rats that has contributed to our understanding of the mechanisms underlying stress vulnerability and stress resilience. Given that PTSD is more common in females than males, we also highlight the need for increased focus on inclusion of both males in females in future studies.

Key words Anxiety, Immunoregulation, Inflammasome, Inflammation, Microbiome, Old Friends, PTSD, Resilience, Stress, Trauma

1 Introduction: An Integrative Model for Endophenotypes Relevant to PTSD

1.1 Posttraumatic Stress Disorder (PTSD)

Posttraumatic stress disorder (PTSD) is a disorder afflicting hundreds of millions worldwide annually [1], with evidence for its prevalence dating back over 4000 years [2–4]. Clinical diagnosis of PTSD is characterized by: (1) chronic intrusive thoughts related to re-experiencing the traumatic event; (2) avoidance of situations that lead to distressing memories, thoughts, feelings, or external reminders of the event; (3) negative cognitions and mood; and (4) heightened arousal, which can include aggressive, reckless, or self-destructive behavior, sleep disturbances, hypervigilance, or related problems [5]. These enduring symptom clusters can be extremely disruptive to an individual's physical and psychosocial well-being. They also may serve as guides in designing and applying animal models that help us to understand specific symptoms and therefore the etiology and pathophysiology of PTSD as a whole. Symptoms often include extreme sleep disruption [6–9], increased incidence and rates of long-term pain [10–15], increased psychological and physical comorbidities [16–20], and relational, familial, and professional difficulties [21–26]. Emotion- and intrusion-based reasoning [27–29], chronic avoidance [30–32] (as seen with many psychiatric disorders [33–36]), and chronic sleep-disruption [9, 37–39] have been linked to symptom perpetuation. What is more, hyperarousal has been linked to increased sleep disruption [40], while sleep disruption, in turn, is associated with increased inflammation [41], which has been increasingly linked to PTSD [42–46]. PTSD has also been observed to undergo "intergenerational transmission" [47–50], including through higher rates of intergenerational domestic violence [51–54]. Some idea of the level of distress in individuals with PTSD is indicated by high rates of comorbidity with major depressive disorder (MDD) and high rates of suicide where PTSD is evaluated as the cause. Comorbid MDD has been observed in 45–70% [55, 56] of those who develop PTSD following experiencing natural disasters. Nearly two decades after the World Trade Center terrorist attack approximately 70% of those suffering from PTSD had comorbid MDD [57]. Among a United States (US) military Veteran population diagnosed with MDD, 36% were found to have undiagnosed PTSD [58]. A meta-analysis, as well as compelling narrative and systematic reviews [59, 60] have found PTSD to be associated with increased suicidal ideation, suicidal behaviors, and suicide attempts, and that the association between depression and suicide was stronger in those with PTSD than those without [61]. Two recent meta-analyses observed a positive association between suicidal attempt in adults with childhood trauma exposure [62], and a "highly significant positive association" between PTSD and suicidality in adolescent populations [63]. Of particular note, the increased risk for suicide from PTSD is far more pronounced in women [64], who constitute the majority of PTSD diagnoses [65–67].

One of the most insidious aspects of PTSD is its high rate of treatment resistance [68]. PTSD is marked by high rates of treatment dropout [69–71] and unwillingness to pursue treatment altogether [72–76]. This is even more pronounced within some populations where rates of PTSD are significantly higher, including military Veterans [77, 78]. Interestingly, some of these same treatment-resistant populations are also more open to nontraditional or "complementary and alternative medicine (CAM)" approaches [79, 80], which are increasing in popularity [81–86]. CAM therapies may even have higher rates of success for some populations [87]. Increased understanding of the etiology and underlying pathophysiology of PTSD will help direct promising new areas of research, many of which may lend themselves well to nontraditional prophylactic and therapeutic interventions. One promising example of this is the possible role of the microbiota-gut-brain axis in the risk of development of PTSD and the persistence of PTSD symptoms [43, 46, 88–93].

The persistent and perpetuating nature of PTSD symptoms combined with their severity and high rates of treatment resistance sum to an enormous cost. Though the true "cost-of-illness" or "burden of disease" has not been comprehensively examined through epidemiological study [94], the disease is well understood to take a profound psychosocial, familial, and societal toll [94–96]. The economic burden is understood to be on the order of at least several billion US dollars per year for the US in individual and societal treatment costs, increased non-PTSD comorbid healthcare costs, and high lost opportunity cost [26, 97–101]. The disorder is most costly when untreated, but even in subpopulations where treatment has a high rate of success, the economic cost is high. In Germany, successful treatment is about 30,000 euros over the course of 3 years before an individual's healthcare costs return to baseline [102].

Given the severity of symptoms, high propensity for them to persist and perpetuate, high rates of treatment resistance, and resulting economic and societal costs, there exists opportunity to effect significant positive change on a global scale through better addressing PTSD both prophylactically and therapeutically. Increased understanding of disease etiology and pathophysiology, and the ability to readily explore new and alternative preventative and therapeutic interventions is essential in realizing this opportunity and meeting this need. The utilization of translational models is essential in this endeavor, allowing for use of invasive procedures to rapidly assess potential mechanisms underlying endophenotypes of PTSD.

1.2 Biological Relevance of Translational Models

It is important to highlight the significance of translational models in the biomedical research of PTSD. The earliest recorded animal research dates back over 2300 years to ancient Greece. Countless biomedical discoveries owe their direct and indirect origins to animal models; a few highlights amongst a vast and growing list include scientific validation of germ theory, vaccination, insulin, antivenin, transplants, and validation of virtually all new surgical techniques. Our understanding of stress-related physiological function is credited directly to an animal model utilizing rats; following hospital rotations in his first year in medical school, Dr. Hans Selye postulated the "syndrome of just being sick" [103]. This was later dubbed "biological stress" after observing common symptoms and necropsy results in rats despite the fact that different groups of rats had received injections of distinct extracts from cow ovaries [103, 104]. Translational models have and continue to serve an invaluable role in biomedical research.

Nonhuman animals cannot display PTSD symptoms or have a diagnosis of PTSD, both of which are based on validated clinician administered surveys. PTSD is a complex psychiatric disorder with established diagnostic criteria that are impossible to assess in nonhumans in any evidenced-based manner. Any claim or attempt otherwise hinges on anthropomorphism, rather than evidence. Even so, as our understanding of disease etiology increases with rapid increases in neuroscientific understanding and omics-based analysis, we have an ever-increasing understanding of the physiological profile that determines risk of developing PTSD following exposure to trauma. A potential future direction that may have merit would be to categorize various stress models that involve assessment of endophenotypes that are based on the symptom clusters of PTSD, rather than a claim of being a model for PTSD as a whole. Within this chapter, we define stressors as actual or perceived threats to the homeostasis of the organism [105]. Physiological responses to stressors are hallmarked by activation of the hypothalamic-pituitary-adrenal (HPA) axis, assessable through biomarkers including an increase in corticosterone (or cortisol in humans) secretion and the resulting physiological response cascade. Endophenotypes are here defined as "measurable components unseen by the unaided eye along the pathway between disease and distal genotype" which could be assessed at various levels of analysis including neurophysiology, biochemistry, neurocognition, self-report, or other outcomes that must "mark genetic vulnerability and be amenable to more precise measurement than overt phenotypes" in order to be useful, as outlined by Dr. Irving I. Gottesman and Dr. Todd D. Gould [106] This was updated to include "markers of genetic liability to transdiagnostic vulnerability traits (e.g., impulsivity, irritability, anhedonia)" by Dr. Theodore P. Beauchaine and Dr. John N. Constantino [107]. These clusters have distinct, translatable phenotypes that can be observed.

The degree to which elements of biology have been evolutionarily conserved is core to the value of translational models. Though directionality here may be biased, the majority of genes associated with human disease have counterparts within the rat genome with over 75% calculated to have a 1:1 ortholog and the remaining ~25% believed to likely have a genetic ortholog [108]. In fact, most of these genes are highly conserved across all mammalian evolution

[108, 109]. This accounts for much of the essential role animal models have played and continue to play in contributing to countless medical advances [110–115].

Imperfect though models are, in addition to maintaining high biological relevance, they also afford us the ability to "peek behind the curtain" physiologically. In other words, we can explore through observation, genetic, and direct physical manipulation, the underlying mechanisms associated with endophenotypes, including physiological and behavioral responses, that are relevant to a human condition such as PTSD. This includes but is not limited to our rapidly expanding ability to study and directly manipulate the central nervous system, including specific neural circuits within the brain. Technological advances include: (1) region-specific brain lesions; (2) genetic lineage tracing models such as the Cre-*lox* system allowing the manipulation of global, tissue, and cell-specific gene expression and calcium imaging, allowing direct, real-time, in vivo imaging of neuron activation; and (3) genetically encoded actuators including optogenetics and designer receptors exclusively activated by designer drugs (DREADDs), allowing for the direct activation and inhibition of specific neural circuits. These techniques are yielding increased understanding of neural circuits that control endophenotypes relevant to PTSD in the journey to better understand and treat this condition.

1.3 Inescapable Tail Shock Stress (IS)

1.3.1 History

Electric shock has been utilized in stress research since the 1950s. It was an ideal target in the quest for better tools to explore and understand stress due to the high degree of accurate and precise controllability of timing, duration, voltage and amperage, relative cost-effectiveness as compared to other stressors, and ease of reproducibility. Early apparatuses largely used bar or grid floors, allowing for the animal's feet to be contact points for shock delivery. However, this design quickly yielded problems. Animals were observed jumping into the air, receiving no shock for that duration [116]. Additionally, animals would by chance, and even by learning, stand partially or completely on bars exclusively of like charge, thus altering or avoiding shock altogether [117, 118]. The appeal of electric shock as a highly controllable stressor was still high enough that this quickly led to innovation in the technology and apparatuses, including development of the inescapable tail shock stress (IS) paradigm, largely in the form that it is used today. To resolve inconsistent shock delivery, resisters were used to verify and ensure consistent current. Perhaps most importantly, "shock scramblers" were added to the design and greatly improved over the decades, disallowing the ability to lessen or prevent circuit completion with foot placement. Though these technologies provided a solution to the problem and are still used today, researchers also pursued avenues for attaching electrodes directly to the animals in order to circumvent the issue entirely as well as to prevent any jumping escape. A number of creative approaches were explored across the 1960s including electrode implantation [119, 120], attachment of a tail electrode to a freely moving rat [121], and combinations of tail electrode attachment with complete [122] or partial restraint [116].

Partial restraint has advantages over other methods of shock-induced stress that are easy to take for granted from today's perspective, which is so far from the design's inception. Inherent to its design was affording the animal enough mobility to take part in an operant paradigm (lever pressing, wheel turning, etc.) while disallowing any ability to avoid complete delivery of the intended shock. Additionally, complete restraint is a well-established stressor in itself [123, 124], compromising the accuracy and precision intended with electrical shock stress methods. By making the chambers out of a smooth, hard, slippery plastic, it allows the animal to struggle freely while preventing the animal from being able to exert its full strength to escape its tail restraint. This is an essential detail if designing modifications to existing apparatuses [116]. A valuable aspect of this paradigm is that tail shock does not increase nociceptive signaling. It has been well established that afferent signaling by C-fibers, which transmit nociceptive signals, are either not activated by tail shock or are inhibited as a result of stress-induced analgesia (SIA) [125–133]. Aδ-fibers, which provide precise localization of nociceptive signals, are also presumably inhibited, especially in SIA that is mediated by opioids, as Aδ-fibers are also inhibited by opioids [134, 135], if differentially from C-fibers [136, 137], the differential mechanisms are mediated, at least in part, via opioid receptor subtypes [138, 139]. There is also evidence for an endocannabinoid mechanism inhibiting both C- and Aδ-fibers underlying non-opioid SIA [133, 140, 141]. Oxytocin, which has been shown to inhibit both C- and Aδ-fibers via GABAergic interneurons [142], may also be involved in non-opioid SIA. Overall, this allows for the isolated application of stress without nociceptive signaling, removing a significant potential confound and allowing a more detailed exploration of underlying mechanisms.

The model, in its current form, has contributed to our understanding of PTSD [143, 144], learned helplessness [145–151], stress [152, 153], stress-induced analgesia [129, 150, 154], and mechanisms underlying behavioral and physiological responses to exposure to aversive stimuli, including anxiety-like defensive behavioral responses [129, 150, 154], nociceptive signaling [154–157], addiction [127, 158], and depressive-like behavioral responses [159–164].

1.3.2 Overview

IS consists of a single stress "session" in which experimental animals are brought to a procedure room and partially restrained in a Plexiglas® tube or partial tube. Tails protrude from the back and are attached to or placed in contact with electrodes and secured into place with tape. This allows for some mobility of the animal, but complete restraint of the tail ensuring proper and complete delivery of shock. Shock is administered for 5 s semi-randomly with 30–90 s intervals (averaging 60 s) between each shock or "inter-trial interval (ITI)" for a total of 100 shocks at 1.6 mA with variable voltage to maintain consistent amperage based on measured resistance (some studies have used an program that ramps the current over time, with 33 trials at 1.0 mA, 33 trials at 1.3 mA, and 34 trials at 1.6 mA, also for a total of 100 trials [165]). Animals are immediately returned to their home cages and the colony following IS. Control animals remain in their home cages for the entirety, undisturbed, as "home cage controls."

1.3.3 Utility and Limitations (Advantages/ Disadvantages of IS as a Model of Endophenotypes Relevant to PTSD)

IS has distinct strengths as compared to other models for stress and endophenotypes relevant to PTSD. The accurate and precise controllability of shock duration and intensity used in IS is ideal for research. It is reliable, yielding highly replicable results across many labs and over several decades [150, 154, 166–168]. The model yields highly consistent results across several rat strains and lines, highlighting the model's potential versatility in the greater context of genetic models [169–173]. Additionally, IS yields similar but distinct physiological responses in female versus male rats, potentially allowing the model to help explore sex differences in endophenotypes relevant to PTSD [174–181]. The distinction between endophenotypes that are (1) "conditioned," when responses are paired with context cues [182]; (2) "trans-situational" or "generalized," when responses are generalized to other contexts [148, 183–188]; (3) "sensitized," when response severity is increased or exaggerated [185, 189–192]; and (4) characterized by "incubation," when sensitization and or generalization increase over time is potentially important in translational research given the stressor generalization, sensitization, and incubation observed in some psychiatric disorders that may have similar etiology [193–196]. As such, the fact that the shuttle-box escape task [185, 197, 198], auditory startle response [193], and juvenile social exploration (JSE)/social avoidance test [199] all show clear trans-situationality or generalized responses to IS underscore the potential utility of IS. In fact, trans-situationality has been a hallmark feature of uncontrollable stress effects from early theoretical conception, planning, and execution [186, 200, 201].

IS, of course, also has limitations. Though restraint is partial, partial restraint is still a moderate stressor to the animals [202]. By placing the animal in a small, confined space it is not possible to have classical methods of context conditioning outside of tone-, light-, or scent-paired cues, so measuring startle or freezing in the same context is not possible. In some stressor paradigms, there is an observed incubation time over the course of which endophenotypes relevant to PTSD will worsen in regard to response to ensitization and or degree of generalization, i.e., symptom incubation [203]. This is not unlike what is observed in individuals suffering from PTSD, as it manifests and worsens over time from trauma onset and thus symptom incubation is an ideal characteristic in stressor models. Though IS has recently been studied in conjunction with a foot-shock paradigm one week following IS in an adapted mouse model yielding significant sensitization [204], IS in its traditional form has not been adequately studied in regards to sensitized phenotypes and the incubation of sensitized and generalized phenotypes, though some findings suggest this could be a promising future direction [193, 205]. This in no way precludes, however, its potential utility in exploring sensitized and incubated phenotypes further, and indeed some early studies suggest there may be promise [185, 193, 197, 198]. All of these studies, however, lack adequate differentiation between fear network priming, a short-term phenomenon with effects that will only last for 24–48 h and an acquired sensitization to fear, a more enduring effect measured over longer time periods, with only one study measuring beyond 48 h, but only up to 10 days post-stress. Other validated models have observed sensitized responses on a scale of 7–28 days post-stress [206], and incubation effects that take up to 31 days to fully manifest and are present 61 days post-stress [207–210]. This distinction may be better defined as there being evidence for short-term, but not long-term sensitization from IS, and an overall lack of investigation into incubation effects. Overall, sensitization and incubation effects are understudied in IS, but constitute potentially promising future directions.

Given its strengths and weaknesses, neither IS nor any other model should be viewed as the "be-all and end-all" model for endophenotypes relevant to PTSD. Rather, it should fit into a larger scope of research within the field with researchers continuing to utilize several well-established models as well as continuing to explore and develop new models where specific as-of-yet unanswered questions remain hindered by the limitations of established models. That said, IS continues to hold a useful place within the field moving forward.

1.4 The Juvenile Social Exploration (JSE) Test for Assessing a Stress-Induced Exaggeration of Anxiety-Like Behavior

1.4.1 History

The quantitative evaluation of anxiety-like defensive behavioral responses relevant to anxiety disorders in humans, and avoidance behaviors in trauma- and stressor-related disorders, such as PTSD, has long been a goal of science and medicine. Within animal models, early methodologies date back to 1941, at which time Dr. William Estes and Dr. Burrhus Skinner introduced an operant and fear conditioning paradigm, the conditioned emotional response test. Food-deprived animals were first trained to press a lever at a specific rate in order to receive food. Following two weeks of periodic reinforcement animals were repeatedly subjected to a tone followed by shock for four days. The animal's hunger-driven lever-response during and following tone presentation, preceding shock, decreased consistently over that time. The conclusion drawn was that anticipation of shock, rather than shock itself, caused quantifiable decreases in hunger-motivated behavior [211].

Creating and then ameliorating "neurotic behavior" was also an early target of anxiety research. Several models utilized the punishing of food-rewarded behavior in food-deprived animals to induce neurotic behavior [212–214]. This structure was proposed as a standardized means for evaluating the effects of drugs on neurotic behavior [215]. These methods were criticized for having only subjective means of evaluation, however, so Dr. Knut Naess and Dr. Ernst Rasmussen designed a paradigm in which animals were deprived of water for 24 h and then trained to a brief period (5 min) of access to water once per day. Following 2 or more days of training ("once the animals had become accustomed in this way"), the animals, still water-deprived, were placed back into the apparatus but with the grid floor having one pole of a current and a wire going into the water dish with the opposite pole. They were able to quantify "nervous" behavior by counting the number of "approach-withdrawals" the animal had wherein they did not actually drink over a 20-min period [216]. This design was further improved with the utilization of operant chambers tying reward and punishment to a single lever press with a tone cue [217].

High tribute should be paid to these early innovations while behavioral science was relatively new. Even so, these paradigms were limited. Something all designs had in common was their reliance on both deprivation and punishment to evaluate anxiety-like behavioral responses. Seeking to address deficits in current methodology, drawing on the foundational 1972 work of Dr. Bibb Latané and Dr. David Hothersall on social attraction in animals [218], a revolutionary approach utilizing social interaction as a measure of anxiety-like behavior was first introduced in 1978 by Dr. Sandra File and Dr. J.R.G. Hyde [219]. The social interaction test has proven to have several advantages over the aforementioned conditioned emotional response test [211] and the rat conflict test [217], which had become the standard paradigms of the time. By quantifying time spent engaged in social interaction as the measure of anxiety-like defensive behavioral responses [220], the model removed the need for the stressful confounds of food or water deprivation and an added stressor, as well as extensive training of the animals. It also allowed for baseline measurements of anxiety-like behavioral responses; it therefore allowed not only the observation of anxiolytic effects of interventions, as noted by the original paper [219], but also the observation of anxiogenic effects of interventions as well [221]. Variations of the social interaction test have been successfully implemented to assess both anxiolytic and anxiogenic effects in rats [219, 221, 222], mice [223, 224], gerbils [225, 226], prairie voles [227–229], and meadow voles [227, 228], and are currently being utilized in the study of *Octodon degus* [230, 231].

The social interaction test marked vast improvements in translational models by reducing confounds. Though the early version of this test was not without a few drawbacks of its own, it has continued to evolve and improve, addressing most if not all of these early drawbacks, continuing to play a vital role in translational research decades later. Despite the original design utilizing a distinct testing chamber it is worth noting that some early adaptations conducted the "social interaction test" by introducing conspecific, same-sex rats in the experimental animal's home cage [232]. Such adaptations might better be considered hybrids of the social interaction test and the "resident intruder [233]," "homecage intruder [234]," or "homecage aggression [235]" test, giving rise to concern for confounds of aggression and social dominance competition. These were not concerns in the original paradigm, which utilizes a novel testing environment, and which has subsequently been more universally adopted. In 1987, a modification of the social interaction test utilizing juvenile social stimulus "partner rats" was put forth [236], later adapted to observe the social effects of vasopressin [237, 238], which has since become known as the JSE test.

The JSE test, in its current form, when paired with IS, has played a vital role in our understanding of endophenotypes relevant to PTSD [143, 144], learned helplessness [145–151], physiological stress responses [152, 153], stress-induced analgesia [129, 150, 154], and mechanisms underlying resilience to behavioral and physiological responses to exposure to aversive stimuli, including as assessed by anxiety-like behavioral responses [129, 150, 154], nociceptive signaling [154–157, 239], addiction [127, 158], and depressive-like behavior [159–164], as well as developmental stress exposure [240], the impacts of the time of day of stress exposure (diurnal rhythm mediated) [165], and the impact of circadian misalignment on stress responses [241].

1.4.2 Overview of IS-Induced JSE Deficits

JSE entails the exposure of both control and experimental rats to novel juveniles in a clean standard housing cage with fresh bedding and devoid of food or water. The animal being tested is immediately transferred to its individual testing cage upon entry into the JSE procedure room and allowed to acclimate to the testing environment for 60 min prior to the introduction of the juvenile [199]. As per the social interaction test, a number of studies have used low lighting (40 lux) to minimize stress and maximize social interaction [219]. It is important to note that lighting levels during IS/JSE have not been systematically assessed to determine the impacts on social avoidance. Though it may have an impact on the magnitude of IS-induced JSE deficits by altering baseline social interaction times (as it does in the social interaction test), the effects of IS on JSE are robust enough to be significant even with lighting levels well above 40 lux. Many successful IS/JSE experiments have used procedure room lighting described as "brightly lit" (measured to be 150 lux at the level of the animal) [179, 180, 242]. Social interaction is scored by an observer blind to treatment groups by timing the duration and recording the number of social exploratory behaviors (sniffing, chasing, pinning, allogrooming) initiated by the adult, for a 3-min session [180, 199, 243–245]. This procedure is conducted twice for each animal. The first JSE session provides a baseline social interaction score. An IS session is conducted after the baseline JSE session, typically 24 h after [199, 180]. The JSE test procedure is then repeated 12–48 h post IS, typically 24 h [199, 244], and each animal's social interaction score is compared to its baseline (i.e., expressed as percent of baseline [179, 180, 245]). In some paradigms a baseline measure may not be available, in which case absolute values can be compared [199, 244]. As outlined in Subheading 1.3, IS-induced effects on JSE diminish after 48 h. Though typical baseline interaction levels have previously been reported to be approximately 50–80 s [246], others have excluded animals with fewer than 60 s of social interaction as outliers [245]. There are so many factors that will impact absolute interaction that it is recommended to establish typical baseline interaction levels in-house for a given age, strain, and sex. It should be noted that if typical baseline interaction levels are too low in certain strains, lines, or paradigms, it would be difficult to observe IS-induced reductions in social interaction, yielding a floor effect.

1.4.3 Utility and Limitations

In its present form, JSE has many well established advantages [244] making it ideal for use in translational PTSD research and pairing with IS. It is a well validated, highly reproducible measure of anxiety-like defensive behavioral responses [220] utilized for decades, and highlights social avoidance [221]. It is also sensitive to stress-related neuroinflammation [247], and reflects robust changes in the form of reduced social interaction time resulting from IS [244, 248]. The original social interaction test paradigm called for singly housing animals for 5 days prior to testing to maximize social interaction [219], with a peak maximization subsequently validated as taking place after 4–7 days of single housing prior to testing [249]. As previously noted, single housing is a stressor and can confound a stress paradigm. Though this may seem to a be a considerable limitation within the context of stress research, over the evolution of both the social interaction test and the JSE variant thereof, single housing has been reduced to acclimation to the test cage, 60 min, preceding assessment [199]. In addition to their utility in exploring the etiology and pathophysiology of PTSD, well-established models for endophenotypes relevant to PTSD such as IS-JSE lend themselves well to the exploration of potential interventions. Prophylactic interventions have been and continue to be explored with this model seeking a better understanding of stress-resilience factors, some of which have had robust results [244, 250–253].

Again, IS-JSE is only one valuable tool in our arsenal for expanding our understanding of PTSD. Limitations include that it has not yet been adequately validated in a stress-induced prolonged incubation of a phenotype, as done in a variety of other models [207–210, 254, 255] (though this may be an area worthy of future exploration), and that the drive to investigate conspecific same-sex juveniles is so strong that it may mask some stress effects. IS-induced JSE deficits, however, are robust and highly reproducible, so, based on empirical findings, this would seem to be a minor limitation. Overall, as with the social interaction test as a whole, there are few noteworthy limitations, and the paradigm continues to evolve to address any that are highlighted within the research. Altogether, these characteristics of the IS-JSE paradigm support the claim that it is useful in exploring stress-related endophenotypes relevant to PTSD.

2 IS/JSE materials

2.1 Animals: Sprague Dawley Rats

2.1.1 Experimental and Control Animals

For details on selecting the rat stock and ordering, including selecting the n for each group, see a detailed description in the Notes section below (4.1.1). In brief, adult male or female Sprague Dawley (SD) rats are used. They are notably calm and easy to handle and have a well-established, robust stress response in many models including IS/JSE [179, 199, 180, 242, 244, 245, 248, 256–261]. SD rats were first generated in 1925 from the breeding of a Wistar female and genetic background unknown hooded male by Sprague-Dawley farms, later Sprague-Dawley Animal Company [262]. NIH acquired them in 1945. They are a closed-colony outbred stock, meaning that they are partially bred to maintain genetic diversity within the stock. A breeding scheme engineered to maintain a foundation colony, minimize inbreeding, and minimize genetic drift was instituted as of 1992 [263].

- Adult male Sprague Dawley® rats (Hsd:Sprague Dawley® SD®; Envigo, Indianapolis, IN, USA) weighing 250–265 g

2.1.2 Juveniles (For Use in JSE)

Juveniles are same-sex SD rats, from the same vendor. Though JSE takes place when juveniles are postnatal day (PND) 28–32; they must be acclimated for 7 days prior to JSE, so they must arrive by PND 21–25, on experimental day –7. Juveniles may be reused in JSE, but never with the same experimental rat. A default of one juvenile is recommended for every four adult animals, though some protocols may merit a larger ratio.

- Juvenile Sprague Dawley® rats (Hsd:Sprague Dawley® SD®; Envigo, Indianapolis, IN, USA) PND 21 on arrival

2.2 Housing Environment

All animal facilities should be overseen by the research institution's Office of Animal Resources or local equivalent and all procedures should be approved by the Institutional Animal Care and Use Committee (IACUC) or local equivalent. The institution must be approved by the Office of Laboratory Animal Welfare (OLAW) or local equivalent. Ideally, the research institution should meet the requirements for American Association for Accreditation of Laboratory Animal Care (AAALAC) accreditation. AAALAC is a private international nonprofit group dedicated to promoting the humane treatment of animals in scientific environments. All accreditation and assessment programs are voluntary. This level of care in animal housing is not only for the well-being of the animals, but equally so, and causally linked to animal well-being, it is for the integrity of experimental design. Unsanitary conditions, inconsistent lighting, noise, and inconsistent feeding schedules have all been used as stressors because they are so disruptive to the animals [264–267]. In other words, if the animals are unhealthy or stressed, it will confound stress research and potentially other areas of research as well.

Animals should be pair-housed upon arrival in cages with approximately 2.5 cm-deep bedding and ad libitum access to food and water.

- Allentown micro-isolator filter-topped caging [259 mm (W) × 476 mm (L) × 209 mm (H); cage model #PC10198HT, cage top #MBT1019HT; Allentown, NJ, USA]
- Cat. no. 7090; Teklad Sani-Chips; Harlan Laboratories, Indianapolis, IN, USA
- Harlan Teklad 2918 Irradiated Rodent Chow, Envigo, Huntingdon, United Kingdom

2.3 Behavioral Test Administrators

The scientist applying the IS procedure should remain consistent throughout the entire experimental design. It is noteworthy that a female scientist produces less potentially compounding stress in the animals [268, 269], though this has not been explored in the context of IS, and IS-induced JSE deficits are robust when conducted by both male and female experimenters. One consistent experimenter should also carry out JSE. In order to minimize the impact of shared stimuli between IS and JSE, different experimenters should conduct each procedure. One person, blind to treatment group, should carry out all JSE scoring to control for interindividual reliability. Behavioral analysis can be done at the time of the procedure by the experimenter conducting JSE or afterwards using video recording.

2.4 Behavioral Assessment Environment (JSE)

All stressor application must take place in an appropriate testing facility for experimental validity. Ideally this will be a dedicated procedure room for a single species (in this case rat) that is clean, well ventilated, quiet, sound-insulated, dimly lit, warm, and near the housing colony room to avoid experimental confounds produced by any stress associated with the movement of animal cages around the animal facility or between buildings. This is not always able to be the case, and in other cases extra care must be taken to minimize confounds, e.g., putting a sheet over cages being transferred greater distances. That being said, care should always be taken in any behavioral paradigm to minimize any potential stress or stimulation to the animals that may serve to confound the validity of or disrupt the success of a paradigm [270, 271]. The equipment and space should be meticulously cleaned after each use to prevent significant confounding factors including scents and sanitation.

2.5 Behavioral Assessment Equipment (JSE)

JSE is conducted in a standard rat housing cage with standard bedding and wire hopper but lacking food and water and without a filter cage top. Either a video recording or direct observation (in each case conducted while the experimenter is blind to the nature of treatment groups) are recommended for scoring.

- Allentown micro-isolator filter-topped caging [259 mm (W) × 476 mm (L) × 209 mm (H); cage model #PC10198HT, cage top #MBT1019HT; Allentown, NJ, USA]
- Cat. no. 7090; Teklad Sani-Chips; Harlan Laboratories, Indianapolis, IN, USA
- Digital video camera
- Stand for camera (able to view the entire cage for all interaction)
- Scoring sheet (or scoring software)
  - Printed or computer spreadsheet
- Behavioral scoring software, e.g. in-house Python program allowing for depression of a key during interaction periods to function as a timer (optional, but helpful)

2.6 Stressor Environment (IS)

IS is conducted in a similar procedure room that is separate from the room in which JSE is conducted.

2.7 Stressor Equipment (IS)

The device used for stressing animals in IS has gone through a number of iterations with origins of its modern form dating back to 1967 [116, 121, 272, 273]. In brief, there is a restraining portion that the animal is secured in. The tail is able to protrude freely outside the restraining chamber through a hole or slit, where it is then attached to electrodes and separately restrained. The electrodes are connected to an unscrambled DC circuit, controlled manually or by a computer, for shock application that is able to be controlled with precision in terms of amperage, voltage, and duration. Over the course of that time some things may have come to be taken for granted that are, in fact, quite essential to the paradigm.

IS device: The stressor environment consists of a Plexiglas® half tube or tube (17.5 cm length × 6.0 cm diameter × 0.5 cm walls) attached to a base in which the animal is restrained, with a small hole through which the tail passes to be secured to the base or an attached rectangular rod.

- Restraint apparatus
    - In some cases this will come with integrated electrode/contact points:
    Harvard Apparatus Item #E93-91R
- Unscrambled shock delivery device
    - Manual
    - Programmable
    - Wires with alligator clips of appropriate gage for amperage utilized
- Electrodes
    - Copper tape (2.5 cm width)
    - Conductive cream
        (e.g., SignaCreme Electrode Cream, Parker Labs Inc, supplied by Medline, Northfield, IL, USA)
- Computer
    - Standard digital camera
        An infrared camera can allow for lower lighting
- Software
    - In-house Python programs can be generated for recording
- Medical tape
- Clydox or 70% EtOH

3 IS/JSE Methods

3.1 Housing

All animals should be pair housed in an appropriately stress-minimizing environment as outlined in detail above. Single housing is a known stressor for rats and can confound any stress paradigm that does not specifically call for single housing [274, 275]. The ARRIVE (Animals in Research: Reporting In

Vivo Experiments) guidelines 2.0 (or since updated) should be followed from guiding initial experimental design through to publication. The ARRIVE guidelines, first introduced in 2010 [276], updated in 2020 [277] serve to increase transparency and reproducibility in in vivo animal research.

3.2 Acclimation

All animals should be acclimated to their new housing conditions for a minimum of seven days from arrival. If they are bred in-house, they need to reach the correct age and or weight for the desired experimental conditions.

3.3 Behavioral Assessment (Baseline and Post-IS JSE)

Ensure all equipment is clean and functional preceding testing. If appropriate, ensure the procedure room is reserved for the correct date and time that it is required.

3.3.1 Setup

3.3.2 Transportation

Transportation of the animals should be quiet and gentle. Begin by flipping water bottles upside down in the cage. This prevents drinking water from dripping onto the bedding during transport which is a known stressor [265–267]. Gently place all cages of a given session onto a cart and bring them to the procedure room. If the distance is long or the lighting conditions vary significantly (e.g., passing by exterior windows), a sheet can be used over the cages on the cart. Take care not to jostle or bump the cages or cart in transport and avoid all unnecessary excess noises, all of which can be compounding stressors [264].

3.3.3 Testing

It is important to test animals in a completely randomized block design in order to take into account any potential time of day, sleep, or circadian confounds. All animals being tested are placed into their own novel clean cage with bedding and a wire hopper with no food or water for 60 min to acclimate to the testing environment. After this time, a novel juvenile is introduced and the encounter is recorded. Each session is timed at 3 min [243].

Manual scoring takes place observing the live interaction or video. The procedure for JSE is conducted as a baseline assessment prior to IS. Though the time interval between baseline JSE testing and IS has not been systematically tested, where possible, establishing baseline JSE from no more than a few days prior to IS [165, 241, 278] to the morning (2 h) before IS [279] is recommended in order to minimize developmental, environmental, and other factors that might alter JSE. Within 48 h is recommended [165, 241, 278], with 24 h pre-IS as a default [245], on experimental day −1 (i.e., baseline JSE), and again 24 h post-IS, on experimental day 1 (24 h is again a default, with IS-induced effects on JSE having been consistently observed 12–48 h post IS [179, 180, 243]).

3.3.4 Scoring

Total interaction time is recorded with a computer program, which can be generated in-house using open source software such as Python to track what percentage of a 3-min testing session a key has been depressed with the scorer depressing the key for interactions. Interactions include: (1) sniffing, when the adult rat is investigating the juvenile by sniffing; (2) chasing, when the adult rat is chasing the juvenile; (3) pinning, when the adult rat uses its front paws to hold in place the juvenile; and (4) allogrooming, when the adult rat licks and grooms the juvenile in a stereotypical social fashion.

3.3.5 In Brief

Preparation

1. Test all (video and scoring) equipment
   (a) Ensure enough memory on the recording device
2. Print scoring sheets if done by hand

Procedure

3. Bring each session's cohort or block of animals to the stressor procedure room (block size for JSE is not limited to apparatuses as is the case for IS). All animals for a given study can be brought to a single session, acclimate at the same time, and be tested serially (as always, order of treatment group must be randomized)
4. Begin recording
5. Note the animal number on the video recording
6. Transfer an experimental rat to a clean cage with fresh bedding with an empty wire hopper and no filter cage top
7. Let the rat acclimate for 60 min
8. Add a juvenile rat to the cage beginning a 3-min JSE session
   (a) Observe and score the interaction
9. It is recommended that there always be one dedicated individual to scoring JSE, thus avoiding the need for extensive validation of inter-rater reliability
10. After five min remove the juvenile and replace the experimental or control animal in its colony cage
11. Remove the used cage and replace with a clean, unused cage with fresh bedding
12. Repeat **steps 6–11** for each group until complete

3.3.6 Data Analysis

Data, in the form of percent of baseline social interaction time (or total interaction time if baseline values are not available within a paradigm), are analyzed utilizing an unpaired t-test or ANOVA in the case of more than two groups.

3.4 Stressor Application (IS)

3.4.1 Setup

All equipment should be cleaned and tested prior to each use. A voltmeter can be used on the electrodes that attach to the copper strips to test for current presence and accuracy of voltage and amperage. If there are more experimental animals than there are stressor apparatuses, the animals should be put into cohorts/randomized blocks equal in size to the number apparatuses. One group is transported into the stress procedure room at any given time. This prevents animals from hearing a shock session separately and in addition to experiencing one. As always, and important to note once again, consistency of timing is imperative in controlling for potential sleep and circadian rhythm confounds. All animals exposed to IS are stressed within *2–5 h into the light cycle*, before plasma concentrations of endogenous glucocorticoids begin to increase [248, 280, 281].

3.4.2 Transportation

Transportation is to be quiet and gentle, as outlined above Subheading (3.4.2).

3.4.3 Procedure

Rats are then picked up and placed head-first into the Plexiglas® restraining tube. Generally, once the head is put into position the rats will crawl the rest of the way into the tube on their own. The rear of the chamber is then secured with the tail remaining exterior through a hole in the rear plate. Electrodes are formed by placing 2.5-cm wide copper tape wrapped completely around the tail over a layer of electrode cream (with enough cream to allow for uniform contact). The first electrode is placed approximately 2.5–3.5-cm from the base of the tail and the second electrode is placed distally with a 2.5 cm gap from first. The tail is then secured with medical tape. Electrodes are connected to the current source via amperage-appropriate gauged wires with alligator clips emanating from the shock generator. Animals receive 100, 5-s tail shocks on a variable 1-min inter-shock interval. The shock intensity is 1.6 mA. Rats are immediately returned to their cages and to their colony room upon termination of the final tail shock.

Just as with JSE, it is important to test animals simultaneously as much as possible in pairs, at minimum per-cage, thus avoiding animals being isolated prior to or after an IS session as their cage mate is undergoing IS [221]. Animals must be periodically monitored either remotely or in-person throughout the duration of each shock session, a minimum of once every 15 min. This ensures that animals remain properly restrained and do not experience any unanticipated adverse events.

Control rats are to remain in the colony room for the entire duration, known as "home cage controls."

3.4.4 In Brief

Preparation

1. Test all equipment
2. Clean all equipment
   (a) Clydox or 70% ETOH

Procedure

3. Bring each block of animals to the stressor procedure room (ensuring use of a randomized block design)
4. Transfer all rats into stressor restraint tubes
5. Apply electrode cream
6. Attach electrodes
7. Secure tail
8. Run programed stressor session
   (a) Check on animals either remotely or in person every 15 min to verify proper restraint is maintained
9. Transfer animals back to their cages
10. Clean all equipment
11. Repeat for subsequent groups

4 IS/JSE Notes

4.1 General

4.1.1 Ordering Rats

Choosing the correct stock (outbred) or strain (inbred) of animal can be essential when selecting a known stress paradigm as it has been shown that stress and specific stressors have stock/strain dependent phenotypes. Rats that have undergone IS include outbred stocks, i.e., Wistar, Sprague Dawley (formerly Sprague-Dawley), RNU Nude, and inbred strains, i.e., Fisher344 (F344), Brown Norway, Dark Agouti, Albino Oxford, Lewis, and spontaneously rat (SHRhypertensive rat (SHR *advertised as an outbred stock, though generated from a combination of inbred strains) [173, 282–286]). It should be noted that Lewis and SHR rats proved to be unresponsive to IS as assessed by increased adrenocorticotropic hormone (ACTH) secretion, suggesting potential HPA deficits and are therefore not considered ideal strains to be used in the model [287]. As noted previously, differences between lines within this model are remarkably small, highlighting the model's utility. The majority of IS experiments utilize SD or F344 rats. F344 rats have been shown to be more active than other lines [288], including specifically in stressor paradigms [289]. Though for some paradigms this strain is ideal, here with the present model we will focus solely on use of SDs.

Age, weight, and sex of the animals are important factors within any stress paradigm and should be kept consistent.

An essential note on animal ordering: as much as possible, one must make sure all animals are ordered from the same company, building, colony, down to the same room. This ensures some consistency not only in the genetics of an outbred line that has some inherent variability but also in microbiomes between animals and across cohorts as much as possible. Researchers should be cognizant of the phase advance/phase delay of the light cycle and altitude when animals are shipped from the vendor to the local research facility, as well as details related to conventions for diet, water, and housing conditions at the vendor.

How many animals required for a given experiment depends on a number of factors. A power calculation given preliminary effect size estimations should be used to calculate an estimated minimum n. For our procedures an n of 8/12 animals per group is generally desired.

4.1.2 Handling Rats

As noted previously, laboratory rats are intelligent, social, and gentle creatures. They are sensitive to handling and to the temperament and biological sex of the handlers [269], and guidelines have been outlined previously [290]. We recommend new trainees imagine that they are calm or even bored before handling the animals. Adopting a calm movement style (while retaining true care) can help the animals to stay calm themselves, avoid unnecessarily stressing the animals, and make handling easier.

This may be obvious to some and should be obvious to all, but an easy thing to overlook when utilizing behavioral stress models is the improper use of identifying notes on cage-cards which can compromise the study's integrity by unblinding researchers.

4.1.3 Timing

Some assessments must begin well prior to the IS/JSE paradigm immediately after acclimation (microbiome sample collection baseline, etc.)

It is worth noting here that not only does the timing of the stress and assessment paradigm matter, but the timing of any potential intervention is also essential. This includes careful thought in designing and strict adherence to the experimental timeline, as well as taking into account sleep and circadian impacts by utilizing a completely randomized block design and remaining consistent in time-of-day application of all procedures and assessments.

4.2 IS

4.2.1 Notes Worth Noting

1. While setting up the IS procedure, it is essential that any animals in preparation for the IS procedure are not habituating near the stressor behavior rooms as they may hear vocalizations from rats currently undergoing the procedure.

2. Any adjustments to shock mA, shock protocol, shock duration, etc., should be made prior to placing rats within the stressor tubes to prevent any nontrial shocks.

3. IS sessions should occur at the same time of day relative to the animal's light/dark cycle to prevent any circadian effects. If multiple rounds of IS are necessary due to a large number of animals, a randomized block design should be used to ensure random distribution of experimental and control animals throughout sessions.

4.2.2 Common Errors (and Fixes)

Throughout the set up and execution of the IS procedure it is possible that minor problems may arise resulting in a need to temporarily interrupt shock administration. In these cases, it is essential that the experimenter resolves the issue quickly to minimize abnormal breaks between the standard 60 s inter-trial intervals. Such problems may include:

1. The animal is not receiving the tail shock. This could be due to a multitude of reasons, but often it is due to the electrode cream from the front copper electrode contacting the electrode cream from the back copper electrode thus allowing the shock to pass through the cream instead of the tail. This can be fixed by simply cleaning the cream off the tail and ensuring a third small piece of medical tape is used to separate the two electrodes and prevent any contact between the two. Other possible causes are that the animal's tail has been taped too tightly to the Plexiglas® rod and blood flow has been cut off or that the copper electrodes are not fully contacting the tail. If blood flow is cut off, remove the tape from the animal's tail to allow regular blood flow, clean the tail of any excess electrode cream and re-tape the tail with less pressure to allow blood flow. A third possibility is that the copper electrodes are coming loose or not fully contacting the tail. In this case simply move the alligator clips on the electrodes closer to the tail to tighten the copper electrode and ensure that there is sufficient electrode cream on all contact points.

2. The animal pulls its tail out of the tape. This issue should not arise if the IS tube has been adjusted to provide insufficient room for the animal to move forward. However, it is possible that the animal could push the adjustable perforated Plexiglas® door backwards to move farther into the tube and pull its tail free. In this instance simply clean the animal's tail of any excess electrode cream (to prevent contact between electrodes and the animal pulling free again due to the cream preventing the tape from securing the tail), re-tape the animal's tail, ensure the perforated door is adjusted to prevent any forward movement from the animal, and continue the procedure.

4.2.3 Pro Tips

1. Complete all steps that do not require the animal prior to placing animals in the IS tubes. For example, all IS tubes can be pre-taped and labeled with the animal number. This allows for the experimenter to place animals into the IS tubes one after another without pausing to tape up the next tail rod and minimizes differences in the length of time the animals are spending in the tube.

2. Label the IS tubes with the animal IDs. This prevents confusion when transferring animals between rooms, as the tape holding the base of the tail down will most likely cover the animal's identification number.

3. Stagger the protocol initiation time for each shock station. This allows for the experimenter to correct any of the potential issues listed in the above section prior to initiating the next animal's protocol as most common issues will arise within the first ~10 trials. This also allows for consistent lengths of time between any treatment (i.e., drug injection/micro-infusion) and initiation of the IS procedure.

4. Throughout the procedure we also recommend the experimenter checks on each animal every 5–10 trials to ensure there are no issues with shock administration.

4.3 JSE

4.3.1 Notes Worth Noting

1. Animals should be placed individually into cages for habituation and testing.

2. Animals should be habituated and tested solely with an empty wire food hopper to prevent exploration of the cage top, food, and water over the juvenile animal.

3. If you are not using a scoring software, scoring of JSE should be conducted by a single experimenter either using a video or live scored in the testing room.

4. When running the test, it is recommended that the experimenter sit quietly within view of the testing cage for the duration of the test to limit any distracting movement or noises that would influence JSE scores. We have found that any variable stimulus (i.e., coughing, talking, walking around, looking into the cage from above, etc.) can change interaction times and should be avoided.

5. Males and females should be run on separate days and in separate cages to ensure there is no effect of male scent on female interaction times and vice versa.

6. Limit scoring "interactions" to instances when the experimental animal is interacting with the juvenile and not when the juvenile is attempting to interact with the experimental animal.

7. JSE testing should take place at the same time of day relative to the animal's light/dark cycle to prevent any circadian effects.

8. JSE testing should take place in a novel room to avoid any effect of previous experiences in the testing room.

4.3.2 Common Errors (and Fixes)

As with most behavioral tasks, any small deviation from protocol could have an unanticipated effect on JSE times. Common issues that can influence JSE times include:

1. The experimental animal becomes more interested in the cage than the juvenile, typically expressed in the form of rearing to investigate the cage top or wire hopper. This can occur if the cage conditions are changed at all between habituation and testing. For example, if the animal is habituated with an empty wire food hopper and a cage top, the JSE testing should be carried out with both an empty wire food hopper and the same cage top. Any deviation from habituation conditions (i.e., using a see-through cage top to record from above, removing the wire hopper, etc.) provides novel stimuli that can distract the animal from the juvenile. We recommend that an empty wire food hopper is used without a filter cage top to contain the animal during both habituation and testing to minimize distraction.

2. Noises from other behavioral rooms are distracting for the experimental animal. This can occur if your behavior room is near another behavior room (e.g., stressor room) or a vivarium room that produces a lot of noise (cage wash, animal housing room, etc.). We recommend that JSE is conducted in a space that limits all potentially distracting stimuli to ensure the animal is solely focused on the juvenile.

4.3.3 Pro Tips

1. If using a recording device and not live scoring, ensure that the camera is positioned within view of the entire cage and connected to a power source to prevent loss of data if the battery dies.

2. Placing a small marking on the tail of each juvenile after testing ensures that the same juvenile is not used more frequently than the others.

3. If training a new lab member on JSE scoring, we recommend using old videos to avoid any effect of a second experimenter in the room during JSE and to prevent any effects on inter-rater reliability.

References

1. Sareen J (2018) Posttraumatic stress disorder in adults: epidemiology, pathophysiology, clinical manifestations, course, assessment, and diagnosis. Up to date [online]. Version May 2023
2. Sandars NK (1972) The epic of Gilgamesh. Penguin, London
3. Ben-Ezra M (2004) Trauma in antiquity: 4000 year old post-traumatic reactions? Stress Health 20:121–125
4. Geppert C (2021) An anniversary postponed and a diagnosis delayed: Vietnam and PTSD. Fed Pract 38:200–201
5. American Psychiatric Association (2013) Diagnostic and Statistical Manual of Mental Disorders: (5th ed). Arlington, VA
6. Ross RJ, Ball WA, Sullivan KA et al (1989) Sleep disturbance as the hallmark of posttraumatic stress disorder. Am J Psychiatry 146(6):697–707
7. Inman DJ, Silver SM, Doghramji K (1990) Sleep disturbance in post-traumatic stress disorder: a comparison with non-PTSD insomnia. J Trauma Stress 3:429–437
8. Lamarche LJ, De Koninck J (2007) Sleep disturbance in adults with posttraumatic stress disorder: a review. J Clin Psychiatry 68(8):1257–1270
9. Pace-Schott EF, Germain A, Milad MR (2015) Sleep and REM sleep disturbance in the pathophysiology of PTSD: the role of extinction memory. Biol Mood Anxiety Disord 5:1–19
10. Otis JD, Keane TM, Kerns RD (2003) An examination of the relationship between chronic pain and post-traumatic stress disorder. J Rehabil Res Dev 40:397–406
11. Scioli-Salter ER, Forman DE, Otis JD et al (2015) The shared neuroanatomy and neurobiology of comorbid chronic pain and PTSD: therapeutic implications. Clin J Pain 31:363–374
12. Sharp TJ, Harvey AG (2001) Chronic pain and posttraumatic stress disorder: mutual maintenance? Clin Psychol Rev 21:857–877
13. Asmundson GJ, Hadjistavropolous HD (2006) Addressing shared vulnerability for comorbid PTSD and chronic pain: a cognitive-behavioral perspective. Cogn Behav Pract 13:8–16
14. Shipherd JC, Keyes M, Jovanovic T et al (2007) Veterans seeking treatment for posttraumatic stress disorder: what about comorbid chronic pain? J Rehabil Res Dev 44:153–166
15. Moeller-Bertram T, Keltner J, Strigo IA (2012) Pain and post traumatic stress disorder – review of clinical and experimental evidence. Neuropharmacology 62:586–597
16. Boudreaux E, Kilpatrick G, Resnick HS et al (1998) Criminal victimization, posttraumatic stress disorder, and comorbid psychopathology among a community sample of women. J Trauma Stress 11:665–678
17. Swart S, Wildschut M, Draijer N et al (2020) Dissociative subtype of posttraumatic stress disorder or PTSD with comorbid dissociative disorders: comparative evaluation of clinical profiles. Psychol Trauma 12:38–45
18. Britvić D, Antičević V, Kaliterna M et al (2015) Comorbidities with Posttraumatic Stress Disorder (PTSD) among combat veterans: 15 years postwar analysis. Int J Clin Health Psychol 15:81–92
19. Pace TW, Heim CM (2011) A short review on the psychoneuroimmunology of posttraumatic stress disorder: from risk factors to medical comorbidities. Brain Behav Immun 25:6–13
20. Xu Y, Vandeleur C, Müller M et al (2021) Retrospectively assessed trajectories of PTSD symptoms and their subsequent comorbidities. J Psychiatr Res 136:71–79
21. McFarlane AC, Bookless C (2001) The effect of PTSD on interpersonal relationships: issues for emergency service workers. Sex Relatsh Ther 16:261–267
22. Laffaye C, Cavella S, Drescher K et al (2008) Relationships among PTSD symptoms, social support, and support source in veterans with chronic PTSD. J Trauma Stress 21:394–401
23. Fox J, Desai MM, Britten K et al (2012) Mental-health conditions, barriers to care, and productivity loss among officers in an urban police department. Conn Med 76:525–531
24. Milligan-Saville J, Choi I, Deady M et al (2018) The impact of trauma exposure on the development of PTSD and psychological distress in a volunteer fire service. Psychiatry Res 270:1110–1115
25. Chopko BA, Palmieri PA, Adams RE (2018) Relationships among traumatic experiences, PTSD, and posttraumatic growth for police officers: a path analysis. Psychol Trauma 10:183–189
26. Dams J, Rimane E, Steil R et al (2020) Health-related quality of life and costs of posttraumatic stress disorder in adolescents and

young adults in Germany. Front Psychiatry 11:697
27. Ehlers A, Clark DM (2000) A cognitive model of posttraumatic stress disorder. Behav Res Ther 38:319–345
28. Engelhard IM, Macklin ML, McNally RJ et al (2001) Emotion- and intrusion-based reasoning in Vietnam veterans with and without chronic posttraumatic stress disorder. Behav Res Ther 39:1339–1348
29. Engelhard IM, Hout MA van den, Arntz A et al (2002) A longitudinal study of "intrusion-based reasoning" and posttraumatic stress disorder after exposure to a train disaster. Behav Res Ther 40:1415–1424
30. Fortier MA, DiLillo D, Messman-Moore TL et al (2009) Severity of child sexual abuse and revictimization: the mediating role of coping and trauma symptoms. Psychol Women Q 33:308–320
31. Bistricky SL, Gallagher MW, Roberts CM et al (2017) Frequency of interpersonal trauma types, avoidant attachment, self-compassion, and interpersonal competence: a model of persisting posttraumatic symptoms. J Aggress Maltreat Trauma 26:608–625
32. Campbell SB, Renshaw KD, Kashdan TB et al (2017) A daily diary study of posttraumatic stress symptoms and romantic partner accommodation. Behav Ther 48:222–234
33. Lawrence JW, Fauerbach J, Munster A (1996) Early avoidance of traumatic stimuli predicts chronicity of intrusive thoughts following burn injury. Behav Res Ther 34:643–646
34. Boeding SE, Paprocki CM, Baucom DH et al (2013) Let me check that for you: symptom accommodation in romantic partners of adults with obsessive–compulsive disorder. Behav Res Ther 51:316–322
35. Eisma MC, Stroebe MS, Schut HA et al (2013) Avoidance processes mediate the relationship between rumination and symptoms of complicated grief and depression following loss. J Abnorm Psychol 122:961–970
36. Rapee RM, Peters L, Carpenter L et al (2015) The Yin and Yang of support from significant others: influence of general social support and partner support of avoidance in the context of treatment for social anxiety disorder. Behav Res Ther 69:40–47
37. Belleville G, Guay S, Marchand A (2011) Persistence of sleep disturbances following cognitive-behavior therapy for posttraumatic stress disorder. J Psychosom Res 70:318–327
38. Liempt S van (2012) Sleep disturbances and PTSD: a perpetual circle? Eur J Psychotraumatol 3:19142
39. Jaoude P, Vermont LN, Porhomayon J et al (2015) Sleep-disordered breathing in patients with post-traumatic stress disorder. Ann Am Thorac Soc 12:259–268
40. Wyk M van, Thomas KG, Solms M et al (2016) Prominence of hyperarousal symptoms explains variability of sleep disruption in posttraumatic stress disorder. Psychol Trauma 8:688–696
41. Hurtado-Alvarado G, Domínguez-Salazar E, Pavon L et al (2016) Blood-brain barrier disruption induced by chronic sleep loss: low-grade inflammation may be the link. J Immunol Res 2016:4576012
42. Groer MW, Kane B, Williams SN et al (2015) Relationship of PTSD symptoms with combat exposure, stress, and inflammation in American soldiers. Biol Res Nurs 17:303–310
43. Loupy KM, Lowry CA (2019) Posttraumatic stress disorder and the gut microbiome. In: The Oxford Handbook of the Microbiome-Gut-Brain Axis. Oxford University Press, Oxford
44. Langgartner D, Lowry CA, Reber SO (2019) Old Friends, immunoregulation, and stress resilience. Pflugers Arch 471:237–269
45. Kim TD, Lee S, Yoon S (2020) Inflammation in post-traumatic stress disorder (PTSD): a review of potential correlates of PTSD with a neurological perspective. Antioxidants (Basel) 9:107
46. Bersani FS, Mellon SH, Lindqvist D et al (2020) Novel pharmacological targets for combat PTSD – metabolism, inflammation, the gut microbiome, and mitochondrial dysfunction. Mil Med 185:311–318
47. Lev-Wiesel R (2007) Intergenerational transmission of trauma across three generations: a preliminary study. Qual Soc Work Res Pract 6:75–94
48. Dekel R, Goldblatt H (2008) Is there intergenerational transmission of trauma? The case of combat veterans' children. Am J Orthop 78:281–289
49. Aguiar W, Halseth R (2015) Aboriginal Peoples and Historic Trauma: The Processes of Intergenerational Transmission. National Collaborating Centre for Aboriginal Health. Prince George, BC
50. Yehuda R, Lehrner A (2018) Intergenerational transmission of trauma effects: putative role of epigenetic mechanisms. World Psychiatry 17:243–257
51. Lünnemann M, Van der Horst F, Prinzie P et al (2019) The intergenerational impact of trauma and family violence on parents and their children. Child Abuse Negl 96:104134

52. Anderson RE, Edwards L-J, Silver KE et al (2018) Intergenerational transmission of child abuse: predictors of child abuse potential among racially diverse women residing in domestic violence shelters. Child Abuse Negl 85:80–90

53. Hoffart R, Jones NA (2018) Intimate partner violence and intergenerational trauma among Indigenous women. Int Crim Justice Rev 28:25–44

54. Menzies P (2010) Intergenerational trauma from a mental health perspective. Native Social Work Journal 7:63–85

55. Başoğlu M, Kiliç C, Şalcioğlu E et al (2004) Prevalence of posttraumatic stress disorder and comorbid depression in earthquake survivors in Turkey: an epidemiological study. J Trauma Stress 17:133–141

56. Salcioglu E, Basoglu M, Livanou M (2007) Post-traumatic stress disorder and comorbid depression among survivors of the 1999 earthquake in Turkey. Disasters 31:115–129

57. Adams SW, Bowler RM, Russell K et al (2019) PTSD and comorbid depression: social support and self-efficacy in World Trade Center tower survivors 14–15 years after 9/11. Psychol Trauma Theory Res Pract Policy 11:156–164

58. Campbell DG, Felker BL, Liu C-F et al (2007) Prevalence of depression–PTSD comorbidity: implications for clinical practice guidelines and primary care-based interventions. J Gen Intern Med 22:711–718

59. Panagioti M, Gooding P, Tarrier N (2009) Post-traumatic stress disorder and suicidal behavior: a narrative review. Clin Psychol Rev 29:471–482

60. Krysinska K, Lester D (2010) Post-traumatic stress disorder and suicide risk: a systematic review. Arch Suicide Res 14:1–23

61. Panagioti M, Gooding PA, Tarrier N (2012) A meta-analysis of the association between posttraumatic stress disorder and suicidality: the role of comorbid depression. Compr Psychiatry 53:915–930

62. Zatti C, Rosa V, Barros A et al (2017) Childhood trauma and suicide attempt: a meta-analysis of longitudinal studies from the last decade. Psychiatry Res 256:353–358

63. Panagioti M, Gooding PA, Triantafyllou K et al (2015) Suicidality and posttraumatic stress disorder (PTSD) in adolescents: a systematic review and meta-analysis. Soc Psychiatry Psychiatr Epidemiol 50:525–537

64. Fox V, Dalman C, Dal H et al (2021) Suicide risk in people with post-traumatic stress disorder: a cohort study of 3.1 million people in Sweden. J Affect Disord 279:609–616

65. Charak R, Armour C, Elklit A et al (2014) Factor structure of PTSD, and relation with gender in trauma survivors from India. Eur J Psychotraumatol 5:25547

66. Meer CA van der, Bakker A, Smit AS et al (2017) Gender and age differences in trauma and PTSD among Dutch treatment-seeking police officers. J Nerv Ment Dis 205:87–92

67. Olff M (2017) Sex and gender differences in post-traumatic stress disorder: an update. Eur J Psychotraumatol 8:1351204

68. Dunlop BW, Kaye JL, Youngner C et al (2014) Assessing treatment-resistant post-traumatic stress disorder: the Emory treatment resistance interview for PTSD (E-TRIP). Behav Sci (Basel) 4:511–527

69. Imel ZE, Laska K, Jakupcak M et al (2013) Meta-analysis of dropout in treatments for posttraumatic stress disorder. J Consult Clin Psychol 81:394–404

70. Najavits LM (2015) The problem of dropout from "gold standard" PTSD therapies. F1000prime Rep 7:43

71. Lewis C, Roberts NP, Gibson S et al (2020) Dropout from psychological therapies for post-traumatic stress disorder (PTSD) in adults: systematic review and meta-analysis. Eur J Psychotraumatol 11:1709709

72. Simpson TL (2002) Women's treatment utilization and its relationship to childhood sexual abuse history and lifetime PTSD. Subst Abus 23:17–30

73. Najavits LM, Sullivan TP, Schmitz M et al (2004) Treatment utilization by women with PTSD and substance dependence. Am J Addict 13:215–224

74. Peltan JR, Cellucci T (2011) Childhood sexual abuse and substance abuse treatment utilization among substance-dependent incarcerated women. J Subst Abus Treat 41:215–224

75. Nobles CJ, Valentine SE, Zepeda ED et al (2017) Usual course of treatment and predictors of treatment utilization for patients with posttraumatic stress disorder. J Clin Psychiatry 78:e559–e566

76. Artime TM, Buchholz KR, Jakupcak M (2019) Mental health symptoms and treatment utilization among trauma-exposed college students. Psychol Trauma Theory Res Pract Policy 11:274–282

77. Hoge CW, Grossman SH, Auchterlonie JL et al (2014) PTSD treatment for soldiers after combat deployment: low utilization of mental health care and reasons for dropout. Psychiatr Serv 65:997–1004

78. Goetter EM, Bui E, Ojserkis RA et al (2015) A systematic review of dropout from

79. Schottenbauer MA, Glass CR, Arnkoff DB et al (2008) Nonresponse and dropout rates in outcome studies on PTSD: review and methodological considerations. Psychiatry 71:134–168
80. Libby DJ, Pilver CE, Desai R (2012) Complementary and alternative medicine in VA specialized PTSD treatment programs. Psychiatr Serv 63:1134–1136
81. Eisenberg DM, Davis RB, Ettner SL et al (1998) Trends in alternative medicine use in the United States, 1990-1997: results of a follow-up national survey. JAMA 280:1569–1575
82. Kessler RC, Davis RB, Foster DF et al (2001) Long-term trends in the use of complementary and alternative medical therapies in the United States. Ann Intern Med 135:262–268
83. Barnes PM, Powell-Griner E, McFann K et al (2004) Complementary and alternative medicine use among adults: United States, 2002. In: Seminars in integrative medicine. Elsevier, pp 54–71
84. Barnes PM, Bloom B, Nahin RL (2008) Complementary and alternative medicine use among adults and children; United States, 2007. Natl Health Stat Report 10(12):1–23
85. Su D, Li L (2011) Trends in the use of complementary and alternative medicine in the United States: 2002–2007. J Health Care Poor Underserved 22:296–310
86. Clarke TC, Black LI, Stussman BJ et al (2015) Trends in the use of complementary health approaches among adults: United States, 2002–2012. Natl Health Stat Report 79:1–16
87. Williams JW Jr, Gierisch JM, McDuffie J et al (2012) An overview of complementary and alternative medicine therapies for anxiety and depressive disorders: supplement to efficacy of complementary and alternative medicine therapies for posttraumatic stress disorder. Department of Veterans Affairs, Washington, DC
88. Hemmings SM, Malan-Muller S, Heuvel LL van den et al (2017) The microbiome in posttraumatic stress disorder and trauma-exposed controls: an exploratory study. Psychosom Med 79:936–946
89. Malan-Muller S, Valles-Colomer M, Raes J et al (2018) The gut microbiome and mental health: implications for anxiety-and trauma-related disorders. OMICS 22:90–107
90. Hoisington AJ, Billera DM, Bates KL et al (2018) Exploring service dogs for rehabilitation of veterans with PTSD: a microbiome perspective. Rehabil Psychol 63:575–587
91. Bajaj JS, Sikaroodi M, Fagan A et al (2019) Posttraumatic stress disorder is associated with altered gut microbiota that modulates cognitive performance in veterans with cirrhosis. Am J Physiol Gastrointest Liver Physiol 317:G661–G669
92. Nikolova VL, Hall MR, Hall LJ et al (2021) Perturbations in gut microbiota composition in psychiatric disorders: a review and meta-analysis. JAMA Psychiatry 78:1343–1354
93. Malan-Muller S, Valles-Colomer M, Foxx CL et al (2022) Exploring the relationship between the gut microbiome and mental health outcomes in a posttraumatic stress disorder cohort relative to trauma-exposed controls. Eur Neuropsychopharmacol 56:24–38
94. McGowan I (2019) The economic burden of PTSD. A brief review of salient literature. Int J Psychiatry Med 1:20–26
95. Ferry FR, Brady SE, Bunting BP et al (2015) The economic burden of PTSD in Northern Ireland. J Trauma Stress 28:191–197
96. McCrone P, Knapp M, Cawkill P (2003) Post-traumatic stress disorder (PTSD) in the Armed Forces: health economic considerations. J Trauma Stress 16:519–522
97. Surís A, Lind L, Kashner TM et al (2004) Sexual assault in women veterans: an examination of PTSD risk, health care utilization, and cost of care. Psychosom Med 66:749–756
98. Chan D, Cheadle AD, Reiber G et al (2009) Health care utilization and its costs for depressed veterans with and without comorbid PTSD symptoms. Psychiatr Serv 60:1612–1617
99. McGeary D, Moore M, Vriend CA et al (2011) The evaluation and treatment of comorbid pain and PTSD in a military setting: an overview. J Clin Psychol Med Settings 18:155–163
100. Vyas KJ, Fesperman SF, Nebeker BJ et al (2016) Preventing PTSD and depression and reducing health care costs in the military: a call for building resilience among service members. Mil Med 181:1240–1247
101. Hardner K, Wolf MR, Rinfrette ES (2018) Examining the relationship between higher educational attainment, trauma symptoms, and internalizing behaviors in child sexual abuse survivors. Child Abuse Negl 86:375–383
102. Bothe T, Jacob J, Kröger C et al (2020) How expensive are post-traumatic stress disorders? Estimating incremental health care and

economic costs on anonymised claims data. Eur J Health Econ 21:917–930
103. Tan SY, Yip A (2018) Hans Selye (1907–1982): founder of the stress theory. Singap Med J 59:170
104. Selye H (1936) A syndrome produced by diverse nocuous agents. Nature 138:32–32
105. Del Giudice M, Buck CL, Chaby LE et al (2018) What is stress? A systems perspective. Integr Comp Biol 58:1019–1032
106. Gottesman II, Gould TD (2003) The endophenotype concept in psychiatry: etymology and strategic intentions. Am J Psychiatry 160: 636–645
107. Beauchaine TP, Constantino JN (2017) Redefining the endophenotype concept to accommodate transdiagnostic vulnerabilities and etiological complexity. Biomark Med 11: 769–780
108. Mullins LJ, Mullins JJ (2004) Insights from the rat genome sequence. Genome Biol 5:1–3
109. Gibbs RA, Pachter L (2004) Genome sequence of the Brown Norway rat yields insights into mammalian evolution. Nature 428:493–521
110. Tabakoff B, Hoffman PL (2000) Animal models in alcohol research. Alcohol Res Health 24:77
111. Van Den Buuse M, Garner B, Gogos A et al (2005) Importance of animal models in schizophrenia research. Aust N Z J Psychiatry 39:550–557
112. Mogil JS, Davis KD, Derbyshire SW (2010) The necessity of animal models in pain research. Pain 151:12–17
113. Swearengen JR (2012) Biodefense research methodology and animal models. CRC Press, Boca Raton, Florida
114. Abdullahi A, Amini-Nik S, Jeschke M (2014) Animal models in burn research. Cell Mol Life Sci 71:3241–3255
115. Budhu S, Wolchok J, Merghoub T (2014) The importance of animal models in tumor immunity and immunotherapy. Curr Opin Genet Dev 24:46–51
116. Azrin N, Hopwood J, Powell J (1967) A rat chamber and electrode procedure for avoidance conditioning 1. J Exp Anal Behav 10: 291–298
117. Skinner B, Campbell S (1947) An automatic shocking-grid apparatus for continuous use. J Comp Physiol Psychol 40:305–37
118. Sloane H (1964) Scramble patterns and escape learning. J Exp Anal Behav 7:336
119. Ulrich RE, Azrin NH (1962) Reflexive fighting in response to aversive stimulation 1. J Exp Anal Behav 5:511–520
120. Silver MP, Schoenfeld WN, Snapper AG et al (1964) Impedance-voltage functions in the white rat with chronic body electrode implants. Psychon Sci 1:61–62
121. Weiss J (1967) A tail electrode for unrestrained rats. J Exp Anal Behav 10:85–86
122. Hall RD, Clayton RJ, Mark RG (1966) A device for the partial restraint of rats in operant conditioning studies. J Exp Anal Behav 9: 143–145
123. Paré WP, Glavin GB (1986) Restraint stress in biomedical research: a review. Neurosci Biobehav Rev 10:339–370
124. Seewoo BJ, Hennessy LA, Feindel KW et al (2020) Validation of chronic restraint stress model in young adult rats for the study of depression using longitudinal multimodal MR imaging. Eneuro 7: ENEURO.0113-20.2020
125. Clutton-Brock J (1960) Some pain threshold studies with particular reference to thiopentone. Anaesthesia 15:71–72
126. Spiaggia A, Bodnar RJ, Kelly DD et al (1979) Opiate and non-opiate mechanisms of stress-induced analgesia: cross-tolerance between stressors. Pharmacol Biochem Behav 10: 761–765
127. Grau JW, Hyson RL, Maier SF et al (1981) Long-term stress-induced analgesia and activation of the opiate system. Science 213: 1409–1411
128. Maier SF, Sherman JE, Lewis JW et al (1983) The opioid/nonopioid nature of stress-induced analgesia and learned helplessness. J Exp Psychol Anim Behav Process 9:80–90
129. Drugan RC, Ader DN, Maier SF (1985) Shock controllability and the nature of stress-induced analgesia. Behav Neurosci 99: 791–801
130. Maier SF (1986) Stressor controllability and stress-induced analgesia. Ann N Y Acad Sci 467:55–72
131. Maier SF, Watkins LR (1991) Conditioned and unconditioned stress-induced analgesia: stimulus preexposure and stimulus change. Anim Learn Behav 19:295–304
132. Grisel JE, Fleshner M, Watkins LR et al (1993) Opioid and nonopioid interactions in two forms of stress-induced analgesia. Pharmacol Biochem Behav 45:161–172
133. Butler RK, Finn DP (2009) Stress-induced analgesia. Prog Neurobiol 88:184–202
134. Heinke B, Gingl E, Sandkühler J (2011) Multiple targets of μ-opioid receptor-mediated presynaptic inhibition at primary afferent Aδ- and C-fibers. J Neurosci 31:1313–1322

135. Koga A, Fujita T, Piao L-H et al (2019) Inhibition by O-desmethyltramadol of glutamatergic excitatory transmission in adult rat spinal substantia gelatinosa neurons. Mol Pain 15:1744806918824243
136. Jones TL, Sweitzer SM, Wilson SP et al (2003) Afferent fiber-selective shift in opiate potency following targeted opioid receptor knockdown. Pain 106:365–371
137. Ikoma M, Kohno T, Baba H (2007) Differential presynaptic effects of opioid agonists on Aδ-and C-afferent glutamatergic transmission to the spinal dorsal horn. J Neurosci 107:807–812
138. Lu Y, Sweitzer SM, Laurito CE et al (2004) Differential opioid inhibition of C-and A-δ-fiber mediated thermonociception after stimulation of the nucleus raphe magnus. Anesth Analg 98:414–419
139. Brederson J-D, Honda CN (2015) Primary afferent neurons express functional delta opioid receptors in inflamed skin. Brain Res 1614:105–111
140. Hohmann AG, Suplita RL, Bolton NM et al (2005) An endocannabinoid mechanism for stress-induced analgesia. Nature 435:1108–1112
141. Suplita RL II, Gutierrez T, Fegley D et al (2006) Endocannabinoids at the spinal level regulate, but do not mediate, nonopioid stress-induced analgesia. Neuropharmacology 50:372–379
142. Rash JA, Aguirre-Camacho A, Campbell TS (2014) Oxytocin and pain: a systematic review and synthesis of findings. Clin J Pain 30:453–462
143. Van der Kolk B, Greenberg M, Boyd H et al (1985) Inescapable shock, neurotransmitters, and addiction to trauma: toward a psychobiology of post traumatic stress. Biol Psychiatry 20:314–325
144. Koba T, Kodama Y, Shimizu K et al (2001) Persistent behavioural changes in rats following inescapable shock stress: a potential model of posttraumatic stress disorder. World J Biol Psychiatry 2:34–37
145. Thornton JW, Jacobs PD (1971) Learned helplessness in human subjects. J Exp Psychol 87:367–372
146. Seligman ME (1972) Learned helplessness. Annu Rev Med 23:407–412
147. Seligman ME, Beagley G (1975) Learned helplessness in the rat. J Comp Physiol Psychol 88:534–541
148. Maier SF, Seligman ME (1976) Learned helplessness: theory and evidence. J Exp Psychol Gen 105:3–46
149. Maier SF, Jackson RL (1979) Learned helplessness: all of us were right (and wrong): inescapable shock has multiple effects. In: Psychology of Learning and Motivation. Elsevier, pp 155–218
150. Jackson RL, Maier SF, Coon DJ (1979) Long-term analgesic effects of inescapable shock and learned helplessness. Science 206:91–93
151. Vollmayr B, Henn FA (2001) Learned helplessness in the rat: improvements in validity and reliability. Brain Res Protocol 8:1–7
152. Dess NK, Minor TR, Brewer J (1989) Suppression of feeding and body weight by inescapable shock: modulation by quinine adulteration, stress reinstatement, and controllability. Physiol Behav 45:975–983
153. Cassens G, Roffman M, Kuruc A et al (1980) Alterations in brain norepinephrine metabolism induced by environmental stimuli previously paired with inescapable shock. Science 209:1138–1140
154. Maier SF, Davies S, Grau JW et al (1980) Opiate antagonists and long-term analgesic reaction induced by inescapable shock in rats. J Comp Physiol Psychol 94:1172–1183
155. Drugan RC, Maier SF (1983) Analgesic and opioid involvement in the shock-elicited activity and escape deficits produced by inescapable shock. Learn Motiv 14:30–47
156. Stuckey J, Marra S, Minor T et al (1989) Changes in mu opiate receptors following inescapable shock. Brain Research 476:167–169
157. Li B, Yang C-J, Yue N et al (2013) Clomipramine reverses hypoalgesia/hypoesthesia and improved depressive-like behaviors induced by inescapable shock in rats. Neurosci Lett 541:227–232
158. Will MJ, Watkins LR, Maier SF (1998) Uncontrollable stress potentiates morphine's rewarding properties. Pharmacol Biochem Behav 60:655–664
159. Weiss J, Bailey W, Pohorecky L et al (1980) Stress-induced depression of motor activity correlates with regional changes in brain norepinephrine but not in dopamine. Neurochem Res 5:9–22
160. Moraska A, Fleshner M (2001) Voluntary physical activity prevents stress-induced behavioral depression and anti-KLH antibody suppression. Am J Phys Regul Integr Comp Phys 281:R484–R489
161. Greenwood BN, Foley TE, Day HE et al (2003) Freewheel running prevents learned helplessness/behavioral depression: role of

dorsal raphe serotonergic neurons. J Neurosci 23:2889–2898

162. Vollmayr B, Gass P (2013) Learned helplessness: unique features and translational value of a cognitive depression model. Cell Tissue Res 354:171–178

163. Landgraf D, Long J, Der-Avakian A et al (2015) Dissociation of learned helplessness and fear conditioning in mice: a mouse model of depression. PLoS One 10: e0125892

164. Cheng Y, Desse S, Martinez A et al (2018) TNFα disrupts blood brain barrier integrity to maintain prolonged depressive-like behavior in mice. Brain Behav Immun 69:556–567

165. Daut RA, Ravenel JR, Watkins LR et al (2020) The behavioral and neurochemical effects of an inescapable stressor are time of day dependent. Stress 23:405–416

166. Biederman GB, Furedy JJ (1973) Preference-for-signaled-shock phenomenon: effects of shock modifiability and light reinforcement. J Exp Psychol 100:380

167. Kelsey JE (1977) Escape acquisition following inescapable shock in the rat. Anim Learn Behav 5:83–92

168. Jackson RL, Maier SF, Rapaport PM (1978) Exposure to inescapable shock produces both activity and associative deficits in the rat. Learn Motiv 9:69–98

169. Nick A, Alexander Z, others (1990) Handling habituation and chlordiazepoxide have different effects on GABA and 5-HT function in the frontal cortex and hippocampus. Eur J Pharmacol 190:229–234

170. Petty F, Kramer G, Wilson L (1992) Prevention of learned helplessness: in vivo correlation with cortical serotonin. Pharmacol Biochem Behav 43:361–367

171. Paré WP (1996) Enhanced retrieval of unpleasant memories influenced by shock controllability, shock sequence, and rat strain. Biological Psychiatry 39:808–813

172. Campisi J, Leem TH, Fleshner M (2003) Stress-induced extracellular Hsp72 is a functionally significant danger signal to the immune system. Cell Stress Chaperones 8: 272

173. Campisi J, Fleshner M (2003) Role of extracellular HSP72 in acute stress-induced potentiation of innate immunity in active rats. J Appl Physiol 94:43–52

174. Kirk RC, Blampied NM (1985) Activity during inescapable shock and subsequent escape avoidance learning: female and male rats. N Z J Psychol 14:9–14

175. Jenkins J, Williams P, Kramer G et al (2000) 249. The effects of gender and the estrous cycle on learned helplessness in the rat. Biol Psychiatry 47:S76

176. Jenkins JA, Williams P, Kramer GL et al (2001) The influence of gender and the estrous cycle on learned helplessness in the rat. Biol Psychol 58:147–158

177. Nickerson M, Kennedy SL, Johnson JD et al (2006) Sexual dimorphism of the intracellular heat shock protein 72 response. J Appl Physiol (1985) 101:566–575

178. Fonken LK, Frank MG, Gaudet AD et al (2018) Neuroinflammatory priming to stress is differentially regulated in male and female rats. Brain Behav Immun 70:257–267

179. Baratta MV, Leslie NR, Fallon IP et al (2018) Behavioural and neural sequelae of stressor exposure are not modulated by controllability in females. Eur J Neurosci 47:959–967

180. Baratta MV, Gruene TM, Dolzani SD et al (2019) Controllable stress elicits circuit-specific patterns of prefrontal plasticity in males, but not females. Brain Struct Funct 224:1831–1843

181. Tanner MK, Fallon IP, Baratta MV et al (2019) Voluntary exercise enables stress resistance in females. Behav Brain Res 369: 111923

182. Vul'fson S (1897) O psikhicheskom vliianii v rabote sliunnykh zhelez [On psychic influence in the work of the salivary glands]. Trudy obshchestva russkikh vrachei 65:110–113

183. Minor TR, LoLordo VM (1984) Escape deficits following inescapable shock: the role of contextual odor. J Exp Psychol Anim Behav Process 10:168–181

184. Minor TR, Jackson RL, Maier SF (1984) Effects of task-irrelevant cues and reinforcement delay on choice-escape learning following inescapable shock: evidence for a deficit in selective attention. J Exp Psychol Anim Behav Process 10:543–556

185. Maier SF (1990) Role of fear in mediating shuttle escape learning deficit produced by inescapable shock. J Exp Psychol Anim Behav Process 16:137–149

186. Maier SF, Watkins LR (2005) Stressor controllability and learned helplessness: the roles of the dorsal raphe nucleus, serotonin, and corticotropin-releasing factor. Neurosci Biobehav Rev 29:829–841

187. Chen C-FF, Barnes DC, Wilson DA (2011) Generalized vs. stimulus-specific learned fear differentially modifies stimulus encoding in primary sensory cortex of awake rats. J Neurophysiol 106:3136–3144

188. Ghosh S, Chattarji S (2015) Neuronal encoding of the switch from specific to generalized fear. Nat Neurosci 18:112–120

189. Harris JD (1943) Studies on Nonassociative Factors Inherent in conditioning. Williams & Wilkins
190. Mackintosh N (1974) Classical conditioning: basic operations. The psychology of animal learning, pp 8–40
191. Kamprath K, Wotjak CT (2004) Nonassociative learning processes determine expression and extinction of conditioned fear in mice. Learn Mem 11:770–786
192. Perusini JN, Meyer EM, Long VA et al (2016) Induction and expression of fear sensitization caused by acute traumatic stress. Neuropsychopharmacology 41:45–57
193. Servatius RJ, Ottenweller JE, Natelson BH (1995) Delayed startle sensitization distinguishes rats exposed to one or three stress sessions: further evidence toward an animal model of PTSD. Biol Psychiatry 38:539–546
194. Mirshekar M, Abrari K, Goudarzi I et al (2013) Systemic administrations of β-estradiol alleviate both conditioned and sensitized fear responses in an ovariectomized rat model of post-traumatic stress disorder. Neurobiol Learn Mem 102:12–19
195. Thompson R, Strong P, Clark P et al (2014) Repeated fear-induced diurnal rhythm disruptions predict PTSD-like sensitized physiological acute stress responses in F 344 rats. Acta Physiol (Oxf) 211:447–465
196. Rajbhandari AK, Baldo BA, Bakshi VP (2015) Predator stress-induced CRF release causes enduring sensitization of basolateral amygdala norepinephrine systems that promote PTSD-like startle abnormalities. J Neurosci 35:14270–14285
197. Maier SF (1990) Diazepam modulation of stress-induced analgesia depends on the type of analgesia. Behav Neurosci 104:339–347
198. Maier SF, Watkins LR (1998) Stressor controllability, anxiety, and serotonin. Cognit Ther Res 22:595–613
199. Christianson JP, Benison AM, Jennings J et al (2008) The sensory insular cortex mediates the stress-buffering effects of safety signals but not behavioral control. J Neurosci 28:13703–13711
200. Weiss JM (1968) Effects of coping responses on stress. J Comp Physiol Psychol 65:251–260
201. Maier SF, Seligman MEP, Solomon RL (1969) Pavlovian fear conditioning and learned helplessness: Effects on escape and avoidance behavior of (a) the CS-US contingency and (b) the independence of the US and voluntary responding. In: Campbell, B. A., and Church, R.M. (Eds.), Punishment. Appleton-Century-Crofts, New York, pp 299–343
202. Traini C, Evangelista S, Girod V et al (2016) Changes of excitatory and inhibitory neurotransmitters in the colon of rats underwent to the wrap partial restraint stress. Neurogastroenterol Motil 28:1172–1185
203. Davidson JR, Stein DJ, Shalev AY et al (2004) Posttraumatic stress disorder: acquisition, recognition, course, and treatment. J Neuropsychiatr Clin Neurosci 16:135–147
204. Jeong M-J, Lee C, Sung K et al (2020) Fear response-based prediction for stress susceptibility to PTSD-like phenotypes. Mol Brain 13:1–9
205. Watkins L, Drugan R, Hyson R et al (1984) Opiate and non-opiate analgesia induced by inescapable tail-shock: effects of dorsolateral funiculus lesions and decerebration. Brain Res 291:325–336
206. Khan S, Liberzon I (2004) Topiramate attenuates exaggerated acoustic startle in an animal model of PTSD. Psychopharmacology 172:225–229
207. Wang W, Liu Y, Zheng H et al (2008) A modified single-prolonged stress model for post-traumatic stress disorder. Neurosci Lett 441:237–241
208. Pickens CL, Golden SA, Adams-Deutsch T et al (2009) Long-lasting incubation of conditioned fear in rats. Biol Psychiatry 65:881–886
209. Pickens CL, Navarre BM, Nair SG (2010) Incubation of conditioned fear in the conditioned suppression model in rats: role of food-restriction conditions, length of conditioned stimulus, and generality to conditioned freezing. Neuroscience 169:1501–1510
210. Pamplona F, Henes K, Micale V et al (2011) Prolonged fear incubation leads to generalized avoidance behavior in mice. J Psychiatr Res 45:354–360
211. Estes WK, Skinner BF (1941) Some quantitative properties of anxiety. J Exp Psychol 29:390–400
212. Masserman JH, Yum K (1946) An analysis of the influence of alcohol on experimental neuroses in cats. Psychosom Med 8:36–52
213. Conger JJ (1951) The effects of alcohol on conflict behavior in the albino rat. Q J Stud Alcohol 12:1–29
214. Jacobsen E (1957) The effect of psychotropic drugs under psychic stress. In: Psychotropic drugs. Elsevier Publishing Co, Amsterdam, pp 119–124

215. Miller NE, Angell JR (1957) Objective techniques for studying motivational effects of drugs on animals. Sympotium on psychotropic drugs
216. Naess K, Rasmussen EW (1958) Approach-withdrawal responses and other specific behaviour reactions as screening test for tranquillizers. Acta Pharmacol 15:99–114
217. Geller I, Seifter J (1960) The effects of meprobamate, barbiturates, d-amphetamine and promazine on experimentally induced conflict in the rat. Psychopharmacologia 1:482–492
218. Latane B, Hothersall D (1972) Social attraction in animals. In: Dodwell PC (ed) New Horizons in Psychology 2. Penguin, Baltimore
219. File SE, Hyde J (1978) Can social interaction be used to measure anxiety? Br J Pharmacol 62:19–24
220. Blanchard RJ, Blanchard DC (1989) Attack and defense in rodents as ethoexperimental models for the study of emotion. Prog Neuro-Psychopharmacol Biol Psychiatry 13: S3–S14
221. File SE, Seth P (2003) A review of 25 years of the social interaction test. Eur J Pharmacol 463:35–53
222. Gong Z-H, Li Y-F, Zhao N et al (2006) Anxiolytic effect of agmatine in rats and mice. Eur J Pharmacol 550:112–116
223. Lapin IP, Mutovkina LG, Ryzov IV et al (1996) Anxiogenic activity of quinolinic acid and kynurenine in the social interaction test in mice. J Psychopharmacol 10:246–249
224. Kita A, Furukawa K (2008) Involvement of neurosteroids in the anxiolytic-like effects of AC-5216 in mice. Pharmacol Biochem Behav 89:171–178
225. File SE, Cheeta S, Akanezi C (2001) Diazepam and nicotine increase social interaction in gerbils: a test for anxiolytic action. Brain Res 888:311–313
226. Cheeta S, Tucci S, Sandhu J et al (2001) Anxiolytic actions of the substance P (NK1) receptor antagonist L-760735 and the 5-HT1A agonist 8-OH-DPAT in the social interaction test in gerbils. Brain Res 915: 170–175
227. Greenberg GD, Westerhuyzen JA van, Bales KL et al (2012) Is it all in the family? The effects of early social structure on neural–behavioral systems of prairie voles (*Microtus ochrogaster*). Neuroscience 216:46–56
228. Lee NS, Goodwin NL, Freitas KE et al (2019) Affiliation, aggression, and selectivity of peer relationships in meadow and prairie voles. Front Behav Neurosci 13:52
229. Normann M (2020) Investigating the behavioral effects of juvenile stress in the prairie vole model. Northern Illinois University ProQuest Dissertations Publishing
230. Rivera DS, Lindsay CB, Codocedo JF et al (2018) Long-term, fructose-induced metabolic syndrome-like condition is associated with higher metabolism, reduced synaptic plasticity and cognitive impairment in *Octodon degus*. Mol Neurobiol 55:9169–9187
231. Rivera DS, Lindsay CB, Oliva CA et al (2021) A multivariate assessment of age-related cognitive impairment in *Octodon degus*. Front Integr Neurosci 15:719076
232. Thor D (1979) Olfactory perception and inclusive fitness. Physiol Psychol 7:303–306
233. Miczek KA (1979) A new test for aggression in rats without aversive stimulation: differential effects of d-amphetamine and cocaine. Psychopharmacology 60:253–259
234. File SE, Deakin J, Longden A et al (1979) An investigation of the role of the locus coeruleus in anxiety and agonistic behaviour. Brain Res 169:411–420
235. File SE, Hyde J, MacLeod N (1979) 5,7-dihydroxytryptamine lesions of dorsal and median raphe nuclei and performance in the social interaction test of anxiety and in a home-cage aggression test. J Affect Disord 1: 115–122
236. Maier DM, Pohorecky LA (1987) The effect of ethanol treatment on social behavior in male rats. Physiol Behav 13:259–268
237. Bluthe R-M, Schoenen J, Dantzer R (1990) Androgen-dependent vasopressinergic neurons are involved in social recognition in rats. Brain Res 519:150–157
238. Dantzer R, Bluthe R-M, Kelley KW (1991) Androgen-dependent vasopressinergic neurotransmission attenuates interleukin-1-induced sickness behavior. Brain Res 557: 115–120
239. Patel HP (2016) Investigating the sensitivity of juvenile social exploration at detecting the affective consequences accompanying chronic neuropathic pain. Mol Pain 12: 1744806916656635
240. Jacobson-Pick S, Audet M-C, Nathoo N et al (2011) Stressor experiences during the juvenile period increase stressor responsivity in adulthood: transmission of stressor experiences. Behav Brain Res 216:365–374
241. Daut RA, Hartsock MJ, Tomczik AC et al (2019) Circadian misalignment has differential effects on affective behavior following exposure to controllable or uncontrollable stress. Behav Brain Res 359:440–445

242. Frank MG, Baratta MV, Zhang K et al (2020) Acute stress induces the rapid and transient induction of caspase-1, gasdermin D and release of constitutive IL-1β protein in dorsal hippocampus. Brain Behav Immun 90:70–80
243. Bilbo SD, Yirmiya R, Amat J et al (2008) Bacterial infection early in life protects against stressor-induced depressive-like symptoms in adult rats. Psychoneuroendocrinology 33:261–269
244. Christianson JP, Paul ED, Irani M et al (2008) The role of prior stressor controllability and the dorsal raphe nucleus in sucrose preference and social exploration. Behav Brain Res 193:87–93
245. Christianson JP, Ragole T, Amat J et al (2010) 5-hydroxytryptamine 2C receptors in the basolateral amygdala are involved in the expression of anxiety after uncontrollable traumatic stress. Biol Psychiatry 67:339–345
246. Christianson JP, Thompson BM, Watkins LR et al (2009) Medial prefrontal cortical activation modulates the impact of controllable and uncontrollable stressor exposure on a social exploration test of anxiety in the rat. Stress 12:445–450
247. Goshen I, Yirmiya R (2009) Interleukin-1 (IL-1): a central regulator of stress responses. Front Neuroendocrinol 30:30–45
248. Loupy KM, Cler KE, Marquart BM et al (2021) Comparing the effects of two different strains of mycobacteria, *Mycobacterium vaccae* NCTC 11659 and *M. vaccae* ATCC 15483, on stress-resilient behaviors and lipid-immune signaling in rats. Brain Behav Immun 91:212–229
249. Niesink RJ, Van Ree JM (1982) Short-term isolation increases social interactions of male rats: a parametric analysis. Physiol Behav 29:819–825
250. Drugan RC, Basile AS, Ha J-H et al (1997) Analysis of the importance of controllable versus uncontrollable stress on subsequent behavioral and physiological functioning. Brain Res Brain Res Protoc 2:69–74
251. Fleshner M, Maier SF, Lyons DM et al (2011) The neurobiology of the stress-resistant brain. Stress 14:498–502
252. Rozeske RR, Greenwood BN, Fleshner M et al (2011) Voluntary wheel running produces resistance to inescapable stress-induced potentiation of morphine conditioned place preference. Behav Brain Res 219:378–381
253. Frank MG, Fonken LK, Dolzani SD et al (2018) Immunization with *Mycobacterium vaccae* induces an anti-inflammatory milieu in the CNS: attenuation of stress-induced microglial priming, alarmins and anxiety-like behavior. Behav Brain Res 73:352–363
254. Siegmund A, Wotjak CT (2007) A mouse model of posttraumatic stress disorder that distinguishes between conditioned and sensitised fear. J Psychiatr Res 41:848–860
255. Golub Y, Mauch CP, Dahlhoff M et al (2009) Consequences of extinction training on associative and non-associative fear in a mouse model of Posttraumatic Stress Disorder (PTSD). Behav Brain Res 205:544–549
256. McNay E Sprague Dawley. https://www.albany.edu/mcnaylab/sd.html
257. Pan Y, Zhang W-Y, Xia X et al (2006) Effects of icariin on hypothalamic-pituitary-adrenal axis action and cytokine levels in stressed Sprague-Dawley rats. Biol Pharm Bull 29:2399–2403
258. Fediuc S, Campbell JE, Riddell MC (2006) Effect of voluntary wheel running on circadian corticosterone release and on HPA axis responsiveness to restraint stress in Sprague-Dawley rats. J Appl Physiol (1985) 100:1867–1875
259. Shi G, Ku B, Yao H (2007) Effects of jieyuwan on HPA axis and immune system in chronic stress models in rats. Zhongguo Zhong Yao Za Zhi 32:1551–1554
260. Knaepen L, Rayen I, Charlier TD et al (2013) Developmental fluoxetine exposure normalizes the long-term effects of maternal stress on post-operative pain in Sprague-Dawley rat offspring. PLoS One 8:e57608
261. Caruso M, McClintock M, Cavigelli S (2014) Temperament moderates the influence of periadolescent social experience on behavior and adrenocortical activity in adult male rats. Horm Behav 66:517–524
262. Gileta AF, Fitzpatrick CJ, Chitre AS et al (2021) Genetic characterization of outbred Sprague Dawley rats and utility for genome-wide association studies. bioRxiv 412924
263. White W, Lee C (1998) The development and maintenance of the Crl: CD (SD) IGS BR rat breeding system. Biol Ref Data CD (SD) IGS Rats 8–14
264. De Boer S, Van der Gugten J, Slangen J (1989) Plasma catecholamine and corticosterone responses to predictable and unpredictable noise stress in rats. Physiol Behav 45:789–795
265. Muscat R, Willner P (1992) Suppression of sucrose drinking by chronic mild unpredictable stress: a methodological analysis. Neurosci Biobehav Rev 16:507–517
266. Papp M, Muscat R, Willner P (1993) Subsensitivity to rewarding and locomotor stimulant

effects of a dopamine agonist following chronic mild stress. Psychopharmacology (Berl) 110:152–158

267. Gouirand AM, Matuszewich L (2005) The effects of chronic unpredictable stress on male rats in the water maze. Physiol Behav 86:21–31

268. Katsnelson A (2014) Male researchers stress out rodents. . Nature doi:10.1038/nature.2014.15106

269. Sorge RE, Martin LJ, Isbester KA et al (2014) Olfactory exposure to males, including men, causes stress and related analgesia in rodents. Nat Methods 11:629–632

270. Terashvili MN, Kozak KN, Gebremedhin D et al (2020) Effect of nearby construction activity on endothelial function, sensitivity to nitric oxide, and potassium channel activity in the middle cerebral arteries of rats. J Am Assoc Lab Anim Sci 59:411–422

271. Dallman MF, Akana SF, Bell ME et al (1999) Warning! Nearby construction can profoundly affect your experiments. Endocrine 11(2):111–113

272. Bitinas IA (1967) Aggression caused by withdrawal from morphine. [Doctoral dissertation, Western Michigan University], ProQuest Dissertations & Theses Global 1301396

273. Lal H (1967) Operant control of vocal responding in rats. Psychon Sci 8:35–36

274. Baker S, Bielajew C (2007) Influence of housing on the consequences of chronic mild stress in female rats. Stress 10:283–293

275. Azar T, Sharp J, Lawson D (2011) Heart rates of male and female Sprague–Dawley and spontaneously hypertensive rats housed singly or in groups. J Am Assoc Lab Anim Sci 50: 175–184

276. McGrath J, Drummond G, McLachlan E et al (2010) Guidelines for reporting experiments involving animals: the ARRIVE guidelines. Br J Pharmacol 160:1573–1576

277. Percie du Sert N, Hurst V, Ahluwalia A et al (2020) The ARRIVE guidelines 2.0: updated guidelines for reporting animal research. J Cereb Blood Flow Metab 40:1769–1777

278. Frank MG, Fonken LK, Watkins LR et al (2020) Acute stress induces chronic neuroinflammatory, microglial and behavioral priming: a role for potentiated NLRP3 inflammasome activation. Brain Behav Immun 89:32–42

279. Worley NB (2019) Prefrontal circuit selection in stress and resilience. [Doctoral dissertation, Boston College], ProQuest Dissertations & Theses Global 13885882

280. Horseman ND, Ehret CF (1982) Glucocorticosteroid injection is a circadian zeitgeber in the laboratory rat. Am J Phys 243:R373–R378

281. Xu R, Liu Z, Zhao Y (1991) A study on the circadian rhythm of glucocorticoid receptor. Neuroendocrinology 53:31–36

282. Deak T, Nguyen KT, Fleshner M et al (1999) Acute stress may facilitate recovery from a subcutaneous bacterial challenge. Neuroimmunomodulation 6:344–354

283. Stanojević S, Mitić K, Vujić V et al (2007) The influence of stress and methionine-enkephalin on macrophage functions in two inbred rat strains. Life Sci 80:901–909

284. Fleshner M, Campisi J, Deak T et al (2002) Acute stressor exposure facilitates innate immunity more in physically active than in sedentary rats. Am J Phys Regul Integr Comp Phys 282:R1680–R1686

285. Campisi J, Leem TH, Fleshner M (2002) Acute stress decreases inflammation at the site of infection: a role for nitric oxide. Physiol Behav 77:291–299

286. Campisi J, Leem TH, Greenwood BN et al (2003) Habitual physical activity facilitates stress-induced HSP72 induction in brain, peripheral, and immune tissues. Am J Physiol Regul Integr Comp Physiol 284:R520–R530

287. Gómez F, De Kloet ER, Armario A (1998) Glucocorticoid negative feedback on the HPA axis in five inbred rat strains. Am J Phys Regul Integr Comp Phys 274:R420–R427

288. Paré WP (1989) Stress ulcer susceptibility and depression in Wistar Kyoto (WKY) rats. Physiol Behav 46:993–998

289. Paré WP (1992) Learning behavior, escape behavior, and depression in an ulcer susceptible rat strain. Integr Physiol Behav Sci 27: 130–141

290. Machholz E, Mulder G, Ruiz C et al (2012) Manual restraint and common compound administration routes in mice and rats. J Vis Exp 67:e2771

Chapter 8

Apolipoprotein E Isoform-Related Translational Measures in PTSD Research

Eileen Ruth Samson Torres, Andrea E. DeBarber, and Jacob Raber

Abstract

Rodent models have been extensively used to study mechanisms underlying post-traumatic stress disorder (PTSD), a complex mental health disorder which can often be directly linked to a traumatic event or events. Previous research suggests that in humans *Apolipoprotein E* (*APOE*) genotype may influence the susceptibility to develop PTSD as well as ensuing symptom severity. Apolipoprotein E is of particular interest due to its association with cognitive health and its role in cholesterol transport and metabolism, which may underlie its association with PTSD. Here, we describe a liquid chromatography and gas chromatography method to measure levels of cholesterol and related metabolites in small amounts of mouse tissue. Specifically, we provide methods to measure cholesterol and related metabolites in the brains of targeted replacement mice that express human E2, E3, or E4 under the control of the murine apoE promotor that were either subjected to the chronic variable stress exposure model of PTSD or a control condition. Facilitating more researchers to measure sterols and oxysterols in rodent models of PTSD will likely increase understanding of cholesterol metabolism in PTSD and related disorders and hopefully aid in the development of novel therapeutic strategies for these conditions.

Key words Post-traumatic stress disorder, Apolipoprotein E, Oxysterols, 7-ketocholesterol

1 Introduction

Estimates suggest that upwards of 70% of the general population in the United States experience at least one traumatic event in their lifetimes, while many individuals experience more than three [1]. Such trauma exposure may result in trauma- and stressor-related disorders, most notably post-traumatic stress disorder (PTSD). While originally characterized in terms of its psychological burden, it cannot be overemphasized that biological research and the identified markers have demonstrated that PTSD truly affects an individual on a whole-body biological level (for review *see* Refs. [2, 3]). Those with PTSD are also highly likely to present with comorbid health disorders, including cardiovascular disease, diabetes, chronic pain, and other mental health disorders [4]. PTSD patients are also more likely to develop dementia, adding to the concerns of cognitive dysfunction, even after excluding potential confounding risk factors such as head injury, depression, and substance abuse [5, 6].

Given the symptoms seen in PTSD and that genetic risk factors influence susceptibility and symptom severity, *APOE* has become a gene of high interest in better understanding and treating PTSD. Apolipoprotein E (apoE) is involved in lipid transport and metabolism and is synthesized throughout the body, with the liver generating most apoE. In the healthy brain under physiological conditions, astrocytes primarily produce apoE [7], although neurons, microglia, oligodendrocytes, and ependymal cells have been demonstrated to generate apoE under distress [8–13]. Considering the high percentage of cholesterol found in the brain, apoE has a critical role in healthy brain function. Cholesterol is the precursor of steroid hormones, including those regulating stress, cortisol/corticosterone, which can be modulated by apoE [14]. ApoE knockout mice, for example, show age-dependent increases in anxiety-like behavior in the elevated plus maze and have higher plasma corticosterone levels after an acute restraint stress [14] as well as a greater acoustic startle response [15]. In addition, direct application of glucocorticoids increases expression of apoE in cultured macrophages [16], and apoE mRNA levels inversely correlate with steroidogenesis [17].

ApoE's potential role in PTSD might relate to recent theories that metabolic dysfunction underlies the disorder [2, 18]. In adult men, an acute, mental task (cognitive, not emotional and devised to be mildly stress-provoking) increased total serum cholesterol by 0.10 mmol/L, and this increase did not reverse during a brief recovery period [19]. In a prospective study, patients with PTSD showed significantly lower high-density lipoprotein cholesterol (HDL-C) and higher triglycerides at baseline compared to those that did not develop PTSD, suggesting that lipid profile at baseline may help determine who is at risk of developing PTSD [20]. In contrast, in a cross-sectional study, lipid profiles and stress hormones in PTSD patients and controls revealed a more favorable lipoprotein profile (i.e., lower LDL-C) in males and females with PTSD than in controls [21]. In other studies, there was a lack of association between a favorable lipoprotein profile and PTSD [22, 23]. Despite the conflicting studies thus far, this research suggests that cholesterol and related lipid metabolism may be predictive factors as well as altered in response to trauma.

In humans, apoE (E) is found in three major isoforms, E2, E3, and E4, with allele frequencies of 8.4%, 77.9%, and 13.7%, respectively [24]. These isoforms have been differentially associated with human disease risk, initially with cardiovascular disease risk [25, 26]. Subsequent to the cardiovascular disease risk, apoE isoforms have been shown to be strong predictors of age-related cognitive decline and Alzheimer's disease (AD): E4 is the strongest genetic risk factor for late-onset AD compared to E3 [27]. In comparison, E2 has been demonstrated to confer protection against Alzheimer's disease as compared to E3 [28]. To explore the association between *APOE* genotype and PTSD, Freeman et al. (2005) assessed 54 male Caucasian combat veterans and found a higher number of E2 homozygotes with PTSD than expected based on the Hardy-Weinberg equilibrium. E2 carriers in this study also showed higher CAPS-2 re-experiencing scores (i.e., more severe re-experiencing symptoms) compared to non-E2 carriers but there was no difference in avoidance or arousal symptom clusters [29]. This increase in susceptibility was supported in a later, larger study assessing Korean male combat veterans from the Vietnam War with and without PTSD [30]. In a separate cohort of male combat veterans from the Vietnam War era, those with E2 showed greater general symptom severity on the CAPS compared to E3 and E4 carriers [31]. However, the association between E2 and PTSD has been less distinct in other studies [29, 32], and in some studies, E4 appeared to be associated with worse outcomes after trauma exposure [33]. In many subsequent studies, E4 carriers versus non-carriers were specifically assessed, while ignoring E2 carriers altogether [34–39]. In some studies, there was no significant association between *APOE* genotype and PTSD incidence or severity [39, 40]. The association between PTSD and *APOE* genotype is complex and likely influenced by sex as well as ethnicity [28, 41–44]. These studies are summarized in Table 1.

Considering the likely role of cholesterol metabolism and the influence of *APOE* genotype in PTSD, it is imperative that we increase our understanding of cholesterol metabolism in response to environmental stressors. Importantly, cholesterol and apoE are unable to cross the blood-brain barrier (BBB). Thus, cerebral cholesterol synthesis as well as its metabolism is carefully regulated. Figure 1 depicts cholesterol synthesis and metabolism pathways of interest within the brain. In brief, cholesterol synthesis occurs via 3-hydroxy-3-methyl-glutaryl-coenzyme A reductase, primarily from astrocyte-secreted apoE-cholesterol complexes (lipoproteins). Cerebral cholesterol can be enzymatically and nonenzymatically converted into oxysterols, which can cross the blood-brain barrier (BBB) into circulation. The major oxysterols thought to be formed enzymatically from cholesterol are 24S-hydroxycholesterol and 27-hydroxycholesterol, which can be further metabolized within the brain. 7-ketocholesterol may be formed as an auto-oxidation product of cholesterol [45]. It can also be formed through enzymatic pathways [45], and there is evidence from human liver microsomes that 7-dehydrocholesterol can be converted to 7-ketocholesterol via P450 7A1 [46]. Previous work has shown the apoE isoforms may influence whole brain levels of oxysterols, including 7-ketocholesterol [47]. Oxysterols have been

Table 1
Studies assessing PTSD symptom severity and/or susceptibility depending apoE genotype, including studies published as recently as April 2, 2020

Article	Population	Sex (n)	PTSD	apoE	Findings
E2 is associated with greater PTSD severity or susceptibility					
Freeman et al. (2005)	Caucasian veterans	Male (54)	All	First E4+ vs. E4−, then E2+ vs. E2−	High rate of E2 homozygotes (16.7%), E2+ had higher reexperiencing scores
Kim et al. (2013)	Korean veterans	Male (256)	With and without	All genotypes	Greater # of E2 in PTSD group than non-PTSD, E2 results in lowered risk of substance use
Johnson et al. (2015)	Caucasian veterans	Male (104)	All	All genotypes	E2 had higher symptom severity (CAPS and PCL), no difference in prevalence
E4 is associated with greater PTSD severity or susceptibility					
Lyons et al. (2013)	Twin pairs, Vietnam veterans	Male (172)	With and without	E4+ vs. E4−	E4 associated with worse PTSD outcomes
Peterson et al. (2015)	Veterans	Male (53), Female (6)	With and without (49 with, 10 with subthreshold PTSD)	All genotypes, cysteine residues	Lower cysteine residue (i.e, E4 carriers) corresponded to greater symptom severity
Kimbrel et al. (2015)	Non-Hispanic White and Non-Hispanic Black Iran/Afghanistan veterans	Male (1291), Female (333)	With and without	E4+ vs. E4−	For those with high exposure, E4 increased susceptibility in non-Hispanic blacks but not whites
James et al. (2017)	US veterans	Male (309), Female (34)	With and without	All genotypes, cysteine residue	E2 genotype related to higher resilience
Mota et al. (2018)	Two cohorts from European American veterans	2011 cohort: Male (1260), Female (126) 2013 cohort: Male (457), Female (52)	With and without	E4+ vs. E4−	E4 carriers showed greater PTSD symptoms in main sample, but not in replicate sample

Study	Population	Sample	PTSD	Genotype	Findings
Merritt et al. (2018)	Military veterans	Male (106), Female (27)	With and without	E4+ vs. E4−	Trend for E4+ greater PCL score in those with TBI
Nielsen et al. (2019)	Veterans	Male, female (total = 87, breakdown NA)	With and without	E4+ vs. E4−, excluded all E2 carriers	Plasma apoE levels correlated to PTSD symptom severity, and E4+ corresponded to increased susceptibility
Huguenard et al. (2020)	Active duty soldiers	Male (120)	With and without	E4+ vs. E4−	E4 carriers showed significant interaction with PTSD diagnosis and lipid profile
APOE genotype is *not* associated with greater PTSD severity or susceptibility					
Yesavage et al. (2012)	Primarily not Hispanic or Latino, Vietnam veterans	Male (105)	All	E4+ vs. E4−	No difference in E4 prevalence or cognitive tests, PTSD severity not directly assessed
Dretsch et al. (2015)	Soldiers	Male (221), Female (9)	With and without	All genotypes	No differences in allele frequency compared to hardy-Weinberg, severity was not assessed
Hayes et al. (2018)	White, non-Hispanic US combat Iraq and Afghanistan veterans	Male (149), Female (11)	With and without	All genotypes	Focused primarily on BDNF and mTBI, did not directly test apoE and PTSD, but no differences in allele frequencies between those with and without PTSD PTSD
Averill et al. (2019)	Two cohorts: European American veterans	2011 cohort: Male (1260), Female (126) 2013 cohort: Male (457), Female (52)	With and without	E4+ vs. E4−	Assessed cognitive performance, did not assess PTSD severity or prevalence

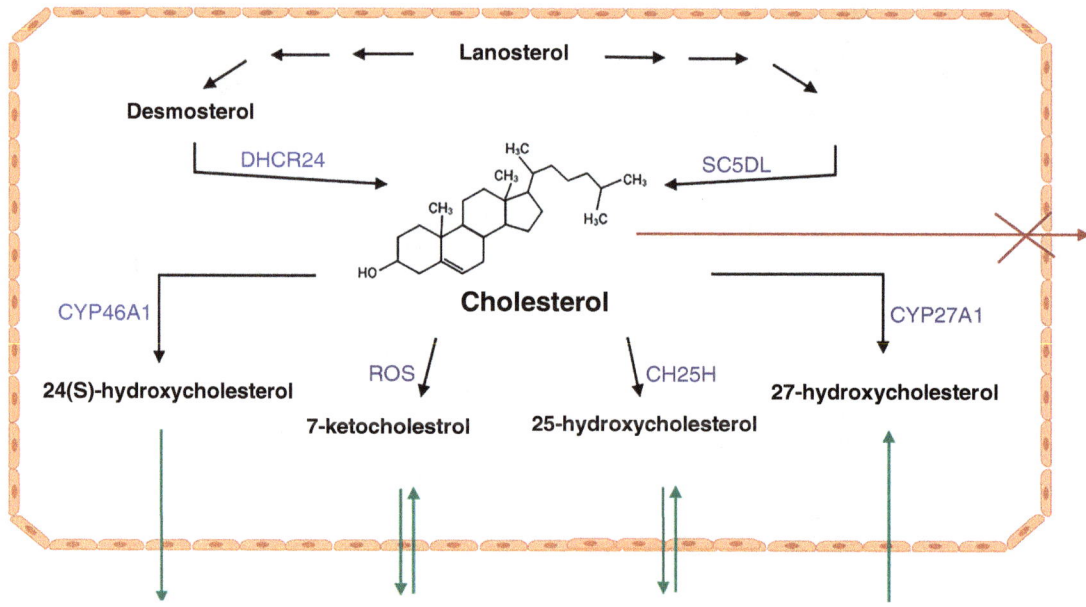

Fig. 1 Overview of cholesterol metabolism in the brain. Most cholesterol is metabolized from astrocyte-secreted apoE-cholesterol complexes. Cholesterol does not cross the BBB, although oxysterols do. Not shown here is the conversion of cholesterol to pregnenolone, the rate-liming step in steroid synthesis. (Created with BioRender.com)

demonstrated to exert a number of biological effects. For example, 7-ketocholesterol inhibits glucocorticoid action in adipocytes by reducing glucocorticoid receptor (GR) activation via substrate competition at the enzyme activity level and subsequently influences GR transcriptional activity [48]. 7-ketocholesterol can also serve as a ligand for oxysterol binding protein receptors, which attenuate glucocorticoid synthesis, in the adrenal gland [49]. However, brain and circulating cholesterol and metabolites such as oxysterols have been little studied in the context of PTSD.

In this chapter, we describe methods to sensitively measure cholesterol and related sterols and oxysterols in a PTSD mouse model expressing human apoE isoforms. As animal models such as this can recapitulate aspects of this complex human disorder, this methodology provides a way to assess biochemical changes that are associated with changes in stress-related behavior and cognition under environmentally controlled conditions.

2 Materials

2.1 Animal Model and Stressor Exposure

Mice do not possess multiple isoforms of apoE as humans do. Murine apoE is described to be most similar to E3, although there remain clear differences in binding patterns between mouse apoE and human E3 [50, 51]. In order to better understand the functional differences of human apoE isoforms, apoE mice that express human apoE isoforms via targeted gene replacement (TR) under the mouse apoE promoter were created by Dr. Patrick Sullivan [52–54] (available for experimental use from Taconic models 001547, 001548, 001549). Like in humans, apoE mRNA levels in apoE TR mice are similar across the different isoforms and also replicate the asymmetric protein levels with E2 > E3 > E4 [55, 56].

Researchers have induced PTSD-like changes in behavior and cognition in rodents using numerous methods including the Single Prolonged Stress (SPS) exposure model, in which rats experience a series of stressors—cold water swim, restraint stress, and induced unconsciousness by ether dosing—originally described by Drs. Liberzon and Young and resulting in a multitude of variations in the length and timing of stressors [57, 58]. This model includes a "sensitization" or "incubation" period in which animals are left untouched for seven days after the stress exposure day in order to facilitate memory of the event.

Similar to the SPS model is the chronic variable or unpredictable stress (CVS) paradigm. This model incorporates several of the stressors in the SPS exposure model in addition to others but extends exposure over 5+ days, depending on the protocol of each lab. During this time, rodents are exposed to one or more stressors each day at different times throughout the day. The perceived "randomness" and unpredictability ensure the animal will not habituate to the stressors, leading to long-term upregulation of the hypothalamic-pituitary-adrenal (HPA) axis. It has been used extensively in both depression and PTSD research [59]. We used this exposure paradigm to induce PTSD-like changes in male and female apoE TR mice based on previous research from our group [31]. All housing and experimental procedures were approved by the OHSU Institutional Animal Care and Use Committee (IACUC).

Exposure to CVS took place over the course of five days and included social deprivation (i.e., single housing), a 30° cage tilt for three h, a wet home cage for three h, overnight food deprivation, a three-min cold swim (10–12 °C), and a 15-minute restraint. Mice were exposed to two unique stressors at random times throughout each day. Stressors within the home cage (white noise, wet cage, cage tilt, and food deprivation) took place in the housing room while stressors that required mice be moved from the home cage (cold swim and restraint) occurred in the testing room adjacent to the housing room. The exact timing and order of specific stressors used in this study [60] are shown in Fig. 2.

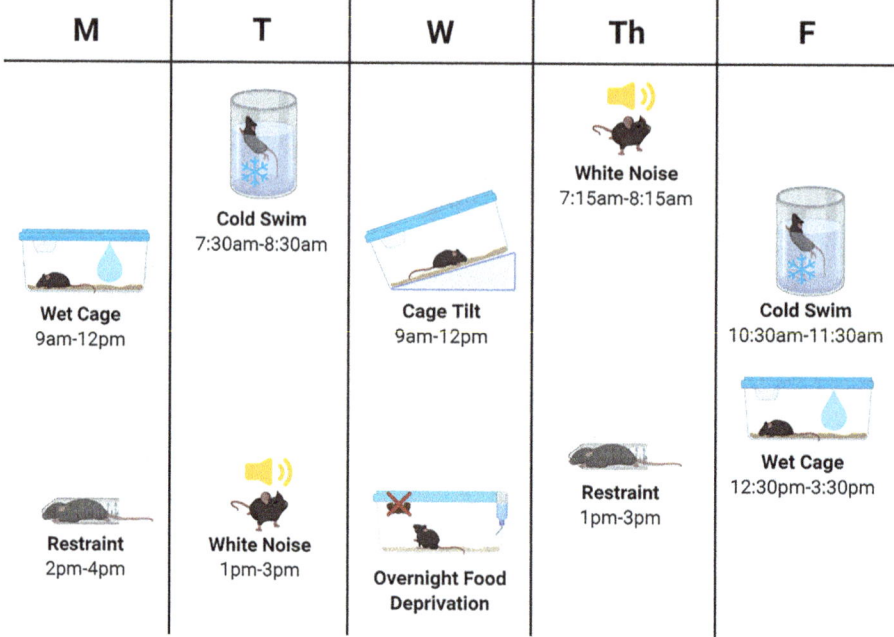

Fig. 2 Details of CVS schedule. (Created with BioRender.com)

2.2 Plasma and Tissue Collection

After mice were exposed to the CVS paradigm, they were assessed for behavioral and cognitive changes [60]. After the final behavioral test, the elevated zero maze, blood was collected from the mandibular vein in unanesthetized mice as per the approved IACUC protocol. Whole blood (~50–200 μL depending on the speed of blood flow and the time to apply pressure to stop bleeding) was collected with 0.5 M EDTA and centrifuged at 4000G for 10 min at 4 °C to obtain plasma.

Three hours after blood collection, mice were anesthetized with a lethal ketamine cocktail and then perfused with 1x phosphate-buffered saline (PBS). PBS perfusion was performed to remove blood from the brain for sterol and oxysterol analyses, described further below. Brain regions, including the frontal cortex, medial prefrontal cortex, and hippocampus [61, 62], along with the adrenal gland, liver, and kidney were dissected and flash frozen using liquid nitrogen. Tissues were kept at −80 °C until further use.

3 Methods

3.1 Oxysterol Measures in Brain Tissue and Plasma

3.1.1 Oxysterol Panel in Cortical Tissue

Total (free plus esterified) 7-ketocholesterol, 24S-hydroxycholesterol, 25-hydroxycholesterol, and 27-hydroxycholesterol were determined in homogenized samples by LC-MS/MS following sample saponification, extraction with hexane, and derivatization with N,N-dimethylglycine (DMG) [63, 64] as follows:

1. Cortical tissues (15 mg) were homogenized in 150 μL 1× PBS using homogenization beads.

2. Samples (75 μL) were spiked with 5 μL of an internal standard mixture containing 7-ketocholesterol-d7 1ng/μL, 25-hydroxycholesterol-d6 2 ng/μL, and 27-hydroxycholesterol-d6 20 ng/μL in methanol.

3. Standards were prepared in homogenization buffer. Saponification was accomplished by diluting the sample with 2 mL of ethanol followed by 0.120 mL of 33% KOH (w/w).

4. Samples were vortexed and then heated at 37 °C for 1 h. After saponification, each sample was diluted with 2 mL of water and extracted twice with 4 mL of hexane.

5. The combined hexane extracts were dried under vacuum. After drying, the tubes were rinsed with 0.4 mL of hexane and dried again.

6. The dried sample was treated with 25 μL mixture of DMG at 0.5 M and 4-(N,N-dimethylamino)pyridine at 1 M in chloroform and 25 μL 1-ethyl-3-(3-dimethylaminopropyl)carbodiimide at 1 M chloroform, then heated at 45 °C.

7. After one hour, 50 μL of methanol was added to deactivate the excess derivatizing agent. Samples were dried, suspended in 100 μL of methanol, vortexed, centrifuged, and filtered prior to analysis of 5 μL injection with LC-MS/MS.

8. DMG derivatives were analyzed using a 4000 QTRAP hybrid/triple quadrupole linear ion trap mass spectrometer (SCIEX, Framingham, MA) with electrospray ionization (ESI) in positive mode. The mass spectrometer was interfaced to a Shimadzu (Columbia, MD) SIL-20 AC XR auto-sampler followed by 2 LC-20 AD XR LC pumps. The instrument was operated in triple-quadrupole mode with the following settings: source voltage 4000 kV, GS1 40, GS2 30, CUR 40, TEM 500, and CAD gas medium.

9. Compounds transitions were quantified with multiple reaction monitoring (MRM) with peak retention times as described in Table 2.

10. Separation was achieved using an ACE Excel 3 μm C18-PFP 100 × 2.1 mm (ACE, part # EXL-1110-1002 U) column kept

Table 2
Oxysterol MRM transitions. Quantifying (quan) and qualifying (qual) ion information

Q1 mass	Q3 mass	Retention time (min)	Analyte	DP (V)	EP (V)	CE (V)	CXP (V)
488.4	385.4	4.4	24SOH-C-DMG quan[a]	81	10	23	12
488.4	367.4	4.4	24SOH-C-DMG qual[a]	81	10	29	12
488.4	104.1	4.7	27OH-C-DMG quan	101	10	39	20
488.4	58.2	4.7	27OH-C-DMG qual	101	10	91	10
494.4	58.2	4.7	27OH-C-d6-DMG quan	101	10	91	10
488.4	367.4	4.0	25OH-C-DMG quan	81	10	29	12
488.4	385.4	4.0	25OH-C-DMG qual	81	10	23	12
494.4	373.4	4.0	25OH-C-d6-DMG quan	81	10	29	12
486.4	58.2	7.2	7-KC-DMG quan	76	10	87	10
486.4	104.1	7.2	7-KC-DMG qual	76	10	37	18
493.4	390.2	7.2	7-KC-d7-DMG quan	76	10	27	12

Values also reported in Torres et al. [60]
[a]Note that 25OH-C-d6-DMG quan internal standard was used for quantification of 24SOH-C-DMG quan

at 18 °C using a Shimadzu CTO-20 AC column oven. The gradient mobile phase was delivered at a flow rate of 0.4 mL/min in 0–6.5 min, 0.8 mL/min in 6.6–10 min, and 0.4 mL/min in 10.1–12 min and consisted of two solvents: A: 0.1% formic acid, 2 mM ammonium acetate in water:methanol at 95:5 v/v, B: 0.1% formic acid, 2 mM ammonium acetate in methanol:acetonitrile at 10:90 v/v. The initial concentration of solvent B was 55%, followed by a linear increase to 70% B in 3 min, then to 100% B in 2.5 min, held for 4 min, decreased back to starting 55% B over 0.1 min, and then held for 2.5 min.

11. Data were acquired using Analyst software and analyzed with Multiquant software (SCIEX, Framingham, MA). Sample values were calculated from standard curves generated from the peak area ratio of the analyte to internal standard versus the analyte concentration that was fit to a linear equation with $1/x$ weighting. The analytical measurement range was 5–1000 ng/mL homogenate.

3.1.2 7-Ketocholesterol Analysis in Liver Tissue and Plasma

After determining that the 7-ketocholesterol analyte was of primary interest from the cortical tissue results, liver tissue samples and plasma samples were assayed for free 7-ketocholesterol levels using a simplified LC-MS/MS method.

1. Liver samples were homogenized (20 s) and briefly sonicated (5 s) using approximately 15 mg of tissue and 500 μL of 1× PBS. Plasma samples were assayed undiluted.
2. Internal standard solution (5 μL, at 0.01 ng/μL 7-ketochol-d7 in ethanol) was added to 50 μL calibrant or unknown tissue sample homogenate/plasma.
3. Calibrants were generated by spiking 45 μL 1× PBS with 5 μL 7-ketocholesterol methanol solution for final concentrations from 1–100 ng/mL.
4. Acetonitrile (250 μL) was added to calibrants/samples, which were then vortexed for 10 s to precipitate any protein present and centrifuged using a microfuge for 10 min at 14,000 rpm.
5. Supernatant were transferred to disposable 13 × 100 mm glass test tubes and dried using a speed vacuum.
6. Amplifex Keto Reagent® (AB Sciex) was mixed 1:1 with Amplifex Keto diluent provided by the vendor and then was diluted 1:4 with 5% acetic acid in methanol.
7. After vortexing, samples were then filtered with centrifugal PVDF 0.22 μm filters and transferred to autosampler vials to react at room temperature for two days.
8. QAO-7-ketocholesterol derivatives were analyzed using a 4000 Q-TRAP hybrid/triple quadrupole linear ion trap mass spectrometer (SCIEX) with electrospray ionization (ESI) in positive mode. The mass spectrometer was interfaced to a Shimadzu (Columbia, MD) SIL-20 AC XR auto-sampler followed by 2 LC-20 AD XR LC pumps. The instrument was operated with the following settings: source voltage 4500 kV, GS1 50, GS2 50, CUR 20, TEM 550, and CAD gas medium. Compounds were quantified with MRM and transitions were optimized by infusion of pure derivatized compounds as presented in Table 3.
9. Separation was achieved using a Gemini 3 μ C6-phenyl 110 Å, 100 × 2 mm column (Phenomenex) kept at 35 °C using a Shimadzu CTO-20 AC column oven. The gradient mobile phase was delivered at a flow rate of 0.5 mL/min and consisted of two solvents: A: 0.1% formic acid in water and B: 0.1%

Table 3
Oxysterol MRM transitions. Quantifying (quan) and qualifying (qual) ion information

Q1 mass	Q3 mass	Analyte	DP (V)	EP (V)	CE (V)	CXP (V)
515.5	58.8	QAO-7-ketocholesterol quan	106	10	99	8
515.5	456.3	QAO-7-ketocholesterol qual	106	10	43	12
522.5	463.4	QAO-d7-ketocholesterol quan	61	10	45	14

formic acid in acetonitrile. The initial concentration of solvent B was 20%, followed by a linear increase to 60% B in 10 min, then to 95% B in 0.1 min, held for 3 min, decreased back to starting 20% B over 0.1 min, and then held for 4 min. The retention time for 7-ketocholesterol was 8.46 min.

10. Data were acquired using Analyst 1.6.2 and analyzed with Multiquant 3.0.1 software.

11. Sample values were calculated from standard curves generated from the peak area ratio of the analyte to internal standard versus the analyte concentration that was fit to a linear equation with $1/x$ weighting. The lower limit of quantification was 1 ng/mL with an accuracy of 102% and precision (relative standard deviation) of 8.5% and the signal to noise (S/N) was 19:1. At a concentration of 100 ng/mL the accuracy was 98% and precision was 0.5% with a S/N of 24:1.

3.2 Sterol Analysis in Brain Tissue

3.2.1 Hippocampal Cholesterol Analyses

1. Gas chromatography-flame ionization detection (GC-FID) was used to measure cholesterol. Hippocampal tissue (~12 mg) was homogenized in 150 μL 1× PBS using homogenization beads.

2. Lipids were then extracted from hippocampal homogenates (0.5 mL) by chloroform: methanol (2:1) extraction (10 mL). The lower layer was dried under nitrogen and residue reconstituted in hexane (2 mL) and 0.1 mL dried for cholesterol analysis.

3. For each sample, 50 μL 0.2 mg/mL EPIC solution in n-propyl alcohol was added followed by 2 mL of ethyl alcohol and then 120 μL of potassium hydroxide (33% w/v). Samples were securely capped and vortexed briefly. They were placed in a heating block set to 37 °C to incubate for 1 h.

4. After incubation, samples were removed from the heating block and uncapped to add 2 mL HPLC grade water. Hexane (4 mL) was then added. Samples were vortexed for 20 s to allow the organic layer to separate. This organic layer was then transferred to a second labeled 15 mL tube.

5. An additional 4 mL of hexane was added to the samples, which were then vortexed for 20 s. The organic layer was then added to the second 15 mL tube.
   *Note that sample preparation may be stopped at this point and samples can be stored in hexane extraction tubes overnight at 4 °C.

6. The organic layer was dried at 40 °C under a nitrogen flow and then rinsed by adding 2 mL of hexane to the sides of the tube.
   *Note that sample preparation may be stopped at this point and samples can be stored in hexane extraction tubes overnight at 4 °C.

7. Again, the organic layer was dried at 40 °C under nitrogen flow. Afterward, 50–70 μL BSTFA reagent was added, and samples were vortexed briefly. Samples were then capped and incubated at 80 °C for 30 min.

   *Note that the sealed ampoule of BSTFA should be carefully opened in the hood. Store any remaining BSFTA at 4 °C, cap tightly, and use within 14 days from the open date.

8. The BSTFA reagent was then dried at 80 °C and samples were cooled down for 1 min.

9. Hexane (100 μL) was added to sample tubes, which were then vortexed and transferred immediately into labeled GC vials.

   *To prevent solvent evaporation, add hexane to three sample tubes at a time and transfer to GC vials with 200 μL inserts.

10. For each batch, a blank sample was prepared by adding hexane to 2 mL auto-sampler amber tubes.

11. Cholesterol concentration was measured by capillary column gas chromatography on an Agilent (Santa Clara, CA) gas chromatograph (Model 6890 N) with a ZB1701 column (30 m, 0.25 mm ID, 0.25 μm film; Phenomenex, Torrance, CA) and a FID detector. An internal standard (epicoprostanol; Sigma, St. Louis, MO) and an authentic cholesterol standard (Steraloids, Newport, RI) were used for calibration.

3.2.2 Sterol Panel in Hippocampal Tissue

1. Gas chromatography-mass spectrometry (GC-MS) was used to measure cholestanol, desmosterol, and lanosterol. For hippocampal sterol analyses, hippocampal tissues from one hemisphere were homogenized. Tissue (~12 mg) was homogenized in 1× PBS (150 μL) using homogenization beads.

2. Lipids were then extracted from hippocampal homogenates (0.5 mL) by chloroform:methanol (2:1) extraction (10 mL). The lower layer was dried under nitrogen and residue reconstituted in hexane (2 mL) and 1.9 mL dried for sterol analysis.

3. All samples were then treated as described in Subheading 3.2.1, **steps 3–10** *except* 50 μL 0.004 mg/mL EPIC solution in n-propyl alcohol was used instead. (Note the difference in concentration.)

4. Concentrations of the trimethylsilyl ether derivatives of sterols were measured using GC performed with a ZB1701 column (Phenomenex, Torrance, CA) coupled to a mass spectrometer (Agilent GC 6890 N and MS 5975; Santa Clara, CA).

5. Mass spectra were collected in selected ion mode with m/z = 355 and 370 ions monitored for epicoprostanol internal standard (quantifying and qualifying ions, respectively), m/z = 393.2 and 498.2 ions for lanosterol, m/z = 343.3

and 441.5 ions for desmosterol, and m/z = 458.5 and 255.3 ions for lathosterol.

6. Calibrants were generated using authentic standards (cholestanol, desmosterol, and lathosterol from Avanti Polar Lipids, Alabaster, AL).

7. Analyte concentrations were calculated across the range 0.04–3.2 mg/dL using calibration curves generated by performing a least-squares linear regression for peak area ratios plotted against specified calibrant concentration. The lower limit of quantification was determined as the lowest spiked concentration in the matrix for which the signal-to-noise ratio was ≥5. The between-run precision of the assay was determined to be <20% relative standard deviation. Note: Due to method sensitivity and sample amount limitations, sterols were assessed only in male hippocampal tissues and oxysterols were only analyzed in female cortical tissues.

3.3 Results from Sterol and Oxysterol Analyses

To examine the effect of apoE genotype and the interaction of CVS on cholesterol metabolism, we assessed cholesterol as well as different sterols (cholestanol, desmosterol, and lathosterol) and oxysterols (24S-hydroxycholesterol, 25-hydroxycholetserol, and 27 hydroxycholesterol) in brain tissue of these mice (*see* Figs. 3 and 4). This was done to better understand part of the changes in cholesterol metabolism associated with CVS exposure. Sterols were assessed only in male hippocampal tissues and oxysterols were only analyzed in female cortical tissues due to technical limitations. Of these sterols and oxysterols, 7-ketocholesterol was the only one significantly affected by CVS exposure (*see* Fig. 3a). A two-way ANOVA revealed a significant main effect of CVS exposure in which CVS corresponded to higher 7-ketocholesterol cortical levels ($F_{1,26} = 6.53$, $p = 0.017$). There was also an interaction between genotype and group ($F_{2,26} = 4.72$, $p = 0.018$), which when evaluated using Sidak's multiple comparisons showed that E4 mice exposed to CVS had higher 7-ketocholesterol levels than their genotype-matched controls ($p = 0.013$). This CVS-related difference was absent in E2 and E3 mice ($p = 0.11$).

This striking genotype × CVS interaction led us to explore 7-ketocholesterol concentrations in the liver, a major organ for cholesterol metabolism and for the generation of apoE, and in the plasma. Exposure to CVS corresponded to higher levels of free 7-ketocholesterol in the liver, regardless of genotype or sex (*see* Fig. 3b. $F_{1,36} = 16.13$, $p < 0.001$). This effect of CVS exposure was not seen in plasma samples; however, E2 mice regardless of sex or CVS exposure showed greater levels of 7-ketcholesterol compared to E3 and E4 mice (*see* Fig. 3c. $F_{2,34} = 24.42$, $p < 0.001$). There were no significant effects of sex or CVS exposure on plasma levels of free 7-ketcholesterol.

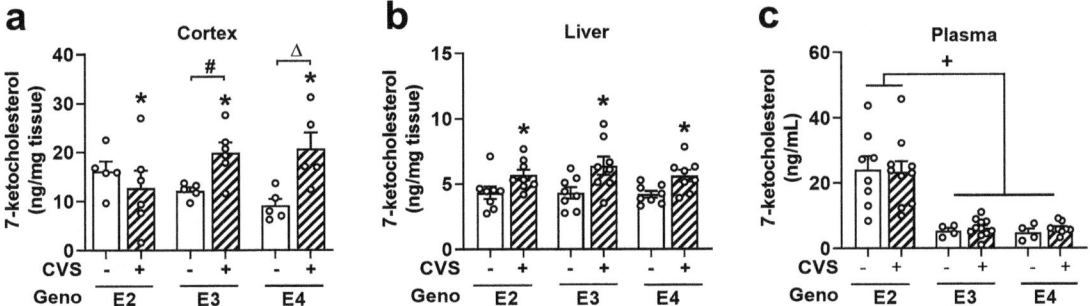

Fig. 3 Levels of 7-ketocholesterol throughout the body. (**a**). Female cortical tissue showed a genotype × group interaction in which only E3 and E4 mice exposed to CVS showed higher levels of 7-ketocholesterol compared to controls ($p < 0.05$). CVS groups demonstrated higher 7-ketocholesterol concentrations (*$p < 0.05$). E4 mice exposed to CVS had higher 7-ketocholesterol levels than genotype-matched controls ($^\Delta p = 0.013$, Sidak's multiple comparisons), which was absent in E2 mice ($p = 0.72$) and E3 mice ($^\# p = 0.11$). (**b**) CVS exposure was associated with higher 7-ketocholesterol levels (*$p < 0.05$) regardless of genotype or sex (shown collapsed). (**c**) Plasma levels were highest in E2 mice (+$p < 0.05$, shown with sexes collapsed)

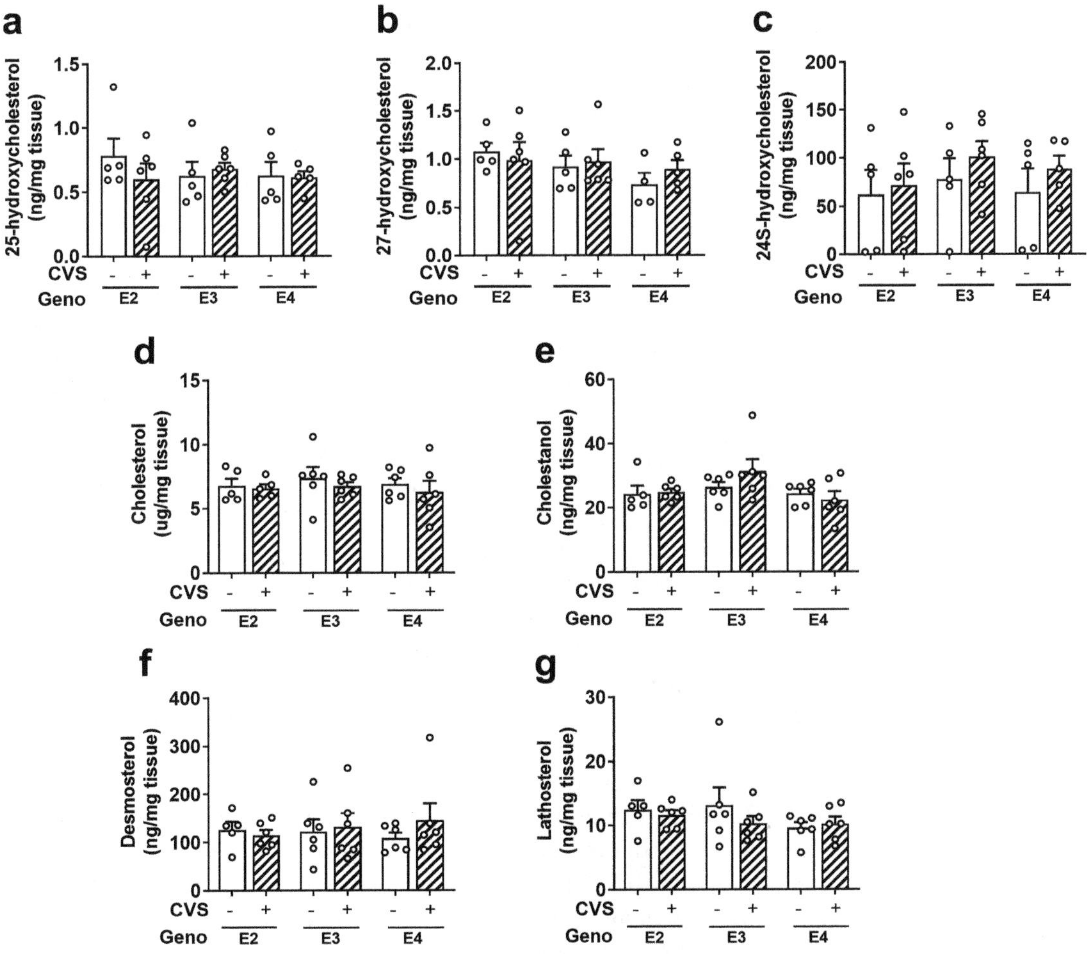

Fig. 4 Additional sterol and oxysterol measures. No significant differences due to genotype or CVS exposure were found in oxysterols in cortical tissue (**a**–**c**) nor in sterols in hippocampal tissue (**d**–**g**)

4 Notes and Conclusion

The human apoE TR mouse models provide valuable information about the differential roles of human apoE isoforms in human disorders [65]. However, there are certain limitations regarding the use of these mouse models that should be kept in mind. For example, based on our data and those of others [54], E2 mice have far higher plasma cholesterol levels than E3 and E4 mice. These differences in cholesterol are paradoxical to what is observed in humans. In fact, human E2 carriers show lower plasma cholesterol levels and lowered risk of developing atherosclerosis unless they have an additional health condition such as diabetes or obesity resulting in Type III hyperlipidemia [66], suggesting that another apolipoprotein or metabolic agent might be dissimilar enough in the mice to cause this discrepancy.

It is important to note that the CVS paradigm has been used to induce both depression- and PTSD-like behavioral changes in rodents. The individual stressors can vary widely. They are largely straightforward to incorporate with very little specialized equipment needed. However, this customizability does require experimenters to take extra care between groups to ensure that the timing of exposures and other environmental conditions remain consistent, especially if the experimental group needs to be tested in separate cohorts. Careful consideration should also be taken when deciding housing conditions. In the study described here, mice were pair housed with a litter mate for the duration of the study except when exposed to the CVS paradigm during which mice were singly housed. Social housing was chosen to avoid potentially worsening the effects of CVS and reaching a ceiling effect. Ros-Simó and Valverde have shown that social isolation in young (three-week-old) CD1 male mice results in greater anxiety-like behavior, as well as hyperactivity and greater increases in plasma corticosterone levels after a stressor compared to mice that were group housed [67]. Thus, pair housing after CVS may have acted as "community support" and mitigated some behavioral effects of CVS exposure. Previous studies show that social buffering can occur when social support lessens the effects of stressors [68]. More recent work shows evidence that if the conspecific spectator is fearful, there will be increased responses in the naïve-exposed animal to the stressor, but if both animals are exposed to the same stressor, then the shared experience results in attenuated fear-related behaviors [69].

Recent studies in humans support the role of oxysterols in mental health disorders [70, 71] although there is much to be still understood. In parallel, there is mounting evidence for their importance in aging [72] as well as neurodegenerative diseases such as AD. For example, brain 7-ketocholesterol levels increased in correspondence with disease progression in postmortem frontal cortical tissue of AD patients [73]. The effects associated between 7-ketocholesterol and brain disorders may be attributed to the pro-inflammatory effects of 7-ketocholesterol. In a mouse model of age-related macular degeneration, 7-ketocholesterol was readily internalized by retinal microglia and 7-ketocholesterol application led to microglial activation [74]. Considering the purported role of neuroinflammation in PTSD [75], studying oxysterols such as 7-ketocholesterol becomes more pressing.

Measuring sterols and oxysterols in rodent tissue requires equipment and expertise not always available within a laboratory or research center. Our experience with institutional bioanalytical shared resource core facilities at OHSU proved invaluable in providing the technical expertise to develop measurements of these sterols and oxysterols in mouse plasma and tissues. From our initial sterol and oxysterol panels, we identified 7-ketocholesterol as altered uniquely in an apoE isoform by CVS exposure interaction that we plan to explore further.

Methods are available to measure oxysterols in small tissue amounts and even specifically in organelles such as mitochondria [76]. With the technology rapidly expanding our ability to assess lipids in small tissues, the knowledge of how lipid metabolism is associated with health and disease will hopefully grow in parallel. Along with measuring oxysterol with methods such as those described in this chapter, we would like to point out that genetic mouse models also exist to explore the roles of oxysterols [45]. Ultimately, the knowledge generated with these methods and models will help elucidate novel pathways and thus potential with identification of novel therapeutic targets in disorders such as PTSD.

Acknowledgments

This work was partially supported by the OHSU Dean's Fund (ERT), T32DA007262-25 (ERT), OHSU Core Pilot Grant (AD, JR), DOD grant W81XWH-17-1-0193 (JR), NIH grant RF1 AG059088 (JR), and JR's development account. We acknowledge the OHSU Bioanalytical Shared Resource core facility for providing technical assistance and access to analytical instrumentation.

References

1. Benjet C et al (2016) The epidemiology of traumatic event exposure worldwide: results from the world mental health survey consortium. Psychol Med 46:327–343. https://doi.org/10.1017/S0033291715001981
2. Levine AB, Levine LM, Levine TB (2014) Posttraumatic stress disorder and cardiometabolic disease. Cardiology 127:1–19. https://doi.org/10.1159/000354910
3. Pitman RK et al (2012) Biological studies of post-traumatic stress disorder. Nat Rev Neurosci 13:769–787. https://doi.org/10.1038/nrn3339

4. Boscarino JA (2004) Posttraumatic stress disorder and physical illness: results from clinical and epidemiologic studies. Ann N Y Acad Sci 1032:141–153. https://doi.org/10.1196/annals.1314.011

5. Flatt JD, Gilsanz P, Quesenberry CP Jr, Albers KB, Whitmer RA (2018) Post-traumatic stress disorder and risk of dementia among members of a health care delivery system. Alzheimers Dement 14:28–34. https://doi.org/10.1016/j.jalz.2017.04.014

6. Yaffe K et al (2010) Posttraumatic stress disorder and risk of dementia among US veterans. Arch Gen Psychiatry 67:608–613. https://doi.org/10.1001/archgenpsychiatry.2010.61

7. Pitas RE, Boyles JK, Lee SH, Foss D, Mahley RW (1987) Astrocytes synthesize apolipoprotein E and metabolize apolipoprotein E-containing lipoproteins. Biochim Biophys Acta 917:148–161. https://doi.org/10.1016/0005-2760(87)90295-5

8. Boyles JK, Pitas RE, Wilson E, Mahley RW, Taylor JM (1985) Apolipoprotein E associated with astrocytic glia of the central nervous system and with nonmyelinating glia of the peripheral nervous system. J Clin Invest 76:1501–1513. https://doi.org/10.1172/jci112130

9. Xu P-T et al (1999) Sialylated human apolipoprotein E (apoEs) is preferentially associated with neuron-enriched cultures from APOE transgenic mice. Neurobiol Dis 6:63–75. https://doi.org/10.1006/nbdi.1998.0213

10. Xu Q et al (2006) Profile and regulation of apolipoprotein E (ApoE) expression in the CNS in mice with targeting of green fluorescent protein gene to the ApoE locus. J Neurosci 26:4985–4994. https://doi.org/10.1523/jneurosci.5476-05.2006

11. Zhao N, Liu CC, Qiao W, Bu G (2018) Apolipoprotein E, receptors, and modulation of Alzheimer's disease. Biol Psychiatry 83:347–357. https://doi.org/10.1016/j.biopsych.2017.03.003

12. Stoll G, Meuller HW, Trapp BD, Griffin JW (1989) Oligodendrocytes but not astrocytes express apolipoprotein E after injury of rat optic nerve. Glia 2:170–176. https://doi.org/10.1002/glia.440020306

13. Poirier J, Hess M, May PC, Finch CE (1991) Astrocytic apolipoprotein E mRNA and GFAP mRNA in hippocampus after entorhinal cortex lesioning. Brain Res Mol Brain Res 11:97–106. https://doi.org/10.1016/0169-328x(91)90111-a

14. Raber J et al (2000) Hypothalamic-pituitary-adrenal dysfunction in Apoe(−/−) mice: possible role in behavioral and metabolic alterations. J Neurosci 20:2064–2071

15. Raber J (2007) Role of apolipoprotein E in anxiety. Neural Plast 2007:91236–91236. https://doi.org/10.1155/2007/91236

16. Trusca VG et al (2017) Differential action of glucocorticoids on apolipoprotein E gene expression in macrophages and hepatocytes. PLoS One 12:e0174078. https://doi.org/10.1371/journal.pone.0174078

17. Nicosia M, Prack MM, Williams DL (1992) Differential regulation of apolipoprotein-E messenger RNA in zona fasciculata cells of rat adrenal gland determined by in situ hybridization. Mol Endocrinol (Baltimore, Md.) 6:288–298. https://doi.org/10.1210/mend.6.2.1373819

18. Michopoulos V, Vester A, Neigh G (2016) Posttraumatic stress disorder: a metabolic disorder in disguise? Exp Neurol 284:220–229. https://doi.org/10.1016/j.expneurol.2016.05.038

19. Muldoon MF et al (1992) Acute cholesterol responses to mental stress and change in posture. Arch Intern Med 152:775–780

20. Hamazaki K et al (2014) The role of high-density lipoprotein cholesterol in risk for posttraumatic stress disorder: taking a nutritional approach toward universal prevention. Eur Psychiatry 29:408–413. https://doi.org/10.1016/j.eurpsy.2014.05.002

21. Vries GJ, Mocking R, Assies J, Schene A, Olff M (2017) Plasma lipoproteins in posttraumatic stress disorder patients compared to healthy controls and their associations with the HPA- and HPT-axis. Psychoneuroendocrinology 86:209–217. https://doi.org/10.1016/j.psyneuen.2017.09.020

22. Dennis PA et al (2014) Behavioral health mediators of the link between posttraumatic stress disorder and dyslipidemia. J Psychosomatic Res 77:45–50. https://doi.org/10.1016/j.jpsychores.2014.05.001

23. Gill JM, Saligan L, Lee H, Rotolo S, Szanton S (2013) Women in recovery from PTSD have similar inflammation and quality of life as non-traumatized controls. J Psychosomatic Res 74:301–306. https://doi.org/10.1016/j.jpsychores.2012.10.013

24. Liu C-C, Kanekiyo T, Xu H, Bu G (2013) Apolipoprotein E and Alzheimer disease: risk, mechanisms and therapy. Nat Rev Neurol 9:106–118

25. Eichner JE et al (1993) Relation of apolipoprotein E phenotype to myocardial infarction and mortality from coronary artery disease. Am J

26. Wilson PWF, Schaefer EJ, Larson MG, Ordovas JM (1996) Apolipoprotein E alleles and risk of coronary disease. Arterioscler Thromb Vasc Biol 16:1250–1255. https://doi.org/10.1161/01.ATV.16.10.1250

27. Strittmatter WJ, Roses AD (1996) Apolipoprotein E and Alzheimer's disease. Annu Rev Neurosci 19:53–77. https://doi.org/10.1146/annurev.ne.19.030196.000413

28. Farrer LA et al (1997) Effects of age, sex, and ethnicity on the association between apolipoprotein E genotype and Alzheimer disease. A meta-analysis APOE and Alzheimer Disease Meta Analysis Consortium. JAMA 278:1349–1356

29. Freeman T, Roca V, Guggenheim F, Kimbrell T, Griffin WS (2005) Neuropsychiatric associations of apolipoprotein E alleles in subjects with combat-related posttraumatic stress disorder. J Neuropsychiatry Clin Neurosci 17:541–543. https://doi.org/10.1176/jnp.17.4.541

30. Kim TY et al (2013) Apolipoprotein E gene polymorphism, alcohol use, and their interactions in combat-related posttraumatic stress disorder. Depress Anxiety 30:1194–1201. https://doi.org/10.1002/da.22138

31. Johnson LA et al (2015) ApoE2 exaggerates PTSD-related behavioral, cognitive, and neuroendocrine alterations. Neuropsychopharmacology 40:2443–2453. https://doi.org/10.1038/npp.2015.95

32. Peterson CK, James LM, Anders SL, Engdahl BE, Georgopoulos AP (2015) The number of cysteine residues per mole in apolipoprotein E is associated with the severity of PTSD re-experiencing symptoms. J Neuropsychiatry Clin Neurosci 27:157–161. https://doi.org/10.1176/appi.neuropsych.13090205

33. Lyons MJ et al (2013) Gene-environment interaction of ApoE genotype and combat exposure on PTSD. Am J Med Genet B Neuropsychiatry Genet 162b:762–769. https://doi.org/10.1002/ajmg.b.32154

34. Emmerich T et al (2015) Plasma lipidomic profiling in a military population of mTBI and PTSD with APOE epsilon4 dependent effect. J Neurotrauma 33:1331. https://doi.org/10.1089/neu.2015.4061

35. Mota NP et al (2018) Apolipoprotein E gene polymorphism, trauma burden, and posttraumatic stress symptoms in U.S. military veterans: results from the National Health and resilience in veterans study. Depress Anxiety 35:168–177. https://doi.org/10.1002/da.22698

36. Nielsen DA, Spellicy CJ, Harding MJ, Graham DP (2019) Apolipoprotein E DNA methylation and posttraumatic stress disorder are associated with plasma ApoE level: a preliminary study. Behav Brain Res 356:415–422. https://doi.org/10.1016/j.bbr.2018.05.013

37. Averill LA et al (2019) Apolipoprotein E gene polymorphism, posttraumatic stress disorder, and cognitive function in older U.S. veterans: results from the National Health and resilience in veterans study. Depress Anxiety 36:834–845. https://doi.org/10.1002/da.22912

38. Hayes JP et al (2017) Mild traumatic brain injury is associated with reduced cortical thickness in those at risk for Alzheimer's disease. Brain J Neurol 140:813–825. https://doi.org/10.1093/brain/aww344

39. Yesavage JA et al (2012) Sleep-disordered breathing in Vietnam veterans with posttraumatic stress disorder. Am J Geriatr Psychiatry 20:199–204. https://doi.org/10.1097/JGP.0b013e3181e446ea

40. Dretsch MN et al (2016) Brain-derived neurotropic factor polymorphisms, traumatic stress, mild traumatic brain injury, and combat exposure contribute to postdeployment traumatic stress. Brain Behav 6:e00392. https://doi.org/10.1002/brb3.392

41. Marini S et al (2019) Association of apolipoprotein E with intracerebral hemorrhage risk by race/ethnicity: a meta-analysis. JAMA Neurol 76:480–491. https://doi.org/10.1001/jamaneurol.2018.4519

42. Rajan KB et al (2017) Racial differences in the association between apolipoprotein E risk alleles and overall and total cardiovascular mortality over 18 years. J Am Geriatr Soc 65:2425–2430. https://doi.org/10.1111/jgs.15059

43. Pole N, Best SR, Metzler T, Marmar CR (2005) Why are Hispanics at greater risk for PTSD? Cult Divers Ethn Minor Psychol 11:144–161. https://doi.org/10.1037/1099-9809.11.2.144

44. Kimbrel NA et al (2015) Effect of the APOE epsilon4 allele and combat exposure on ptsd among Iraq/Afghanistan-era veterans. Depress Anxiety 32:307–315. https://doi.org/10.1002/da.22348

45. Mutemberezi V, Guillemot-Legris O, Muccioli GG (2016) Oxysterols: from cholesterol metabolites to key mediators. Prog Lipid Res 64:152–169. https://doi.org/10.1016/j.plipres.2016.09.002

46. Shinkyo R et al (2011) Conversion of 7-dehydrocholesterol to 7-ketocholesterol is

catalyzed by human cytochrome P450 7A1 and occurs by direct oxidation without an epoxide intermediate. J Biol Chem 286:33021–33028. https://doi.org/10.1074/jbc.M111.282434

47. Jenner AM et al (2010) The effect of APOE genotype on brain levels of oxysterols in young and old human APOE epsilon2, epsilon3 and epsilon4 knock-in mice. Neuroscience 169: 109–115. https://doi.org/10.1016/j.neuroscience.2010.04.026

48. Wamil M et al (2008) 7-oxysterols modulate glucocorticoid activity in adipocytes through competition for 11β-hydroxysteroid dehydrogenase type. Endocrinology 149:5909–5918. https://doi.org/10.1210/en.2008-0420

49. Escajadillo T, Wang H, Li L, Li D, Sewer MB (2016) Oxysterol-related-binding-protein related protein-2 (ORP2) regulates cortisol biosynthesis and cholesterol homeostasis. Mol Cell Endocrinol 427:73–85. https://doi.org/10.1016/j.mce.2016.03.006

50. Nguyen D et al (2014) Influence of domain stability on the properties of human apolipoprotein E3 and E4 and mouse apolipoprotein E. Biochemistry 53:4025–4033. https://doi.org/10.1021/bi500340z

51. Raffai RL, Dong LM, Farese RV Jr, Weisgraber KH (2001) Introduction of human apolipoprotein E4 'domain interaction' into mouse apolipoprotein E. Proc Natl Acad Sci U S A 98:11587–11591. https://doi.org/10.1073/pnas.201279298

52. Sullivan PM et al (1997) Targeted replacement of the mouse apolipoprotein E gene with the common human APOE3 allele enhances diet-induced hypercholesterolemia and atherosclerosis. J Biol Chem 272:17972–17980

53. Sullivan PM, Mezdour H, Quarfordt SH, Maeda N (1998) Type III hyperlipoproteinemia and spontaneous atherosclerosis in mice resulting from gene replacement of mouse Apoe with human Apoe*2. J Clin Invest 102: 130–135. https://doi.org/10.1172/JCI2673

54. Sullivan PM, Mace BE, Maeda N, Schmechel DE (2004) Marked regional differences of brain human apolipoprotein E expression in targeted replacement mice. Neuroscience 124: 725–733. https://doi.org/10.1016/j.neuroscience.2003.10.011

55. Riddell DR et al (2008) Impact of apolipoprotein E (ApoE) polymorphism on brain ApoE levels. J Neurosci 28:11445–11453. https://doi.org/10.1523/jneurosci.1972-08.2008

56. Bales KR et al (2009) Human APOE isoform-dependent effects on brain β-amyloid levels in PDAPP transgenic mice. J Neurosci 29:6771–6779. https://doi.org/10.1523/jneurosci.0887-09.2009

57. Liberzon I, Young EA (1997) Effects of stress and glucocorticoids on CNS oxytocin receptor binding. Psychoneuroendocrinology 22:411–422. https://doi.org/10.1016/s0306-4530(97)00045-0

58. Yamamoto S et al (2009) Single prolonged stress: toward an animal model of posttraumatic stress disorder. Depress Anxiety 26: 1110–1117. https://doi.org/10.1002/da.20629

59. Goswami S, Rodriguez-Sierra O, Cascardi M, Pare D (2013) Animal models of post-traumatic stress disorder: face validity. Front Neurosci 7:89. https://doi.org/10.3389/fnins.2013.00089

60. Torres ERS et al (2022) Apolipoprotein E isoform-specific changes related to stress and trauma exposure. Transl Psychiatry 12:125. https://doi.org/10.1038/s41398-022-01848-7

61. Pflibsen L et al (2015) Executive function deficits and glutamatergic protein alterations in a progressive 1-methyl-4-phenyl-1,2,3,6-tetrahydropyridine mouse model of Parkinson's disease. J Neurosci Res 93:1849–1864. https://doi.org/10.1002/jnr.23638

62. Raber J et al (2019) Combined effects of three high-energy charged particle beams important for space flight on brain, behavioral and cognitive endpoints in B6D2F1 female and male mice. Front Physiol 10:179. https://doi.org/10.3389/fphys.2019.00179

63. Jiang X, Ory DS, Han X (2007) Characterization of oxysterols by electrospray ionization tandem mass spectrometry after one-step derivatization with dimethylglycine. Rapid Commun Mass Spectrom 21:141–152. https://doi.org/10.1002/rcm.2820

64. Pataj Z, Liebisch G, Schmitz G, Matysik S (2016) Quantification of oxysterols in human plasma and red blood cells by liquid chromatography high-resolution tandem mass spectrometry. J Chromatogr A 1439:82–88. https://doi.org/10.1016/j.chroma.2015.11.015

65. Balu D et al (2019) The role of APOE in transgenic mouse models of AD. Neurosci Lett 707:134285. https://doi.org/10.1016/j.neulet.2019.134285

66. Mahley RW, Rall SC Jr (2000) Apolipoprotein E: far more than a lipid transport protein. Annu Rev Genomics Hum Genet 1:507–537. https://doi.org/10.1146/annurev.genom.1.1.507

67. Ros-Simo C, Valverde O (2012) Early-life social experiences in mice affect emotional behaviour and hypothalamic-pituitary-adrenal axis function. Pharmacol Biochem Behav 102:434–441. https://doi.org/10.1016/j.pbb.2012.06.001
68. Davitz JR, Mason DJ (1955) Socially facilitated reduction of a fear response in rats. J Comp Physiol Psychol 48:149–151. https://doi.org/10.1037/h0046411
69. Lee H, Noh J (2016) Pair exposure with conspecific during fear conditioning induces the link between freezing and passive avoidance behaviors in rats. Neurosci Res 108:40–45. https://doi.org/10.1016/j.neures.2016.01.005
70. Sun Z et al (2021) Brain-specific oxysterols and risk of schizophrenia in clinical high-risk subjects and patients with schizophrenia. Front Psych 12:711734. https://doi.org/10.3389/fpsyt.2021.711734
71. Freemantle E, Chen GG, Cruceanu C, Mechawar N, Turecki G (2013) Analysis of oxysterols and cholesterol in prefrontal cortex of suicides. Int. J. Neuropsychopharmacol 16:1241–1249. https://doi.org/10.1017/s1461145712001587
72. Zarrouk A et al (2014) Involvement of oxysterols in age-related diseases and ageing processes. Ageing Research Reviews 18:148–162. https://doi.org/10.1016/j.arr.2014.09.006
73. Testa G et al (2016) Changes in brain oxysterols at different stages of Alzheimer's disease: their involvement in neuroinflammation. Redox Biol 10:24–33. https://doi.org/10.1016/j.redox.2016.09.001
74. Indaram M et al (2015) 7-Ketocholesterol increases retinal microglial migration, activation, and angiogenicity: a potential pathogenic mechanism underlying age-related macular degeneration. Sci Rep 5:9144. https://doi.org/10.1038/srep09144
75. Zass LJ, Hart SA, Seedat S, Hemmings SM, Malan-Müller S (2017) Neuroinflammatory genes associated with post-traumatic stress disorder: implications for comorbidity. Psychiatr Genet 27:1–16. https://doi.org/10.1097/ypg.0000000000000143
76. Borah K et al (2020) A quantitative LC-MS/MS method for analysis of mitochondrial-specific oxysterol metabolism. Redox Biol 36:101595. https://doi.org/10.1016/j.redox.2020.101595

Chapter 9

Stress Immobilization Inducing Fear Extinction Deficits in Male and Female Mice

Eric Raul Velasco, Antonio Florido, Ignacio Javier Marin-Blasco, Patricia Molina, Laura Perez-Caballero, and Raul Andero

Abstract

Fear extinction alterations are a core symptom of post-traumatic stress disorder(PTSD). Traditionally, animal models of fear extinction mainly use male subjects. Recently, more studies on fear extinction are including sex as a biological variable and thus also test females. We describe here how to induce impairments in fear extinction in both male and female mice by previously exposing them to acute stress immobilization.

Key words Acute stress, Immobilization, Fear extinction, Sex as a biological variable, Animal model

1 Introduction

Exposure to acute traumatic stressors such as car accidents or sexual abuse is an important environmental factor modifying brain function. Most individuals exposed to trauma recover while others have long-term trauma-related symptoms such as fear extinction deficits [1]. In this regard, many studies have highlighted the importance of biological factors like genes and hormones [2] whereas others emphasize the role of environmental factors such as trauma severity [3]. In this regard, being a woman is one of the main sociodemographic risk factors associated with post-traumatic stress disorder (PTSD) [3], and it is still uknown why they present a higher prevalence of this disorder. There is raising awareness for the need to include sex as a biological variable [4]. However, when modeling traumatic stress in animals, most studies use male rodents.

The main advantages of immobilization stress (IMO) as an animal model of traumatic stress is its high face, predictive, and construct validity in both male and female mice [5, 6]. IMO evokes the highest hypothalamic-pituitary-adrenal (HPA) response in comparison with other models such as exposure to restraint in a tube, cat urine, or intense footshocks [7–9]. Moreover, IMO has negative consequences on depressive-like and anxiety-like behavior [6, 10], and fear extinction [11].

The study of fear extinction in both humans and animal models is a useful tool to investigate intermediate phenotypes of PTSD [12]. In mice, this procedure consists of pairing a neutral stimulus (e.g., tone or light), or context (e.g., experimental box), with an unconditioned stimulus (e.g., mild electric footshocks). After this pairing, the neutral stimulus becomes a conditioned stimulus (CS) eliciting fear responses, like freezing [13] (see *Gafford* et al. for more details on fear conditioning and fear extinction procedures in mice [14]). For studies with female mice that require monitoring the internal hormonal states, the vaginal smear cytology is a valid and reliable technique. In comparison to the measurement of vaginal impedance, this procedure does not require complex equipment and it is more accurate than the inspection of the vaginal aperture [15]. It is necessary to acknowledge the potential limitations when translating fear extinction findings from rodents into humans, since there are specific procedures for each species and there is a much bigger cortex in humans [16].

We have established an animal model of impaired fear extinction by submitting adult wild-type mice previously exposed to acute IMO to an auditory-cued fear conditioning paradigm [10, 11, 17]. After IMO, mice are kept undisturbed in the vivarium and trained 6 days later in cued-fear conditioning. Twenty-four hours later animals are exposed to the same experimental box but changing cues such as odors and lighting to avoid contextual conditioning. In this fear extinction session, mice receive 15 or 30 CS presentations. This fear extinction session is then repeated for three more days or until animals extinguish their fear response to the CS. Our results show that animals previously exposed to IMO present impaired fear extinction across these different sessions (*see* Fig. 1 for a graphical representation of a typical experiment).

2 Equipment Materials and Setup

2.1 Mice

Use adult (8–12 weeks old) wild-type C57BL/6 J male and female mice (Charles River). Mice must be housed in groups with constant environmental conditions, for example, temperature (22 ± 1 °C) and humidity (~ 40%) in a 12-h light/12-h dark schedule (lights on at 8:00 h), with ad libitum food and water intake. Behavioral procedures should start early in the light phase. Avoid external stressors, monitor for fights in the home cage, and minimize manipulations. Ensure proper home cage conditions before IMO stress procedures. Once mice undergo IMO, they must be left undisturbed in their home cages. Note that freezing rates during fear conditioning, fear extinction, and other behavioral tests may

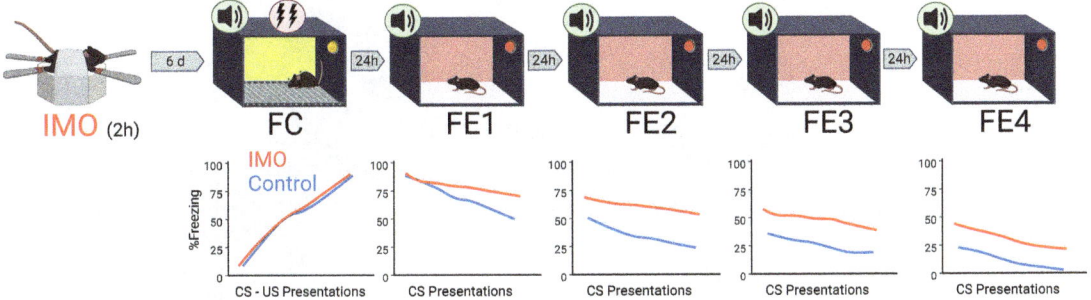

Fig. 1 Expected freezing levels in auditory-cued fear conditioning in mice previously exposed to acute IMO. Control animals have compensatory handling the same day the IMO group is exposed to stress. *IMO* acute stress immobilization, *FC* fear conditioning, *FE* fear extinction, *CS* conditioned stimulus (tone), *US* unconditioned stimulus (mild electric footshocks)

vary according to the mouse strain used [18–21]. The materials and procedures described here have been developed and established in adult male and female mice, but if younger mice are used, smaller IMO boards can be built, adjusting the rest of the parameters.

2.2 Stress Immobilization Boards

1. To build an IMO board (Fig. 2), use an acrylonitrile butadiene styrene (ABS) plastic block cut in an octagonal shape with the dimensions shown in Fig. 3a and 5 cm of height. Edges and borders must be rounded off.

2. On the top side, drill four holes (x) as shown in Fig. 3a. On the bottom side, drill two holes (z) as shown in Fig. 3b.

3. Attach four metal arms (preferably made of stainless steel) of dimensions shown in Fig. 3c to the holes at the top (x) using 4× flathead 6 mm screws and 4× 12 mm metal washers (Fig. 3d, e).

4. Attach a ~4 cm wide hook-and-loop fastener (Fig. 3f) (e.g., Velcro®) to the two bottom holes (z) using 2× flathead screws and 2× 12 mm metal washers.

5. Stick four adhesive rubber legs (Fig. 3g) to the bottom side for stabilization.

6. Use duct tape (e.g., American Tape®) for animal immobilization.

2.3 Vaginal Cytology

1. Use adult cycling female mice (~P40 first estrous cycle; ~9–12 months reproductive senescence) [22, 23].

2. Solutions: Fresh NaCl 0.9%, 0.1% cresyl violet acetate, distilled water.

3. Materials: Pipette with 10 ul capacity and appropriate pipette tips, paper towel, microscopy adhesion slides, coverslips, and DPX mounting medium (Sigma).

4. Equipment: Brightfield microscope with 10×–20× objectives, slide warmer (optional).

Fig. 2 IMO board after assembly

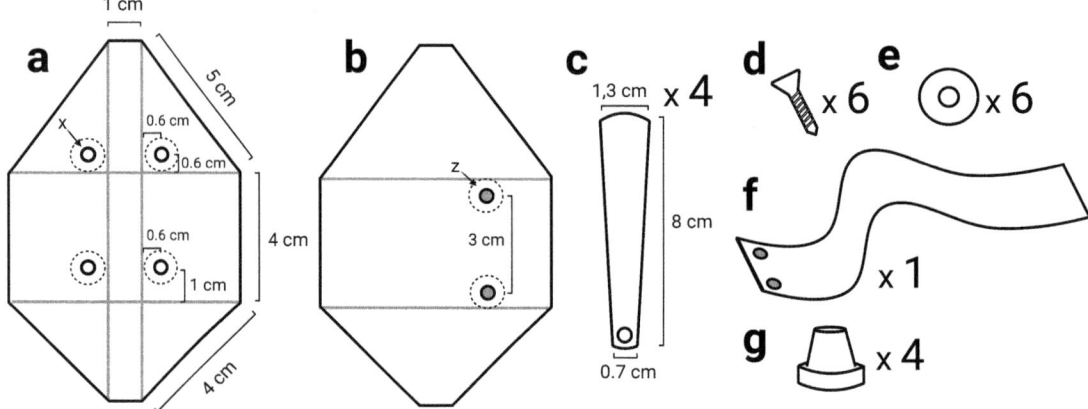

Fig. 3 Detailed graphical representation of an IMO board. (**a**) Topside, (**b**) bottom side, (**c**) metal arms, (**d**) flathead screws, (**e**) metal washers, (**f**) hook-and-loop fastener, and (**g**) adhesive rubber leg

2.4 Cued-Fear Conditioning and Fear Extinction

1. Fear chambers and software: Fear conditioning (FC) and fear extinction (FE) procedures are carried out using a StartFear system (Panlab-Harvard, Barcelona, Spain). The fear chamber consists of a black methacrylate box with a transparent front door (25 × 25 × 25 cm) inside a sound-attenuating chamber (67 × 53 × 55 cm). The same boxes are used for FC and FE. These chambers are equipped with a speaker for delivering white noise or tones, a fan, and a lightbulb for visual cues. Footshocks are used as the unconditioned stimuli, and they are administered through a grid floor coupled with a shocker. Animals' behavior is recorded with an analog pinhole camera suitable for low illumination conditions. Tones and shocks are delivered and controlled using Freezing software (Panlab-Harvard, Barcelona, Spain).

2. Stimuli: A 30 s, 6 kHz, 75 dB tone is used as the Conditioned Stimulus (CS), which co-terminates with a footshock of 1 s, 0.3 mA, the unconditioned stimulus (US).
3. Context: Different contexts are used for FC and FE procedures. Habituation and FC context consist of a yellow light source (~10 lux), a grid floor of 25 bars (3 mm Ø and 10 mm between bars), the background noise of 60 dB produced by a ventilation fan, and a solution of 70% ethanol as odor. FE context consists of a red-light source (~10 lux), a gray plexiglass floor covering the bars, no background noise, and CR36 (bronopol 0.26%, benzalkonium chloride 0.08%, and isopropyl alcohol 41%) (José Collado, Barcelona, Spain) as odor.
4. Transportation: White methacrylate boxes (30 × 15 × 5 cm; with a transparent removable top lid) are needed for animal transportation from the vivarium to the testing rooms.
5. Materials to score freezing: Freezing behavior, defined as the absence of any movement except respiration, is scored by a high-sensitivity weight transducer system located at the bottom of the experimental chambers. The system records and analyzes the signal generated by the movement of the mice. An episode of freezing is scored using Freezing software (Panlab-Harvard, Barcelona, Spain). As an alternative, videos can be recorded and stored for further manual or automated analyses when necessary.

3 Methods

3.1 Stress Immobilization Procedure

The immobilization procedure involves fixing the four limbs of the mouse in the prone position on the plain board with adhesive tape.

1. Immobilization board preparation: Cut four strips of duct tape approximately 5 cm long and 2 cm broad. It is recommended to stick one end on itself to facilitate its removal once the immobilization session has finished. Then, stick the strips to the bottom of each metal arm (*see* Fig. 4a for further details).
2. Take the animal from the home cage by gently grasping the tail at the base with your dominant hand and place it on the wire bar lid or a rough surface. Restrain the mice by placing your non-dominant hand in the trunk and advancing your first and second fingers rostrally to firmly grasp the mouse by the scruff of the neck. The mouse must be correctly restrained so that its movements do not interfere with the procedure.
3. Place the mouse on top of the board and carefully fix each forepaw with tape (*see* Fig. 4a). When both forepaws are tied, grasp the animal by the base of the tail and fix each of the hind

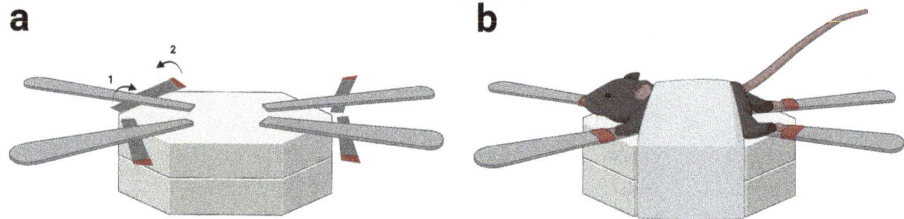

Fig. 4 Graphical representation of the stress immobilization procedure. (**a**) Preparation of the immobilization board with tape strips on each metal arm. (**b**) Mouse placed in the stress immobilization board

paws by gently pulling and taping them so the animal is slightly stretched. Note that the metal surfaces and the paws must be completely dry to ensure the sticking of the adhesive tape.

4. Finally, gently surround the animal with the hook-and-loop fastener and leave it in position for 2 h (Fig. 4b).

5. To release, remove the hook-and-loop fastener and carefully untie the hind paws. Then, grasp the animal by the tail and untie the forepaws. Animals are returned to their home cage until further testing.

3.2 Vaginal Cytology

1. Timing: Perform the vaginal cytology early in the morning. Obtain vaginal cytologies 8–12 days before testing to ensure estrous cycle regularity. If behavioral tests are carried out on the same day, collect vaginal cytologies at least 2 h before starting behavioral procedures.

2. Restraint technique: Use your non-dominant hand to restrain the mouse by the scruff on the neck, turn it to a supine position, and hold its tail between your fourth and fifth fingers. Clean any secretions or urine in the vaginal opening.

3. Vaginal cytology: With your dominant hand, load 10 μL of NaCl 0.9% and insert the pipette tip into the vaginal opening (max 2–3 mm). Flush gently five times ensuring that all the fluid is collected in the last flush. Spread the solution into an adhesion slide and return the animal to its home cage. Each slide can be used to accurately determine the cycle of up to eight females. To avoid complications during the following steps, make sure there are no debris of urine or feces before collecting the sample. A poor performance during the collection will cause low cellularity during the reading.

4. Slide staining: Let the slides dry completely at room temperature or use a hot plate at 37 °C. For staining, soak the slides for 5 min in 1% cresyl violet acetate and then rinse twice in clean distilled water. Dry at room temperature.

5. Slide visualization and interpretation: Use a brightfield microscope at 10× or 20× to visualize cells. In-depth instructions to determine the estrous cycle are covered elsewhere [24]. In short, determine the proportion of leucocytes, cornified and nucleated cells to determine the estrous cycle phase. In the proestrus stage, nucleated cells predominate with few leukocytes or cornified cells. Estrus consists mostly of cornified cells. Metestrus shows predominantly cornified cells mixed with leukocytes, whereas in diestrus leukocytes are predominant.

6. Slide storage: To preserve slides, cover them with DPX mounting medium (Sigma) or equivalent, and store at room temperature.

3.3 Cued-Fear Conditioning and Fear Extinction Procedures

1. Preparation: Ensure proper animal identification with either ear punches or tail marks. Calibrate and test the equipment before starting the experiment (e.g., shocker, tone calibration, and video recording). All behavioral procedures are performed at the same time each day within the animal's light cycle.

2. Estrous cycle monitorization: If the estrous cycle is going to be monitored, start taking vaginal cytologies 8–12 days before behavioral testing to ensure cycle regularity. Perform vaginal cytologies at the same time every day and at least 2 h before any behavioral test.

3. Habituation: To minimize anxiety related to transportation or the fear conditioning chamber, mice must be habituated to the fear chambers for 5 min/day in at least two consecutive days. Habituation must precede IMO stress. Transport the animals individually in a methacrylate box and use the same transportation route and contextual cues as the ones used for the fear conditioning day. Clean the chamber and grid floor with watery soap between each animal.

4. IMO: Immobilize animals 24 h after habituation, and 2 h after taking vaginal cytologies. Use a separate room, different from the fear conditioning room. After the procedure, animals must be left undisturbed in their home cages.

5. Fear conditioning: After 6 days, transport animals to the fear chambers individually. Allow for 5 min of habituation to the chamber, and then present five trials of a tone (30 s, 6 kHz, 75 dB) that co-terminate with a footshock (1 s, 0.3 mA). Use intertrial intervals (ITI) of 3 min, with 3 additional min following the last trial. Different stimulus intensities and duration may be adapted according to the experimental hypotheses [14]. Return animals to their home cage until further testing. Clean the chamber and grid floor with watery soap between each animal.

6. Fear extinction: After 24 h, transport animals to the fear chambers individually using a different route and contextual cues than the ones used for fear conditioning day. For each fear extinction test, start with 5 min of habituation to the chamber, and then deliver 15 trials of the tone alone (30 s) with a 30 s ITI, with an additional 30 s following the last trial. Return animals to their home cage until further testing. Clean the chamber and grid floor with watery soap between each animal.

7. Repeat fear extinction tests every 24 h according to the protocol (four fear extinction sessions) or until animals reach baseline freezing levels.

8. Freezing episodes are scored when a mouse remains immobile for more than 1 s, or adapt as necessary, using Freezing software (Panlab-Harvard, Barcelona, Spain). Extract freezing data and prepare it for analysis by averaging freezing time into 30-s slots.

4 Data Analysis

Repeated measures ANOVA tests are used to analyze fear conditioning and fear extinction sessions. For fear conditioning, freezing during CS presentations (CS1, CS2, CS3, CS4, and CS5) is used as the within-subjects factor and experimental groups (control vs. IMO) and/or estrous cycle (proestrus vs. metestrus) as between-subjects factor. For fear extinction, freezing levels across time and between extinction sessions can be analyzed by introducing mean freezing in fear extinction sessions (FE1, FE2, FE3, and FE4) as the within-subjects factor and experimental groups (control vs. IMO) and/or estrous cycle (proestrus vs. metestrus) as between-subjects factor. However, freezing performance during each fear extinction session may also be analyzed by introducing the average freezing for 5 CS presentations (CS 1–5, CS 6–10, CS 11–15) as the within-subjects factor and the desired between-subjects factor. In some protocols where fear extinction is not evident, it may be necessary to compare freezing levels between early (CS 1–5) and late (CS 11–15) trials.

5 Troubleshooting

Environmental factors play a major role in the modulation of the consequences of exposure to a stressful event. We suggest grouping the housed animals whether or not they have been stressed. Animals that have been under stressful conditions may recover easily in the presence of other animals with intact social abilities with whom they are familiar [25]. Further, environmental enrichment of the home cages has shown to increase basal adrenal function as well as to facilitate the extinction of novel light stressors such as handling [26]. Moreover, the manipulation of the animal has shown to have a relieving effect on animals when the handling comes from a familiar source. As an example, two-week handling increases resiliency in the forced swimming test [27]. Environmental conditions might mitigate or enhance the consequences of stress exposure, so they must be taken into account when performing a study.

Stressed mice present similar fear of extinction to controls. There could be multiple reasons for this issue. It is key to make sure that control mice are not exposed to any kind of stressor during the procedure which may impair their fear of extinction, such as changes in the temperature or humidity or unusual noise in the vivarium. It is also crucial to ensure that IMO is performed step by step as indicated in the methods. Any slight variation of the methods described here could lead to different behavioral results.

Mice release themselves from the stress immobilization board. Mice should not be able to release themselves from the stress procedure. However, they release when the limbs are not properly taped onto the metal attachments. It is critical to test several tape brands before starting the experiment to find out which one holds the animals better to the immobilization board.

Troubleshooting with estrous cycles. Animals should be checked for correct or regular cycling before starting the experiment. Different factors such as pseudopregnancy, the age of the animal, or stress may alter the normal cycling of the animals. Some of these issues may be addressed by simple changes in the housing of the animal, for example, by including males in the room where the females are housed or by reducing environmental stress to facilitate correct cycling.

Troubleshooting for vaginal smear cytology. Make sure to use clean reagents and materials (e.g., fresh 0.9% NaCl, slides) and remove urine and detritus from the vaginal opening. High cellularity in the sample can be avoided by spreading evenly the vaginal lavage, whereas low cellularity may arise from not letting the slides dry and losing cells in the staining. Moreover, double-check the positioning of the pipette tip into the vaginal opening. Dirty staining may be fixed by filtering the cresyl violet solution and using clean distilled water to wash the slides.

6 Concluding Remarks

Exposure to IMO may recapitulate impairments in fear extinction that are also observed in people suffering from mental disorders with fear dysregulations such as PTSD. These impairments in fear extinction are consistently observed in male and female mice exposed to IMO, making it an interesting model to study the neurobiology of traumatic stress. There is an increasing interest to include sex as a biological variable when approaching research questions related to fear learning, memory, and fear-related disorders. In this regard, IMO stress and fear conditioning procedures can be easily combined with vaginal smear cytology, thereby allowing to approach research questions investigating the mechanisms of sex hormones on stress processing. These approaches can be further combined with direct hormonal measurements, genetic and chemogenetic techniques, pharmacological challenges, and environmental manipulations which may inform better about plausible interventions for patients suffering from mental disorders with fear dysregulation.

Acknowledgments

RA was supported by ERANET-Neuron JTC 2019 ISCIII AC19/00077, Fundacion Koplowitz, Beca Leonardo BBVA, RETOS-MINECO PID2020-112705RB-I00 funded by MCIN/AEI/10.13039/501100011033 and by "ERDF, A way of making Europe". ERV was supported by the BES-2017-080870 FPI-2017 fellowship from MINECO. AF was a recipient of the Generalitat de Catalunya predoctoral fellowship (2018 FI_B00030). IMB received support from ERANET-Neuron JTC 2019 ISCIII AC19/00077. PM received support from FPU16/05416 from Ministerio de Educación, Cultura y Deporte (MECD). LPC was supported by an FJC2018-037958-I from Ministerio de Ciencia, Innovación y Universidades (MCIyU).

References

1. Maren S, Holmes A (2016) Stress and fear extinction. Neuropsychopharmacology 41:58–79
2. Ressler KJ et al (2011) Post-traumatic stress disorder is associated with PACAP and the PAC1 receptor. Nature 470:492–497
3. Tortella-Feliu M et al (2019) Risk factors for posttraumatic stress disorder: an umbrella review of systematic reviews and meta-analyses. Neurosci Biobehav Rev 107:154–165
4. Shansky RM, Murphy AZ (2021) Considering sex as a biological variable will require a global shift in science culture. Nat Neurosci 24:457–464
5. Belzung C, Lemoine M (2011) Criteria of validity for animal models of psychiatric disorders: focus on anxiety disorders and depression. Biol Mood Anxiety Disord 1
6. Wingo AP et al (2018) Expression of the PPM1F gene is regulated by stress and associated with anxiety and depression. Biol Psychiatry 83:284–295
7. Márquez C, Belda X, Armario A (2002) Post-stress recovery of pituitary-adrenal hormones and glucose, but not the response during exposure to the stressor, is a marker of stress intensity in highly stressful situations. Brain Res 926:181–185
8. Campmany L, Pol O, Armario A (1996) The effects of two chronic intermittent stressors on brain monoamines. Pharmacol Biochem Behav 53:517–523
9. Muñoz-Abellán C, Andero R, Nadal R, Armario A (2008) Marked dissociation between hypothalamic-pituitary-adrenal activation and long-term behavioral effects in rats exposed to immobilization or cat odor. Psychoneuroendocrinology 33:1139–1150

10. Andero R et al (2013) Amygdala-dependent fear is regulated by Oprl1 in mice and humans with PTSD. Sci Transl Med 5:188ra73
11. Andero R et al (2011) Effect of 7,8-dihydroxyflavone, a small-molecule TrkB agonist, on emotional learning. Am J Psychiatry 168:163–172
12. Maeng LY, Milad MR (2017) Post-traumatic stress disorder: the relationship between the fear response and chronic stress. Chronic Stress 1
13. Myers KM, Davis M (2002) Behavioral and neural analysis of extinction. Neuron 36:567–584
14. Gafford GM, Ressler KJ (2011) Fear conditioning and extinction as a model of PTSD in mice. NeuroMethods 63:171–184
15. Florido A et al (2021) Direct and indirect measurements of sex hormones in rodents during fear conditioning. Curr Protoc 1:e102
16. Flores Á, Fullana M, Soriano-Mas C, Andero R (2018) Lost in translation: how to upgrade fear memory research. Mol Psychiatry 23:2122–2132
17. Andero R, Dias BG, Ressler KJ (2014) A role for Tac2, NkB, and Nk3 receptor in normal and dysregulated fear memory consolidation. Neuron 83:444–454
18. Holmes A, Wrenn CC, Harris AP, Thayer KE, Crawley JN (2002) Behavioral profiles of inbred strains on novel olfactory, spatial and emotional tests for reference memory in mice. Genes Brain Behav 1:55–69
19. Siegmund A, Langnaese K, Wotjak CT (2005) Differences in extinction of conditioned fear in C57BL/6 substrains are unrelated to expression of α-synuclein. Behav Brain Res 157:291–298
20. Camp MC et al (2012) Genetic strain differences in learned fear inhibition associated with variation in neuroendocrine, autonomic, and amygdala dendritic phenotypes. Neuropsychopharmacology 37:1534–1547
21. Temme SJ, Bell RZ, Pahumi R, Murphy GG (2014) Comparison of inbred mouse substrains reveals segregation of maladaptive fear phenotypes. Front Behav Neurosci 0:282
22. Koebele SV, Bimonte-Nelson HA (2016) Modeling menopause: the utility of rodents in translational behavioral endocrinology research. Maturitas 87:5–17
23. Schroeder A, Notaras M, Du X, Hill RA (2018) On the developmental timing of stress: delineating sex-specific effects of stress across development on adult behavior. Brain Sci 8
24. Caligioni CS (2009) Assessing reproductive status/stages in mice. Curr Protoc Neurosci. Appendix 4, Appendix 4I
25. Kikusui T, Winslow JT, Mori Y (2006) Social buffering: relief from stress and anxiety. Philos Trans R Soc B 361:2215–2228
26. Moncek F, Duncko R, Johansson BB, Jezova D (2004) Effect of environmental enrichment on stress related systems in rats. J Neuroendocrinol 16:423–431
27. Neely C, Lane C, Torres J, Flinn J (2018) The effect of gentle handling on depressive-like behavior in adult male mice: considerations for human and rodent interactions in the laboratory. Behav Neurol 2018:1

Chapter 10

5-HT Neural System Abnormalities in PTSD Model Rats

Hiroki Shikanai, Hirokazu Matsuzaki, Rina Kasai, Shota Kusaka, Tsugumi Shindo, and Takeshi Izumi

Abstract

To do translational research on illness, a suitable animal model is needed. Since the symptoms of post-traumatic stress disorder (PTSD) exist across a wide range of mental functions such as fear memory, negative operant conditioning, cognition, mood, and vigilance, it is difficult to completely reproduce the pathophysiology of PTSD with a single animal model. Here, we will explain the 3WFS model in rats, in which rats are subjected to aversive footshock during the third week of the postnatal period. 3WFS rats show an abnormality that delays the extinction of contextual fear conditioning, which is antagonized by the administration of D-cycloserine, which in turn is thought to promote the extinction of fear memory in humans. This indicates that the 3WFS model reproduces the pathophysiology of PTSD intrusion symptoms (recollection of morbid fear memory). Based on previous data, we speculated that the impaired fear extinction in 3WFS rats is due to abnormalities in the 5-HT neural system of the hippocampus. In this chapter, we will present how to create a 3WFS model and introduce our research on fear memory and the 5-HT neural system of the hippocampus.

Key words Animal model, Conditioned fear, Cognitive enhancer, Extinction, 5-hydroxytryptamine (5-HT), Hippocampus, Post-traumatic stress disorder, Serotonin

1 Introduction

Post-traumatic stress disorder (PTSD) is a mental illness that develops after exposure to traumatic events such as life-threatening situations, serious injury, or sexual assault [1]. PTSD may present with symptoms like panic disorder and depression, but it is characterized by memory-related symptoms such as involuntary re-experiencing of trauma which triggers nightmares and flashbacks, distress in response to trauma-related cues, and avoidance of those cues [2, 3].

By definition, PTSD is an illness caused by a single trauma, but individuals who are repeatedly stressed over a long period, such as child abuse survivors, are known to exhibit a wider range of symptoms than PTSD, including affective dysregulation, negative self-concept, and disturbed relationships. The concept of complex PTSD (CPTSD) has been proposed to describe such conditions [4]. In addition, these patients have been reported to exhibit impaired autobiographical memory coherence [5].

From the above, it can be seen that pathological conditions related to memory are very important in traumatic disorders. Meanwhile, several animal models of PTSD have been proposed, such as maternal separation, stress-enhanced fear learning, and single-prolonged stress [6–8]. These models employ a variety of physical or psychological stressors, but are commonly characterized by exhibiting memory dysfunction, such as impaired fear memory extinction, while also exhibiting anxiety and depression-like behaviors.

Here, we will explain the 3WFS model in rats, in which rats are subjected to aversive footshock during the third postnatal week period. 3WFS rats show an abnormality that delays the extinction of contextual fear conditioning [9]. Our hypothesis is that the impaired fear extinction in 3WFS rats is due to abnormalities in the 5-HT neural system of the hippocampus. In this chapter, we will present a method for creating a 3WFS model and introduce our research on fear memory and the 5-HT neural system.

2 Materials

2.1 Animals

Male Wistar/ST rats were bred in our laboratory, with the exception of the first-breeder adult rats, which were supplied by Sankyo Labo Service Corporation, INC. (Shizuoka, Japan). The day of birth was denoted postnatal day 0 (PND 0). Gender was confirmed on PND 14, and weaning occurred on PND 21. Only male pups were used in the present experiments. Rats were housed in a room with a 12 h light/dark cycle and a constant temperature (21 ± 2 °C). Food and water were available ad libitum. All animal procedures were performed in accordance with the National Institutes of Health Guide for the Care and Use of Laboratory Animals.

2.2 Aversive Stress During Early Postnatal Periods

The male pups were randomly assigned to a postnatal stress group or a control group. The rats in the postnatal stress groups were placed into a footshock box ($30.5 \times 24.1 \times 21.0$ cm, Med Associates Inc., VT, USA) for 5 min and subjected to five footshocks (shock intensity, 0.5 mA; intershock interval, 30 s; shock duration, 2 s), and they remained in the box for 5 min after the last footshock. Footshocks were repeated for five consecutive days at PND 21–25 (3WFS group). The rats in the control groups were placed in the footshock box for 12.5 min without footshock (non-FS group).

2.3 Contextual Fear Conditioning Test

Behavioral experiments were performed between 11:00 h and 14:00 h to minimize circadian influences. The rats were subjected to the contextual fear conditioning test in the adult period at 10–14 weeks of age. Each rat was acclimated to the footshock box for 5 min and subjected to five footshocks (shock intensity, 0.5 mA; intershock interval, 30 s; shock duration, 2 s; acquisition session). Rats remained in the box for 5 min after the last footshock and then were returned to their home cage. Twenty-four hours later, rats were placed in the footshock box for a 10-min testing period without footshock. Freezing was defined as the lack of any observable movement of the body and the vibrissae, with the exception of movements related to respiration. The presence of freezing was estimated by an automatic system (FreezeFrame, Actimetrics, IL, USA) using a pixel difference method.

2.4 Extinction of Contextual Fear Conditioning

During the adult period at 10–12 weeks of age, rats were subjected to contextual fear conditioning as described in Subheading 2.3. From the day after conditioning, the extinction of fear memory was estimated by measuring freezing behavior during a 5-min exposure to the footshock box without footshock. Extinction trials were repeated for five consecutive days (from the first day [Day 1] to the fifth [Day 5]).

2.5 Surgery for Microinjection

The rats were anesthetized with sodium pentobarbital (50 mg/kg, i.p.) and fixed in a stereotaxic frame (Narishige, Tokyo, Japan). Stainless steel guide cannulae (24 gauge) were implanted 2 mm above the dorsal hippocampus (guide cannula tip; 3.2 mm posterior, ±2.0 mm lateral to bregma, 1.6 mm ventral from the dura) and amygdala (guide cannula tip; 3.2 mm posterior, ±5.0 mm lateral to bregma, 6.8 mm ventral from the dura). After surgery, the rats were housed individually and allowed a one-week recovery period prior to testing. After completion of the experiments, the rats were sacrificed, and the cannula placements were verified histologically according to the rat brain atlas [10].

2.6 Drug Administration for Contextual Fear Conditioning

In the systemic administration experiment, the rats were injected with either R-(+)-8-hydroxy-2-(di-n-propylamino) tetralin hydrobromide (8-OH-DPAT) (0.2 mg/kg, i.p.) (Sigma, MO, USA), a $5-HT_{1A/7}$ agonist, or saline 20 min before re-exposure to the footshock box. In the bilateral microinjection experiment, either 8-OH-DPAT (1 μg in 0.5 μl saline) or vehicle was injected into the hippocampus or amygdala using a 30-gauge stainless steel injector that protruded 2 mm from the tip of the guide cannula. The solution was infused over a period of 1 min at a constant flow by a microinjection pump (CMA100, Carnegie Medicine, Sweden), and the injection cannula was left in place for 1 min after injection to allow for diffusion. Ten minutes later, the rats were re-exposed to the footshock box.

2.7 Drug Administration for Extinction of Contextual Fear Conditioning

D-cycloserine (DCS) (15 mg/kg, i.p.) (Sigma), a partial N-methyl-D-aspartate (NMDA) receptor agonist, diazepam (1 mg/kg, i.p.) (Sigma), escitalopram (10 mg/kg, i.p.) (Tocris Bioscience, MN, USA), a selective serotonin reuptake inhibitor (SSRI), and tandospirone (1 mg/kg, i.p.), a $5-HT_{1A}$ receptor partial agonist, were used to characterize contextual fear conditioning during extinction trials. Tandospirone was provided by Sumitomo Pharmaceuticals Co. Ltd. (Tokyo, Japan). Drugs were administered 20 min before each extinction trial for four consecutive days.

3 Results

Our juvenile stress model (3WFS stress model) showed several abnormalities such as depression-like behavior in adulthood [11]. Aside from this, 3WFS rats had reduced anxiolytic effects of the 5-HT1A agonist compared to controls, which may be caused by abnormalities in the hippocampal 5-HT nervous system [12]. In addition, the 3WFS stress model shows PTSD-like impaired fear memory extinction, which was improved by the administration of DCS and $5-HT_{1A}$ agonist (the so-called cognitive enhancer effect) [9]. The benzodiazepine and SSRI used in our study did not show this effect. Data are shown below.

3.1 Effects of 3WFS on the Anxiolytic Effect of 8-OH-DPAT Systemic Administration [12]

3WFS rats showed reduced response to the anxiolytic effect of the $5-HT_{1A}$ receptor agonist in the contextual fear conditioning test (Fig. 1). The systemic administration of 8-OH-DPAT (0.2 mg/kg i.p.) attenuated freezing behavior in the non-FS group (non-FS/saline group vs. non-FS/8-OH-DPAT group, $P < 0.05$) but not in the 3WFS group (3WFS/saline group vs. 3WFS/8-OH-DPAT group, N.S.). This suggests that 3WFS influences the anxiolytic effect via the $5-HT_{1A}$ receptor in the contextual fear conditioning test.

3.2 Effects of 3WFS on the Anxiolytic Effect of 8-OH-DPAT Local Injection into the Brain [12]

We have already reported that local administration of $5-HT_{1A}$ agonists to both the dorsal hippocampus and amygdala shows anxiolytic effects in the contextual fear conditioning test [13]. To investigate which brain region abnormality caused the attenuation of $5-HT_{1A}$ agonist anxiolytic effect in 3WFS, the $5-HT_{1A}$ agonist was locally injected into either the dorsal hippocampus or amygdala of 3WFS rats, and its anxiolytic effect in the contextual fear conditioning test was investigated.

The bilateral local injection of 8-OH-DPAT (1 μg/side) into the dorsal hippocampus attenuated freezing behavior in the non-FS group (non-FS/saline group vs. non-FS/8-OH-DPAT group, $P < 0.05$) but not in the 3WFS group (3WFS/saline group vs. 3WFS/8-OH-DPAT group, N.S.) (Fig. 2a). However, in the amygdala, 8-OH-DPAT injection attenuated freezing behavior in both the non-FS and 3WFS groups ($P < 0.01$) (Fig. 2b).

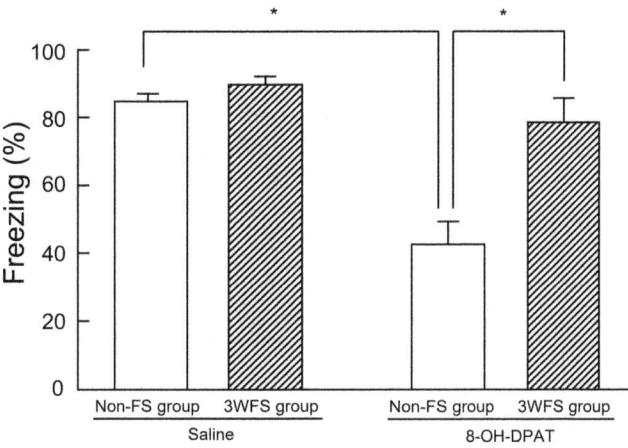

Fig. 1 Effects of 3WFS stress on anxiolytic effect of 8-OH-DPAT systemic administration in conditioned fear. 8-OH-DPAT or saline was injected 20 min before re-exposure to the footshock box. Data are expressed as mean ± S.E.M. of %freezing for a 5-min observation period. $N = 5$–7. *$P < 0.05$. FS, footshock; 3WFS, footshock at the third postnatal week. (This figure was modified with permission from a figure in the author's published work [12])

These suggest that the 5-HT_{1A} receptor-mediated anxiolytic effect in the amygdala of 3WFS rats remains unchanged, but the 5-HT_{1A} receptor-mediated anxiolytic effect in the dorsal hippocampus is diminished. Moreover, the Bmax value of the [^{3}H] WAY-100635 binding, which reflects the number of 5-HT_{1A} receptors, was significantly lower in the dorsal hippocampus of 3WFS than in the control group (non-FS group, 68.8 ± 8.1; 3WFS group, 45.9 ± 3.7 fmol/mg protein). There was no difference in the amygdala.

3.3 Effects of Early Postnatal Stress on Freezing Behavior During Extinction [9]

We investigated the effect of 3WFS on the extinction of contextual fear conditioning (Fig. 3). There was no significant difference in freezing behavior during the initial extinction trial (Day 1) between the 3WFS group and controls. In the subsequent extinction trials (Day 2–5), freezing behavior gradually decreased in both groups. However, the decrease in freezing behavior was attenuated in the 3WFS group compared to controls, and there were significant differences in freezing between the groups on each day (Day 2–5, all $P < 0.05$).

3.4 Effects of DCS on Freezing Behavior During Extinction in 3WFS Rats [9]

Animal studies have revealed that DCS enhanced the extinction of conditioned fear [14]. The effects of DCS plus trauma-focused psychotherapies have been investigated in several clinical trials with promising findings [15], though subsequent studies have shown that the effects of DCS were not as initially expected. However, as there are no other clinically effective drugs, DCS continues to be considered a "cognitive enhancer" for validating animal models of PTSD.

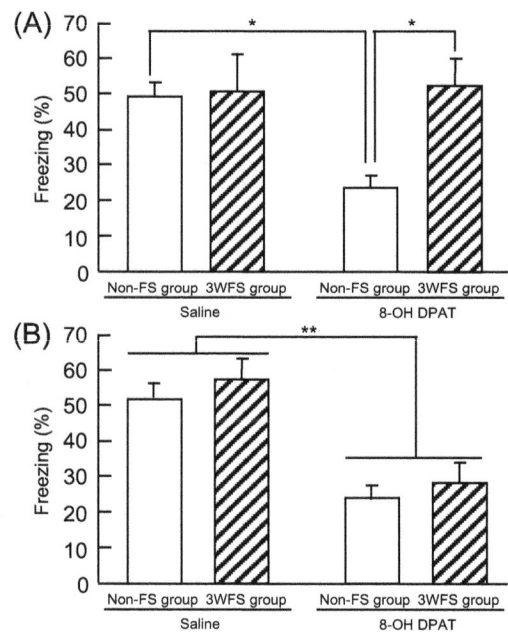

Fig. 2 Effects of 3WFS stress on the anxiolytic effect of 8-OH-DPAT local injection. 8-OH-DPAT or saline was locally injected 10 min before re-exposure to the footshock box. Data are expressed as mean ± S.E.M. of %freezing for a 5-min observation period. (**a**) Effect of local injection into the dorsal hippocampus. $N = 6–9$. *$P < 0.05$. (**b**) Effect of local injection into the basolateral amygdala. $N = 4–5$. **$P < 0.01$. 8-OH-DPAT, R-(+)-8-hydroxy-2-(di-n-propylamino) tetralin hydrobromide; FS, footshock; 3WFS, footshock at the third postnatal week. (This figure was modified with permission from a figure in the author's published work [12])

We examined the effects of DCS on the extinction of conditioned fear (Fig. 4). There was no significant difference in freezing behavior during the initial extinction trial (Day 1) between the 3WFS/saline group and the 3WFS/DCS group. In the subsequent extinction trials (Day 2–5), freezing behavior gradually decreased in both groups. However, the decrease in freezing behavior was enhanced in the 3WFS/DCS group compared to the 3WFS/saline group, and there were significant differences in freezing between the groups on each day (Day 2–5, all $P < 0.05$).

3.5 Effects of Escitalopram, an SSRI, on Freezing Behavior During Extinction in 3WFS Rats

SSRIs are drugs used to treat PTSD in humans, but there are few studies using animal experiments. Pedraza et al. (2019) reported that fluoxetine was effective for the extinction of conditioned fear [16], while Matthew et al. (2018) reported that both fluoxetine and citalopram were ineffective [17]. Thus, we investigated the effect of escitalopram, an SSRI, on the extinction of conditioned fear in 3WFS rats (Fig. 5). There was no significant difference in freezing behavior during the initial extinction trial (Day 1) between

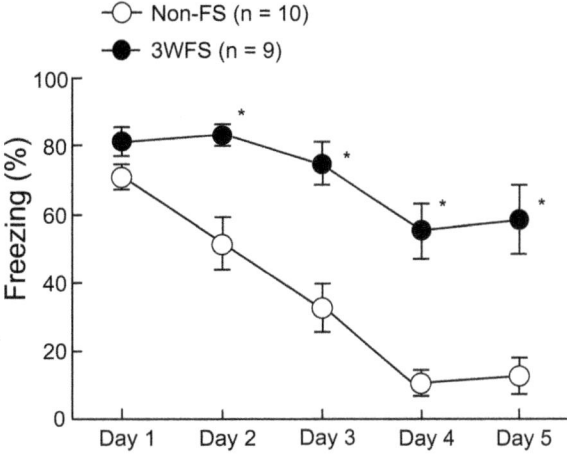

Fig. 3 Effects of 3WFS stress on freezing behavior during extinction Extinction trials were repeated for five consecutive days (from the first [Day 1] to the fifth [Day 5] extinction trial). Data are expressed as mean ± S.E.M. of % freezing for a 5-min observation period. Non-FS group, $n = 10$; 3WFS group, $n = 12$. *$P < 0.05$. FS, footshock; 3WFS, footshock at the third postnatal week. (This figure was modified with permission from a figure in the author's published work [9])

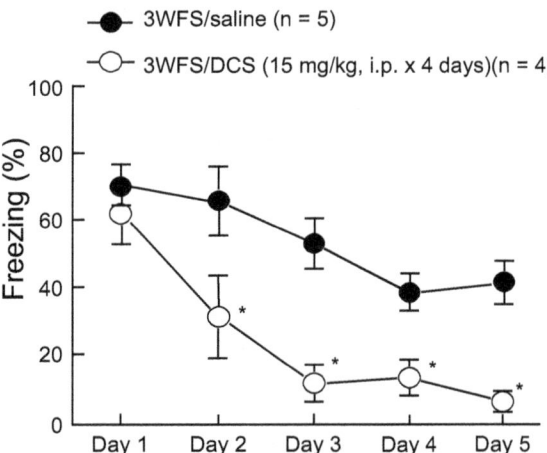

Fig. 4 Effects of DCS on freezing behavior during extinction in 3WFS rats. DCS or saline was administered 20 min before each extinction trial for four consecutive days (from Day 2 to Day 5). Data are expressed as mean ± S.E.M. of %freezing for a 5-min observation period. Saline group, $n = 5$; DCS group, $n = 4$. *$P < 0.05$. DCS, D-cycloserine; 3WFS, footshock at the third postnatal week. (This figure was modified with permission from a figure in the author's published work [9])

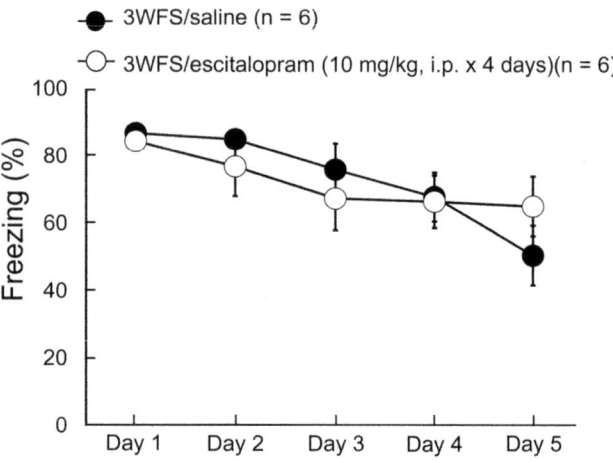

Fig. 5 Effects of escitalopram on freezing behavior during extinction in 3WFS rats. Escitalopram or saline was administered 20 min before each extinction trial for four consecutive days (from Day 2 to Day 5). Data are expressed as mean ± S.E.M. of %freezing for a 5-min observation period. Saline group, $n = 6$; escitalopram group, $n = 6$. 3WFS, footshock at the third postnatal week. (This figure was created by the authors for this publication)

the 3WFS/saline group and the 3WFS/escitalopram group. In the subsequent extinction trials (Day 2–5), freezing behavior gradually decreased in both groups, and there was no significant difference in freezing between the groups.

3.6 Effects of Tandospirone and Diazepam on Freezing Behavior During Extinction in 3WFS Rats [9]

We also investigated the effects of two anxiolytics, tandospirone (5-HT$_{1A}$ partial agonist) and diazepam (benzodiazepine), on the extinction of conditioned fear (Fig. 6). There was no significant difference in freezing behavior during the initial extinction trial (Day 1) between the 3WFS/saline group, the 3WFS/tandospirone group, and the 3WFS/diazepam group. In the subsequent extinction trials (Day 2–5), freezing behavior gradually decreased in all three groups. Although there was no difference in freezing between the 3WFS/diazepam group and the 3WFS/saline group, the decrease in freezing behavior was enhanced in the 3WFS/tandospirone group compared to the 3WFS/saline group, and there were significant differences in freezing between these groups on each day (Day 2–5, all $P < 0.05$).

4 Conclusion

A number of double-blind, placebo-controlled randomized clinical trials (RCTs) have indicated the effectiveness of SSRIs, SNRIs, and atypical antipsychotics for PTSD, although their effects are relatively small [18]. Accordingly, drugs that enhance the effects of

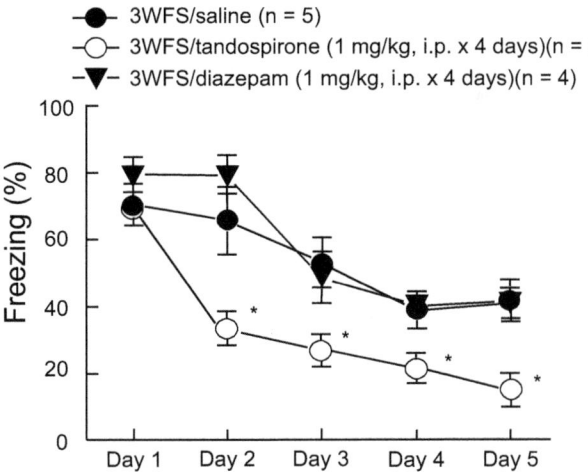

Fig. 6 Effects of diazepam and tandospirone on freezing behavior during extinction in 3WFS rats. Drugs or saline were administered 20 min before each extinction trial for four consecutive days (from Day 2 to Day 5). Data are expressed as mean ± S.E.M. of %freezing for a 5-min observation period. Saline group, $n = 5$; diazepam group, $n = 4$; tandospirone group, $n = 5$. *$P < 0.05$. 3WFS, footshock at the third postnatal week. (This figure was modified with permission from a figure in the author's published work [9])

trauma-focused psychotherapies (cognitive enhancers) are the focus of much research [19, 20]. A recent meta-analysis did not support the cognitive enhancer effects of SSRIs, SNRIs, or DCS, while the effects of the hallucinogen 3,4-methylenedioxymethamphetamine (MDMA) were found to be statistically superior to placebo, though the number of participants included was too small to be conclusive [21]. In our data presented here, DCS and 5-HT$_{1A}$ agonists were effective in improving the delayed extinction of 3WFS rats, while an SSRI and benzodiazepine were ineffective. It cannot be said that the delayed extinction in 3WFS rats completely reproduces the traumatic memory symptoms of PTSD, but in the future, we would like to further elucidate the pathological memory-related condition of PTSD using 3WFS rats as a clue, while searching for a clinically effective cognitive enhancer.

Acknowledgments

This work was supported by JSPS KAKENHI Grant Number JP21K07506. The authors are grateful to Dr. Hiroko Togashi (former professor, Health Sciences University of Hokkaido) and Dr. Machiko Matsumoto (former associate professor, Health Sciences University of Hokkaido) who founded and developed juvenile stress research. The authors also express our appreciation to everyone who has participated in juvenile stress research.

References

1. Javidi H, Yadollahie M (2012) Post-traumatic stress disorder. Int J Occup Environ Med 3(1): 2–9
2. Careaga MBL, Girardi CEN, Suchecki D (2016) Understanding posttraumatic stress disorder through fear conditioning, extinction and reconsolidation. Neurosci Biobehav Rev 71:48–57. https://doi.org/10.1016/j.neubiorev.2016.08.023
3. Liberzon I, Abelson JL (2016) Context processing and the neurobiology of post-traumatic stress disorder. Neuron 92(1):14–30. https://doi.org/10.1016/j.neuron.2016.09.039
4. Herman JL (1992) Complex PTSD: a syndrome in survivors of prolonged and repeated trauma. J Trauma Stress 5(3):377–391
5. Shin YJ, Kim SM, Hong JS, Han DH (2021) Correlations between cognitive functions and clinical symptoms in adolescents with complex post-traumatic stress disorder. Front Public Health 9:586389. https://doi.org/10.3389/fpubh.2021.586389
6. Whitaker AM, Gilpin NW, Edwards S (2014) Animal models of post-traumatic stress disorder and recent neurobiological insights. Behav Pharmacol 25(5–6):398–409. https://doi.org/10.1097/FBP.0000000000000069
7. Deslauriers J, Toth M, Der-Avakian A, Risbrough VB (2018) Current status of animal models of posttraumatic stress disorder: behavioral and biological phenotypes, and future challenges in improving translation. Biol Psychiatry 83(10):895–907. https://doi.org/10.1016/j.biopsych.2017.11.019
8. Richter-Levin G, Stork O, Schmidt MV (2019) Animal models of PTSD: a challenge to be met. Mol Psychiatry 24(8):1135–1156. https://doi.org/10.1038/s41380-018-0272-5
9. Matsumoto M, Togashi H, Konno K, Koseki H, Hirata R, Izumi T, Yamaguchi T, Yoshioka M (2008) Early postnatal stress alters the extinction of context-dependent conditioned fear in adult rats. Pharmacol Biochem Behav 89(3):247–252. https://doi.org/10.1016/j.pbb.2007.12.017
10. Paxinos G, Watson C (2007) The rat brain in stereotaxic coordinates, 6th edn. Elsevier, Amsterdam
11. Lyttle K, Ohmura Y, Konno K, Yoshida T, Izumi T, Watanabe M, Yoshioka M (2015) Repeated fluvoxamine treatment recovers juvenile stress-induced morphological changes and depressive-like behavior in rats. Brain Res 1616:88–100. https://doi.org/10.1016/j.brainres.2015.04.058
12. Matsuzaki H, Izumi T, Horinouchi T, Boku S, Inoue T, Yamaguchi T, Yoshida T, Matsumoto M, Togashi H, Miwa S, Koyama T, Yoshioka M (2011) Juvenile stress attenuates the dorsal hippocampal postsynaptic 5-HT1A receptor function in adult rats. Psychopharmacology 214(1):329–337. https://doi.org/10.1007/s00213-010-1987-4
13. Li X, Inoue T, Abekawa T, Weng S, Nakagawa S, Izumi T, Koyama T (2006) 5-HT1A receptor agonist affects fear conditioning through stimulations of the postsynaptic 5-HT1A receptors in the hippocampus and amygdala. Eur J Pharmacol 532(1–2):74–80. https://doi.org/10.1016/j.ejphar.2005.12.008
14. Parnas AS, Weber M, Richardson R (2005) Effects of multiple exposures to D-cycloserine on extinction of conditioned fear in rats. Neurobiol Learn Mem 83(3):224–231. https://doi.org/10.1016/j.nlm.2005.01.001
15. Attari A, Rajabi F, Maracy MR (2014) D-cycloserine for treatment of numbing and avoidance in chronic post traumatic stress disorder: a randomized, double blind, clinical trial. J Res Med Sci 19(7):592–598
16. Pedraza LK, Sierra RO, Giachero M, Nunes-Souza W, Lotz FN, de Oliveira AL (2019) Chronic fluoxetine prevents fear memory generalization and enhances subsequent extinction by remodeling hippocampal dendritic spines and slowing down systems consolidation. Transl Psychiatry 9(1):53. https://doi.org/10.1038/s41398-019-0371-3
17. Young MB, Norrholm SD, Khoury LM, Jovanovic T, Rauch SAM, Reiff CM, Dunlop BW, Rothbaum BO, Howell LL (2017) Inhibition of serotonin transporters disrupts the enhancement of fear memory extinction by 3,4-methylenedioxymethamphetamine (MDMA). Psychopharmacology 234(19): 2883–2895. https://doi.org/10.1007/s00213-017-4684-8
18. Steckler T, Risbrough V (2012) Pharmacological treatment of PTSD - established and new approaches. Neuropharmacology 62(2): 617–627. https://doi.org/10.1016/j.neuropharm.2011.06.012
19. McGuire JF, Lewin AB, Storch EA (2014) Enhancing exposure therapy for anxiety disorders, obsessive-compulsive disorder and post-traumatic stress disorder. Expert Rev Neurother 14(8):893–910. https://doi.org/10.1586/14737175.2014.934677
20. Singewald N, Schmuckermair C, Whittle N, Holmes A, Ressler KJ (2015) Pharmacology

of cognitive enhancers for exposure-based therapy of fear, anxiety and trauma-related disorders. Pharmacol Ther 149:150–190. https://doi.org/10.1016/j.pharmthera.2014.12.004

21. Hoskins MD, Sinnerton R, Nakamura A, Underwood JFG, Slater A, Lewis C, Roberts NP, Bisson JI, Lee M, Clarke L (2021) Pharmacological-assisted psychotherapy for post-traumatic stress disorder: a systematic review and meta-analysis. Eur J Psychotraumatol 12(1):1853379. https://doi.org/10.1080/20008198.2020.1853379

Chapter 11

Animal Models of PTSD: The Role of Fear Conditioning

Mariella B. L. Careaga, Carlos Eduardo Neves Girardi, and Deborah Suchecki

Abstract

Adverse situations that challenge an individual's physical or psychological integrity are frequent throughout the lifespan. However, some situations go beyond the adaptive capacity and are considered traumatic, leading, in some individuals, to the development of post-traumatic stress disorder (PTSD), a condition characterized by persistent recollection of the trauma, avoidance of trauma-related cues, increased arousal, and fear generalization and sensitization. Some of these symptoms indicate that fear conditioning (cue or context-based) plays a major role in this disorder. Because individual variability is a major feature of PTSD, it is crucial to understand the psychological and biological factors that confer vulnerability and resilience to the development of this disorder. Animal models based on fear conditioning, which incorporates individual variability and sex differences, could, therefore, increase the translational value and validity of these models for testing of potential pharmacological and non-pharmacological treatments. In the present chapter, we will present the behavioral and neurobiological outcomes of animal models of PTSD based on paradigms of fear conditioning and the putative systems that may be involved with vulnerability and resilience. We will close the chapter by presenting the gaps in the literature and propose future directions on how to fill them in.

Key words Traumatic stress, Animal models, Fear conditioning, Behavior, Fear extinction, Individual variability

1 Introduction: Animals Models for the Study of PTSD

Animal models are one of the major scientific tools currently available to gain insight into the physiological and neural mechanisms underlying human pathologies. They allow scientists to reproduce features of a specific disease under controlled settings, using a large number of subjects. In addition, these models are useful to monitor the progression of the disease and the manifestation of symptoms, which can be limited in clinical studies. Animal models are also suited for the development and testing of new treatments, which are crucial in the search for more effective treatment strategies. Like the difficulties faced for other psychiatric disorders, the development of animal models for PTSD is no easy task. There is a clear limitation to modeling some of the symptoms, not only because they are essentially reported by the patients, but also because they seem to occur exclusively in humans (e.g., flashbacks and nightmares). Despite these limitations, animal models of PTSD may still capture behavioral and physiological changes important to this disorder, allowing its in-depth study.

The complexity of PTSD psychopathology makes it clear how unrealistic it is to expect a "one size fits all" model. Instead, different animal models should be interpreted within the features of PTSD that they can capture. To evaluate whether an animal model resembles the core features of this disorder, researchers proposed a set of criteria to be met [1, 2]. First, long-term behavioral and neurochemical alterations should arise after the animals are exposed to a brief severe stressor, capturing the sudden nature of several trauma-inducing events such as violent assaults and traffic accidents. Second, biological alterations induced by the traumatic situation should be modulated by its intensity. Thus, the greater the magnitude of the stressor, the greater the number and/or intensity of the sequelae observed in animals. According to the Diagnostic and Statistical Manual of Mental Disorders 5th edition (DSM-5), PTSD symptoms must endure for at least a month following a traumatic event and they usually have a delayed onset. Thus, relevant animal models of PTSD should identify long-term biological changes following stress exposure. Additionally, the behavioral phenotype needs to include signs of hyper- and hypo-responsiveness as seen in PTSD patients, who might exhibit exaggerated fear reactions to trauma-related cues (i.e., hyper-response) along with emotional blunting and social withdrawal. Finally, as not all people exposed to traumatic events develop PTSD symptoms, individual variability in response to trauma should be present in the animal model. The detection of this variability might allow researchers to investigate important etiological issues related to PTSD such as vulnerability and resilience to stress. Embedded in the individual variability is sexual dimorphism, as women are twice as much vulnerable as men to develop PTSD [3–5], but this aspect seems to be overlooked by scholars who work in this field.

In the past decades, several animal models for the study of PTSD have been developed, and a summary of some of the most used models is presented in Table 1. Among them, those aiming to model the re-experiencing and avoidance symptoms are of particular interest, as fear learning is considered a core aspect of PTSD development [6–9].

Table 1
Commonly used animal models for the study of PTSD

Model	General protocol	Aims	Advantages	Disadvantages	References
Predator-based models (e.g., predator-stress, predator-scent, predator-based psychosocial stress)	Involuntary exposure to a predator (e.g., cat) or a predator scent, in combination or not with other stressors	To recreate a highly naturalistic stressful event for rodents	Ecologically relevant. Predator scent exposure does not result in physical injury	Predator exposure may result in physical injury. Variable predator-rodent interaction. Predator scent is challenging to control. Predator-exposure models may require additional facilities	[10–16]
Underwater trauma	One-min swim sessions for 5 days, followed by a forced underwater submersion for 45 s using a metal net	To model a brief, inescapable, traumatic experience	Ecologically relevant	May result in physical injuries	[17–20]
Single prolonged stress (SPS)	Exposure to 2 h restraint, 20 min forced swimming, and anesthesia (ether) until loss of consciousness	Long-lasting PTSD-like changes due to the combination of severe stressors	Combine effects of different stressors	Uncontrollable intensity. Some stressors may not be ecologically relevant (i.e., ether)	[21–24]
Stress-enhanced fear learning (SEFL)	Pre-exposure to a series of foot shocks enhances conditional fear responding to a single, mild context-shock pairing	To model the effects of a pre-trauma stressful event on trauma outcomes	Stressor's intensity (i.e., foot shock) is controllable. Reproducible context	May result in injury	[25–27]

(continued)

Table 1
(continued)

Model	General protocol	Aims	Advantages	Disadvantages	References
Fear conditioning	An environmental cue is paired with a foot shock and conditioned responses, often freezing response, are measured as an index of trauma-related memory	To model PTSD-related enhancement of fear response in face of trauma-related cues	Use of controllable parameters (e.g., current intensity, duration, and number of shocks) Reproducible context and cues	May result in physical injury Foot shock is not ecologically relevant	[28–35]

2 Role of Fear Conditioning in PTSD

Fear learning is an evolutionary advantageous response mechanism that allows individuals to use information that was learned from a previous threatening experience so they might predict and avoid dangerous encounters in the future. Fear learning, also known as fear or threat conditioning, is a form of associative learning in which a neutral stimulus (e.g., tone, light, odor, context as a whole) is paired with an aversive, unconditioned stimulus (US) such as foot shocks. As a result of this pairing, the presentation of the neutral stimulus, now known as conditioned stimulus (CS), acquires aversive properties and induces behavioral and physiological changes, including freezing behavior [36], potentiated startle [37], ultrasonic distress vocalization [38], and changes in blood pressure and respiratory and heart rates [39, 40]. These physiological and behavioral responses are collectively known as conditioned responses (CRs) as their expression is conditioned to the presentation of the CS after conditioning.

When a person experiences a traumatic event, the physiological responses during the trauma can be associated with previously neutral environmental cues. These cues can, in turn, still serve as triggers of similar physiological responses even though these may now be presented in a safe environment. The person may be consciously aware of the environmental triggers and the physiological responses may include autonomic changes (e.g., increased heart and respiratory rates), increased startle, hypervigilance, recollection of the trauma by involuntary memories (i.e., intrusive memories) and, in some cases, reexperiencing of the traumatic event by flashbacks, during which the person cannot temporarily separate the past traumatic experience from the present moment. Additionally, individuals affected by PTSD often avoid reminders of the trauma, like places, people, sounds, and smells. Brain imaging studies with PTSD patients support the critical role of memory processes in this disorder. Increased activity in brain regions essential for the formation and expression of conditioned fear (e.g., amygdala, posterior cingulate cortex) and reduced activity of structures related to fear inhibition processes such as extinction (e.g., prefrontal cortex) have been reported both during fear conditioning and in response to trauma-related cues [41, 42]. When the CS is repeatedly presented in the absence of the US, fear extinction is induced, leading to a reduction of the CR expression [43]. Fear extinction is thought to represent an inhibitory process, in which a new memory associating the CS with the absence of the US (i.e., CS no-US association) is formed, competing with the original CS-US association [44–46]. Human studies with PTSD patients often report reduced extinction of conditioned fear, suggesting that impaired fear extinction is a hallmark of PTSD [9, 47–50]. Additionally, fear extinction has been proposed to be the main process underlying exposure therapy, a non-pharmacological intervention frequently employed to treat PTSD [51].

Over the past decades, the neurobiological underpinnings of fear learning have been extensively studied in male rodents. Therefore, there are well-defined brain circuits and molecular events crucial for the acquisition, storage, retrieval, and extinction of fear memories [43, 44, 52, 53]. As fear dysregulation seems to be critical for PTSD, it is expected that some—if not most—of these neural circuits and cellular events are altered in this disorder. In the methods section of this chapter, we provide a description of how fear conditioning paradigms have been used as traumatic events in animal models of PTSD and discuss the PTSD-like changes observed in these models.

3 Methods

3.1 Fear Conditioning Models of PTSD

Fear conditioning paradigms are widely used to unravel the molecular and cellular events essential for fear learning and memory [52, 53]. Even though electric foot shocks might not be considered as ethologically relevant as natural predators, fear conditioning models of PTSD present several advantages when compared to other animal models of this disorder. First, the traumatic stress session consists of a single stress exposure, reducing the chance of habituation due to multiple stress exposures. Second, different aspects of the stressor (i.e., electric foot shock) such as current intensity, duration, and number of exposures can be well controlled and defined, making the stress exposure easier to reproduce among different laboratories. Third, fear conditioning apparatus are widely available and can be easily accommodated in standard laboratory rooms. Lastly, increasing data indicate that PTSD is, at least in part, a disorder of fear systems and, therefore, the use of fear conditioning paradigms may advance our understanding of PTSD pathophysiology. In the following sections we provide a description of how fear conditioning can be used to study PTSD in rodents.

3.2 Traumatic Stress Procedure

Before starting fear conditioning experiments, rodents are allowed to acclimate to the facility for at least 1 week. To date, most PTSD studies using fear conditioning as a traumatic experience employed adult (8–16 weeks old) mice and rats. Animals can be grouped (3–5 animals/group) or single-housed, and they are kept under controlled light and temperature conditions. In experiments with mice, fear conditioning and long-term behavioral and physiological measurements are conducted in the active phase of the rodent's daily cycle (i.e., dark phase) [28, 29], whereas studies using rats perform the assessments during the inactive phase of the cycle (i.e., light phase) [30–32]. The findings indicate that there are no significant differences when fear conditioning and behavioral outcomes are assessed during the active and inactive phases of the rodent's light cycle.

Most fear conditioning models of PTSD employ a version of contextual fear conditioning. In this paradigm, a whole context—that incorporates the visual appearance, tactile cues, background noise, and scent present in an environment—is used as CS. This context, also known as the shock chamber, is commonly made of acrylic walls and stainless-steel rods as floor. The electric foot shock is generated by a shock controller and delivered through the rods. The duration of CS presentation before shock delivery ranges from 120 s to 198 s [28–32]. The US is a high-intensity electric foot shock ranging from 1.5 to 2 mA with durations of 1 s [30, 32], 2 s [28, 29], and 10 s [35]. It is worth mentioning that the foot shocks in these models use electric currents higher than those commonly employed in learning and memory studies of fear (e.g., 0.4–1.0 mA) [54–57]. Therefore, the aversive event reaches a threshold that can be considered a traumatic stress event [29, 30, 32–34]. Animals can be removed from the shock chamber immediately after shock delivery [30, 32] or remain there for an additional 60 s [28, 29]. After the fear conditioning training is over, the shock chamber is thoroughly cleaned using a disinfectant solution (e.g., 70% ethanol) and prepared for the next animal.

To assess long-term behavioral changes, conditioned fear is commonly evaluated at least 2 weeks after the training session of fear conditioning. During testing of fear memory, animals are brought back to the same training room and placed inside the same chamber for 3–5 min in the absence of electric foot shock. Conditioned responses are then measured continuously until the end of the trial. When the trial is over, animals are removed from the chamber and placed back in their home-cages. The chamber is then thoroughly cleaned and prepared for the next animal.

3.3 Scoring of Freezing Behavior as a Conditioned Response

Conditioned fear responses in rodents can be measured by a variety of physiological indicators. In fear conditioning models of PTSD, the freezing response is the most commonly assessed CR. Freezing is defined as the absence of all movements except for those related to respiration [36] and it is a well-documented behavior associated with fear expression in rodents. Freezing behavior can be manually or automatically scored. In a manual scoring procedure, the experimental session is videotaped and later the experimenter reviews the recordings and meticulously scores freezing time using a stopwatch. Software for behavioral analysis are useful to perform automatic freezing scoring. As most fear conditioning models of PTSD employ contextual fear conditioning paradigms, freezing scoring needs to start immediately after the animal is placed in the shock chamber and continues until the end of the trial. The time of freezing during the trial is then converted to a percentage score in relation to the total duration of the trial.

3.4 Behavioral Test Batteries

In order to establish a PTSD-like phenotype, animals exposed to the trauma should be submitted to further behavioral tests (i.e., test battery), unrelated to fear conditioning, in an attempt to capture distinct features related to PTSD pathophysiology. Behavioral testing should take place a few weeks after the traumatic experience so long-term changes in behavior can be evaluated. An interval between tests should also be used to minimize the interference of one behavioral test on the following one [29, 30, 32]. Figure 1 shows the basic protocol of the behavioral test battery used by our group to assess different symptom-related changes in the animal model [30–32].

3.5 Main Behavioral and Physiological Changes Observed in Fear Conditioning Models of PTSD

Fear conditioning models of PTSD recapitulate several features of this disorder. Shock-exposed rodents show conditioned fear response upon re-exposure to the context of the shock even 1 month after the traumatic experience [28–32, 35]. In male mice, the traumatic procedure induces social withdrawal, neophobia, and depression-like phenotypes, as well as sensitization of fear response, which is ameliorated by chronic fluoxetine treatment [29]. Blunted exploratory activity, exacerbated startle response, and fear sensitization are observed only in animals allowed to explore the context before foot shock delivery, indicating a crucial role of context exploration, rather than simply shock intensity, for the development of long-term outcomes induced by the trauma [30, 32]. Moreover, trauma-exposed rodents exhibit high corticosterone levels when exposed to a novel socioenvironmental stimulus (i.e., a new cage filled with sawdust from a cage of unfamiliar

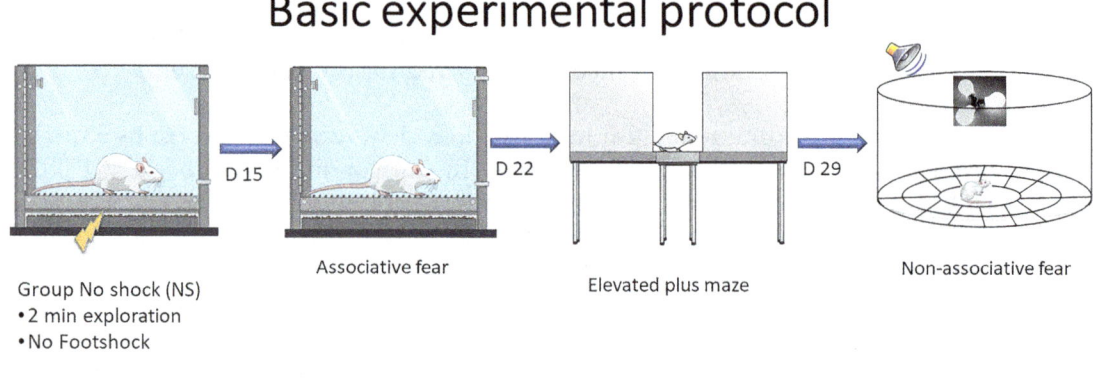

Fig. 1 The protocol of context fear conditioning (CFC) and subsequent tests used to assess behaviors that are relevant to symptoms of PTSD. Different groups of rats are exposed to one of three conditions: (1) exploration of the conditioning box for 2 min, without any foot shock delivery (control group or non-shock [NS]); (2) animals receive an intense and very brief shock immediately upon placement in the conditioning box and are removed immediately after the shock delivery, not being allowed to explore the box (immediate shock [IS]); (3) animals are allowed to explore the conditioning box for 2 min, after which an intense and very brief shock is delivered and the animal is removed from the box. Two weeks after the CFC protocol, all animals are placed back in the same box, without shock delivery, and freezing time is recorded as an index of associative learning. One week later, animals are exposed to the elevated plus maze to assess behaviors relevant to anxiety, followed 1 week later by a test of non-associative fear. In this test, animals are placed in a brightly lit open field arena (novel environment) and allowed to explore it for 3 min. After this period, two 90 db tones, 10 s long and 50 s apart, are presented and freezing time is recorded. This test permits the assessment of fear generalization (fear from an unknown environment) and sensitization (measured by the increase in freezing time from the first to the second tone). After this basic protocol, we evaluated other parameters 50 days after CFC, such as startle response [32], corticosterone response to an odor stimulus [35], and freezing time in a second test of associative fear [30]

male rats), indicating that trauma exposure induces a hyper-responsiveness of the HPA axis [35]. Individual variability in response to a stressor is a relevant feature to be included in an animal model of PTSD [11, 19, 58]. Several behavioral or physiological criteria can be used alone or in combination to classify animals as more or less responsive to a traumatic event [10, 11, 30, 59]. Using the conditioned fear response as the criterion, rats exposed to the traumatic event and classified as highly reactive, that is, expressing high levels of freezing behavior, not only express blunted exploratory activity and fear sensitization to a novel stimulus but also display anxiety-like phenotype and brain co-activity patterns relevant to the behavioral outcomes [30]. Fear-sensitization responses can also be used to segregate animals. In mice classified as high responders according to their fear-sensitized response, chronic treatment with a high dose of fluoxetine reduces fear sensitization [29].

Fear conditioning models of PTSD have also been used to assess the effects of extinction on associative and non-associative fear (i.e., fear sensitization). Mice undergoing extinction close to (1 day after shock) but not far (1 month after shock) from the traumatic event show reduced contextual and generalized fear as well as hyperarousal (i.e., fear sensitization), suggesting that early post-trauma interventions may be effective in reducing some PTSD-like outcomes [60].

3.6 Sex Differences in the Response to PTSD Animal Models

Clinical and epidemiologic studies show that PTSD is about twice as much prevalent in women than in men [3–5]. Despite that, for many years, the neurobiological and endocrine factors related to PTSD were mostly studied in sex-mixed samples or only in men [61]. In the last few decades, an attempt has been made to investigate sex-related differences in PTSD, and more studies with women, as well as with female animals, have been published [21, 27, 34, 62–65].

Some of the animal models of PTSD mentioned earlier in this chapter have been used to investigate sex-differences in the aftermath of a traumatic event. In the single prolonged stress model, male rats exhibit deficits in extinction recall while females do not [21]. Females seem to be more resilient to changes induced by the predator stress model, inasmuch as they do not show increased startle response and exaggerated negative feedback of the HPA axis as males do [66, 67]. However, these neuroendocrine changes might be related to the type of trauma, as female rats exposed to a foot shock followed by situational reminders of the trauma exhibit increased negative feedback of the HPA axis when compared to males [34]. After exposure to the predator scent stress, both females and males show a robust increase in heart rate. When animals are further classified according to a cut-off behavioral criterion, extreme-behavioral responder females display faster autonomic recovery than the corresponding subgroup of males [68].

Even though the abovementioned studies indicate the existence of sex-differences after exposure to a traumatic event, there are preclinical studies that do not report such effects. For instance, females exposed to the predator-based psychosocial stress model exhibit anxiety-like behaviors comparable to those observed in males [69]. The same incidence of PTSD-like behavioral responses is also observed in both sexes after exposure to predator scent stress [70].

Perhaps, the most challenging aspect of modeling fear conditioning and extinction in female rats is the choice of the appropriate behavioral endpoint. Freezing is traditionally used as the index of fear in tests of cued or contextual fear conditioning. However, several studies report that females do not express fear by freezing as much as males, but by using active coping behaviors, such as escape [71], darting [72], and rearing [73]. Despite the distinct behavioral expression of fear response, activation of the HPA axis was observed in both sexes after training and testing of contextual fear conditioning [73], indicating that the situation is similarly stressful, but the behavioral expression is different between males and females.

In addition to behavioral and HPA axis-related hormonal changes, the effect of sex hormones should also be further investigated in animal models of PTSD as previous clinical studies indicate an association between low estrogen levels and PTSD development in women with trauma histories [50].

4 Concluding Remarks

Animal models of PTSD based on fear conditioning and fear extinction paradigms are extremely useful for the study of the neural basis of this disorder. Nonetheless, there are still some gaps that have not been fulfilled by these models and should be addressed in future studies. The first one is the clear existence of individual differences in vulnerability or resilience to develop PTSD, which may be related to previous history, such as exposure to early life adversities, to the genetic and epigenetic background, as some polymorphisms were shown to be associated to increased cortisol negative feedback in PTSD patients [74]. This individual variability requires that a large number of animals be tested in order to reach statistical power for the establishment of neurotransmitters, neuromodulators, and neuropeptides potentially involved with vulnerability and resilience to PTSD. Unfortunately, ethical committees show enormous resistance to approve studies with large animal sample sizes [75]. The second is the failure to replicate the sex differences in the incidence of PTSD, given that epidemiological and clinical data indicate that being women doubles the risk to develop this disorder. As mentioned above, the choice of behavior to be measured in the tests of associative and non-associative learning may be the major obstacle to evoke the clinical findings and should be selected differentially for each sex [76]. These caveats in the experimental approach to PTSD using animal models of fear conditioning have hindered the findings of effective treatments for PTSD, inasmuch as psychotherapy, exposure, and behavioral cognitive therapies are more effective than pharmacological treatments, but still are far from ideal [77].

In this chapter, we presented the usefulness of paradigms of fear conditioning to study the neurobiological underpinnings of PTSD and the existing problems to propose animal models necessary to meet the pressing demands in this field. By considering these aspects, animal models will be strong enough to shed light on the potential treatments for this devastating disorder.

References

1. Yehuda R, Antelman SM (1993) Criteria for rationally evaluating animal models of posttraumatic stress disorder. Biol Psychiatry 33(7):479–486. https://doi.org/10.1016/0006-3223(93)90001-t
2. Siegmund A, Wotjak CT (2006) Toward an animal model of posttraumatic stress disorder. Ann N Y Acad Sci 1071:324–334. https://doi.org/10.1196/annals.1364.025
3. Kessler RC, Sonnega A, Bromet E, Hughes M, Nelson CB (1995) Posttraumatic stress disorder in the National Comorbidity Survey. Arch Gen Psychiatry 52(12):1048–1060. https://doi.org/10.1001/archpsyc.1995.03950240066012
4. Ditlevsen DN, Elklit A (2010) The combined effect of gender and age on post traumatic stress disorder: do men and women show differences in the lifespan distribution of the disorder? Ann General Psychiatry 9:32. https://doi.org/10.1186/1744-859X-9-32
5. McLean CP, Asnaani A, Litz BT, Hofmann SG (2011) Gender differences in anxiety disorders: prevalence, course of illness, comorbidity and burden of illness. J Psychiatr Res 45(8):1027–1035. https://doi.org/10.1016/j.jpsychires.2011.03.006
6. Pitman RK (1989) Post-traumatic stress disorder, hormones, and memory. Biol Psychiatry 26(3):221–223. https://doi.org/10.1016/0006-3223(89)90033-4
7. Pitman RK, Orr SP, Shalev AY (1993) Once bitten, twice shy: beyond the conditioning model of PTSD. Biol Psychiatry 33(3):145–146. https://doi.org/10.1016/0006-3223(93)90132-w
8. Elzinga BM, Bremner JD (2002) Are the neural substrates of memory the final common pathway in posttraumatic stress disorder (PTSD)? J Affect Disord 70(1):1–17. https://doi.org/10.1016/s0165-0327(01)00351-2
9. Jovanovic T, Ressler KJ (2010) How the neurocircuitry and genetics of fear inhibition may inform our understanding of PTSD. Am J Psychiatry 167(6):648–662. https://doi.org/10.1176/appi.ajp.2009.09071074
10. Cohen H, Zohar J, Matar M (2003) The relevance of differential response to trauma in an animal model of posttraumatic stress disorder. Biol Psychiatry 53(6):463–473
11. Cohen H, Zohar J, Matar MA, Zeev K, Loewenthal U, Richter-Levin G (2004) Setting apart the affected: the use of behavioral criteria in animal models of post traumatic stress disorder. Neuropsychopharmacology 29(11):1962–1970. https://doi.org/10.1038/sj.npp.1300523
12. Dopfel D, Perez PD, Verbitsky A, Bravo-Rivera H, Ma Y, Quirk GJ, Zhang N (2019) Individual variability in behavior and functional networks predicts vulnerability using an animal model of PTSD. Nat Commun 10(1):2372. https://doi.org/10.1038/s41467-019-09926-z
13. Zoladz PR, Conrad CD, Fleshner M, Diamond DM (2008) Acute episodes of predator exposure in conjunction with chronic social instability as an animal model of post-traumatic stress disorder. Stress 11(4):259–281. https://doi.org/10.1080/10253890701768613
14. Zoladz PR, Fleshner M, Diamond DM (2012) Psychosocial animal model of PTSD produces a long-lasting traumatic memory, an increase in general anxiety and PTSD-like glucocorticoid abnormalities. Psychoneuroendocrinology 37(9):1531–1545. https://doi.org/10.1016/j.psyneuen.2012.02.007
15. Zoladz PR, Park CR, Fleshner M, Diamond DM (2015) Psychosocial predator-based animal model of PTSD produces physiological and behavioral sequelae and a traumatic memory four months following stress onset. Physiol Behav 147:183–192. https://doi.org/10.1016/j.physbeh.2015.04.032
16. Mendes-Gomes J, Paschoalin-Maurin T, Donaldson LF, Lumb BM, Blanchard DC, Coimbra NC (2020) Repeated exposure of naive and peripheral nerve-injured mice to a snake as an experimental model of post-traumatic stress disorder and its co-morbidity with neuropathic pain. Brain Res 1744:146907. https://doi.org/10.1016/j.brainres.2020.146907

17. Ardi Z, Albrecht A, Richter-Levin A, Saha R, Richter-Levin G (2016) Behavioral profiling as a translational approach in an animal model of posttraumatic stress disorder. Neurobiol Dis 88:139–147. https://doi.org/10.1016/j.nbd.2016.01.012

18. Ardi Z, Ritov G, Lucas M, Richter-Levin G (2014) The effects of a reminder of underwater trauma on behaviour and memory-related mechanisms in the rat dentate gyrus. Int J Neuropsychopharmacol 17(4):571–580. https://doi.org/10.1017/S1461145713001272

19. Richter-Levin G (1998) Acute and long-term behavioral correlates of underwater trauma—potential relevance to stress and post-stress syndromes. Psychiatry Res 79(1):73–83

20. Ritov G, Richter-Levin G (2017) Pre-trauma Methylphenidate in rats reduces PTSD-like reactions one month later. Transl Psychiatry 7(1):e1000. https://doi.org/10.1038/tp.2016.277

21. Keller SM, Schreiber WB, Staib JM, Knox D (2015) Sex differences in the single prolonged stress model. Behav Brain Res 286:29–32. https://doi.org/10.1016/j.bbr.2015.02.034

22. Knox D, Stanfield BR, Staib JM, David NP, DePietro T, Chamness M, Schneider EK, Keller SM, Lawless C (2018) Using c-Jun to identify fear extinction learning-specific patterns of neural activity that are affected by single prolonged stress. Behav Brain Res 341:189–197. https://doi.org/10.1016/j.bbr.2017.12.037

23. Knox D, Stanfield BR, Staib JM, David NP, Keller SM, DePietro T (2016) Neural circuits via which single prolonged stress exposure leads to fear extinction retention deficits. Learn Mem 23(12):689–698. https://doi.org/10.1101/lm.043141.116

24. Serova LI, Laukova M, Alaluf LG, Pucillo L, Sabban EL (2014) Intranasal neuropeptide Y reverses anxiety and depressive-like behavior impaired by single prolonged stress PTSD model. Eur Neuropsychopharmacol 24(1):142–147. https://doi.org/10.1016/j.euroneuro.2013.11.007

25. Rau V, DeCola JP, Fanselow MS (2005) Stress-induced enhancement of fear learning: an animal model of posttraumatic stress disorder. Neurosci Biobehav Rev 29(8):1207–1223. https://doi.org/10.1016/j.neubiorev.2005.04.010

26. Poulos AM, Zhuravka I, Long V, Gannam C, Fanselow M (2015) Sensitization of fear learning to mild unconditional stimuli in male and female rats. Behav Neurosci 129(1):62–67. https://doi.org/10.1037/bne0000033

27. Sillivan SE, Joseph NF, Jamieson S, King ML, Chevere-Torres I, Fuentes I, Shumyatsky GP, Brantley AF, Rumbaugh G, Miller CA (2017) Susceptibility and resilience to posttraumatic stress disorder-like behaviors in inbred mice. Biol Psychiatry 82(12):924–933. https://doi.org/10.1016/j.biopsych.2017.06.030

28. Siegmund A, Wotjak CT (2007) Hyperarousal does not depend on trauma-related contextual memory in an animal model of Posttraumatic Stress Disorder. Physiol Behav 90(1):103–107. https://doi.org/10.1016/j.physbeh.2006.08.032

29. Siegmund A, Wotjak CT (2007) A mouse model of posttraumatic stress disorder that distinguishes between conditioned and sensitised fear. J Psychiatr Res 41(10):848–860. https://doi.org/10.1016/j.jpsychires.2006.07.017

30. Careaga MBL, Girardi CEN, Suchecki D (2019) Variability in response to severe stress: highly reactive rats exhibit changes in fear and anxiety-like behavior related to distinct neuronal co-activation patterns. Behav Brain Res 373:112078. https://doi.org/10.1016/j.bbr.2019.112078

31. Careaga MBL, Girardi CEN, Suchecki D (2021) Propranolol failed to prevent severe stress-induced long-term behavioral changes in male rats. Prog Neuro-Psychopharmacol Biol Psychiatry 108:110079. https://doi.org/10.1016/j.pnpbp.2020.110079

32. Girardi CE, Tiba PA, Llobet GB, Levin R, Abilio VC, Suchecki D (2013) Contextual exploration previous to an aversive event predicts long-term emotional consequences of severe stress. Front Behav Neurosci 7:134. https://doi.org/10.3389/fnbeh.2013.00134

33. Pynoos RS, Ritzmann RF, Steinberg AM, Goenjian A, Prisecaru I (1996) A behavioral animal model of posttraumatic stress disorder featuring repeated exposure to situational reminders. Biol Psychiatry 39(2):129–134. https://doi.org/10.1016/0006-3223(95)00088-7

34. Louvart H, Maccari S, Lesage J, Leonhardt M, Dickes-Coopman A, Darnaudery M (2006) Effects of a single footshock followed by situational reminders on HPA axis and behaviour in the aversive context in male and female rats. Psychoneuroendocrinology 31(1):92–99. https://doi.org/10.1016/j.psyneuen.2005.05.014

35. Girardi CEN, Llobet GB, Suchecki D, Tiba PA (2018) High corticosterone after olfactory social stimuli in a rodent model of traumatic stress. Psychol Neurosci 11(1):105–115. https://doi.org/10.1037/pne0000128

36. Blanchard DC, Blanchard RJ (1972) Innate and conditioned reactions to threat in rats with amygdaloid lesions. J Comp Physiol Psychol 81(2):281–290. https://doi.org/10.1037/h0033521
37. Hitchcock J, Davis M (1986) Lesions of the amygdala, but not of the cerebellum or red nucleus, block conditioned fear as measured with the potentiated startle paradigm. Behav Neurosci 100(1):11–22. https://doi.org/10.1037//0735-7044.100.1.11
38. Blanchard RJ, Blanchard DC, Agullana R, Weiss SM (1991) Twenty-two kHz alarm cries to presentation of a predator, by laboratory rats living in visible burrow systems. Physiol Behav 50(5):967–972. https://doi.org/10.1016/0031-9384(91)90423-l
39. Iwata J, LeDoux JE, Reis DJ (1986) Destruction of intrinsic neurons in the lateral hypothalamus disrupts the classical conditioning of autonomic but not behavioral emotional responses in the rat. Brain Res 368(1):161–166. https://doi.org/10.1016/0006-8993(86)91055-3
40. Kapp BS, Frysinger RC, Gallagher M, Haselton JR (1979) Amygdala central nucleus lesions: effect on heart rate conditioning in the rabbit. Physiol Behav 23(6):1109–1117. https://doi.org/10.1016/0031-9384(79)90304-4
41. Bremner JD (2005) Effects of traumatic stress on brain structure and function: relevance to early responses to trauma. J Trauma Dissociation 6(2):51–68. https://doi.org/10.1300/J229v06n02_06
42. Bremner JD (2002) Neuroimaging studies in post-traumatic stress disorder. Curr Psychiatry Rep 4(4):254–263. https://doi.org/10.1007/s11920-996-0044-9
43. Quirk GJ, Mueller D (2008) Neural mechanisms of extinction learning and retrieval. Neuropsychopharmacology 33(1):56–72. https://doi.org/10.1038/sj.npp.1301555
44. Myers KM, Davis M (2002) Behavioral and neural analysis of extinction. Neuron 36(4):567–584. https://doi.org/10.1016/s0896-6273(02)01064-4
45. Bouton ME (2004) Context and behavioral processes in extinction. Learn Mem 11(5):485–494. https://doi.org/10.1101/lm.78804
46. Lissek S, van Meurs B (2015) Learning models of PTSD: theoretical accounts and psychobiological evidence. Int J Psychophysiol 98(3 Pt 2):594–605. https://doi.org/10.1016/j.ijpsycho.2014.11.006
47. Peri T, Ben-Shakhar G, Orr SP, Shalev AY (2000) Psychophysiologic assessment of aversive conditioning in posttraumatic stress disorder. Biol Psychiatry 47(6):512–519. https://doi.org/10.1016/s0006-3223(99)00144-4
48. Blechert J, Michael T, Vriends N, Margraf J, Wilhelm FH (2007) Fear conditioning in posttraumatic stress disorder: evidence for delayed extinction of autonomic, experiential, and behavioural responses. Behav Res Ther 45(9):2019–2033. https://doi.org/10.1016/j.brat.2007.02.012
49. Jovanovic T, Norrholm SD, Sakoman AJ, Esterajher S, Kozaric-Kovacic D (2009) Altered resting psychophysiology and startle response in Croatian combat veterans with PTSD. Int J Psychophysiol 71(3):264–268. https://doi.org/10.1016/j.ijpsycho.2008.10.007
50. Glover EM, Jovanovic T, Mercer KB, Kerley K, Bradley B, Ressler KJ, Norrholm SD (2012) Estrogen levels are associated with extinction deficits in women with posttraumatic stress disorder. Biol Psychiatry 72(1):19–24. https://doi.org/10.1016/j.biopsych.2012.02.031
51. Choi DC, Rothbaum BO, Gerardi M, Ressler KJ (2010) Pharmacological enhancement of behavioral therapy: focus on posttraumatic stress disorder. Curr Top Behav Neurosci 2:279–299
52. Johansen JP, Cain CK, Ostroff LE, LeDoux JE (2011) Molecular mechanisms of fear learning and memory. Cell 147(3):509–524. https://doi.org/10.1016/j.cell.2011.10.009
53. Maren S (2011) Seeking a spotless mind: extinction, deconsolidation, and erasure of fear memory. Neuron 70(5):830–845. https://doi.org/10.1016/j.neuron.2011.04.023
54. Debiec J, Ledoux JE (2004) Disruption of reconsolidation but not consolidation of auditory fear conditioning by noradrenergic blockade in the amygdala. Neuroscience 129(2):267–272. https://doi.org/10.1016/j.neuroscience.2004.08.018
55. Hefner K, Whittle N, Juhasz J, Norcross M, Karlsson RM, Saksida LM, Bussey TJ, Singewald N, Holmes A (2008) Impaired fear extinction learning and cortico-amygdala circuit abnormalities in a common genetic mouse strain. J Neurosci 28(32):8074–8085.

https://doi.org/10.1523/JNEUROSCI.4904-07.2008
56. Muravieva EV, Alberini CM (2010) Limited efficacy of propranolol on the reconsolidation of fear memories. Learn Mem 17(6):306–313. https://doi.org/10.1101/lm.1794710
57. Careaga MB, Tiba PA, Ota SM, Suchecki D (2015) Pre-test metyrapone impairs memory recall in fear conditioning tasks: lack of interaction with β-adrenergic activity. Front Behav Neurosci 9:51. https://doi.org/10.3389/fnbeh.2015.00051
58. Verbitsky A, Dopfel D, Zhang N (2020) Rodent models of post-traumatic stress disorder: behavioral assessment. Transl Psychiatry 10(1):132. https://doi.org/10.1038/s41398-020-0806-x
59. Ritov G, Boltyansky B, Richter-Levin G (2016) A novel approach to PTSD modeling in rats reveals alternating patterns of limbic activity in different types of stress reaction. Mol Psychiatry 21(5):630–641. https://doi.org/10.1038/mp.2015.169
60. Golub Y, Mauch CP, Dahlhoff M, Wotjak CT (2009) Consequences of extinction training on associative and non-associative fear in a mouse model of Posttraumatic Stress Disorder (PTSD). Behav Brain Res 205(2):544–549. https://doi.org/10.1016/j.bbr.2009.08.019
61. Dell'Osso L, Carmassi C, Del Debbio A, Catena Dell'Osso M, Bianchi C, da Pozzo E, Origlia N, Domenici L, Massimetti G, Marazziti D, Piccinni A (2009) Brain-derived neurotrophic factor plasma levels in patients suffering from post-traumatic stress disorder. Prog Neuro-Psychopharmacol Biol Psychiatry 33(5):899–902. https://doi.org/10.1016/j.pnpbp.2009.04.018
62. Bremner JD, Vermetten E, Schmahl C, Vaccarino V, Vythilingam M, Afzal N, Grillon C, Charney DS (2005) Positron emission tomographic imaging of neural correlates of a fear acquisition and extinction paradigm in women with childhood sexual-abuse-related post-traumatic stress disorder. Psychol Med 35(6):791–806. https://doi.org/10.1017/s0033291704003290
63. Bremner JD, Vythilingam M, Vermetten E, Southwick SM, McGlashan T, Staib LH, Soufer R, Charney DS (2003) Neural correlates of declarative memory for emotionally valenced words in women with posttraumatic stress disorder related to early childhood sexual abuse. Biol Psychiatry 53(10):879–889. https://doi.org/10.1016/s0006-3223(02)01891-7
64. Louvart H, Maccari S, Ducrocq F, Thomas P, Darnaudery M (2005) Long-term behavioural alterations in female rats after a single intense footshock followed by situational reminders. Psychoneuroendocrinology 30(4):316–324. https://doi.org/10.1016/j.psyneuen.2004.09.003
65. Milad MR, Pitman RK, Ellis CB, Gold AL, Shin LM, Lasko NB, Zeidan MA, Handwerger K, Orr SP, Rauch SL (2009) Neurobiological basis of failure to recall extinction memory in posttraumatic stress disorder. Biol Psychiatry 66(12):1075–1082. https://doi.org/10.1016/j.biopsych.2009.06.026
66. Pooley AE, Benjamin RC, Sreedhar S, Eagle AL, Robison AJ, Mazei-Robison MS, Breedlove SM, Jordan CL (2018) Sex differences in the traumatic stress response: the role of adult gonadal hormones. Biol Sex Differ 9(1):32. https://doi.org/10.1186/s13293-018-0192-8
67. Pooley AE, Benjamin RC, Sreedhar S, Eagle AL, Robison AJ, Mazei-Robison MS, Breedlove SM, Jordan CL (2018) Sex differences in the traumatic stress response: PTSD symptoms in women recapitulated in female rats. Biol Sex Differ 9(1):31. https://doi.org/10.1186/s13293-018-0191-9
68. Koresh O, Kaplan Z, Zohar J, Matar MA, Geva AB, Cohen H (2016) Distinctive cardiac autonomic dysfunction following stress exposure in both sexes in an animal model of PTSD. Behav Brain Res 308:128–142. https://doi.org/10.1016/j.bbr.2016.04.024
69. Zoladz PR, D'Alessio PA, Seeley SL, Kasler CD, Goodman CS, Mucher KE, Allison AS, Smith IF, Dodson JL, Stoops TS, Rorabaugh BR (2019) A predator-based psychosocial stress animal model of PTSD in females: influence of estrous phase and ovarian hormones. Horm Behav 115:104564. https://doi.org/10.1016/j.yhbeh.2019.104564
70. Mazor A, Matar MA, Kaplan Z, Kozlovsky N, Zohar J, Cohen H (2009) Gender-related qualitative differences in baseline and post-stress anxiety responses are not reflected in the incidence of criterion-based PTSD-like behaviour patterns. World J Biol Psychiatry 10(4 Pt 3):856–869. https://doi.org/10.1080/15622970701561383
71. Ribeiro AM, Barbosa FF, Godinho MR, Fernandes VS, Munguba H, Melo TG, Barbosa

MT, Eufrasio RA, Cabral A, Izidio GS, Silva RH (2010) Sex differences in aversive memory in rats: possible role of extinction and reactive emotional factors. Brain Cogn 74(2):145–151. https://doi.org/10.1016/j.bandc.2010.07.012

72. Gruene TM, Flick K, Stefano A, Shea SD, Shansky RM (2015) Sexually divergent expression of active and passive conditioned fear responses in rats. elife 4. https://doi.org/10.7554/eLife.11352

73. Daviu N, Andero R, Armario A, Nadal R (2014) Sex differences in the behavioural and hypothalamic-pituitary-adrenal response to contextual fear conditioning in rats. Horm Behav 66(5):713–723. https://doi.org/10.1016/j.yhbeh.2014.09.015

74. Yehuda R, Flory JD, Pratchett LC, Buxbaum J, Ising M, Holsboer F (2010) Putative biological mechanisms for the association between early life adversity and the subsequent development of PTSD. Psychopharmacology 212(3): 405–417. https://doi.org/10.1007/s00213-010-1969-6

75. Richter-Levin G, Sandi C (2021) Title: "labels matter: is it stress or is it trauma?". Transl Psychiatry 11(1):385. https://doi.org/10.1038/s41398-021-01514-4

76. Shansky RM (2015) Sex differences in PTSD resilience and susceptibility: challenges for animal models of fear learning. Neurobiol Stress 1:60–65. https://doi.org/10.1016/j.ynstr.2014.09.005

77. Richter-Levin G, Stork O, Schmidt MV (2019) Animal models of PTSD: a challenge to be met. Mol Psychiatry 24(8):1135–1156. https://doi.org/10.1038/s41380-018-0272-5

Chapter 12

Combining Virtual Reality Exposure Therapy with Non-invasive Brain Stimulation for the Treatment of Post-traumatic Stress Disorder and Related Syndromes: A Perspective

Carmelo M. Vicario, Mohammad A. Salehinejad, Chiara Lucifora, Gabriella Martino, Alessandra M. Falzone, G. Grasso, and Michael A. Nitsche

Abstract

The current literature supports the therapeutic effectiveness of virtual reality exposure therapy and non-invasive brain stimulation as stand-alone approaches for the treatment of post-traumatic stress disorder and related syndromes. However, studies exploring the combined use of such therapeutic approaches are extremely limited. In this chapter, we provide a complete review of research exploring the effect of combining these two types of treatment in PTSD and related syndromes. We also provide a brief discussion about the main limitations and future directions of this emerging research field which has a relevant potential in boosting the treatment of these mental disorders.

Key words Non-invasive brain stimulation, Virtual reality exposure therapy, Post-traumatic stress disorder and related syndromes

1 Introduction

Post-traumatic Stress Disorder (PTSD) is a clinical condition caused by exposure to a severe traumatic event (e.g., a bereavement, a car accident, an assault), which is estimated to involve about 70% of the population during their lifetime [1]. The conditional risk to develop PTSD after such an event is around 4%, but varies significantly by trauma type [1]. The most prevalent symptoms include memory and attention deficits (i.e., attentional biases to threat); temporal processing deficits [2, 3]; hyperarousal [4]; and dissociation, which is related with the experience of depersonalization and derealisation [5]. Although 25–40% of cases of PTSD recover within 1 year [1], about 25% of patients do not recover within 10 years with standard treatment approaches [6]

PTSD-related syndromes include a wide range of clinical conditions – including anxiety disorders (ADs) – that refer to different clinical profiles such as *separation anxiety disorder* (related to the pediatric population), *selective mutism, specific phobia, social anxiety disorder, panic disorder, agoraphobia*, and *generalized anxiety disorder* [7]. ADs are the most common mental disorders as they affect around 3.6% of the population, according to the World Health Organization. The specific prevalence, however, varies widely, depending on the country, population, and measurement tools used [8]. The most prevalent common symptoms across different clinical profiles include panic, social awkwardness, and obsessiveness [9]. Moreover, ADs often overlap with PTSD, such as in the case of social phobias (SPs) [10]. The recovery rate of people affected by AD is estimated to be around 48% following conventional cognitive-behavioral therapy [11].

Overall, the literature examined above shows that several individuals with PTSD and related disorders remain symptomatic despite the state-of-the-art treatment. Therefore, it is more necessary than ever to develop more effective treatment protocols that help to increase overall recovery rates.

Exposure therapy (ET) is the gold standard for the treatment of PTSD and related syndromes [12, 13] and is highly effective in reducing respective symptoms. An evolution of this ET approach is virtual reality exposure therapy (VRET). Compared to standard ET, VRET enables a higher emotional engagement of patients during exposure due to multiple sensory stimuli provided by the virtual environment [14]. Moreover, it helps to bypass symptoms of avoidance [e.g., 15].

Recently it was shown that also non-invasive brain stimulation (NIBS) methods, such as transcranial direct current stimulation (tDCS) and repetitive transcranial magnetic stimulation (rTMS), are promising for the treatment of several mental disorders affecting adults [16–21], as well as pediatric patients [*see* 22–27]. This includes the treatment of PTSD and AD, as suggested by two recently published systematic reviews [28–30]. The physiological foundation of respective effects is the modulation of synaptic plasticity accomplished by these approaches [31], which is pathologically altered in several mental disorders, including PTSD [32] and AD [e.g., 33]. In this chapter, we provide a short section about the mechanisms of action of NIBS and an overview of the state-of-the-art treatment with respect to the effects of combined NIBS-VRET treatment of PTSD and related syndromes.

2 Basic Mechanisms of Action of NIBS

Repetitive transcranial magnetic stimulation (rTMS) and tDCS are the most common NIBS techniques for the modulation of brain activity, excitability, and plasticity.

rTMS modulates the activity and excitability of neural tissue by applying trains of magnetic pulses over the target cortical region. Magnetic pulses are applied via a copper coil over the scalp of the patient. This induces electrical currents in the target area and activates neurons via the induction of an electric field. If these stimuli are applied repetitively within specific frequencies, beyond the acute activation of neuronal populations, prolonged neuroplastic excitability alterations are accomplished. These alterations are excitability-enhancing or excitability-reducing, depending on the specific stimulation protocol. With conventional protocols, decreased cortical excitability is accomplished by low-frequency rTMS (1–5 Hz), while increased cortical excitability is induced by high-frequency rTMS (>5 Hz) [34, 35].

For the effects of rTMS on brain plasticity, evidence for a contribution of the glutamatergic system is available [36], and beyond functional long-term potentiation (LTP)-like plasticity induced by high-frequency rTMS, also evidence for structural plasticity induction via rTMS is available [37, 38]. New protocols emerged in recent years, which allow faster induction of plasticity. Theta Burst Stimulation (TBS), a rapid rTMS paradigm, allows induction of plasticity within a few minutes [39]. Continuous TBS (cTBS) produces a reduction of cortical excitability [40], while intermittent TBS (iTBS) enhances cortical excitability.

tDCS accomplishes stimulation of the cerebral cortex via two or more electrodes with opposite polarities (i.e., anodal and cathodal) placed on the scalp and connected with a battery-driven constant current stimulator. These electrodes release a relatively weak electrical direct current (usually 1 ∼ 2 mA) which, in difference to TMS, does not induce neuronal activity, but modulates neuronal resting membrane potentials at a subthreshold level, thus enhancing or reducing spontaneous cortical activity and excitability, depending on the stimulation polarity. Beyond acute effects, prolonged stimulation for a few minutes induces identically directed neuroplastic aftereffects [41–43]. These changes share similarities with mechanisms of LTP and long-term depression (LTD) [44, 45], which are known to involve N-methyl D-aspartate receptors and glutamate release (for details *see* [46, 47]), and are mediated by calcium-dependent processes [47]. Generally speaking, anodal stimulation increases cortical excitability, whereas cathodal stimulation decreases it at the macroscale level [47]. These effects are valid for conventional dosages, but also nonlinear effects have been documented for higher stimulation intensities and durations which exceed these windows. For instance, there is evidence [48, 49] that both anodal and cathodal tDCS at the intensity of 2 mA increase corticospinal excitability, whereas 1 mA cathodal tDCS decreases it. These nonlinear effects are thought to be caused by calcium dynamics of plasticity where LTD and LTP are induced by specific calcium concentrations but exceeding respective concentration windows results in null or conversion of effects [50].

Overall, the application of these NIBS methods to PTSD and related disorders is based on their role in boosting exposition-related learning and memory processes, which are considered central to promote the successful treatment of these mental disorders [51], by acting in a synergistic way on learning-related plasticity.

3 An Overview of the Therapeutic Effectiveness of VRET and NIBS

Virtual reality exposure therapy (VRET) has become increasingly established as an alternative to in-vivo exposition in the treatment of pathological fear. Its effectiveness is mainly due to its imaginative power, which allows individuals to experience reality-like situations, as well as the possibility of actively perceiving one's own body within a simulated environment [52–57]. Recent studies show that this approach provides better results compared to traditional therapies of mental disorders [58]. This also includes the treatment of PTSD and several forms of phobias [59, 60].

Application of NIBS is an emerging concept for the treatment of PTSD and related disorders with promising initial results (for reviews in the field *see* [28, 29]). In most available studies so far, the clinical effectiveness of stand-alone NIBS was suggested by surrogate markers in laboratory studies, including psychophysiological measures (e.g., heart rate, skin conductance) and respective scores from specific questionnaires before and after the intervention. In other cases, the clinical effectiveness of NIBS has been investigated by coupling this method and respective behavioral and psychophysiological measures with virtual scenarios displayed on a computer screen as part of the intervention. The use of computer screen–displayed scenarios for therapy has the advantage of exposing the patient to clinically relevant stimuli, but may have a limited impact caused by relatively low ecological validity, as compared to VRET. Overall, the current literature supports the therapeutic effectiveness of VRET and NIBS as stand-alone approaches in the clinical field, including the treatment of PTSD and related disorders. In the next section, we discuss available research exploring the combined effect of these two types of treatment in respective clinical populations.

4 Combining VRET and NIBS in PTSD Treatment

Despite growing evidence for the therapeutic effects of NIBS [see 29] and VRET [e.g., 14, 61] in PTSD, the combined effect of these two treatments has been explored in just one recently published study [62].

In that study, 12 male veterans (mean age 40.5 years, SD 8.8, range 30–53 years) with warzone-related PTSD received six combat-related VRET sessions over 2 weeks (single-blind design). During these VRET sessions, participants were randomized to receive in addition 25 min of 2 mA anodal or sham tDCS over the left ventromedial prefrontal cortex (vmPFC) (AF3 according to the 10–20 EEG system). The vmPFC is a crucial region for fear extinction learning and recall [63, 64], and its relevance to promote fear extinction learning via NIBS has been documented in several reports [e.g., 16, 65, 66]. Outcome measures included psychophysiological arousal (skin conductance reactivity – SCR) during each virtual reality (VR) session and self-reported PTSD symptoms (PTSD checklist for DSM-5) [67] obtained at baseline, after all VRET sessions, and 1 month later. The results show a greater decrease of physiological arousal (i.e., reduced SCR) in the active tDCS group, as compared to the sham intervention group. Furthermore, a clinically meaningful reduction of PTSD symptom severity, as suggested by the PTSD checklist of the DSM-5 (pcl-5) score, was reported, which was larger in the active group, as compared to the sham stimulation group. Although preliminary, this study provides promising evidence for the feasibility and therapeutic effectiveness of a NIBS-VRET combination for the treatment of PTSD.

5 Overview of the Available Studies Combining NIBS and VRET in PTSD-Related Syndromes

Regarding PTSD-related syndromes, three studies with combined NIBS and VRET have been published so far. Two of these involved patients were affected by spider phobia (SP) [68, 69]. The other study was conducted on patients affected by acrophobia (fear of heights) [70].

In the study by Notzon et al. [68], 41 participants (mean age 27.51 years, SD 9.45) affected by SP and 42 healthy controls (mean age 25.43 years, SD 7.37) (total age range between 18 and 65 years) were exposed to a spider scenario in VR after one session of intermittent Theta Burst Stimulation (iTBS) or sham treatment over the left dorsolateral prefrontal cortex (dlPFC). This target was selected due to evidence for its reduced activation in patients affected by other types of phobias such as panic disorder

[71]. Moreover, the left dlPFC is a relevant cortical target to downregulate negative reactions to emotional stimuli/outcomes via excitatory NIBS [28]. The primary outcome measure was the fear of spiders questionnaire (FSQ; [72]). Participants completed the anxiety sensitivity index (ASI; [73]) and the questionnaire for the assessment of disgust sensitivity (disgust scale: DS; [74]) at baseline. Moreover, heart rate (HR) and SCR measurements were conducted to monitor vegetative signs of fear. Overall, the results showed that a single session of iTBS had no effect on the self-reported measures of fear of spiders and anxiety. An increase in HR in response to VR scenarios was described for the phobic group during the presentation of spider scenes, as compared to baseline. However, no significant main or interaction effects of iTBS on vegetative measures were found. In a separate publication, this research team explored the impact of this intervention on cerebral blood flow during the conduction of an emotional Stroop paradigm in the same group of participants [69]. After iTBS/VR, no significant differences in cortical activation between the phobic and control groups remained. However, verum-iTBS did not augment this intervention effect, as compared to sham stimulation, which might not necessarily show the inadequacy of this method, but could have been also caused by a ceiling effect.

Herrmann et al. [70] performed a double-blind, sham-controlled study to test the effect of two sessions of high frequency (10 Hz) rTMS over the vmPFC with the coil positioned over FPz (according to the 10–20 EEG system) in 39 patients (average age 44.9, SD = 13.1) affected by acrophobia to boost the efficacy of VRET. This cortical site was chosen according to evidence that this region is highly relevant for fear extinction learning [63]. rTMS was applied immediately before VRET sessions. Outcome measures included the Acrophobia Questionnaire (AQ) [75], the Attitude Towards Heights Questionnaire (ATHQ) [76], the State Trait Anxiety Inventory (STAI) [77], the Anxiety Sensitivity Index (ASI) [78], and the Behavioral Avoidance Test (BAT) [79]. The results showed that high-frequency rTMS improved the efficacy of VRET immediately after the intervention. For the AQ and BAT questionnaires, the reduction of symptoms was significantly higher in the active, as compared to the sham group for both anxiety and avoidance outcomes. However, no significant difference between active and sham stimulation conditions were reported at the 3-month follow-up.

Overall, the examined literature provides heterogeneous results with respect to the therapeutic effectiveness of NIBS-VRET combinations for the treatment of PTSD-related syndromes. In the next paragraph, we provide a brief discussion of possible reasons for these heterogeneities.

6 Discussion and Future Directions

In this chapter, we examined the available literature ($N = 4$) exploring the combined effect of NIBS-VRET for the treatment of PTSD and related syndromes (see Table 1 for details). Overall, the results of these studies provide preliminary evidence in support of the suggestion that NIBS-VRET combinations can be effective for the treatment of these mental disorders. Two [68, 69] of the four examined reports do, however, not show clinically meaningful fear-related effects. However, these might be considered to represent a single study, as they refer to the same sample and treatment protocol, although separate measures are presented in these two articles.

The low number of studies is a central limitation of this overview and highlights the necessity to conduct more (and systematic) investigations in this field. Nevertheless, the available studies differ not only with respect to outcomes but also the specific protocols applied, which might thus help to gain information about more or less promising approaches.

One relevant difference between these studies is the chosen target region. The left dlPFC was the cortical target in two studies [68, 69] which, however, did not reveal a beneficial effect of NIBS on symptoms. In the remaining studies, the cortical target was the vmPFC [62, 70], although the authors chose two different cortical regions (i.e., AF3, FPz) referring to the left hemisphere or the midline position, respectively. Nevertheless, in these latter two reports, the NIBS-VRET combination provided clinically meaningful results.

The results of these four studies thus suggest that vmPFC may be the preferential cortical target to be considered primarily for NIBS-VRET combinations to provide clinical benefits in PTSD and related disorders.

The vmPFC is known to be relevant for the downregulation of negative affective responses [80] and contributes to the processing of valence, with regionally distinct processing of reward and punishment [81]. Moreover, the vmPFC seems to play a causal role in the regulation of physiological arousal, according to evidence showing that increased vmPFC activity leads to decreased skin conductance [82]. Excitatory NIBS over the left vmPFC downregulates the output of the central nucleus of the amygdala, which is important mainly for vegetative components of fear, via activation of the intercalated nucleus, which in turn upregulates the hippocampus, which is involved in memory consolidation [29]. This mechanistic aspect makes the vmPFC relevant for fear extinction learning. However, the results are too preliminary to consider the dlPFC as a non-relevant target for the described purpose. First, the absence of clinical benefits in the treatment of SP through iTBS over the left dlPFC might be due to the limited

Table 1

SCR (skin conductance response), HR (heart rate), ASI (anxiety sensitivity index), FSQ (fear of spiders questionnaire), DS (disgust sensitivity), fNIRs (functional near infrared spectroscopy), AQ (acrophobia questionnaire), ATHQ (attitude towards heights questionnaire), STAI (state trait anxiety inventory), BAT (behavioral avoidance test), SUDS (subjective units of discomfort scale), IPQ (Igroup Presence Questionnaire)

Article	N	Age (mean)	Target	Return	NIBS	Intensity	Duration	Polarity	Control	Measures	Outcome
Post-Traumatic Stress Disorder											
van 't Wout-Frank et al. (2019)	12	40.5	Left vmPFC (AF3)	Contralateral mastoid (PO8)	tDCS	2 mA	25 min	Anodal	Sham	SCR, PTSD checklist for DSM-5; PCL	Reduced SCR and PTSD symptom severity
PTSD-related disorders											
Spider phobia											
Notzon et al. (2015)	83	26.46	Left dlPFC (F3)	NA	rTMS	80% rMT	3 min (iTBS – 600 pulses, 15 pulses per second via 2 s trains, every 10 s)	NA	Sham/healthy participants	HR, SCR, ASI, FSQ, SUDS, IPQ, DS	No effect of iTBS on clinically relevant measures
Deppermann et al. (2016)	83	26.46	Left dlPFC (F3)	NA	rTMS	80% rMT	3 min (iTBS – 600 pulses, 15 pulses per second via 2 s trains, every 10 s)	NA	Sham/healthy participants	fNIRs emotional Stroop paradigm	No significant differences in cortical activation between the phobic and control group
Acrophobia											
Herrmann et al. (2017)	39	44.9	vmPFC (FPz)	NA	rTMS	100% rMT	20 min 40 trains of 4 s duration (1560 pulses) with a 10 Hz. Inter train intervals were 26 s	NA	Sham	AQ, ATHQ, STAI, ASI, Bat	Higher reduction of phobic anxiety and avoidance after active NIBS-VRET

number of stimulation sessions (only one session applied in these studies, as compared with two – [70] – and six – [62] – sessions). Moreover, the heterogeneous outcomes between studies might be related to the type of VR scenario used and/or the specific NIBS (i.e., iTBS) protocol. Regarding the latter aspect, however, a recent report comparing the clinical effects of iTBS with the effects of 10 Hz rTMS for the treatment of depression did report an equivalence of clinical efficacy of these protocols [83], which are both enhancing cortical excitability. One might thus speculate that the absence of clinically meaningful effects in these iTBS studies [i.e., 68, 69] is not caused by a limited capacity of iTBS to modulate cortical excitability. Finally, the potential relevance of dlPFC stimulation for the treatment of (at least) PTSD-related syndromes is suggested by a recent systematic review of our group [28], showing that NIBS (via tDCS and rTMS) of the dlPFC is a promising cortical target for the treatment of AD in combination with standard psychotherapy. The potential mechanism of action of NIBS over the dlPFC is based on the connectivity between this cortical region and several remote brain structures of the meso-cortico-limbic reward circuit – including the amygdala – which is known to be dysfunctional in AD and PTSD (for a review see [84]). Further evidence for a relevant role of the dlPFC in fear-related psychological processes is available, which stresses the potential relevance of this area for intervention approaches. A recent study [85] involving six patients with left dlPFC lesions showed that these were able to acquire conditioned threat, but in these patients, cognitive regulation training to regulate subjective fear to a threatening stimulus was inefficient. This suggests that the dlPFC might be essential for providing cognitive regulation of subjective fear to threatening stimuli. Moreover, Liu et al. [86] reported that an attenuated functional connectivity between the left dlPFC and the hippocampus was predictive of effective suppression of overnight consolidation of aversive memories. In line with the suggestion that the dlPFC is critically involved in top-down regulation of the cortico-meso-limbic network [28], and evidence of functional connectivity between the dlPFC and hippocampus [87], this brain region might be relevant to counteract the process of fear memory consolidation. Overall, the literature examined above suggests that different prefrontal areas might be relevant to boost the treatment of PTSD and related disorders by acting on different physiological mechanisms.

Although the results discussed in this chapter provide preliminary support for the efficacy of NIBS-VRET combinations for the treatment of PTSD and related syndromes, several limitations do apply. First, the number of investigations in the field is surprisingly low, with no studies performed for most types of PTSD-related syndromes. Second, only one study adopted a double-blind/sham-controlled protocol (the others were single blind/sham controlled). Third, none of the available studies has explored right hemispheric interventions, although inhibitory stimulation of the right dlPFC improves symptomatology in several types of ADs by downregulating reactions to negative emotional stimuli/outcomes (*see* [28] for a discussion). These results are in line with suggestions that the right dlPFC is a cortical region relevant for upregulation (in terms of emotional control) of reactions to negative emotional outcomes [e.g., 28, 88]. Therefore, future investigations that consider the right hemisphere as a target of NIBS-VRET treatments are highly recommended.

Finally, NIBS and VRET were applied simultaneously only in one study [62]. While no clinical study has systematically investigated differences in effectiveness of online and offline NIBS directly in clinical studies, one might expect a superior efficacy of online stimulation to modulate memory/learning processes via synergistic effects of learning and stimulation on plasticity, as shown for motor learning [89, 90] and other cognitive processes. With respect to the online application of stimulation, tDCS and other non-invasive electrical stimulation methods (e.g., transcranial alternating current stimulation) may be a better option, since in difference to rTMS these modulate, but do not disrupt, spontaneous cerebral activity, and thus can work synergistically with spontaneous activity via online application. Furthermore, non-invasive electrical stimulation allows an easier management/application of VRET and is associated with lower discomfort, compared with rTMS, which might be especially annoying when applied over the ventromesial prefrontal cortex, because of its effects on head muscle contraction [91].

7 Conclusion and Perspectives

In conclusion, the current state of research suggests that the combination of NIBS-VRET may be effective and promising for the treatment of PTSD and related syndromes. The low number of investigations in this field, however, does not allow to make definite conclusions in this regard. Future work should increase the number of systematic investigations that combine these two treatment methods. These studies should be designed according to high-quality standards, that is, double-blind, sham-controlled protocols with a sufficient number of participants. Additionally, systematic titration of stimulation parameters such as intensity, duration, repetition rate/intervals, and cortical targets for optimization is required to identify optimal protocols. Finally, it would be relevant to explore the combination of NIBS-VRET treatment with pharmacotherapy to explore the relative efficacy of these protocols and the collection of physiological measures, including data from functional imaging methods (e.g., fMRI, EEG), and neurovegetative parameters would be important to explore mechanisms of action of this treatment approach [92].

Acknowledgments

This work was supported by Deutscher Akademischer Austauschdienst (DAAD), BIAL foundation (Prot. Number 160/18), and the Deutsche Forschungsgemeinschaft (DFG, German Research Foundation) – Projektnummer 316803389 – SFB 1280. All authors declare no competing interests.

References

1. Kessler RC, Aguilar-Gaxiola S, Alonso J, Benjet C, Bromet EJ, Cardoso G, Degenhardt L, de Girolamo G, Dinolova RV, Ferry F, Florescu S, Gureje O, Haro JM, Huang Y, Karam EG, Kawakami N, Lee S, Lepine JP, Levinson D, Navarro-Mateu F, Pennell BE, Piazza M, Posada-Villa J, Scott KM, Stein DJ, Ten Have M, Torres Y, Viana MC, Petukhova MV, Sampson NA, Zaslavsky AM, Koenen KC (2017) Trauma and PTSD in the WHO World Mental Health Surveys. Eur J Psychotraumatol 8(sup5):1353383
2. Vicario CM, Felmingham KL (2018) Slower Time estimation in Post-Traumatic Stress Disorder. Sci Rep 8(1):39
3. Correa R, Rodriguez N, Bortolaso M (2021) What is the nature of the alteration of temporality in Trauma-Related Altered States of Consciousness? A neuro-phenomenological analysis. Eur J Trauma & Dissociation. https://doi.org/10.1016/j.ejtd.2021.100227
4. American Psychiatric Association diagnostic and statistical manual of mental disorders, 5th edn. Author, Washington, DC (2013)
5. Lanius RA, Brand B, Vermetten E, Frewen PA, Spiegel D (2012) The dissociative subtype of posttraumatic stress disorder: rationale, clinical and neurobiological evidence, and implications. Depress Anxiety 8:701–708
6. Rosellini AJ, Liu H, Petukhova MV, Sampson NA, Aguilar-Gaxiola S, Alonso J et al (2018) Recovery from DSM-IV post-traumatic stress disorder in the WHO World Mental Health surveys. Psychol Med 48(3):437–450
7. American Psychiatric Association, DSM-5 Task Force (2013) Diagnostic and statistical manual of mental disorders: DSM-5™ (5th ed.). American Psychiatric Publishing, Inc. https://doi.org/10.1176/appi.books.9780890425596
8. Chen J T-H, Belcher J, Zagic D, Wuthrich VM (2020) Anxiety disorders in later life. Reference module in neuroscience and biobehavioral psychology. https://doi.org/10.1016/B978-0-12-818697-8.00020-0
9. Bystritsky A, Khalsa SS, Cameron ME, Schiffman JPT (2013) Current diagnosis and treatment of anxiety disorders. P & T 38:30–57
10. Zayfert C, DeViva JC, Hofmann SG (2005) Comorbid PTSD and social phobia in a treatment-seeking population: an exploratory study. J Nerv Ment Dis 193(2):93–101
11. Springer KS, Levy HC, Tolin DF (2018) Remission in CBT for adult anxiety disorders: a meta-analysis. Clin Psychol Rev 61:1–8
12. Rauch SA, Eftekhari A, Ruzek JI (2012) Review of exposure therapy: a gold standard for PTSD treatment. J Rehabil Res Dev 49(5):679–687
13. Weisman JS, Rodebaugh TL (2018) Exposure therapy augmentation: a review and extension of techniques informed by an inhibitory learning approach. Clin Psychol Rev 59:41–51
14. Gonçalves R, Pedrozo AL, Coutinho ES, Figueira I, Ventura P (2012) Efficacy of virtual reality exposure therapy in the treatment of PTSD: a systematic review. PLoS One 7: e48469
15. Ready DJ, Pollack S, Rothbaum BO, Alarcon RD (2006) Virtual reality exposure for veterans with posttraumatic stress disorder. J Aggress Maltreat Trauma 12:199–220
16. Vicario CM, Nitsche MA, Hoysted I, Yavari F, Avenanti A, Salehinejad MA, Felmingham KL (2020a) Anodal transcranial direct current stimulation over the ventromedial prefrontal cortex enhances fear extinction in healthy humans: a single blind sham-controlled study. Brain Stimul 13(2):489–491
17. Vicario CM, Salehinejad MA, Mosayebi-Samani M, Maezawa H, Avenanti A, Nitsche MA (2020b) Transcranial direct current stimulation over the tongue motor cortex reduces appetite in healthy humans. Brain Stimul 13(4):1121–1123
18. Rostami M, Mosallanezhad Z, Ansari S, Ehsani F, Kidgell D, Nourbakhsh MR et al (2020) Multi-session anodal transcranial direct

current stimulation enhances lower extremity functional performance in healthy older adults. Exp Brain Res 238(9):1925–1936

19. Alizadehgoradel J, Nejati V, Movahed FS et al (2020) Repeated stimulation of the dorsolateral-prefrontal cortex improves executive dysfunctions and craving in drug addiction: a randomized, double-blind, parallel-group study. Brain Stimul 13:582–593

20. Salehinejad MA, Paknia N, Hosseinpour AH, Yavari F, Vicario CM, Nitsche MA, Nejati V (2021) Contribution of the right temporoparietal junction and ventromedial prefrontal cortex to theory of mind in autism: a randomized, sham-controlled tDCS study. Autism Res. Online ahead of print https://doi.org/10.1002/aur.2538

21. Jafari E, Alizadehgoradel J, Pourmohseni Koluri F, Nikoozadehkordmirza E, Refahi M, Taherifard M, Nejati V, Hallajian AH, Ghanavati E, Vicario CM, Nitsche MA, Salehinejad MA (2021) Intensified electrical stimulation targeting lateral and medial prefrontal cortices for the treatment of social anxiety disorder: a randomized, double-blind, parallel-group, dose-comparison study. Brain Stimul 14(4):974–986

22. Vicario CM, Nitsche MA (2013a) Transcranial direct current stimulation: a remediation tool for the treatment of childhood congenital dyslexia? Front Hum Neurosci 7:139

23. Vicario CM, Nitsche MA (2013b) Non-invasive brain stimulation for the treatment of brain diseases in childhood and adolescence: state of the art, current limits and future challenges. Front Syst Neurosci 7:94

24. Vicario CM, Nitsche MA (2018) tDCS in pediatric neuropsychiatric disorders. In: Neurotechnology and brain stimulation in pediatric psychiatric and neurodevelopmental disorders. Academic, London, pp 217–235. https://doi.org/10.1016/B978-0-12-812777-3.00009-X

25. Rivera-Urbina GN, Nitsche MA, Vicario CM, Molero-Chamizo A (2017) Applications of transcranial direct current stimulation in children and pediatrics. Rev Neurosci 28(2): 173–184

26. Salehinejad MA, Nejati V, Mosayebi-Samani M, Mohammadi A, Wischnewski M, Kuo MF, Avenanti A, Vicario CM, Nitsche MA (2020) Transcranial direct current stimulation in ADHD: a systematic review of efficacy, safety, and protocol-induced electrical field modeling results. Neurosci Bull 36(10): 1191–1212

27. Salehinejad MA, Wischnewski M, Nejati V, Vicario CM, Nitsche MA (2019) Transcranial direct current stimulation in attention-deficit hyperactivity disorder: a meta-analysis of neuropsychological deficits. PLoS One 14(4): e0215095

28. Vicario CM, Salehinejad MA, Felmingham K, Martino G, Nitsche MA (2019) A systematic review on the therapeutic effectiveness of non-invasive brain stimulation for the treatment of anxiety disorders. Neurosci Biobehav Rev 96:219–231

29. Marković V, Vicario CM, Yavari F, Salehinejad MA, Nitsche MA (2021) A systematic review on the effect of transcranial direct current and magnetic stimulation on fear memory and extinction. Front Hum Neurosci 15:655947

30. Vicario CM, Salehinejad MA, Avenanti A, Nitsche MA (2020c) Transcranial Direct Current Stimulation (tDCS) in anxiety disorders. In: Non invasive brain stimulation in psychiatry and clinical neurosciences. Springer, Cham, pp 301–317. https://doi.org/10.1007/978-3-030-43356-7_21

31. Kronberg G, Bridi M, Abel T, Bikson M, Parra LC (2017) Direct current stimulation modulates LTP and LTD: activity dependence and dendritic effects. Brain Stimul 10(1):51–58

32. Mahan AL, Ressler KJ (2012) Fear conditioning, synaptic plasticity and the amygdala: implications for posttraumatic stress disorder. Trends Neurosci 35(1):24

33. Lüthi A, Lüscher C (2014) Pathological circuit function underlying addiction and anxiety disorders. Nat Neurosci 17(12):1635–1643

34. Houdayer E, Degardin A, Cassim F, Bocquillon P, Derambure P, Devanne H (2008) The effects of low- and high-frequency repetitive TMS on the input/output properties of the human corticospinal pathway. Exp Brain Res 187(2):207–217

35. Pell GS, Roth Y, Zangen A (2011) Modulation of cortical excitability induced by repetitive transcranial magnetic stimulation: influence of timing and geometrical parameters and underlying mechanisms. Prog Neurobiol 93(1): 59–98

36. Mori F, Ribolsi M, Kusayanagi H, Siracusano A, Mantovani V, Marasco E, Bernardi G, Centonze D (2011) Genetic variants of the NMDA receptor influence cortical excitability and plasticity in humans. J Neurophysiol 106(4):1637–1643

37. Vlachos A, Müller-Dahlhaus F, Rosskopp J, Lenz M, Ziemann U, Deller T (2012) Repetitive magnetic stimulation induces functional and structural plasticity of excitatory postsynapses in mouse organotypic hippocampal

slice cultures. J Neurosci 32(48): 17514–17523

38. Lenz M, Platschek S, Priesemann V, Becker D, Willems LM, Ziemann U, Deller T, Müller-Dahlhaus F, Jedlicka P, Vlachos A (2015) Repetitive magnetic stimulation induces plasticity of excitatory postsynapses on proximal dendrites of cultured mouse CA1 pyramidal neurons. Brain Struct Funct 220(6): 3323–3337

39. Huang YZ, Edwards MJ, Rounis E, Bhatia KP, Rothwell JC (2005) Theta burst stimulation of the human motor cortex. Neuron 45(2): 201–206

40. Li C-T, Chen M-H, Juan C-H, Huang H-H, Chen L-F, Hsieh J-C, Tu P-C, Bai Y-M, Tsai S-J, Lee Y-C, Su T-P (2014) Efficacy of prefrontal theta-burst stimulation in refractory depression: a randomized sham-controlled study. Brain 137:2088–2098

41. Nitsche MA, Cohen LG, Wassermann EM, Priori A, Lang N, Antal A et al (2008) Transcranial direct current stimulation: state of the art 2008. Brain Stimul 1(3):206–223

42. Nitsche MA, Fricke K, Henschke U, Schlitterlau A, Liebetanz D, Lang N, Henning S, Tergau F, Paulus W (2003) Pharmacological modulation of cortical excitability shifts induced by transcranial direct current stimulation in humans. J Physiol 553(Pt 1): 293–301

43. Nitsche MA, Paulus W (2000) Excitability changes induced in the human motor cortex by weak transcranial direct current stimulation. J Physiol 527(3):633–639

44. Monte-Silva K, Kuo MF, Hessenthaler S, Fresnoza S, Liebetanz D, Paulus W, Nitsche MA (2013) Induction of late LTP-like plasticity in the human motor cortex by repeated non-invasive brain stimulation. Brain Stimul 6(3):424–432

45. Ziemann U, Paulus W, Nitsche MA, Pascual-Leone A, Byblow WD, Berardelli A, Siebner HR, Classen J, Cohen LG, Rothwell JC (2008) Consensus: motor cortex plasticity protocols. Brain Stimul 1(3):164–182

46. Hoogendam JM, Ramakers GM, Di Lazzaro V (2010) Physiology of repetitive transcranial magnetic stimulation of the human brain. Brain Stimul 3(2):95–118

47. Stagg CJ, Nitsche MA (2011) Physiological basis of transcranial direct current stimulation. Neuroscientist 17(1):37–53

48. Mosayebi Samani M, Agboada D, Jamil A, Kuo MF, Nitsche MA (2019) Titrating the neuroplastic effects of cathodal transcranial direct current stimulation (tDCS) over the primary motor cortex. Cortex 119:350–361

49. Batsikadze G, Paulus W, Kuo MF, Nitsche MA (2013) Effect of serotonin on paired associative stimulation-induced plasticity in the human motor cortex. Neuropsychopharmacology 38: 2260–2267

50. Mosayebi-Samani M, Melo L, Agboada D, Nitsche MA, Kuo MF (2020) Ca2+ channel dynamics explain the nonlinear neuroplasticity induction by cathodal transcranial direct current stimulation over the primary motor cortex. Eur Neuropsychopharmacol 38:63–72

51. Ressler KJ, Mayberg HS (2007) Targeting abnormal neural circuits in mood and anxiety disorders: from the laboratory to the clinic. Nat Neurosci 10(9):1116–1124

52. Lucifora C, Grasso GM, Perconti P, Plebe A (2021) Moral reasoning and automatic risk reaction during driving. Cogn Tech Work 23: 705

53. Lucifora C, Grasso GM, Perconti P, Plebe A (2020) Moral dilemmas in self-driving cars | Dilemmi morali nelle automobili a guida autonoma. Rivista Internazionale di Filosofia e Psicologia 11(2):238–250

54. Lucifora C, Angelini L, Meteier Q, Vicario CM, Abou Khaled O, Mugellini E, Grasso GM (2021) Cyber-therapy: the use of artificial intelligence in psychological practice. In: International conference on intelligent human systems integration. Springer, Cham, pp 127–132

55. Daher K, Capallera M, Lucifora C et al (2021) Empathic interactions in automated vehicles #EmpathicCHI. Conf Hum Factors Comput Syst Proc 90:1–4. https://doi.org/10.1145/3411763.3441359

56. Grasso GM, Lucifora C, Perconti P, Plebe A (2020) Integrating human acceptable morality in autonomous vehicles. In: Advances in intelligent systems and computing, vol 1131. AISC, Modena, pp 41–45

57. Grasso G, Lucifora C, Perconti P, Plebe, A (2019) Evaluating mentalization during driving. In: Proceedings of the 5th international conference on Vehicle Technology and Intelligent Transport Systems VEHITS 2019, Heraklion, Crete, Greece, 3–5 May 2019

58. Botella C, Fernández-Álvarez J, Guillén V, García-Palacios A, Baños R (2017) Recent progress in virtual reality exposure therapy for phobias: a systematic review. Curr Psychiatry Rep 19:1–13

59. Parsons TD, Rizzo AA (2008) Affective outcomes of virtual reality exposure therapy for anxiety and specific phobias: a meta-analysis. J Behav Ther Exp Psychiatry 39(3):250–261

60. Anderson PL, Zimand E, Hodges LF, Rothbaum BO (2005) Cognitive behavioral therapy for public-speaking anxiety using virtual reality for exposure. Depress Anxiety 22:156–158
61. Freitas JRS, Velosa VHS, Abreu LTN, Jardim RL, Santos JAV, Peres B, Campos PF (2021) Virtual reality exposure treatment in phobias: a systematic review. Psychiatry Q 92:1685. https://doi.org/10.1007/s11126-021-09935-6
62. van 't Wout-Frank M, Shea MT, Larson VC, Greenberg BD, Philip NS (2019) Combined transcranial direct current stimulation with virtual reality exposure for posttraumatic stress disorder: feasibility and pilot results. Brain Stimul 12(1):41–43
63. Phelps EA, Delgado MR, Nearing KI, LeDoux JE (2004) Extinction learning in humans: role of the amygdala and vmPFC. Neuron 43(6):897–905
64. Milad MR, Wright CI, Orr SP, Pitman RK, Quirk GJ, Rauch SL (2007) Recall of fear extinction in humans activates the ventromedial prefrontal cortex and hippocampus in concert. Biol Psychiatry 62:446–454. https://doi.org/10.1016/j.biopsych.2006.10.011
65. van't Wout M, Mariano TY, Garnaat SL, Reddy MK, Rasmussen SA, Greenberg BD (2016) Can transcranial direct current stimulation augment extinction of conditioned fear? Brain Stimul 9(4):529–536
66. van 't Wout M, Longo SM, Reddy MK, Philip NS, Bowker MT, Greenberg BD (2017) Transcranial direct current stimulation may modulate extinction memory in posttraumatic stress disorder. Brain Behav 7(5):e00681
67. Weathers FW, Litz BT, Keane TM, Palmieri PA, Marx BP, Schnurr PP (2013) *The ptsd checklist for dsm-5 (pcl-5)*. Scale available from: the National Center for PTSD at, www.ptsd.va.gov
68. Notzon S, Deppermann S, Fallgatter A, Diemer J, Kroczek A, Domschke K et al (2015) Psychophysiological effects of an iTBS modulated virtual reality challenge including participants with spider phobia. Biol Psychol 112:66–76
69. Deppermann S, Notzon S, Kroczek A et al (2016) Functional co-activation within the prefrontal cortex supports the maintenance of behavioural performance in fear-relevant situations before an iTBS modulated virtual reality challenge in participants with spider phobia. Behav Brain Res 307:208–217
70. Herrmann MJ, Katzorke A, Busch Y et al (2017) Medial prefrontal cortex stimulation accelerates therapy response of exposure therapy in acrophobia. Brain Stim 10:291–297
71. Nishimura Y, Tanii H, Fukuda M, Kajiki N, Inoue K, Kaiya H, Nishida A, Okada M, Okazaki Y (2007) Frontal dysfunction during a cognitive task in drug-naive patients with panic disorder as investigated by multi-channel near-infrared spectroscopy imaging. Neurosci Res 59(1):107–112
72. Szymanski J, O'Donohue WJ (1995) Fear of spiders questionnaire. Behav Ther Exp Psychiatry 26(1):31–34
73. Reiss S, Peterson RA, Gursky DM, McNally RJ (1986) Anxiety sensitivity, anxiety frequency and the prediction of fearfulness. Behav Res Ther 24(1):1–8
74. Haidt J, McCauley C, Rozin P (1994) Individual differences in sensitivity to disgust: a scale sampling seven domains of disgust elicitors. Personal Individ Differ 16:701–716
75. Cohen DC (1977) Comparison of self-report and overt-behavioral procedures for assessing acrophobia. Behav Ther 8:17–23
76. Abelson JL, Curtis GC (1989) Cardiac and neuroendocrine responses to exposure therapy in height phobics: desynchrony within the 'physiological response system'. Behav Res Ther 27:561–567
77. Spielberger CD (1983) State-Trait Anxiety Inventory for Adults (STAI-AD) [Database record]. APA PsycTests. https://doi.org/10.1037/t06496-000
78. Deacon BJ, Abramowitz JS, Woods CM, Tolin DF (2003) The Anxiety Sensitivity Index-Revised: psychometric properties and factor structure in two nonclinical samples. Behav Res Ther 41:1427–1449
79. McGlynn EA, Norquist GS, Wells KB, Sullivan G, Liberman RP (1988) Quality-of-care research in mental health: responding to the challenge. Inquiry 25(1):157–170
80. Diekhof EK, Geier K, Falkai P, Gruber O (2011) Fear is only as deep as the mind allows: a coordinate-based meta-analysis of neuroimaging studies on the regulation of negative affect. NeuroImage 58:275–285
81. Monosov IE, Hikosaka O (2012) Regionally distinct processing of rewards and punishments by the primate ventromedial prefrontal cortex. J Neurosci 32(30):10318–10330
82. Zhang L, Lengersdorff L, Mikus N, Gläscher J, Lamm C (2020) Using reinforcement learning models in social neuroscience: frameworks, pitfalls and suggestions of best practices. Soc Cogn Affect Neurosci 15(6):695–707
83. Blumberger DM, Vila-Rodriguez F, Thorpe KE et al (2018) Effectiveness of theta burst

versus high-frequency repetitive transcranial magnetic stimulation in patients with depression (THREE-D): a randomised non-inferiority trial. Lancet 391:1683–1692

84. Duval ER, Javanbakht A, Liberzon I (2015) Neural circuits in anxiety and stress disorders: a focused review. Ther Clin Risk Manag 11: 115–126

85. Kroes MCW, Dunsmoor JE, Hakimi M, Oosterwaal S, NYU PROSPEC collaboration, Meager MR, Phelps EA (2019) Patients with dorsolateral prefrontal cortex lesions are capable of discriminatory threat learning but appear impaired in cognitive regulation of subjective fear. Soc Cogn Affect Neurosci 14(6):601–612

86. Liu Y, Lin W, Liu C, Luo Y, Wu J, Bayley PJ et al (2016) Memory consolidation reconfigures neural pathways involved in the suppression of emotional memories. Nat Commun 7: 1–12. https://doi.org/10.1038/ncomms13375

87. Wang SH, Morris RG (2010) Hippocampal-neocortical interactions in memory formation, consolidation, and reconsolidation. Annu Rev Psychol 61:49–79. https://doi.org/10.1146/annurev.psych.093008.100523

88. De Raedt R, Leyman L, Baeken C, Van Schuerbeek P, Luypaert R, Vanderhasselt MA, Dannlowski U (2010) Neurocognitive effects of HF-rTMS over the dorsolateral prefrontal cortex on the attentional processing of emotional information in healthy women: an event-related fMRI study. Biol Psychol 85:487–495

89. Nitsche MA, Schauenburg A, Lang N, Liebetanz D, Exner C, Paulus W, Tergau F (2003) Facilitation of implicit motor learning by weak transcranial direct current stimulation of the primary motor cortex in the human. J Cogn Neurosci 15(4):619–626

90. Kuo TY, Van Petten C (2008) Perceptual difficulty in source memory encoding and retrieval: prefrontal versus parietal electrical brain activity. Neuropsychologia 46(8):2243–2257

91. Meteyard L, Holmes NP (2018) TMS SMART – Scalp mapping of annoyance ratings and twitches caused by Transcranial Magnetic Stimulation. J Neurosci Methods 299:34–44

92. Polania R, Nitsche MA, Ruff CC (2018) Studying and modifying brain function with non-invasive brain stimulation. Nat Neurosci 21(2):174–187

Chapter 13

Associating Aversive Task Exposure with Pharmacological Intervention to Model Traumatic Memories in Laboratory Rodents

Lucas Gazarini, Cristina A. J. Stern, and Leandro J. Bertoglio

Abstract

Post-traumatic stress disorder is associated with highly threatening and stressful events. The underlying memory is overconsolidated, leading to generalized fear expression and overall resistance to extinction- and reconsolidation-based interventions. Fear conditioning and avoidance protocols commonly used in laboratory settings induce specific and moderate-intensity aversive memories, but traumatic ones differ in quantitative and qualitative aspects. It would be appropriate to reproduce their abnormal features for studying PTSD neurobiology and assessing potential new therapeutics. After discussing the mnemonic basis of PTSD, its memory-related symptoms, and neurochemical findings underlying the traumatic memory, we aimed to review and discuss studies addressing the abovementioned question in rats and mice. Because of its potential translational value, the focus was on procedures associating an aversive task with single or combined post-training pharmacological interventions. Nearly 200 studies published since 1975 report that this protocol enhances aversive memory strength. The parallel assessment of abnormal features related to traumatic memories, such as altered specificity and susceptibility to extinction and drug-induced reconsolidation blockade, started more recently. Systemically administered drugs potentiating noradrenergic or glucocorticoid mechanisms have predominated, probably because of PTSD's physiopathology. Other options and discrete brain infusions have provided complementary information. Currently available data indicate that aversive task exposure followed by adequate drug interference during consolidation generates more intense and generalized memories, which are less prone to modulation by behavioral and pharmacological strategies. These findings based on the bedside-to-bench approach are instructive for future analyses to advance our understanding of the underlying neurobiological mechanisms and develop more effective treatments for PTSD.

Key words Memory consolidation, Translational research, PTSD model

Abbreviations

ACTH	Adrenocorticotropic hormone
AFC	Auditory fear conditioning
BLA	Basolateral amygdala
BNST	Bed nucleus of the stria terminalis
CB1	Cannabinoid type-1 receptor

CeA	Central amygdala
CFC	Contextual fear conditioning
CS	Conditioned stimulus
dACC	Dorsal anterior cingulate cortex
DH	Dorsal hippocampus
DSM-5	Diagnostic and Statistical Manual of Mental Disorders, 5th edition
eCB	Endocannabinoid
HPA	Hypothalamic–pituitary–adrenal axis
i.c.v.	Intracerebroventricular
IL	Infralimbic cortex
i.p.	Intraperitoneal
LTP	Long-term potentiation
mTORC1	Mammalian target of rapamycin complex 1
ND	Not described or retrieved
NMDA	N-Methyl-D-aspartate
NR	Nucleus reuniens of the thalamus
PFC	Prefrontal cortex
PI3K	Phosphoinositide 3-kinase
PL	Prelimbic cortex
p.o.	via oral, oral route
PTSD	Post-traumatic stress disorder
SAM	Sympathoadrenomedullary axis
s.c.	Subcutaneous
US	Unconditioned stimulus
VH	Ventral hippocampus
vmPFC	Ventromedial prefrontal cortex

1 Introduction

Memories of threatening events are more intense and long-lasting [1], and stressful experiences enhance memory formation [2–5]. Fear learning is an evolutionarily and advantageous mechanism for animal survival [6]. However, when fear is exacerbated, it becomes harmful and is a hallmark symptom of post-traumatic stress disorder (PTSD), panic disorder, and phobias [7].

Fear memory consolidation is an active process modulated by several hormones and neurotransmitters, mainly those released under stressful situations, such as noradrenaline, glucocorticoids, and endocannabinoids, which influence the activity of the brain circuitries [8–12]. Both amygdala and hippocampus play pivotal roles in fear memory consolidation [13, 14]. Basolateral amygdala (BLA) activation appears to integrate sensory information from cortical and information storage from subcortical regions following an aversive event, processing the adequate reaction to environmental threats. In order to differentiate cues with different meanings in different contexts, the interaction of the medial prefrontal cortex and hippocampus takes place, playing a fundamental role in memory encoding and retrieval [15]. Activity and plasticity in several thalamic nuclei, including the reuniens, are also crucial [16].

Over the last 50 years, several studies have contributed to the current understanding of memory consolidation [1, 14, 17, 18]. The memory consolidation theory states that acquiring new and relevant information requires a time-dependent stabilization process for "permanent" information storage [19, 20]. The most accepted theory of memory consolidation is the "systems consolidation memory," stating that the hippocampus integrates information to form a representation of the cortical activity present during the event [21]. This theory has been grounded on the observations of the H.M. patient, whose medial temporal lobe was resected and, consequently, he could not form new episodic memories [22]. Molecularly, memories are consolidated by lasting changes in synaptic strength and connections among neurons, a form of Hebbian learning called long-term potentiation (LTP; [23]). In preclinical studies, it is shown that consolidation of a fear memory lasts for 6 h, and then it becomes stable [24–27].

It was accepted for decades that a memory trace was not susceptible to modifications once consolidated. In 1997, however, it was shown that a glutamatergic N-Methyl-D-aspartate (NMDA) receptor antagonist administered after short retrieval of a consolidated radial maze task led rats to present more errors than controls at the test, which suggests that memories may become again labile and susceptible to interferences. This finding supported early observations and was referred to as reconsolidation. The reconsolidation theory states that retrieving a consolidated memory may render it unstable and susceptible to modification. It has been observed in several animal species and types of memories, including the fearful ones, and suggested as a mechanism for memory maintenance and update [28–31].

The prolonged or repeated retrieval of a fear memory may induce its extinction, a form of inhibitory learning competing with the original one and reducing fear expression. Fear extinction depends on the medial prefrontal cortex activity, mainly the infralimbic (IL) area [32]. Since extinction does not erase the associative fear memory, fear can emerge over time in a process called spontaneous recovery. In addition, stress or context exposures may elicit reinstatement or renewal of fear memory [33]. As a result, it has been shown that impairing fear memory reconsolidation has more long-lasting effects than facilitating fear extinction.

Emotionally relevant stimuli increase the peripheral adrenaline release through sympathoadrenomedullary (SAM) axis activation [12, 34, 35] and the brain noradrenaline release [34, 36, 37]. Activation of this system during a traumatic event or its recall has been associated with potentiated memory consolidation and reconsolidation in rodents and humans [25, 26, 38–40]. Consequently, individuals often present a generalized fear response to cues or environments with a sufficient degree of similarity to the original event [9, 25, 26, 41].

Generalization is a significant feature of anxiety- and stress-related disorders [42, 43]. It can be influenced by previous and/or early life stress [44], the passage of time [45], genetic background [46], and the intensity of the aversive event experienced [9, 25–27], among others. Recent studies have investigated the impact of modulating the memory consolidation strength on the generalization [9, 25–27]. As subsequently discussed, the use of associative memory protocols in rodents or humans, such as Pavlovian conditioning, and the use of passive or active avoidance tasks, which combines Pavlovian contextual fear conditioning and an instrumental response, have been of great value, providing new insights into the molecular and behavioral mechanisms of aberrant fear memories [7, 47, 48].

2 Models and Protocols to Study Fear Memory Processing in Preclinical Research

Pavlovian fear conditioning is a form of implicit learning and memory [49]. Fear conditioning occurs when a neutral and unpaired conditioned stimulus (CS) is contingently associated with an aversive unconditioned stimulus (US) that generates a reflex unconditioned fear response [50]. After CS-US pairings, CS presentation elicits a conditioned response. In studies of Pavlovian fear conditioning, the most common CS is a context or a tone, and an electric foot shock is adopted as the US. The usual laboratory animals employed in these studies are rats and mice, although this associative memory is conserved from crabs to humans [51]. The typical conditioned response evaluated is freezing behavior, a defensive response characterized by the complete absence of body movements, except those breathing-related [52]. However, other parameters such as 22 kHz ultrasonic vocalizations, analgesia, and autonomic responses such as blood pressure and heart rate can also be recorded [27, 53–55]. In the context of modeling PTSD in animal studies, the intensity of US matters since the expression of conditioned response is generally proportional to US intensity [56, 57]. Furthermore, preserving the conditioned response for months is still possible, depending on the intensity of the protocol adopted, contributing to the study of aspects related to recent and remote fear memories. In addition, it allows the study of underlying mechanisms of "acute generalization" (i.e., generalization induced by higher training intensity and overconsolidation) and time-dependent generalization (i.e., generalization induced by the passage of time and associated with systems consolidation) [27, 45, 58].

Other widely used tasks to study the neurobiology of fear memory in rodents involve the measurement of active or inhibitory (passive) avoidance associated with a type of instrumental memory involving fear conditioning. In these protocols, the presentation of an aversive stimulus is conditional to the animal's behavior [59]. Usually, the animal's behavior is stepping down a platform into a grid floor or entering a dark compartment when placed in the adjacent light compartment. In either case, a foot shock follows behavior; thus, the animal learns to present a response that may vary according to the protocol: Active avoidance tasks involve the induction of actions to avoid the aversive stimuli (e.g., transiting from one chamber to another to "escape" a foot shock after a luminous or auditory cue), while inhibitory (or passive) avoidance tasks require the suppression of expected behavior after learning (e.g., keep still on the platform or in the light compartment to avoid a foot shock, even though the expected behavior would tend to be exploratory) [51]. Thus, the fear memory index is the latency to step-down, enter the bright light compartment (step through), or perform active behaviors to avoid the US. For passive avoidance, memory intensity would be proportional to animals' latency during the test, which usually occurs 24 or 48 h later, although the response may last for months [1, 59]. More recently, some authors have highlighted the importance of developing protocols to investigate how specific this type of memory is (e.g., [59]).

In general, fear conditioning and avoidance tasks enable studying each step of memory formation, the participation of specific hormones and neurotransmitters, brain areas involved, and potential treatments to alleviate or adjust aberrant fear memory consolidation and reconsolidation. Box 1 contains additional information about fear conditioning and avoidance tasks.

Box 1: An Overview of Fear Conditioning and Avoidance Tasks on Studying Aversive Memories

Fear conditioning usually uses more than one pairing of CS-US.

The number of CS-US pairings and foot shock intensity influence the memory strength and specificity.

Inhibitory (or passive) avoidance task generally uses one pairing.

Amygdala is the locus of CS-US association.

Hippocampus is a general site for contextual processing.

Increasing training intensities may cause overconsolidation.

Overconsolidation is associated with fear generalization/overgeneralization.

Fear generalization is typically measured in an unpaired (neutral) context.

Since fear conditioning and avoidance protocols' neurobiology are different, they may produce contradictory results.

3 PTSD as a Memory-Related Psychiatric Disorder

PTSD presents a clear etiological factor: a traumatic event experienced directly or indirectly (witnessed) by the victim. That was one of the reasons why it was recently put apart from other anxiety disorders, being one of the now-called "stress- and trauma-related disorders," a classification created in the last edition of the *Diagnostic and Statistical Manual of Mental Disorders* (DSM-5; [60]). However, diagnostic criteria are complex and include excitability signals (hyperarousal, anger, impulsivity, concentration, and sleep perturbations); cognition and humor symptoms (negative emotional state, social distancing, anhedonia, distorted understanding related to trauma, and dissociative amnesia related to trauma details); and persistent avoidance of thoughts, emotions, memories, and stimuli that may elicit memory retrievals such as people, objects, situations, and places related to trauma. Intrusive symptoms, such as invasive memories, anguishing dreams/nightmares, dissociative events, and psychological/physical reactions when confronted with reminders of the trauma, are also present [60] and highlight the involvement of abnormal fear memories in PTSD [61, 62].

PTSD develops in a proportion of people exposed to a traumatic event of great severity that forms a "dysfunctional" memory [63–65]. Such traumatic memory differs from a regular aversive memory since it presumably extrapolates the evolutive function of protecting the subject from future dangers based on previous experience and results in maladaptive features. Indeed, PTSD patients express inadequate levels of anxiety and fear responses to cues related to the traumatic event but also to situations that do not directly relate to the traumatic event or do not offer risks themselves, characterizing fear overgeneralization [66, 67]. Patients are also often disturbed by spontaneous retrieval of trauma-related memories (intrusive memories or "flashbacks" in which the feeling of reviviscence is present) [68]. Moreover, traumatic memory is less prone to therapeutic strategies aiming to "regulate" it, including options focusing on extinction facilitation and reconsolidation blockade [33, 69–79]. The idea that PTSD relies on an overconsolidated aversive memory is supported by studies that correlate the stress level and/or biological features in the trauma aftermath and posterior disease development [63, 80–83], supporting the occurrence of several trauma-related features associated with altered memory processing.

4 PTSD Features Correlated with Memory Overconsolidation

Generalized fear expression (or overgeneralization) is a traumatic memory-related feature described extensively in PTSD patients [84–90] associated with the lack of specificity (or accuracy) in discriminating safe cues from threatening ones [66, 84]. Such patients also present higher threat expectancy during fear conditioning protocols [91]. Neuroimaging studies associate memory specificity loss in PTSD with abnormal recruitment of salience and central executive networks [92]. Dysfunctional frontolimbic activation [84] and significant hippocampal atrophy [93] are present in such cases, reinforcing the neuroanatomical findings usually related to PTSD physiopathology (for additional information, see Box 2). Human neural substrates for overgeneralization are consonant with those observed in animal studies [94], highlighting the relevance of animal modeling focusing on fear generalization as a translational tool for PTSD research [95].

Current treatment options for PTSD are minimal, palliative, and antidepressant-based, mainly aiming at the symptomatic control of fear and anxiety responses [96–98]. As the traumatic memory seems to be the central core of PTSD development and maintenance, recent efforts focus on the possibility of modulating fear memory processing as a more effective—and even cure-related—treatment option for such patients [99–104]. Some potentially useful strategies to mitigate the fear memory underlying PTSD aim to uncouple, attenuate, or "erase" negative valence associated with the traumatic memory in different ways [61, 105–110]. Even though some studies suggest the possibility of prophylactic measures to prevent the consolidation of PTSD-related memory [111–113], this approach raises several ethical concerns [114–117] and therefore is not explored currently as a memory-directed treatment option.

A well-debated strategy for mitigating the traumatic memory is extinction-based therapy, as continuous and/or frequent exposure to specific memory-related cues could lead to further safety association that suppresses the original traumatic memory [33, 118, 119]. Once again, animal models to understand such protocols have also been developed [120]. Even though exposure therapies aiming at extinction learning might be helpful for PTSD, several limitations are possible in the clinical context [119]. PTSD patients are generally less susceptible to extinction-based therapies, with lower learning retention and facilitated renewal and reinstatement of the traumatic memory [33, 69, 71, 73, 74, 77, 79]. Some theories even propose that extinction deficits might underlie the long-lasting psychological distress experienced by PTSD patients following trauma, making it hard to define whether impaired extinction learning would be a cause or consequence of the disorder

[121]. Interestingly, sex differences and the hormonal state could be related to increased extinction resistance in women [122], further supporting the gender-biased incidence of PTSD. Animal models for impaired extinction also provide relevant information concerning the mechanisms underlying that abnormal feature [123], eventually suggesting alternative measures to overcome this limitation. For example, the adoption of extinction-augmentation techniques is currently under investigation, including the use of behavioral or pharmacological interventions [48, 124–132]. Targeting the retrieval-induced memory destabilization as a strategy to enhance extinction learning during the reconsolidation time window is also an option under investigation and debate [133–137].

As the process of destabilization-reconsolidation represents a new opportunity for memory vulnerability and interferences after retrieval, reconsolidation-based therapies targeting the traumatic memory are currently attracting more attention and debate on pros and cons [28, 78, 100, 126, 138–143]. Among the advantages is the long-lasting adjustment of the traumatic memory that leads to PTSD, which eventually could cure patients [101, 117, 144]. In that sense, many safe and FDA-approved drugs have been shown to impair fear memories' reconsolidation in both laboratory animals ([30, 39, 40, 145]; for a review, *see* [28]) and healthy humans [4, 146–149]. However, even though such results may be relatively consistent and robust in the preclinical context, the translation of their usefulness to clinical situations is still limited [70, 72, 75, 76, 78]. Considering that inappropriate memories underlying PTSD may be less prone to interventions (e.g., [150]) the lack of adequate rodent PTSD models may account at least partly for the slow improvements in translational science. Although it seems illogical, modeling stronger memories less prone to interferences may be an essential step to developing strategies to overcome such limitations and, therefore, increase the translational value of preclinical research [151–153]. For example, the lack of retrieval-induced labilization/destabilization of the traumatic memory could explain divergences in translational research as more intense memories may be less prone to destabilization [102, 154, 155]. That idea is supported by preclinical research showing the influence of memory's intensity in allowing reconsolidation-directed interferences and the possibility of adopting drug-induced destabilization as an adjunct alternative to overcome such limitation [26, 156–158].

All such abnormal features seem to be associated with the "overconsolidation" of the traumatic memory, distancing it from "normal" fear memories. Somehow, the interaction of both environmental (e.g., the magnitude of the traumatic event and previous experiences lived by the victim) and biological factors, including genetic predisposition, particularities in neurotransmission systems, and neuroanatomy, may offer the necessary background to allow the modeling of traumatic memories underlying PTSD [47, 159–163].

5 The Challenge of Modeling Complex Conditions Like PTSD

Despite being a very complex condition, many behavioral tests and animal models may help the understanding of aversive memories' dynamics and neurobiology, such as fear conditioning protocols [1, 15, 47, 164, 165]. However, a clear obstacle to translational advances—especially regarding rodents to the human scenario—is that, in most cases, the aversive memory evaluated in experimental conditions are those lying and comprehended in physiological and adaptive levels of learning, which would not necessarily relate to traumatic memories at the PTSD core [166, 167]. Pathological and aberrant features of such memories include fear generalization to neutral cues and increased resistance to strategies to reduce its magnitude (i.e., drug-induced extinction facilitation or reconsolidation impairment) [168], which are not observed commonly or even assessed in preclinical studies. The inability to properly induce dysfunctional features in laboratory conditions may render any interpretation of simple behavioral testing protocols such as fear conditioning as "PTSD models," per se, very simplistic and flawed [160, 169]. For example, besides fear learning being enhanced in PTSD [170], it also represents an adaptive feature of memory processing (learning relevant vs. irrelevant information). In that sense, even though many studies point to the effect of consolidation-enhancement of fear memories induced by systemic drug treatments (Table 1), such neurochemical modulation could not be considered PTSD models per se as they only focus on adaptive features of fear memories (i.e., an increasing intensity in a physiological range).

The optimization of PTSD modeling should aim at inducing the formation of overconsolidated aversive memories that would be as far as possible from the adaptive range, allowing a more plausible translation approach [161, 171–173]. Alternatives to overcome such limitations include manipulating environmental aspects that may strengthen traumatic memory consolidation [47]. Those include the use of stressor agents, such as immobilization [174], confrontation with natural predators [175], or sequential "reminders" of aversive events [176], that may be useful to potentiate fear learning, thus simulating some features of pathological memories. Another approach would be exploring physiopathological features of PTSD in a preclinical context to refine animal modeling. Besides exploring the biological factor underlying PTSD, this approach is exciting since it can be used as a reverse translational measure—from the bedside to the bench—allowing the fine-tuning of animal models of PTSD based on patients' features and biomarkers [47, 177, 178]. Several neurotransmitter systems have been implicated in PTSD physiopathology, along with some neuroanatomic particularities, as the search for possible biomarkers for the disease

Table 1
Systemic, single or combined pharmacological interventions reported to potentiate the strength of the aversive memory during consolidation in rats or mice

Paradigm	Animal			Intervention		
Task, US number, and intensity	Species, strain	Sex	Age in days	Drug, effective dose(s) in mg/kg, administration route	Post-training moment of treatment (h)	References
Step-through inhibitory avoidance, 1 × 0.7 mA	Sprague-Dawley rats	♂	50–55	Adrenaline, 0.01–0.1, s.c.	0 or 0.17	[392]
Step-through inhibitory avoidance, 1 × 0.7 mA	Sprague-Dawley rats	♂	50–55	Adrenaline, 0.01–0.1, s.c.	0	[392]
Step-through inhibitory avoidance, 1 × 0.3 or 0.4 mA	Sprague-Dawley rats	♂	60–80	ACTH, 0.03 and 3.0 IU, i.p.	0	[410]
T-maze active avoidance, 1 × 0.36 mA	ICR (CD-1) mice	♂	42	Hydrocortisone, 30, s.c.	0	[411]
T-maze active avoidance, 1 × 0.36 mA	ICR (CD-1) mice	♂	42	Dexamethasone, 4.0 and 8.0, s.c.	0 or 2.5	[411]
T-maze active avoidance, 1 × 0.36 mA	ICR (CD-1) mice	♂	42	Corticosterone, 30, s.c.	0	[411]
Active avoidance, ND	Rats, strain ND	ND	ND	Substance P, 250, i.p.	0	[448]
Active avoidance, 1 × 1.0 mA	Wistar rats	♀	90–120	Naloxone, 0.3, i.p.	0	[449]
Active avoidance, 1 × 1.0 mA	Wistar rats	♀	90–120	Amphetamine, 2.0, i.p.	0	[449]
Active avoidance, 1 × 1.0 mA	Wistar rats	♀	90–120	Naloxone 0.1 + amphetamine 0.5, i.p.	0	[449]
Step-down inhibitory avoidance, 1 × 0.5 mA	Wistar rats	♂	50–70	Naloxone, 0.4, i.p.	0	[393]
Step-down inhibitory avoidance, 1 × 0.3 mA	Wistar rats	♀	50–70	ACTH 1-24, 0.0002 and 0.002, i.p.	0	[393]
Step-down inhibitory avoidance, 1 × 0.3 mA	Wistar rats	♀	50–70	Adrenaline, 0.005 and 0.05, i.p.	0	[393]
Step-down inhibitory avoidance, 1 × 0.3 mA	Wistar rats	♂	50–70	Adrenaline, 0.05, i.p.	0	[393]
Step-down inhibitory avoidance, 1 × 0.3 mA	Wistar rats	♂	50–70	ACTH 1-24, 0.002, i.p.	0	[393]
Step-down inhibitory avoidance, 1 × 0.3 mA	Wistar rats	♂	50–70	Naloxone, 0.4, i.p.	0	[393]

Task	Subject	Retention (min)	Treatment (mg/kg)	Effect	Ref.
Step-down inhibitory avoidance, 1 × 0.3 mA	Wistar rats	♂	50–70	ACTH 1-24 0.002 + naloxone 0.4, i.p.	0	[393]
Step-down inhibitory avoidance, 1 × 0.3 mA	Wistar rats	♂	50–70	Adrenaline 0.05 + naloxone 0.4, i.p.	0	[393]
Step-down inhibitory avoidance, 1 × 0.3 mA	Wistar rats	♀	50–70	ACTH 1-24, 0.002, i.p.	0	[393]
Step-down inhibitory avoidance, 1 × 0.3 mA	Wistar rats	♀	50–70	Adrenaline, 0.005 and 0.05, i.p.	0	[393]
Step-down inhibitory avoidance, 1 × 0.3 mA	Wistar rats	♂	50–70	ACTH 1-24, 0.002, i.p.	0	[393]
Step-down inhibitory avoidance, 1 × 0.3 mA	Wistar rats	♂	50–70	Adrenaline, 0.05, i.p.	0	[393]
Step-down inhibitory avoidance, 1 × 0.3 mA	Wistar rats	♂	50–70	Naloxone, 0.4, i.p.	0	[393]
Step-down inhibitory avoidance, 1 × 0.3 mA	Wistar rats	♂	50–70	ACTH 1-24 0.002 + naloxone 0.4, i.p.	0	[393]
Step-down inhibitory avoidance, 1 × 0.3 mA	Wistar rats	♂	50–70	Adrenaline 0.05 + naloxone 0.4, i.p.	0	[393]
Inhibitory avoidance, ND	C57BL/6 mice	ND	ND	Morphine 0.5 and 1.0, i.p.	0 or 0.5	[450]
Step-down inhibitory avoidance, 1 × 0.3 mA	Wistar rats	♀	80–130	β-endorphin, 0.001, i.p.	0	[394]
Step-down inhibitory avoidance, 1 × 0.3 mA	Wistar rats	♀	80–130	ACTH, 0.0002, i.p.	0	[394]
Step-down inhibitory avoidance, 1 × 0.3 mA	Wistar rats	♀	80–130	Adrenaline 0.005, i.p.	0	[394]
Step-through inhibitory avoidance, 1 × 0.8 mA	Swiss mice	♂	60	Naloxone, 0.1, i.p.	0	[396]
Step-through inhibitory avoidance, 1 × 0.8 mA	Swiss mice	♂	60	Clenbuterol, 0.01 and 0.03, i.p.	0	[396]
Y-maze active avoidance, 1 × 0.3 mA	CFW mice	♂	ND	Adrenaline, 0.1 and 0.3, i.p.	0	[397]
Step-through inhibitory avoidance, 1 × 0.7 mA	Sprague-Dawley rats	♂	60–70	Adrenaline, 0.1, s.c.	0	[395]
Y-maze discrimination test, 2 × 0.3 mA/10s	ICR (CD-1) mice	♂	ND	Fluoxetine, 5.0–15, s.c.	0 or 1.0	[441]
Step-through inhibitory avoidance, 1 × 0.2 mA	ICR (CD-1) mice	♂	ND	Fluoxetine, 15, s.c.	0	[441]
Step-through inhibitory avoidance, 1 × 0.7 mA	CFW mice	♂	60	Adrenaline, 0.01, i.p.	0	[398]
Step-through inhibitory avoidance, 1 × 0.7 mA	CFW mice	♂	60	Naloxone, 0.3 and 1.0, i.p.	0	[398]
Y-maze discrimination test, 2 × 0.35 mA	CFW mice	♂	60	Naloxone, 1.0 and 3.0, i.p.	0	[398]
Step-through inhibitory avoidance, 1 × 0.7 mA	CFW mice	♂	60	Adrenaline 0.003 + naloxone 0.1, i.p.	0	[398]
Y-maze discrimination test, 2 × 0.35 mA	CFW mice	♂	60	Adrenaline 0.1 + naloxone 0.3, i.p.	0	[398]
Step-through inhibitory avoidance, 1 × 0.2–1.0 mA	CFW mice	♂	60	Picrotoxin, 1.0 and 3.0, i.p.	0	[455]

(continued)

Table 1
(continued)

Paradigm	Animal			Intervention		References
Task, US number, and intensity	Species, strain	Sex	Age in days	Drug, effective dose(s) in mg/kg, administration route	Post-training moment of treatment (h)	
Y-maze discrimination test, 2 × 0.35 mA	CFW mice	♂	60	Picrotoxin, 1.0 and 3.0, i.p.	0	[455]
Step-through inhibitory avoidance, 1 × 0.2–1.0 mA	CFW mice	♂	60	Bicuculline, 0.1–3.0, i.p.	0	[455]
Y-maze discrimination test, 2 × 0.35 mA	CFW mice	♂	60	Bicuculline, 0.1–3.0, i.p.	0	[455]
Step-through inhibitory avoidance, 1 × 0.7 mA	CFW mice	♂	60	Adrenaline, 0.01, i.p.	0	[399]
Step-through inhibitory avoidance, 1 × 0.7 mA	CFW mice	♂	60	Oxotremorine 0.05, i.p.	0	[399]
Step-through inhibitory avoidance, 1 × 0.7 mA	CFW mice	♂	60	Adrenaline 0.003 + oxotremorine 0.005, i.p.	0	[399]
Step-through inhibitory avoidance, 1 × 0.7 mA	CFW mice	♂	60	Adrenaline 0.003 + physostigmine 0.0068, i.p.	0	[399]
Y-maze discrimination test, no footshock description	CFW mice	♂	60	Adrenaline, 0.01, i.p.	0	[399]
Y-maze discrimination test, no footshock description	CFW mice	♂	60	Oxotremorine 0.05, i.p.	0	[399]
Y-maze discrimination test, no footshock description	CFW mice	♂	60	Adrenaline 0.003 + oxotremorine 0.005, i.p.	0	[399]
Y-maze discrimination test, no footshock description	CFW mice	♂	60	Physostigmine, 0.022, i.p.	0	[399]
Y-maze discrimination test, no footshock description	CFW mice	♂	60	Adrenaline 0.003 + physostigmine 0.0068, i.p.	0	[399]
Step-through inhibitory avoidance, 1 × 0.7 mA	ICR (CD-1) mice	♂	ND	Naloxone, 2.0 and 4.0, i.p.	0	[451]
Step-through inhibitory avoidance, 1 × 0.7 mA	ICR (CD-1) mice	♂	ND	Naltrexone, 0.5 and 1.0, i.p.	0	[451]
Step-through inhibitory avoidance, 1 × 0.7 mA	ICR (CD-1) mice	♂	ND	Picrotoxin, 0.5 and 1.0, i.p.	0	[451]
Step-through inhibitory avoidance, 1 × 0.7 mA	ICR (CD-1) mice	♂	ND	Bicuculline, 0.25 and 0.5, i.p.	0	[451]
Step-through inhibitory avoidance, 1 × 0.5 mA	Sprague-Dawley rats	♂	ND	Physostigmine, 0.015 and 0.03, i.p.	0	[434]
Step-through inhibitory avoidance, 1 × 0.7 mA	ICR (CD-1) mice	♂	ND	Ro15-4513, 5.0 and 10, i.p.	0	[456]

Task	Strain	Sex	Age	Drug		Ref
Step-through inhibitory avoidance, 1 × 0.2 mA	Wistar rats	♂	ND	Dynorphin-A, 0.025, s.c.	0	[452]
Step-down inhibitory avoidance, 1 × 0.3 mA	Wistar rats	♂	80–95	Adrenaline, 0.005, i.p.	0	[400]
Step-down inhibitory avoidance, 1 × 0.3 mA	Wistar rats	♂/♀	80–95	Naloxone, 0.8, i.p.	0	[400]
Step-down inhibitory avoidance, 1 × 0.3 mA	Wistar rats	♂/♀	80–95	ACTH 1-24, 0.0002, i.p.	0	[400]
Active avoidance, 1 × 0.4 mA	Wistar rats	♂/♀	80–95	Adrenaline, 0.005, i.p.	0	[400]
Active avoidance, 1 × 0.4 mA	Wistar rats	♂/♀	80–95	Naloxone, 0.8, i.p.	0	[400]
Active avoidance, 1 × 0.4 mA	Wistar rats	♂/♀	80–95	ACTH, 0.0002, i.p.	0	[400]
Step-through inhibitory avoidance, 1 × 0.2 mA	C57BL/6 mice	♂	56	SKF38393, 10 and 20, i.p.	0	[437]
Step-through inhibitory avoidance, 1 × 0.2 mA	C57BL/6 mice	♂	56	LY171555, 0.5 and 1.0, i.p.	0	[437]
Step-through inhibitory avoidance, 1 × 0.5 mA	CFW mice	♂	60	Dipivefrin, 0.001–0.01, i.p.	0	[401]
Y-maze discrimination test, 2 × 0.35 mA	CFW mice	♂	60	Dipivefrin, 0.0003–0.01, i.p.	0	[401]
Step-through inhibitory avoidance, 1 × 0.5 mA	CFW mice	♂	60	Adrenaline, 0.1, i.p.	0	[401]
Y-maze discrimination test, 2 × 0.35 mA	CFW mice	♂	60	Adrenaline, 0.1, i.p.	0	[401]
Active avoidance, 1 × 0.5 mA	Wistar rats	♂	90–97	Adrenaline, 0.05, i.p.	0	[402]
Step-through inhibitory avoidance, 1 × 0.2 mA	DBA2 mice	♂	56	Muscimol, 1.0 and 2.0, i.p.	0	[457]
Step-through inhibitory avoidance, 1 × 0.2 mA	DBA2 mice	♂	56	Baclofen, 10 and 20, i.p.	0	[457]
Step-through inhibitory avoidance, 1 × 0.2 mA	C57BL/6 mice	♂	56	Picrotoxin, 0.5 and 1.0, i.p.	0	[457]
Step-through inhibitory avoidance, 1 × 0.2 mA	C57BL/6 mice	♂	56	Bicuculline, 0.25 and 0.5, i.p.	0	[457]
Step-through inhibitory avoidance, 1 × 0.2 mA	C57BL/6 mice	♂	56	CGP 35348, 200 and 300, i.p.	0	[457]
Step-through inhibitory avoidance, 1 × 0.8 mA	Swiss mice	♂	ND	Glucose, 30, i.p.	0 or 0.17	[474]
Step-through inhibitory avoidance, 1 × 0.35 mA	Swiss mice	♂	42	Amlodipine, 7.0–30, i.p.	0	[475]
Contextual fear conditioning, 2 × 0.15 or 0.3 mA	Swiss mice	♂	42	Amlodipine, 7.0, i.p.	0	[475]
Active avoidance, 1 × 0.16 mA	Swiss mice	♂	42	Amlodipine, 10, i.p.	0	[475]
Y-maze active avoidance, 1 × 0.35 mA	Sprague-Dawley rats	♂	ND	Adrenaline, 0.05, i.p.	0	[403]

(continued)

Table 1
(continued)

Paradigm	Animal			Intervention		
Task, US number, and intensity	Species, strain	Sex	Age in days	Drug, effective dose(s) in mg/kg, administration route	Post-training moment of treatment (h)	References
Active avoidance, 1 × 0.5 mA	Wistar rats	♂	92	Adrenaline, 0.01 and 0.05, i.p.	0	[404]
Step-through inhibitory avoidance, 1 × 0.2 mA	C57BL/6 mice	♂	56	Cocaine, 2.5–10, i.p.	0	[438]
Step-through inhibitory avoidance, 1 × 0.2 mA	C57BL/6 mice	♂	56	Nomifensine, 2.5–10, i.p.	0	[438]
Step-through inhibitory avoidance, 1 × 0.5 mA	ICR (CD-1) mice	♂	56	Minaprine, 10, i.p.	0	[438]
Auditory fear conditioning, 3 × 0.5 mA	Long-Evans rats	♂	ND	Scopolamine, 1.0, i.p.	0	[435]
Step-through inhibitory avoidance, 1 × 0.2 mA	C57BL/6 mice	♂	56	Corticosterone, 0.5 and 1.0, i.p.	0	[412]
Step-through inhibitory avoidance, 1 × 0.2 mA/1s	C57BL/6 mice	♂	56	Cocaine, 2.5 and 5.0, i.p.	0	[439]
Step-through inhibitory avoidance, 1 × 0.2 mA	C57BL/6 mice	♂	56	Physostigmine, 0.1–0.4, i.p.	0	[439]
Step-through inhibitory avoidance, 1 × 0.2 mA	C57BL/6 mice	♂	56	Cocaine, 1.0 or 2.5 + physostigmine 0.05 or 0.1, i.p.	0	[439]
Step-through inhibitory avoidance, 1 × 0.2 mA	DBA2 mice	♂	56	Physostigmine, 0.05–0.4, i.p.	0	[439]
Step-through inhibitory avoidance	ICR (CD-1) mice	ND	ND	Caffeine, 0.25–1.0, i.p.	0	[440]
Step-through inhibitory avoidance	ICR (CD-1) mice	ND	ND	Cocaine, 1.0–5.0, i.p.	0	[440]
Step-through inhibitory avoidance	ICR (CD-1) mice	ND	ND	Caffeine + cocaine, i.p.	0	[440]
Step-through inhibitory avoidance, 1 × 0.45 mA	Sprague-Dawley rats	♂	ND	Dexamethasone, 0.3, s.c.	0	[413]
Step-through inhibitory avoidance, 1 × 0.45 mA	Sprague-Dawley rats	♂	ND	Dexamethasone, 0.3 and 1.0, s.c.	0	[414]
Step-through inhibitory avoidance, 1 × 0.45 mA	Sprague-Dawley rats	♂	ND	Dexamethasone, 0.3, s.c.	0	[415]

Task	Strain	Sex	Age	Drug, dose, route		Ref
Active avoidance, 1 × 0.5 mA	Wistar rats	♂	90	Adrenaline, 0.05, i.p.	0	[405]
Contextual fear conditioning, 3 × 0.2 mA	Wistar rats	♂	ND	Corticosterone, 5.0, i.p.	0	[280]
Active avoidance, 1 × 0.5 mA	Wistar rats	♂	94	Adrenaline, 0.01, i.p.	0	[406]
Step-through inhibitory avoidance, 1 × 0.2 mA	C57BL/6 mice	♂	56	Anandamide, 3.0 and 6.0, i.p.	0	[453]
Step-through inhibitory avoidance, 1 × 0.2 mA	DBA2 mice	♂	56	Naltrexone, 0.2, i.p.	0	[453]
Step-through inhibitory avoidance, 1 × 0.2 mA	DBA2 mice	♂	56	Oxotremorine, 0.01–0.04, i.p.	0	[454]
Step-through inhibitory avoidance, 1 × 0.2 mA	C57BL/6 mice	♂	56	Oxotremorine, 0.04, i.p.	0	[454]
Step-through inhibitory avoidance, 1 × 0.2 mA	C57BL/6 mice	♂	56	Oxotremorine 0.04 + LY171555 0.25, i.p.	0	[454]
Step-through inhibitory avoidance, 1 × 0.2 mA	C57BL/6 mice	♂	56	Oxotremorine 0.04 + SKF 38393 5.0, i.p.	0	[454]
Step-through inhibitory avoidance, 1 × 0.2 mA	DBA2 mice	♂	56	Oxotremorine 0.04 + sulpiride 6.0, i.p.	0	[454]
Step-through inhibitory avoidance, 1 × 0.2 mA	DBA2 mice	♂	56	Oxotremorine 0.04 + SCH23390 0.025, i.p.	0	[454]
Step-through inhibitory avoidance, 1 × 0.4 mA	Swiss mice	♂	ND	Caffeine, 0.3, i.p.	0	[458]
Step-through inhibitory avoidance, 1 × 0.4 mA	Swiss mice	♂	ND	SCH 58261, 0.3, i.p.	0	[458]
Step-through inhibitory avoidance, 1 × 0.35 or 0.45 mA	Sprague-Dawley rats	♂	ND	Dexamethasone, 0.3 and 1.0, s.c.	0	[416]
Step-through inhibitory avoidance, 1 × 0.1 mA	Swiss mice	♂	ND	ACPC 100–400, i.p.	0	[459]
Step-through inhibitory avoidance, 1 × 0.1 mA	Swiss mice	♂	ND	7KYN, 15–30, i.p.	0	[459]
Step-through inhibitory avoidance, 1 × 0.8 mA	CF-1 mice	♂	60–70	Gabapentin, 10, i.p.	0	[476]
Step-through inhibitory avoidance, 1 × 0.8 mA	CF-1 mice	♂	60–70	Gabapentin 5.0 + physostigmine 0.035, i.p.	0	[476]
Step-through inhibitory avoidance, 1 × 0.25 mA	Long-Evans rats	♀	55	Dihydrotestosterone 3.0 and 7.5, s.c.	0	[466]
Step-through inhibitory avoidance, 1 × 0.25 mA	Long-Evans rats	♀	55	5α-androstane-3α,17β-diol, 3.0 and 7.5, s.c.	0	[466]
Step-down inhibitory avoidance, 1 × 0.2 mA	Wistar rats	♂	80	Agmatine, 1.0–20, i.p.	0	[467]
Auditory fear conditioning, 2 × 0.75 mA	F344 rats	♂	90–150	MK-801, 0.3, s.c.	0	[460]
Auditory fear conditioning, ND	Lister-Hooded rats	♂	ND	Dexamethasone, 1.2, s.c.	0	[417]
Auditory fear conditioning, 3 × 0.5 mA	Sprague-Dawley rats	♂	ND	Corticosterone 3.0, s.c.	0	[418]

(continued)

Table 1 (continued)

Paradigm	Animal				Intervention	Post-training moment of treatment (h)	References
Task, US number, and intensity	Species, strain	Sex	Age in days		Drug, effective dose(s) in mg/kg, administration route		
Contextual fear conditioning, 7 × 1.0 mA	Wistar rats	♂	ND		Proxyfan, 0.04 and 0.2, i.p.	0	[477]
Step-down inhibitory avoidance, 1 × 0.4 mA	Wistar rats	♂	ND		Dexamethasone, 0.3, i.p.	0	[419]
Active avoidance, 1 × 0.5 mA	Wistar rats	♂	95		Adrenaline, 0.01, i.p.	0	[407]
Auditory fear conditioning, 5 × 0.5 mA	Sprague-Dawley rats	♂	ND		Corticosterone, 3.0, s.c.	0	[420]
Auditory fear conditioning, 3 × 0.6 mA	Wistar rats	♂	ND		Spermidine, 10 and 100, i.p.	0	[470]
Contextual fear conditioning, 3 × 0.6 mA	Wistar rats	♂	ND		Spermidine, 10 and 100, i.p.	0	[470]
Auditory fear conditioning, 1 × or 2 × 0.6 mA	Long-Evans rats	♂	56		Corticosterone, 3.0, s.c.	0	[284]
Conditioned taste aversion, mildly aversive LiCl solution	Sprague-Dawley rats	♂	ND		Corticosterone, 1.0 and 3.0, s.c.	0	[421]
Step-through inhibitory avoidance, 1 × 0.33 mA	Sprague-Dawley rats	♂	ND		Dexamethasone, 0.3 and 1.0, s.c.	0	[428]
Step-down inhibitory avoidance, 1 × 0.3 mA	Wistar rats	♂	ND		Adrenaline, 0.025, i.p.	0	[408]
Step-down inhibitory avoidance, 1 × 0.3 mA	Wistar rats	♂	ND		Naloxone, 0.4, i.p.	0	[408]
Step-down inhibitory avoidance, 1 × 0.3 mA	Wistar rats	♂	ND		Dexamethasone, 0.3, i.p.	0	[408]
Step-down inhibitory avoidance, 1 × 0.3 mA	Wistar rats	♂	ND		Glucose, 320, i.p.	0	[408]
Contextual fear conditioning, 2 × 0.4 mA	Wistar rats	♂	ND		Corticosterone, 3.0, i.p.	0	[423]
Step-through inhibitory avoidance, 1 × 0.4 mA	NMRI mice	♂	42		DHEA, 6.1 μM/kg, s.c.	0	[471]
Step-through inhibitory avoidance, 1 × 0.4 mA	Sprague-Dawley rats	♂	ND		Oleoylethanolamide, 5.0 and 10, i.p.	0	[464]

Task	Subject	Age (days)	Drug, dose (mg/kg), route	Delay (h)	Ref
Step-through inhibitory avoidance, 1 × 0.4 mA	Sprague-Dawley rats ♂	ND	GW7647, 10, i.p.	0	[464]
Step-through inhibitory avoidance, 1 × 0.4 mA	Sprague-Dawley rats ♂	ND	(R)-1′-methyl-oleoylethanolamide, 50, i.p.	0	[464]
Step-through inhibitory avoidance, 1 × 0.6 mA	Sprague-Dawley rats ♂	ND	Corticosterone, 3.0, s.c.	0	[424]
Contextual fear conditioning, 1 × 0.75 mA	Sprague-Dawley rats ♂	ND	Escitalopram, 5.0, s.c.	0	[442]
Contextual fear conditioning, 2 × 0.25 mA	C57BL/6J mice ♀	70–84	Thioperamide, 10 and 20, i.p.	0	[444]
Step-through inhibitory avoidance, 1 × 0.35 mA	Sprague-Dawley rats ♂	ND	Propofol, 300 and 350, i.p.	0 or 0.5	[472]
Auditory fear conditioning, 2 × 0.3 mA	C57Bl/6J ♂	90	Corticosterone, 0.75–10, i.p.	0	[425]
Contextual fear conditioning, 2 × 0.25 mA	C57BL/6J mice ♀	70–84	Pitolisant, 2.5 and 5.0, i.p.	0	[446]
Step-through inhibitory avoidance, 1 × 0.3 mA	C57BL/6J mice ♀	70–84	Thioperamide, 10 and 20, i.p.	0	[445]
Contextual fear conditioning, 1 × 0.7 mA	Wistar rat ♂	90–120	Yohimbine, 1.0, i.p.	0	[25]
Auditory fear conditioning, 1 × 0.8 mA	Sprague-Dawley rats ♂	ND	Heroin, 0.03, s.c.	0	[454]
Auditory fear conditioning, 1 × 0.8 mA	Sprague-Dawley rats ♂	ND	Amphetamine, 0.5, s.c.	0	[454]
Contextual fear conditioning, 3 × 0.4 mA	Sprague-Dawley rats ♂	ND	Corticosterone, 3.0, i.p.	0	[426]
Contextual fear conditioning, 1 × 0.4 mA	Sprague-Dawley rats ♂	ND	Corticosterone, 10, i.p.	1.0	[428]
Step-through inhibitory avoidance, 1 × 0.5 mA	Wistar rats ♂	ND	CI08297, 20, s.c.	0	[427]
Step-through inhibitory avoidance, 1 × 0.5 mA	Wistar rats ♂	ND	Corticosterone, 1.0, s.c.	0	[427]
Auditory fear conditioning, 8 × 0.75 mA	C57BL/6J mice ♂	56–84	TCB-2, 1.0, i.p.	0	[443]
Contextual fear conditioning, 2 × 0.4 mA	Wistar rats ♀	ND	Corticosterone, 0.3, i.p.	0	[430]
Contextual fear conditioning, 2 × 0.4 mA	Wistar rats ♀	ND	Corticosterone 0.3 + 17β-estradiol 0.001, i.p.	0	[430]
Step-through inhibitory avoidance, 1 × 0.25 mA	ICR (CD-1) mice ♂	40	Oroxylin A, 5.0 and 10, p.o.	0 or 3.0	[447]

(continued)

Table 1
(continued)

Paradigm	Animal			Intervention		Post-training moment of treatment (h)	References
Task, US number, and intensity	Species, strain	Sex	Age in days	Drug, effective dose(s) in mg/kg, administration route			
Step-through inhibitory avoidance, 1 × 0.32 mA	Sprague-Dawley rats	♂	ND	Corticosterone, 3.0, i.p.		0	[429]
Step-through inhibitory avoidance, 1 × 0.3 mA	ICR (CD-1) mice	♂	ND	MMPP, 0.56, i.p.		0	[478]
Step-through inhibitory avoidance, 1 × 0.3 mA	ICR (CD-1) mice	♂	ND	Piracetam, 100, i.p.		0	[478]
Olfactory fear conditioning, 1 × 0.4 mA	Wistar rat	♂	85–105	Fludrocortisone, 1.0 and 3.0, s.c.		0	[431]
Step-through inhibitory avoidance, 1 × 0.38 mA	Sprague-Dawley rats	♂	70–100	Corticosterone, 1.0, s.c.		0	[59]
Contextual fear conditioning, 1 × 0.2 mA	Wistar rats	ND	ND	Corticosterone, 2.0, i.p.		0	[432]
Contextual fear conditioning, 1 × 2.0 mA	Sprague-Dawley rats	♂	90–120	Oxytocin, 0.2, s.c.		0	[468]
Step-through inhibitory avoidance, 1 × 0.2 mA	Swiss mice	♂	ND	AM251, 1.0 and 3.0, i.p.		0	[461]
Step-through inhibitory avoidance, 1 × 0.35 mA	Sprague-Dawley rats	♂	ND	Ketamine, 125, i.p.		0	[462]
Contextual fear conditioning, 5 × 0.8 mA	Sprague-Dawley rats	♂	ND	Propofol, 300, i.p.		0	[462]
Step-through inhibitory avoidance, 1 × 0.35 mA	Sprague–Dawley rats	♂	ND	JZL184, 0.5, i.p.		0	[465]
Step-through inhibitory avoidance, 1 × 0.45 mA	Wistar rats	♂	ND	Corticosterone, 3.0, i.p.		0	[433]
Step-through inhibitory avoidance, 1 × 1.5 mA	Wistar rats	♂	ND	Olanzapine, 2.0 and 5.0, i.p.		0	[473]
Step-through inhibitory avoidance, 1 × 0.35 mA	Sprague-Dawley rats	♂	ND	Yohimbine, 1.0, s.c.		0	[409]

Inhibitory avoidance discrimination test, 1 × 0.35 mA	Sprague-Dawley rats	♂	ND	Corticosterone, 3.0, s.c.	0	[409]
Step-through inhibitory avoidance, 1 × 0.35 mA	Sprague-Dawley rats	♂	ND	D-lactate, 1000, s.c.	0	[469]
Step-through inhibitory avoidance, 1 × 0.35 mA	Sprague-Dawley rats	♂	ND	3,5-DHBA, 60, s.c.	0	[469]
Step-through inhibitory avoidance, 1 × 0.35, 0.6 or 0.65 mA	Sprague-Dawley rats	♂	ND	Ketamine, 125, i.p.	0	[463]
Contextual fear conditioning, 1 × 0.7 mA	Wistar rats	♂	90–120	Adrenaline 0.05 + AM251 1.0 + corticosterone 1.0, i.p.	0	[9]
Contextual fear conditioning, 1 × 0.7 mA	Wistar rats	♂	90–120	Adrenaline 0.1, i.p.	0	[9]
Contextual fear conditioning, 1 × 0.7 mA	Wistar rats	♂	90–120	AM251 3.0, i.p.	0	[9]
Contextual fear conditioning, 1 × 0.7 mA	Wistar rats	♂	90–120	Corticosterone 10, i.p.	0	[9]

Legend: *ND* not described or retrieved, *♂* male, *♀* female, *i.p.* intraperitoneal, *s.c.* subcutaneous, *p.o.* via oral, *ACPC* Glycine B partial agonist, *ACTH* adrenocorticotropic hormone, *Agmatine* multiple molecular targets modulator, *Amlodipine* calcium channel blocker, *Amphetamine* neuronal and vesicular monoamine transporters inhibitor, *AM251* CB1 receptor antagonist/inverse agonist, *Anandamide* CB1/CB2 receptor agonist, *Baclofen* GABA B receptor agonist, *Bicuculline* GABA B receptor antagonist, *Caffeine* adenosine 2A receptor antagonist, phosphodiesterase inhibitor, *Cocaine* neuronal monoamine transporters inhibitor, *C108297* glucocorticoid receptor modulator, *CGP 35348* GABA B receptor antagonist, *Clenbuterol β2*-adrenergic receptor agonist, *Corticosterone* glucocorticoid and mineralocorticoid receptors agonist, *Dexamethasone* glucocorticoid receptor agonist, *DHEA* dehydroepiandrosterone, a metabolic intermediate required for the synthesis of testosterone and estrogen, *Dihydrotestosterone* androgen receptor agonist, *Dipivefrin* adrenaline's dipivalate ester, *D-lactate* hydroxycarboxylic acid 1 receptor partial agonist, *Dynorphin-A* κ-opioid receptor agonist, *Escitalopram* selective serotonin reuptake inhibitor, *Fludrocortisone* mineralocorticoid receptor agonist, *Fluoxetine* selective serotonin reuptake inhibitor, *Gabapentin* calcium influx blocker and potassium efflux activator, *GW 7647* PPAR-alfa agonist, *Heroin* μ-opioid receptor agonist, *Hydrocortisone* glucocorticoid and mineralocorticoid receptors agonist, *JZL184* monoacylglycerol lipase inhibitor, *Ketamine* NMDA receptor antagonist, *LY171555* Dopamine 2/3 receptor agonist (quinpirole), *Minaprine* monoamine oxidase inhibitor, *MK-801* NMDA receptor antagonist, *MMPP* (1-(4-methoxy-2-methylphenyl) piperazine), a nootropic drug, *Morphine* μ-opioid receptor agonist, *Naloxone* μ-opioid receptor antagonist, *Muscimol* GABA A receptor agonist, *Naltrexone* μ-opioid receptor antagonist, *Nomifensine* norepinephrine-dopamine reuptake inhibitor, *Olanzapine* serotonin 2A and dopamine 2 receptors antagonist, *Oleoylethanolamide* PPAR-alfa agonist, *Oroxylin A* increases BDNF levels, *Oxotremorine* muscarinic receptor agonist, *Physostigmine* acetylcholinesterase inhibitor, *Picrotoxin* GABA A receptor antagonist, *Piracetam* nootropic drug, *Pitolisant* histamine 3 receptor antagonist, *Propofol* GABA A receptor agonist, *Proxyfan* histamine 3 receptor agonist, *Ro15-4513* partial inverse agonist of the benzodiazepine receptor, *SCH 23390* Dopamine 1/5 receptor partial agonist, *SCH 58261* Adenosine A2A receptor antagonist, *Scopolamine* muscarinic receptor antagonist, *SKF 38393* Dopamine 1/5 receptor partial agonist, *Spermidine* NMDA receptor modulator, *Substance P* neurokinin type-1 receptor agonist, *Sulpiride* dopamine 2 receptor antagonist, *TCB-2* Serotonin 2A receptor agonist, *Thioperamide* histamine 3 receptor antagonist, *Yohimbine* α2-adrenergic receptor antagonist/inverse agonist, *17β-estradiol* estrogen receptor agonist, *3,5-DHBA* hydroxycarboxylic acid receptor 1 agonist, *5α-androstane-3α,17β-diol* androgen receptor agonist, *7KYN* Glycine B antagonist, *(R)-1'-methyl-oleoylethanolamide* PPAR-alfa agonist, *β-endorphin* μ-opioid receptor agonist follows (for more information, *see* Boxes 2 and 3 below). This reverse translational paradigm would further confirm the relevance of a specific biomarker for PTSD in a preclinical context, increase the construct- and face-validity of current animal models for the disease, and improve research findings' translational value. Based on the above, mimicking neuroendocrine features observed in PTSD patients in laboratory animals would represent an innovative tool to induce traumatic memories due to their overconsolidation [161, 168, 179]. Some classically studied neurotransmission systems that can enhance fear memory consolidation (as seen in Table 1) may be related to PTSD physiopathology. However, the simple fact that a drug or strategy for neurotransmission modulation induced a single PTSD-like feature (e.g., memory consolidation enhancement) does not necessarily make it a plausible "PTSD model," as it may lack face validity [180]. Because of that, knowledge concerning the neurochemical basis of PTSD is crucial.

6 Neural Basis and Neurochemical Signatures of PTSD Related to Memory Processing

Several neuroimaging studies have highlighted neuroanatomical findings in PTSD patients in the past decades. Hippocampal shrinkage is one of the most described findings associated with the disorder. Lower hippocampal volumes could be observed in PTSD patients compared to trauma-exposed or non-trauma-exposed healthy subjects [181–183]. Notably, hippocampal shrinkage does not depend on gender [184]. It could serve as a biomarker for fear generalization [93] because reduced neurogenesis has been correlated with impairments in pattern separation [185]. Prefrontal cortex (PFC) abnormalities include volume reductions in the ventromedial area (vmPFC; [186]) and dorsal anterior cingulate cortex (dACC; [187]), which are also dysfunctional during trauma-related stimuli presentation [188]. Lower neuronal density was also observed in dACC [189]. Aberrant connectivity of dACC and amygdala was also described [190], consonant to the amygdala hyperactivation during exposure to trauma-related cues in PTSD [191]. Overall, many dysfunctional brain areas described in PTSD are the same ones implicated in defensive behavior elaboration and fear memory processing [192, 193] and, thus, could be directly related to specific trauma-related features [194–196]. For further insights on PTSD's neural basis, check Box 2.

Box 2: Further Reading on the Neural Abnormalities Described in PTSD	
Category	**References**
General studies	[161, 197–222]
Prefrontal cortex	[223–233]
Amygdala	[223, 229, 233–243]
Hippocampus	[236, 239, 244–255]
Other brain areas or aspects	[256–265]

When analyzing the neurochemical basis for PTSD, a particular issue is that most studies cannot discriminate if the dysfunctions observed may be the cause or consequence of PTSD [82], a pervasive limitation when trying to establish the basis of this condition. However, there is a large spectrum of possible neurochemical biomarkers for PTSD, including monoamines, amino acid derivatives, neuropeptides, hormones of the hypothalamic-pituitary-adrenal (HPA) axis, neurotrophic factors, and immune mediators [162, 266, 267]. Some of them are more classically associated with the physiopathology of the disease, such as the noradrenergic system and the HPA axis, which have been extensively studied concerning their relation to PTSD development.

The role of the noradrenergic system in modulating adaptive aversive memories is clear [10, 11]. Emotionally relevant stimuli can increase the brain noradrenaline release [34, 36, 37] and the peripheral adrenaline release through SAM axis activation, which also allows a positive feedback loop for brain noradrenaline release [12, 34, 35]. Patients that presented more profound autonomic changes such as sympathetic hyperactivation (i.e., increased heart rate) in the trauma aftermath were more associated with PTSD development and worse symptoms [80, 81, 83], igniting the interpretation that noradrenergic dysfunctions might underlie the traumatic memory. Later, a correlation between brain noradrenergic levels and PTSD symptoms' severity was shown [268], along with increased noradrenaline in the cerebrospinal fluid and catecholamines' metabolites in urine after the retrieval of traumatic memories in PTSD patients [269]. Correspondingly, increased noradrenergic modulation during aversive learning could render fear memories less prone to retrieval-induced destabilization in rodents [36], a feature observed in PTSD patients.

Many particularities in the noradrenergic system have been described in PTSD patients and might explain or support the hyperactivation suggested earlier. A lower activity of enzymes related to catecholamine (including noradrenaline and adrenaline) degradation may underlie the increased noradrenergic tonus [270, 271]; reduced inhibitory control mediated by α_2-adrenergic receptors [272, 273] would increase the release of noradrenaline in normal conditions. Increased α_1-adrenergic receptor sensitivity was also described in such patients [274]. Additionally, PTSD patients also present higher responsiveness to drugs (e.g., yohimbine) that potentiate the release of noradrenaline [275–277], and increased levels of catecholamines' metabolites were detected under various situations, including the trauma aftermath [269, 278, 279].

The glucocorticoid actions in fear memory have also long been described. In rodents, blood corticosterone (cortisol in humans) increases upon aversive stimuli presentation and according to its strength [34, 56, 280, 281]. Blood glucocorticoid is readily available in the brain, enhancing fear memory processing [282–284] and reinforcing its role in memory modulation. For that reason, it seems logical to evaluate the involvement of the HPA axis in PTSD [285]. Indeed, the glucocorticoid system was also one of the first neurochemical modulators to be implicated in those patients, as normal to lower blood and salivary cortisol levels were observed [286, 287]. Even though that feature seems paradoxical given the general potentiating effects of glucocorticoids on memory, an increased glucocorticoid receptors sensitivity or responsiveness, resulting in higher activation and increased negative feedback on cortisol release, was described later [288–291], probably owing to epigenetic modifications that could be transferred across generations [292]. Interestingly, the peripheral lower cortisol levels found in PTSD patients should not be transposed directly to its central action, as increases in cortisol and corticotropin-releasing hormone levels were detected in the brain [293, 294].

Altogether, data supports the glucocorticoid influence on traumatic memory formation and maintenance [295], even though clinical research often presents mixed findings concerning the glucocorticoid involvement in PTSD [296] and, therefore, demands a critical analysis [297]. Some recent studies point to glucocorticoid reduction because of persistent inflammatory stimuli in PTSD [298]. Indeed, increased levels of cytokines and proteins of the immune system transduction pathway were described in reviews and meta-analyses [299–301], supporting the role of inflammatory mechanisms underlying PTSD and collateral to HPA axis involvement. It would be rational to think that other signaling mechanisms that could counteract such stress and inflammatory responses could probably take part in PTSD physiopathology.

Systemic levels of endocannabinoids (eCB) are acutely modified by stress. They are involved in stress-buffering responses, especially concerning the HPA axis activity [302]. This premise was further supported by a human study that described higher eCB salivary concentrations just after a stressful event but not 30 or 45 min later, coinciding with a cortisol increase [303]. Regarding the eCB involvement in PTSD, studies have found lower eCB levels [304] that could be associated with higher PTSD symptoms' severity [305]. PTSD patients also present impaired eCB signaling after psychological stress [306]. Mutations in eCB degradation enzymes that could lead to enhanced anandamide levels are associated with increased extinction learning in healthy adults [307], a result that could be translated to PTSD patients as behavioral-induced augmented anandamide levels correlate with more efficient extinction learning, which is limited in these patients [308]. The abovementioned results suggest an overall reduced/impaired eCB tonus, a premise supported by increased brain cannabinoid type-1 (CB1) receptor expression in PTSD [309]. Altogether, such findings support the theory that PTSD patients may present hypoactive eCB signaling [306, 310].

As mentioned before, one of the theories explaining PTSD development and maintenance is the overconsolidation of fear memory, which would allow the development of an abnormal memory with inappropriate features [63, 64]. The neuroendocrine profile described in PTSD patients could lead to dysfunctional memory consolidation. Increased noradrenergic and glucocorticoid signaling mechanisms are associated with stronger emotional memories. Reduced eCB tonus could also play a part in enhancing fear memories under certain conditions, highlighting the possibility of interaction of various stress-related neurotransmitters in the development of PTSD [311] (for further reading on the neurochemical basis of PTSD, *see* references on Box 3). For that reason, we further point to the usefulness of the reverse translational approach to mimic neurochemical dysfunctions described in PTSD to better model the disease in preclinical conditions.

Box 3: Further Reading on the Neurochemical Dysfunctions Described in PTSD

Category	References
General neurochemical basis	[161, 266, 267, 311–322]
Noradrenergic system and the SAM axis	[323–346]
Glucocorticoids and the HPA axis	[324, 327, 330, 347–378]
Endocannabinoids	[379–391]

Based on the above, we aim to review and discuss animal studies in which aversive memories with traumatic-like features were induced in rats or mice. Because of its potential translational value, the focus was on those associating an aversive task with post-training pharmacological intervention(s), as depicted in Fig. 1.

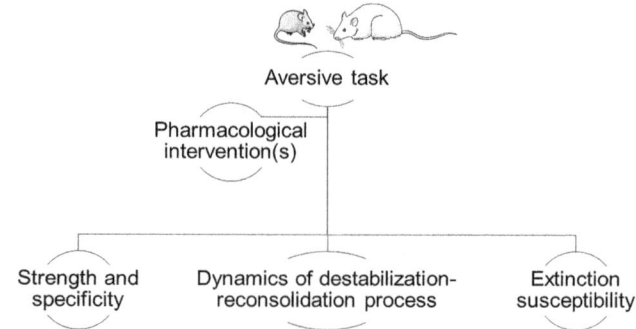

Fig. 1 Scheme of the experimental protocol used and mnemonic aspects potentially investigated in the studies reviewed herein

7 Potentiating Effects of Post-acquisition Interventions on Aversive Memory Strength

Table 1 summarizes the main details from studies in which a single or combined systemic pharmacological intervention has potentiated the strength of the aversive memory during consolidation in rats or mice. Most of the studies have used a drug that stimulates directly or indirectly the neurotransmission via noradrenaline [9, 25, 392–409], adrenocorticotropic hormone (ACTH; [393, 394, 400, 410]), glucocorticoids [9, 59, 284, 408, 411–433], acetylcholine [399, 434–436], dopamine [437–440], serotonin [441–443], histamine [444–447], or substance P [448]. Alternatively, it attenuates the neurotransmission via endogenous opioids [393, 394, 396, 398, 400, 408, 449, 450–454], GABA [451, 455–457], or adenosine [440, 458].

Both glutamatergic and cannabinoid drugs have similarly been used, but whether stimulation or attenuation of these systems potentiates the strength of the memory depends on several factors, including the primary receptor subtype engaged and the level of emotional arousal induced by environmental conditions. For example, the antagonism of glutamate NMDA or CB1 receptors has often enhanced memory intensity ([9, 459–463]; but see [453, 464, 465]).

Drugs targeting other neurotransmitter systems [466–469] and those substances acting on multiple targets [438, 449, 454, 462, 470–473] or producing more overall effects [408, 469, 474–478] have also occasionally been used. Notably, many studies have investigated the effects of combined pharmacological interventions on memory strength. Typically, two or three drugs are associated at subthreshold doses to investigate the potential interaction between/among the neurotransmitter systems involved [9, 393, 398, 399, 440, 449, 453, 476].

In summary, single and combined interventions following aversive experience can potentiate memory strength, an effect relatively comparable across tasks and animal species/strains used. These studies have primarily tested adult male rats or mice, although available female data appear consistent [393, 400, 430, 449, 466]. Though not serving as PTSD models per se, such studies provide helpful insights for future research and animal model refinement.

8 Detrimental Effects of Post-acquisition Interventions on Aversive Memory Specificity, Destabilization, Drug-Induced Reconsolidation Blockade, and Susceptibility to Extinction

Table 2 summarizes the main details from studies in which systemic or central pharmacological interventions have affected the specificity, destabilization upon retrieval, drug-induced reconsolidation blockade, or susceptibility to the extinction of the aversive memory during consolidation in rats or mice. Most studies have focused on memory specificity and systemically administered a drug that stimulates the neurotransmission via glucocorticoids [57, 425, 479, 480, 481], noradrenaline ([25, 26, 158]; but see [482]), or their association with the CB1 receptor antagonist AM251 at subthreshold doses [9]. The studies mentioned above have reported memory specificity deficits. The dopamine type-2 receptor antagonist raclopride [483] and the GABA-B receptor antagonist CGP 36216 [484] similarly impaired memory specificity after post-acquisition systemic treatment. Raclopride infused into the central amygdala or the bed nucleus of the stria terminalis (BNST) [483], CGP 36216 infused along the dorsoventral axis of the hippocampus [484], the allosteric inhibitor of mammalian target of rapamycin complex 1 (mTORC1) rapamycin infused into the BNST [483], the glutamate NMDA receptor agonist NMDA infused into the prelimbic cortex (PL; [485]), the GABA-A receptor agonist muscimol infused into the thalamic nucleus reuniens (NR), or the IL [16, 486] also produced corresponding results.

Post-acquisition systemic administration of the adrenergic α_2-receptor antagonist yohimbine, which enhances the noradrenergic tonus indirectly, impaired memory destabilization upon retrieval [26, 158]. Additionally, either systemically administered yohimbine or intra-NR infusion of muscimol attenuated the drug-induced reconsolidation disruption and the susceptibility to extinction [16, 26, 158]. The selective and cell-permeable phosphoinositide 3-kinase (PI3K) inhibitor LY294002 infused into the basolateral amygdala [487], the NMDA infused into the PL [485], and the protein synthesis inhibitor anisomycin or muscimol infused into the IL [486] also produced a relative extinction resistance.

Table 2
Systemic or brain pharmacological interventions reported to affect the specificity, destabilization upon retrieval, drug-induced reconsolidation impairment, or susceptibility to extinguish the aversive memory during consolidation in rats or mice

Paradigm	Animal			Intervention		Memory feature				References
Task, US number, and intensity	Species, strain	Sex	Age in days	Drug, effective dose(s), route or local of administration	Post-training moment of treatment (h)	Specificity	Destabilization	Drug-induced reconsolidation impairment	Susceptibility to extinction	
AFC, 2 × 0.8 mA	C57Bl/6J mice	♂	90	Corticosterone, 0.75–10 mg/kg, i.p.	0	↓				[425]
CFC, 1 × 0.7 mA	Wistar rats	♂	90–120	Yohimbine, 1.0 mg/kg, i.p.	0	↓				[25]
CFC, 3 × 0.8 mA	Sprague-Dawley rats	♂	60	LY294002, 25 µM/0.5 µL/side, BLA	0				↓	[487]
CFC, 3 × 0.7 mA	Wistar rats	♂	90–120	Yohimbine, 1.0 mg/kg, i.p.	0	↓	↓		↓	[26]
AFC, 5 × 0.6 mA	C57BL/6J mice	♂	60–90	Raclopride, 0.3 mg/kg, i.p.	0	↓				[483]
AFC, 5 × 0.6 mA	C57BL/6J mice	♂	60–90	Raclopride, 0.5 µg/side, CeA	0	↓				[483]
AFC, 5 × 0.6 mA	C57BL/6J mice	♂	60–90	Raclopride, 0.5 µg/side, BNST	0	↓				[483]
AFC, 5 × 0.6 mA	C57BL/6J mice	♂	60–90	Raclopride 0.5 µg, BNST + CeA contralateral	0	↓				[483]
AFC, 5 × 0.6 mA	C57BL/6J mice	♂	60–90	Rapamycin, 1.0 µg/side, BNST	0	↓				[483]
CFC, 5 × 0.8 mA	C57BL/6 mice	♂	45–70	CGP 36216, 3 mM/side, i.c.v.	0	↓				[484]
CFC, 5 × 0.8 mA	C57BL/6 mice	♂	45–70	CGP 36216, 3 mM, DH	0	↓				[484]
CFC, 5 × 0.8 mA	C57BL/6J mice	♂	45–70	CGP 36216, 3 mM, VH	0	↓				[484]

Task	Strain	Sex	Time	Drug			Ref
CFC, 3 × 1.0 mA	Wistar rats	♂	90–105	NMDA, 0.1 nmol/side, PL	0	↓	[485]
CFC, 3 × 0.8 mA	Wistar rats	♂	100–115	Muscimol, 9.2 ng/side, NR	0	↓	[16]
CFC, 1 × 0.8 mA	Wistar rats	♂	100–115	Muscimol, 9.2 ng/side, NR	0	↓	[16]
AFC or unpaired cued CFC, 2 × 1.4 mA	Sprague-Dawley rats	♂	ND	Propranolol, 2.0 mg/kg, s.c.	0	↓	[482]
AFC or unpaired cued CFC, 2 × 1.4 mA	Sprague-Dawley rats	♂	ND	Propranolol, 1.5 and 5 ug/side, DH	0	↓	[482]
AFC, 2 × 0.4 mA	C57Bl/6J mice	♂	90	Corticosterone, 2.5 mg/kg, i.p.	0	↓	[479]
CFC, 3 × 1.0 mA	Wistar rats	♂	90–105	Muscimol, 9.2 ng/side, IL	0	↓	[486]
CFC, 1 × 1.0 mA	Wistar rats	♂	90–105	Muscimol, 9.2 ng/side, IL	0	↓	[486]
CFC, 3 × 1.0 mA	Wistar rats	♂	90–105	Anisomycin, 20 ug/side, IL	0	↓	[486]
AFC, 2 × 1.2 mA	NPSR knockout	♂♀	60–90	Corticosterone, 5.0 mg/kg, i.p.	0	↓	[480]
CFC, 3 × 0.7 mA	Wistar rats	♂	90–120	Yohimbine, 2.0 mg/kg, i.p.	0	↓	[158]
IA discrimination test, 1 × 0.35 mA	Sprague-Dawley rats	♂	ND	Corticosterone, 3.0 mg/kg, s.c.	0	↓	[409]
CFC, 3 × 0.6 mA	Wistar rats	♂	90	Corticosterone, 4.0 mg/kg, s.c.	0	↓	[57]
AFC, 3 × 0.2 mA	C57BL/6J mice	♂	60–90	Corticosterone, 2.0 mg/kg, i.p.	0	↓	[481]
CFC, 1 × 0.7 mA	Wistar rats	♂	90–120	Adrenaline 0.05 + AM251 1.0 + Corticosterone 1.0 mg/kg, i.p.	0	↓	[9]

Legend: *ND* not described or retrieved, ♂ male, ♀ female, *i.p.* intraperitoneal, *s.c.* subcutaneous, *i.c.v.* intracerebroventricular, *AFC* auditory fear conditioning, *CFC* contextual fear conditioning, *Anisomycin* protein synthesis inhibitor, *BNST* bed nucleus of the stria terminalis, *BLA* basolateral amygdala, *CeA* central amygdala, *CGP 36216* GABA B receptor antagonist, *Corticosterone* glucocorticoid and mineralocorticoid receptors agonist, *DH* dorsal hippocampus, *IL* infralimbic cortex, *LY294002* selective and cell-permeable Phosphoinositide 3-kinase (PI3K) inhibitor, *Muscimol* GABA A receptor agonist, *NMDA* glutamate NMDA receptor agonist, *NR* nucleus reuniens of the thalamus, *PL* prelimbic cortex, *Raclopride* dopamine 2 antagonist, *Propranolol* β-adrenergic receptor antagonist, *Rapamycin* allosteric inhibitor of mammalian target of rapamycin complex 1 (mTORC1), *VH* ventral hippocampus, *Yohimbine* α2-adrenergic receptor antagonist/inverse agonist

In summary, systemic or brain interventions during the consolidation of the aversive experience affect memory specificity. Similar impairing effects on destabilization upon retrieval, drug-induced reconsolidation blockade, and susceptibility to extinguish the aversive memory have been reported. However, the studies are still scarce and essentially limited to males. Notably, several of the studies listed used pharmacological interferences described in Table 1 as memory-strengthening treatments, reinforcing the usefulness of revisiting classical studies to optimize animal models for PTSD research.

9 General Discussion

Studies reporting potentiated aversive memory strength (Table 1), an effect usually inferred by an increase in inhibitory/passive or active avoidance and freezing behaviors at test, have used several tasks associated with post-acquisition systemic pharmacological intervention(s). As shown in Fig. 2a, the prevalence of using the step-through, step-down, active avoidance, contextual, and auditory fear conditioning procedures varies over time: Early work has used avoidance-based tasks while the step-through and fear conditioning use has been balanced in the last decade. This situation is opportune because they are complementary and allow the identification of the relative participation of a given neurotransmitter system, differences between two or more neurotransmitter systems, and their potentially varying interactions.

As shown in Fig. 2b, studies reporting deficits in aversive memory specificity (inferred by generalized fear expression at recent time points, usually ≤ a week), destabilization upon retrieval, drug-induced memory reconsolidation blockade, and susceptibility to extinction (all three inferred by the absence of expected between-group differences at test, which varies on the length, for instance) still have mainly used fear-conditioning procedures associated with post-acquisition systemic or local pharmacological intervention(s) (Table 2). Although it is not necessarily an issue, studies using inhibitory/passive or active avoidance tasks are needed to compare their data with relatively consistent yet incipient available ones, suggesting that the conditions mentioned above involve altered functioning of memory-related brain areas during consolidation.

Varying pharmacological interventions can induce abnormal, PTSD-like memory features during consolidation. The ones acting through noradrenergic or glucocorticoid mechanisms still prevail, although other options are numerous, as depicted in Tables 1 and 2. At first glance, one might consider associating an aversive task with administering one or more drugs, irrespective of the primary mechanism of action engaged, relatively "unnatural" to address this question. However, since growing evidence indicate that aversive

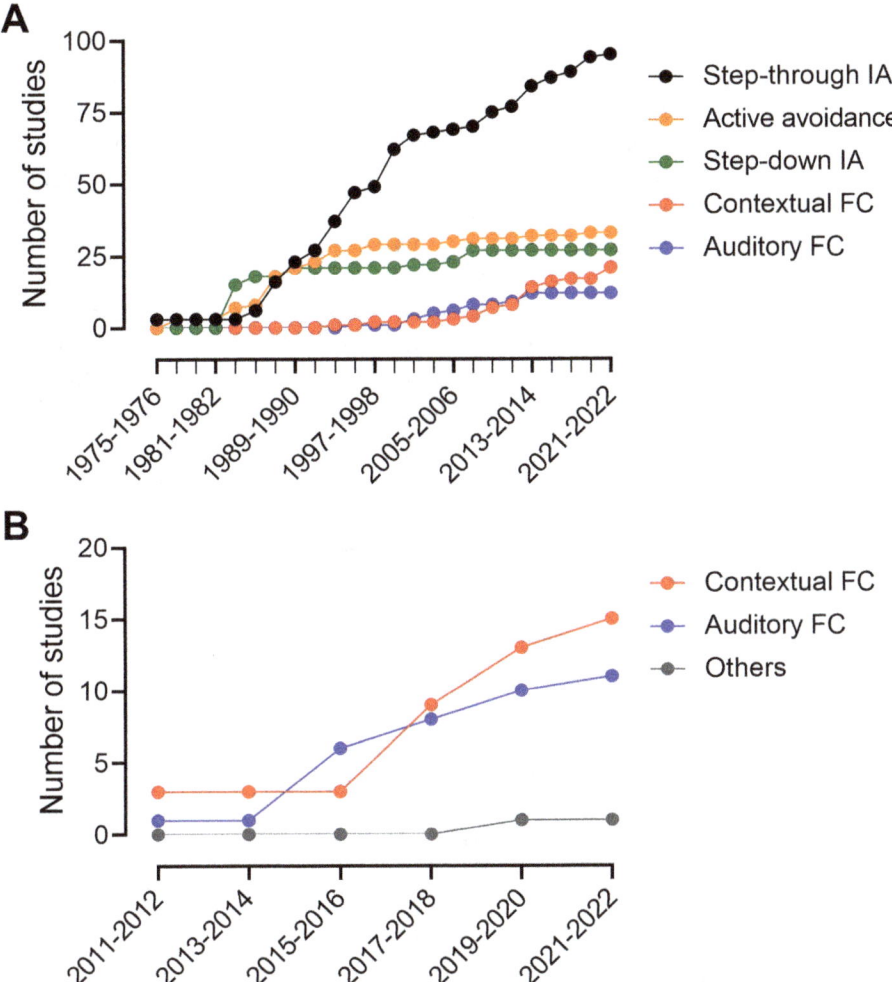

Fig. 2 (**a**) Cumulative curves of main tasks associated with post-acquisition systemic pharmacological interventions used in experiments reporting potentiated aversive memory strength (Table 1). (**b**) Cumulative curves of tasks associated with post-acquisition systemic or local pharmacological interventions used in experiments reporting impairing effects on destabilization upon retrieval, drug-induced reconsolidation blockade, and susceptibility to extinguish the aversive memory (Table 2)

memory-related neuromodulators often interact with each other [433, 488, 489, 490, 491, 492, 493, 494], single and particularly combined systemic interferences may be adequate to replicate altered memory attributes reported following traumatic experiences in humans (e.g., [9, 26, 158]). Studies reviewed here have only occasionally and more recently (from 2012 onwards; *see* Table 2) examined other attributes of abnormal aversive memories besides their strength. Thus, additional studies are necessary to advance the understanding of their neural basis and underlying mechanisms, which might shed light on new potential therapeutic targets to regulate traumatic memories.

It is worth remembering that step-through, step-down, active avoidance, contextual, and auditory fear conditioning tasks have long been used to study normal, adaptive aversive memory. Their integration in more complex experimental designs to study abnormal, potentially pathophysiological aspects of PTSD-like memories is a recent trend. However, whether these aversive tasks are specific to studying PTSD is still a question of debate and investigation [495, 496]. Besides, increased fear expression and/or impaired extinction may occur spontaneously by individual variability [82, 497], and already-stressed animals subjected to aversive tasks may present deficits not only in the process of destabilization-reconsolidation but also in fear extinction [498, 499, 500, 501]. Based on the above, everyone should rely on various behavioral outcomes to ensure that a PTSD-like memory is modeled and assessed.

10 Concluding Remarks

Traumatic memories are often more intense, generalized, inflexible, and resistant to suppression. Most studies that show promising results of post-retrieval behavioral- and pharmacological-based strategies use avoidance- and freezing-associated tasks but focus on the adaptive range of aversive memories. There has been a gradual scientific interest in mimicking and assessing all the altered mnemonic aspects to increase the validity of PTSD modeling and improve the translation value of preclinical findings. The use of commonly aversive tasks (e.g., contextual fear conditioning) associated with post-acquisition administration of one or more drugs that simulate neurochemical changes seen in the trauma aftermath, or disrupt their physiological dynamics, has been considered adequate to achieve this goal. Futures are guaranteed to investigate still-open questions systematically. For instance, are altered aversive memory strength and specificity interrelated? Based on preliminary evidence (e.g., [16, 481]), the site of the pharmacological intervention(s) matters because studies using spatially restricted manipulations into discrete brain areas can dissociate each other. Similarly, as extinction is a context-dependent process, extinguishing less precise fear memories would be relatively more difficult [486].

References

1. Izquierdo I, Furini CR, Myskiw JC (2016) Fear memory. Physiol Rev 96(2):695–750. https://doi.org/10.1152/physrev.00018.2015
2. Aubry AV, Serrano PA, Burghardt NS (2016) Molecular mechanisms of stress-induced increases in fear memory consolidation within the amygdala. Front Behav Neurosci 10:191. https://doi.org/10.3389/fnbeh.2016.00191
3. de Quervain D, Schwabe L, Roozendaal B (2017) Stress, glucocorticoids and memory: implications for treating fear-related

disorders. Nat Rev Neurosci 18(1):7–19. https://doi.org/10.1038/nrn.2016.155

4. Schwabe L, Nader K, Wolf OT, Beaudry T, Pruessner JC (2012) Neural signature of reconsolidation impairments by propranolol in humans. Biol Psychiatry 71(4):380–386. https://doi.org/10.1016/j.biopsych.2011.10.028

5. Wingenfeld K, Wolf OT (2014) Stress, memory, and the hippocampus. Front Neuro Neurosci 34:109–120. https://doi.org/10.1159/000356423

6. Orsini CA, Maren S (2012) Neural and cellular mechanisms of fear and extinction memory formation. Neurosci Biobehav Rev 36(7): 1773–1802. https://doi.org/10.1016/j.neubiorev.2011.12.014

7. Mahan AL, Ressler KJ (2012) Fear conditioning, synaptic plasticity and the amygdala: implications for posttraumatic stress disorder. Trends Neurosci 35(1):24–35. https://doi.org/10.1016/j.tins.2011.06.007

8. Costanzi M, Battaglia M, Rossi-Arnaud C, Cestari V, Castellano C (2004) Effects of anandamide and morphine combinations on memory consolidation in cd1 mice: involvement of dopaminergic mechanisms. Neurobiol Learn Mem 81(2):144–149. https://doi.org/10.1016/j.nlm.2003.09.003

9. Gazarini L, Stern CA, Takahashi RN, Bertoglio LJ (2022) Interactions of noradrenergic, glucocorticoid and endocannabinoid systems intensify and generalize fear memory traces. Neuroscience 497:118–133. https://doi.org/10.1016/j.neuroscience.2021.09.012

10. Kety SS (1972) The possible role of the adrenergic systems of the cortex in learning. Res Publica 50:376–389

11. McGaugh JL, Gold PE, Van Buskirk R, Haycock J (1975) Modulating influences of hormones and catecholamines on memory storage processes. Prog Brain Res 42:151–162. https://doi.org/10.1016/S0079-6123(08)63656-0

12. Dronjak S, Gavrilović L, Filipović D, Radojcić MB (2004) Immobilization and cold stress affect sympatho-adrenomedullary system and pituitary-adrenocortical axis of rats exposed to long-term isolation and crowding. Physiol Behav 81(3):409–415. https://doi.org/10.1016/j.physbeh.2004.01.011

13. van Marle HJ, Hermans EJ, Qin S, Fernández G (2009) From specificity to sensitivity: how acute stress affects amygdala processing of biologically salient stimuli. Biol Psychiatry 66(7):649–655. https://doi.org/10.1016/j.biopsych.2009.05.014

14. McGaugh JL (2015) Consolidating memories. Annu Rev Psychol 66:1–24. https://doi.org/10.1146/annurev-psych-010814-014954

15. Maren S, Phan KL, Liberzon I (2013) The contextual brain: implications for fear conditioning, extinction and psychopathology. Nat Rev Neurosci 14(6):417–428. https://doi.org/10.1038/nrn3492

16. Troyner F, Bicca MA, Bertoglio LJ (2018) Nucleus reuniens of the thalamus controls fear memory intensity, specificity and long-term maintenance during consolidation. Hippocampus 28(8):602–616. https://doi.org/10.1002/hipo.22964

17. McIntyre CK, McGaugh JL, Williams CL (2012) Interacting brain systems modulate memory consolidation. Neurosci Biobehav Rev 36(7):1750–1762. https://doi.org/10.1016/j.neubiorev.2011.11.001

18. Haubrich J, Nader K (2018) Memory reconsolidation. Curr Top Behav Neurosci 37:151–176. https://doi.org/10.1007/7854_2016_463

19. Izquierdo I (1992) Dopamine receptors in the caudate nucleus and memory processes. Trends Pharmacol Sci 13(1):7–8. https://doi.org/10.1016/0165-6147(92)90004-p

20. Dudai Y (2012) The restless engram: consolidations never end. Annu Rev Neurosci 35: 227–247. https://doi.org/10.1146/annurev-neuro-062111-150500

21. Eichenbaum H (2016) Still searching for the engram. Learn Behav 44(3):209–222. https://doi.org/10.3758/s13420-016-0218-1

22. Scoville WB, Milner B (1957) Loss of recent memory after bilateral hippocampal lesions. J Neurol Neurosurg Psychiatry 20(1):11–21. https://doi.org/10.1136/jnnp.20.1.11

23. Bliss TV, Lomo T (1973) Long-lasting potentiation of synaptic transmission in the dentate area of the anaesthetized rabbit following stimulation of the perforant path. J Physiol 232(2):331–356. https://doi.org/10.1113/jphysiol.1973.sp010273

24. Lynch MA (2004) Long-term potentiation and memory. Physiol Rev 84(1):87–136. https://doi.org/10.1152/physrev.00014.2003

25. Gazarini L, Stern CA, Carobrez AP, Bertoglio LJ (2013) Enhanced noradrenergic activity potentiates fear memory consolidation and reconsolidation by differentially recruiting α1- and β-adrenergic receptors. Learn Mem 20(4):210–219. https://doi.org/10.1101/lm.030007.112

26. Gazarini L, Stern CA, Piornedo RR, Takahashi RN, Bertoglio LJ (2014) PTSD-like memory generated through enhanced noradrenergic activity is mitigated by a dual step pharmacological intervention targeting its reconsolidation. Int J Neuropsychopharmacol 18(1):pyu026. https://doi.org/10.1093/ijnp/pyu026

27. Stern CAJ, da Silva TR, Raymundi AM, de Souza CP, Hiroaki-Sato VA, Kato L, Guimarães FS, Andreatini R, Takahashi RN, Bertoglio LJ (2017) Cannabidiol disrupts the consolidation of specific and generalized fear memories via dorsal hippocampus CB1 and CB2 receptors. Neuropharmacology 125: 220–230. https://doi.org/10.1016/j.neuropharm.2017.07.024

28. Besnard A, Caboche J, Laroche S (2012) Reconsolidation of memory: a decade of debate. Prog Neurobiol 99(1):61–80. https://doi.org/10.1016/j.pneurobio.2012.07.002

29. Nader K, Schafe GE, Le Doux JE (2000) Fear memories require protein synthesis in the amygdala for reconsolidation after retrieval. Nature 406(6797):722–726. https://doi.org/10.1038/35021052

30. Stern CA, Gazarini L, Takahashi RN, Guimarães FS, Bertoglio LJ (2012) On disruption of fear memory by reconsolidation blockade: evidence from cannabidiol treatment. Neuropsychopharmacology 37(9):2132–2142. https://doi.org/10.1038/npp.2012.63

31. Pedreira ME, Pérez-Cuesta LM, Maldonado H (2004) Mismatch between what is expected and what actually occurs triggers memory reconsolidation or extinction. Learn Mem 11(5):579–585. https://doi.org/10.1101/lm.76904

32. Milad MR, Quirk GJ (2002) Neurons in medial prefrontal cortex signal memory for fear extinction. Nature 420(6911):70–74. https://doi.org/10.1038/nature01138

33. VanElzakker MB, Dahlgren MK, Davis FC, Dubois S, Shin LM (2014) From Pavlov to PTSD: the extinction of conditioned fear in rodents, humans, and anxiety disorders. Neurobiol Learn Mem 113:3–18. https://doi.org/10.1016/j.nlm.2013.11.014

34. Swenson RM, Vogel WH (1983) Plasma Catecholamine and corticosterone as well as brain catecholamine changes during coping in rats exposed to stressful footshock. Pharmacol Biochem Behav 18(5):689–693. https://doi.org/10.1016/0091-3057(83)90007-2

35. de Diego AM, Gandía L, García AG (2008) A physiological view of the central and peripheral mechanisms that regulate the release of catecholamines at the adrenal medulla. Acta Physiol 192(2):287–301. https://doi.org/10.1111/j.1748-1716.2007.01807.x

36. Haubrich J, Bernabo M, Nader K (2020) Noradrenergic projections from the locus coeruleus to the amygdala constrain fear memory reconsolidation. elife 9:e57010. https://doi.org/10.7554/eLife.57010

37. Seo DO, Zhang ET, Piantadosi SC, Marcus DJ, Motard LE, Kan BK, Gomez AM, Nguyen TK, Xia L, Bruchas MR (2021) A locus coeruleus to dentate gyrus noradrenergic circuit modulates aversive contextual processing. Neuron 109(13):2116–2130.e6. https://doi.org/10.1016/j.neuron.2021.05.006

38. Borrell J, De Kloet ER, Versteeg DH, Bohus B (1983) Inhibitory avoidance deficit following short-term adrenalectomy in the rat: the role of adrenal catecholamines. Behav Neural Biol 39(2):241–258. https://doi.org/10.1016/s0163-1047(83)90910-x

39. Gamache K, Pitman RK, Nader K (2012) Preclinical evaluation of reconsolidation blockade by clonidine as a potential novel treatment for posttraumatic stress disorder. Neuropsychopharmacology 37(13): 2789–2796. https://doi.org/10.1038/npp.2012.145

40. Przybyslawski J, Roullet P, Sara SJ (1999) Attenuation of emotional and nonemotional memories after their reactivation: role of beta adrenergic receptors. J Neurosci 19(15): 6623–6628. https://doi.org/10.1523/JNEUROSCI.19-15-06623.1999

41. Asok A, Kandel ER, Rayman JB (2019) The neurobiology of fear generalization. Front Behav Neurosci 12:329. https://doi.org/10.3389/fnbeh.2018.00329

42. Dunsmoor JE, Prince SE, Murty VP, Kragel PA, LaBar KS (2011) Neurobehavioral mechanisms of human fear generalization. NeuroImage 55(4):1878–1888. https://doi.org/10.1016/j.neuroimage.2011.01.041

43. Lissek S, Rabin S, Heller RE, Lukenbaugh D, Geraci M, Pine DS, Grillon C (2010) Overgeneralization of conditioned fear as a pathogenic marker of panic disorder. Am J Psychiatr 167(1):47–55. https://doi.org/10.1176/appi.ajp.2009.09030410

44. Elliott ND, Richardson R (2019) The effects of early life stress on context fear generalization in adult rats. Behav Neurosci 133(1): 50–58. https://doi.org/10.1037/bne0000289

45. Pedraza LK, Sierra RO, Boos FZ, Haubrich J, Quillfeldt JA, Alvares L (2016) The dynamic

nature of systems consolidation: stress during learning as a switch guiding the rate of the hippocampal dependency and memory quality. Hippocampus 26(3):362–371. https://doi.org/10.1002/hipo.22527

46. Temme SJ, Bell RZ, Pahumi R, Murphy GG (2014) Comparison of inbred mouse substrains reveals segregation of maladaptive fear phenotypes. Front Behav Neurosci 8:282. https://doi.org/10.3389/fnbeh.2014.00282

47. Bienvenu T, Dejean C, Jercog D, Aouizerate B, Lemoine M, Herry C (2021) The advent of fear conditioning as an animal model of post-traumatic stress disorder: learning from the past to shape the future of PTSD research. Neuron 109(15):2380–2397. https://doi.org/10.1016/j.neuron.2021.05.017

48. Lisboa SF, Vila-Verde C, Rosa J, Uliana DL, Stern C, Bertoglio LJ, Resstel LB, Guimaraes FS (2019) Tempering aversive/traumatic memories with cannabinoids: a review of evidence from animal and human studies. Psychopharmacology 236(1):201–226. https://doi.org/10.1007/s00213-018-5127-x

49. Schafe GE, LeDoux JE (2008) Learning and memory: a comprehensive reference, vol 4. Academic, Boston, pp 157–192. https://doi.org/10.1016/B978-012370509-9.00045-0

50. Rescorla RA (1968) Probability of shock in the presence and absence of CS in fear conditioning. J Comp Physiol Psychol 66(1):1–5. https://doi.org/10.1037/h0025984

51. Kim JJ, Jung MW (2006) Neural circuits and mechanisms involved in Pavlovian fear conditioning: a critical review. Neurosci Biobehav Rev 30(2):188–202. https://doi.org/10.1016/j.neubiorev.2005.06.005

52. Blanchard RJ, Blanchard DC (1969) Passive and active reactions to fear-eliciting stimuli. J Comp Physiol Psychol 68(1):129–135. https://doi.org/10.1037/h0027676

53. Fanselow MS (1986) Conditioned fear-induced opiate analgesia: a competing motivational state theory of stress analgesia. Ann N Y Acad Sci 467:40–54. https://doi.org/10.1111/j.1749-6632.1986.tb14617.x

54. Schwarting RK, Wöhr M (2012) On the relationships between ultrasonic calling and anxiety-related behavior in rats. Braz J Med Biol Res 45(4):337–348. https://doi.org/10.1590/s0100-879x2012007500038

55. Stiedl O, Tovote P, Ögren SO, Meyer M (2004) Behavioral and autonomic dynamics during contextual fear conditioning in mice. Auton Neurosci 115(1–2):15–27. https://doi.org/10.1002/da.22903

56. Dos Santos Corrêa M, Vaz B, Grisanti G, de Paiva J, Fornari PA, Tiba RV (2019) Relationship between footshock intensity, post-training corticosterone release and contextual fear memory specificity over time. Psychoneuroendocrinology 110:104447. https://doi.org/10.1016/j.psyneuen.2019.104447

57. Dos Santos Corrêa M, Vaz B, Menezes BS, Ferreira TL, Tiba PA, Fornari RV (2021) Corticosterone differentially modulates time-dependent fear generalization following mild or moderate fear conditioning training in rats. Neurobiol Learn Mem 184:107487. https://doi.org/10.1016/j.nlm.2021.107487

58. da Silva TR, Sohn J, Andreatini R, Stern CA (2020) The role of prelimbic and anterior cingulate cortices in fear memory reconsolidation and persistence depends on the memory age. Learn Mem 27(8):292–300. https://doi.org/10.1101/lm.051615.120

59. Atucha E, Zalachoras I, van den Heuvel JK, van Weert LT, Melchers D, Mol IM, Belanoff JK, Houtman R, Hunt H, Roozendaal B, Meijer OC (2015) A mixed glucocorticoid/mineralocorticoid selective modulator with dominant antagonism in the male rat brain. Endocrinology 156(11):4105–4114. https://doi.org/10.1210/en.2015-1390

60. American Psychiatric Association (2013) Diagnostic and statistical manual of mental disorders, 5th edn. Author, Washington, DC

61. Taylor JR, Torregrossa MM (2015) Pharmacological disruption of maladaptive memory. Handb Exp Pharmacol 228:381–415. https://doi.org/10.1007/978-3-319-16522-6_13

62. Herz N, Bar-Haim Y, Holmes EA, Censor N (2020) Intrusive memories: a mechanistic signature for emotional memory persistence. Behav Res Ther 135:103752. https://doi.org/10.1016/j.brat.2020.103752

63. Pitman RK (1989) Post-traumatic stress disorder, hormones, and memory. Biol Psychiatry 26(3):221–223. https://doi.org/10.1016/0006-3223(89)90033-4

64. Bracha HS (2006) Human brain evolution and the "Neuroevolutionary Time-depth Principle": implications for the Reclassification of fear-circuitry-related traits in DSM-V and for studying resilience to warzone-related posttraumatic stress disorder. Prog Neuro-Psychopharmacol Biol Psychiatry 30(5):827–853. https://doi.org/10.1016/j.pnpbp.2006.01.008

65. Norrholm SD, Jovanovic T (2018) Fear processing, psychophysiology, and PTSD. Harv Rev Psychiatry 26(3):129–141. https://doi.org/10.1097/HRP.0000000000000189

66. Jovanovic T, Kazama A, Bachevalier J, Davis M (2012) Impaired safety signal learning may be a biomarker of PTSD. Neuropharmacology 62(2):695–704. https://doi.org/10.1016/j.neuropharm.2011.02.023
67. Norrholm SD, Jovanovic T, Briscione MA, Anderson KM, Kwon CK, Warren VT, Bosshardt L, Bradley B (2014) Generalization of fear-potentiated startle in the presence of auditory cues: a parametric analysis. Front Behav Neurosci 8:361. https://doi.org/10.3389/fnbeh.2014.00361
68. Ehlers A, Hackmann A, Michael T (2004) Intrusive re-experiencing in post-traumatic stress disorder: phenomenology, theory, and therapy. Memory 12(4):403–415. https://doi.org/10.1080/09658210444000025
69. Wessa M, Flor H (2007) Failure of extinction of fear responses in posttraumatic stress disorder: evidence from second-order conditioning. Am J Psychiatry 164(11):1684–1692. https://doi.org/10.1176/appi.ajp.2007.07030525
70. Muravieva EV, Alberini CM (2010) Limited efficacy of propranolol on the reconsolidation of fear memories. Learn Mem 17(6):306–313. https://doi.org/10.1101/lm.1794710
71. Garfinkel SN, Abelson JL, King AP, Sripada RK, Wang X, Gaines LM, Liberzon I (2014) Impaired contextual modulation of memories in PTSD: an fMRI and psychophysiological study of extinction retention and fear renewal. J Neurosci 34(40):13435–13443. https://doi.org/10.1523/JNEUROSCI.4287-13.2014
72. Wood NE, Rosasco ML, Suris AM, Spring JD, Marin MF, Lasko NB, Goetz JM, Fischer AM, Orr SP, Pitman RK (2015) Pharmacological blockade of memory reconsolidation in post-traumatic stress disorder: three negative psychophysiological studies. Psychiatry Res 225(1–2):31–39. https://doi.org/10.1016/j.psychres.2014.09.005
73. McGuire JF, Orr SP, Essoe JK, McCracken JT, Storch EA, Piacentini J (2016) Extinction learning in childhood anxiety disorders, obsessive compulsive disorder and post-traumatic stress disorder: implications for treatment. Expert Rev Neurother 16(10):1155–1174. https://doi.org/10.1080/14737175.2016.1199276
74. Wicking M, Steiger F, Nees F, Diener SJ, Grimm O, Ruttorf M, Schad LR, Winkelmann T, Wirtz G, Flor H (2016) Deficient fear extinction memory in posttraumatic stress disorder. Neurobiol Learn Mem 136:116–126. https://doi.org/10.1016/j.nlm.2016.09.016
75. Beckers T, Kindt M (2017) Memory reconsolidation interference as an emerging treatment for emotional disorders: strengths, limitations, challenges, and opportunities. Annu Rev Clin Psychol 13:99–121. https://doi.org/10.1146/annurev-clinpsy-032816-045209
76. AlOkda AM, Nasr MM, Amin SN (2019) Between an ugly truth and a perfect lie: wiping off fearful memories using beta-adrenergic receptors antagonists. J Cell Physiol 234(5):5722–5727. https://doi.org/10.1002/jcp.27441
77. Hennings AC, McClay M, Lewis-Peacock JA, Dunsmoor JE (2020) Contextual reinstatement promotes extinction generalization in healthy adults but not PTSD. Neuropsychologia 147:107573
78. Uniyal A, Singh R, Akhtar A, Dhaliwal J, Kuhad A, Sah SP (2020) Pharmacological rewriting of fear memories: a beacon for post-traumatic stress disorder. Eur J Pharmacol 870:172824. https://doi.org/10.1016/j.ejphar.2019.172824
79. Norbury A, Brinkman H, Kowalchyk M, Monti E, Pietrzak RH, Schiller D, Feder A (2021) Latent cause inference during extinction learning in trauma-exposed individuals with and without PTSD. Psychol Med:1–12. https://doi.org/10.1017/S0033291721000647
80. Yehuda R, McFarlane AC, Shalev AY (1998) Predicting the development of posttraumatic stress disorder from the acute response to a traumatic event. Biol Psychiatry 44(12):1305–1313. https://doi.org/10.1016/s0006-3223(98)00276-5
81. Shalev AY, Sahar T, Freedman S, Peri T, Glick N, Brandes D, Orr SP, Pitman RK (1998) A prospective study of heart rate response following trauma and the subsequent development of posttraumatic stress disorder. Arch Gen Psychiatry 55(6):553–559. https://doi.org/10.1001/archpsyc.55.6.553
82. Yehuda R, LeDoux J (2007) Response variation following trauma: a translational neuroscience approach to understanding PTSD. Neuron 56(1):19–32. https://doi.org/10.1016/j.neuron.2007.09.006
83. Ginty AT, Young DA, Tyra AT, Hurley PE, Brindle RC, Williams SE (2021) Heart rate reactivity to acute psychological stress predicts higher levels of posttraumatic stress disorder symptoms during the COVID-19 pandemic. Psychosom Med 83(4):351–357. https://doi.org/10.1097/PSY.0000000000000848
84. Morey RA, Dunsmoor JE, Haswell CC, Brown VM, Vora A, Weiner J, Stjepanovic D,

Wagner HR 3rd, VA Mid-Atlantic MIRECC Workgroup, LaBar KS (2015) Fear learning circuitry is biased toward generalization of fear associations in posttraumatic stress disorder. Transl Psychiatry 5(12):e700. https://doi.org/10.1038/tp.2015.196

85. Besnard A, Sahay A (2016) Adult hippocampal neurogenesis, fear generalization, and stress. Neuropsychopharmacology 41(1): 24–44. https://doi.org/10.1038/npp.2015.167

86. Lopresto D, Schipper P, Homberg JR (2016) Neural circuits and mechanisms involved in fear generalization: implications for the pathophysiology and treatment of posttraumatic stress disorder. Neurosci Biobehav Rev 60: 31–42. https://doi.org/10.1016/j.neubiorev.2015.10.009

87. Hammell AE, Helwig NE, Kaczkurkin AN, Sponheim SR, Lissek S (2020) The temporal course of over-generalized conditioned threat expectancies in posttraumatic stress disorder. Behav Res Ther 124:103513. https://doi.org/10.1016/j.brat.2019.103513

88. Lis S, Thome J, Kleindienst N, Mueller-Engelmann M, Steil R, Priebe K, Schmahl C, Hermans D, Bohus M (2020) Generalization of fear in post-traumatic stress disorder. Psychophysiology 57(1):e13422. https://doi.org/10.1111/psyp.13422

89. Sangha S, Diehl MM, Bergstrom HC, Drew MR (2020) Know safety, no fear. Neurosci Biobehav Rev 108:218–230. https://doi.org/10.1016/j.neubiorev.2019.11.006

90. Lin CC, Liu YP (2021) Psychiatric view of generalization and nonspecific memory after traumatic stress. Biol Psychiatry 90(7): 434–435. https://doi.org/10.1016/j.biopsych.2021.07.022

91. Rabinak CA, Mori S, Lyons M, Milad MR, Phan KL (2017) Acquisition of CS-US contingencies during Pavlovian fear conditioning and extinction in social anxiety disorder and posttraumatic stress disorder. J Affect Disord 207:76–85. https://doi.org/10.1016/j.jad.2016.09.018

92. Berg H, Ma Y, Rueter A, Kaczkurkin A, Burton PC, DeYoung CG, MacDonald AW, Sponheim SR, Lissek SM (2021) Salience and central executive networks track overgeneralization of conditioned-fear in post-traumatic stress disorder. Psychol Med 51(15):2610–2619. https://doi.org/10.1017/S0033291720001166

93. Levy-Gigi E, Szabo C, Richter-Levin G, Kéri S (2015) Reduced hippocampal volume is associated with overgeneralization of negative context in individuals with PTSD. Neuropsychology 29(1):151–161. https://doi.org/10.1037/neu0000131

94. Bian XL, Qin C, Cai CY, Zhou Y, Tao Y, Lin YH, Wu HY, Chang L, Luo CX, Zhu DY (2019) Anterior cingulate cortex to ventral hippocampus circuit mediates contextual fear generalization. J Neurosci 39(29): 5728–5739. https://doi.org/10.1523/JNEUROSCI.2739-18.2019

95. Bergstrom HC (2020) Assaying fear memory discrimination and generalization: methods and concepts. Curr Protoc Neurosci 91(1): e89. https://doi.org/10.1002/cpns.89

96. Akhtar A, Pilkhwal Sah S (2021) Advances in the pharmacotherapeutic management of post-traumatic stress disorder. Expert Opin Pharmacother 22(14):1919–1930. https://doi.org/10.1080/14656566.2021.1935871

97. Martin A, Naunton M, Kosari S, Peterson G, Thomas J, Christenson JK (2021) Treatment guidelines for PTSD: a systematic review. J Clin Med 10(18):4175. https://doi.org/10.3390/jcm10184175

98. Pary R, Micchelli AN, Lippmann S (2021) How we treat posttraumatic stress disorder. Prim Care Compan CNS Disord 23(1): 19nr02572. https://doi.org/10.4088/PCC.19nr02572

99. Brewin CR (2018) Memory and forgetting. Curr Psychiatry Rep 20(10):87. https://doi.org/10.1007/s11920-018-0950-7

100. Waits WM, Hoge CW (2018) Reconsolidation of traumatic memories using psychotherapy. Am J Psychiatry 175(11):1145. https://doi.org/10.1176/appi.ajp.2018.18060646

101. Monfils MH, Holmes EA (2018) Memory boundaries: opening a window inspired by reconsolidation to treat anxiety, trauma-related, and addiction disorders. Lancet Psychiatry 5(12):1032–1042. https://doi.org/10.1016/S2215-0366(18)30270-0

102. Kida S (2019) Reconsolidation/destabilization, extinction and forgetting of fear memory as therapeutic targets for PTSD. Psychopharmacology 236(1):49–57. https://doi.org/10.1007/s00213-018-5086-2

103. Milton AL (2019) Fear not: recent advances in understanding the neural basis of fear memories and implications for treatment development. F1000Research 8:F1000 Faculty Rev-1948. https://doi.org/10.12688/f1000research.20053.1

104. Phelps EA, Hofmann SG (2019) Memory editing from science fiction to clinical practice. Nature 572(7767):43–50. https://doi.org/10.1038/s41586-019-1433-7

105. Cain CK, Maynard GD, Kehne JH (2012) Targeting memory processes with drugs to prevent or cure PTSD. Expert Opin Investig Drugs 21(9):1323–1350. https://doi.org/10.1517/13543784.2012.704020

106. Debiec J (2012) Memory reconsolidation processes and posttraumatic stress disorder: promises and challenges of translational research. Biol Psychiatry 71(4):284–285. https://doi.org/10.1016/j.biopsych.2011.12.009

107. Steckler T, Risbrough V (2012) Pharmacological treatment of PTSD - established and new approaches. Neuropharmacology 62(2): 617–627. https://doi.org/10.1016/j.neuropharm.2011.06.012

108. Parsons RG, Ressler KJ (2013) Implications of memory modulation for post-traumatic stress and fear disorders. Nat Neurosci 16(2):146–153. https://doi.org/10.1038/nn.3296

109. Schwabe L, Nader K, Pruessner JC (2014) Reconsolidation of human memory: brain mechanisms and clinical relevance. Biol Psychiatry 76(4):274–280. https://doi.org/10.1016/j.biopsych.2014.03.008

110. Haider S, Batool Z, Rafiq S (2020) Method for the identification of pharmacological intervention for the disruption of fear memory in PTSD-rat model. MethodsX 7: 101059. https://doi.org/10.1016/j.mex.2020.101059

111. Vaiva G, Ducrocq F, Jezequel K, Averland B, Lestavel P, Brunet A, Marmar CR (2003) Immediate treatment with propranolol decreases posttraumatic stress disorder two months after trauma. Biol Psychiatry 54(9): 947–949

112. Amos T, Stein DJ, Ipser JC (2014) Pharmacological interventions for preventing posttraumatic stress disorder (PTSD). Cochrane Database Syst Rev 7:CD006239. https://doi.org/10.1002/14651858.CD006239.pub2

113. Roque AP (2015) Pharmacotherapy as prophylactic treatment of post-traumatic stress disorder: a review of the literature. Issues Ment Health Nurs 36(9):740–751. https://doi.org/10.3109/01612840.2015.1057785

114. Pope KS (1999) The ethics of research involving memories of trauma. Gen Hosp Psychiatry 21(3):157. https://doi.org/10.1016/s0163-8343(99)00019-5

115. Dossey L (2006) Memory management: a gathering ethical storm. Explore 2(3): 185–188. https://doi.org/10.1016/j.explore.2006.03.001

116. Henry M, Fishman JR, Youngner SJ (2007) Propranolol and the prevention of post-traumatic stress disorder: is it wrong to erase the "sting" of bad memories? Am J Bioeth 7(9):12–20. https://doi.org/10.1080/15265160701518474

117. Donovan E (2010) Propranolol use in the prevention and treatment of posttraumatic stress disorder in military veterans: forgetting therapy revisited. Perspect Biol Med 53(1): 61–74. https://doi.org/10.1353/pbm.0.0140

118. Quirk GJ, Paré D, Richardson R, Herry C, Monfils MH, Schiller D, Vicentic A (2010) Erasing fear memories with extinction training. J Neurosci 30(45):14993–14997. https://doi.org/10.1523/JNEUROSCI.4268-10.2010

119. Markowitz S, Fanselow M (2020) Exposure therapy for post-traumatic stress disorder: factors of limited success and possible alternative treatment. Brain Sci 10(3):167. https://doi.org/10.3390/brainsci10030167

120. Paredes D, Morilak DA (2019) A rodent model of exposure therapy: the use of fear extinction as a therapeutic intervention for PTSD. Front Behav Neurosci 13:46. https://doi.org/10.3389/fnbeh.2019.00046

121. Zuj DV, Palmer MA, Lommen MJ, Felmingham KL (2016) The centrality of fear extinction in linking risk factors to PTSD: a narrative review. Neurosci Biobehav Rev 69: 15–35. https://doi.org/10.1016/j.neubiorev.2016.07.014

122. Pineles SL, Nillni YI, King MW, Patton SC, Bauer MR, Mostoufi SM, Gerber MR, Hauger R, Resick PA, Rasmusson AM, Orr SP (2016) Extinction retention and the menstrual cycle: different associations for women with posttraumatic stress disorder. J Abnorm Psychol 125(3):349–355. https://doi.org/10.1037/abn0000138

123. Singewald N, Holmes A (2019) Rodent models of impaired fear extinction. Psychopharmacology 236(1):21–32. https://doi.org/10.1007/s00213-018-5054-x

124. Marin MF, Lonak SF, Milad MR (2015) Augmentation of evidence-based psychotherapy for PTSD with cognitive enhancers. Curr Psychiatry Rep 17(6):39. https://doi.org/10.1007/s11920-015-0582-0

125. Singewald N, Schmuckermair C, Whittle N, Holmes A, Ressler KJ (2015) Pharmacology of cognitive enhancers for exposure-based therapy of fear, anxiety and trauma-related disorders. Pharmacol Ther 149:150–190. https://doi.org/10.1016/j.pharmthera.2014.12.004

126. Smith NB, Doran JM, Sippel LM, Harpaz-Rotem I (2017) Fear extinction and memory reconsolidation as critical components in behavioral treatment for posttraumatic stress disorder and potential augmentation of these processes. Neurosci Lett 649:170–175. https://doi.org/10.1016/j.neulet.2017.01.006

127. Baker JF, Cates ME, Luthin DR (2018) D-cycloserine in the treatment of posttraumatic stress disorder. Ment Health Clin 7(2):88–94. https://doi.org/10.9740/mhc.2017.03.088

128. Bitencourt RM, Takahashi RN (2018) Cannabidiol as a therapeutic alternative for posttraumatic stress disorder: from bench research to confirmation in human trials. Front Neurosci 12:502. https://doi.org/10.3389/fnins.2018.00502

129. Dittert N, Hüttner S, Polak T, Herrmann MJ (2018) Augmentation of fear extinction by transcranial direct current stimulation (tDCS). Front Behav Neurosci 12:76. https://doi.org/10.3389/fnbeh.2018.00076

130. de Quervain D, Wolf OT, Roozendaal B (2019) Glucocorticoid-induced enhancement of extinction-from animal models to clinical trials. Psychopharmacology 236(1):183–199. https://doi.org/10.1007/s00213-018-5116-0

131. Lebois L, Seligowski AV, Wolff JD, Hill SB, Ressler KJ (2019) Augmentation of extinction and inhibitory learning in anxiety and trauma-related disorders. Annu Rev Clin Psychol 15:257–284. https://doi.org/10.1146/annurev-clinpsy-050718-095634

132. Inslicht SS, Niles AN, Metzler TJ, Lipshitz SL, Otte C, Milad MR, Orr SP, Marmar CR, Neylan TC (2022) Randomized controlled experimental study of hydrocortisone and D-cycloserine effects on fear extinction in PTSD. Neuropsychopharmacology 47(11):1945–1952. https://doi.org/10.1038/s41386-021-01222-z

133. Fiorenza NG, Sartor D, Myskiw JC, Izquierdo I (2011) Treatment of fear memories: interactions between extinction and reconsolidation. An Acad Bras Cienc 83(4):1363–1372. https://doi.org/10.1590/s0001-37652011000400023

134. Auber A, Tedesco V, Jones CE, Monfils MH, Chiamulera C (2013) Post-retrieval extinction as reconsolidation interference: methodological issues or boundary conditions? Psychopharmacology 226(4):631–647. https://doi.org/10.1007/s00213-013-3004-1

135. Post RM, Kegan R (2017) Prevention of recurrent affective episodes using extinction training in the reconsolidation window: a testable psychotherapeutic strategy. Psychiatry Res 249:327–336. https://doi.org/10.1016/j.psychres.2017.01.034

136. Chalkia A, Van Oudenhove L, Beckers T (2020) Preventing the return of fear in humans using reconsolidation update mechanisms: a verification report of Schiller et al. (2010). Cortex 129:510–525. https://doi.org/10.1016/j.cortex.2020.03.031

137. Lancaster CL, Monfils MH, Telch MJ (2020) Augmenting exposure therapy with pre-extinction fear memory reactivation and deepened extinction: a randomized controlled trial. Behav Res Ther 135:103730. https://doi.org/10.1016/j.brat.2020.103730

138. Diergaarde L, Schoffelmeer AN, De Vries TJ (2008) Pharmacological manipulation of memory reconsolidation: towards a novel treatment of pathogenic memories. Eur J Pharmacol 585(2–3):453–457. https://doi.org/10.1016/j.ejphar.2008.03.010

139. Vermetten E, Zhohar J, Krugers HJ (2014) Pharmacotherapy in the aftermath of trauma; opportunities in the 'golden hours'. Curr Psychiatry Rep 16(7):455. https://doi.org/10.1007/s11920-014-0455-y

140. Kindt M, van Emmerik A (2016) New avenues for treating emotional memory disorders: towards a reconsolidation intervention for posttraumatic stress disorder. Ther Adv Psychopharmacol 6(4):283–295. https://doi.org/10.1177/2045125316644541

141. Dunbar AB, Taylor JR (2017) Reconsolidation and psychopathology: moving towards reconsolidation-based treatments. Neurobiol Learn Mem 142(Pt A):162–171. https://doi.org/10.1016/j.nlm.2016.11.005

142. Elsey J, Kindt M (2017) Tackling maladaptive memories through reconsolidation: from neural to clinical science. Neurobiol Learn Mem 142(Pt A):108–117. https://doi.org/10.1016/j.nlm.2017.03.007

143. Astill Wright L, Horstmann L, Holmes EA, Bisson JI (2021) Consolidation/reconsolidation therapies for the prevention and treatment of PTSD and re-experiencing: a systematic review and meta-analysis. Transl Psychiatry 11(1):453. https://doi.org/10.1038/s41398-021-01570-w

144. Pitman RK (2011) Will reconsolidation blockade offer a novel treatment for posttraumatic stress disorder? Front Behav Neurosci 5:11. https://doi.org/10.3389/fnbeh.2011.00011

145. Bustos SG, Maldonado H, Molina VA (2006) Midazolam disrupts fear memory reconsolidation. Neuroscience 139(3):831–842. https://doi.org/10.1016/j.neuroscience.2005.12.064
146. Lonergan MH, Olivera-Figueroa LA, Pitman RK, Brunet A (2013) Propranolol's effects on the consolidation and reconsolidation of long-term emotional memory in healthy participants: a meta-analysis. J Psychiatry Neurosci 38(4):222–231. https://doi.org/10.1503/jpn.120111
147. Brunet A, Thomas É, Saumier D, Ashbaugh AR, Azzoug A, Pitman RK, Orr SP, Tremblay J (2014) Trauma reactivation plus propranolol is associated with durably low physiological responding during subsequent script-driven traumatic imagery. Can J Psychiatr 59(4):228–232. https://doi.org/10.1177/070674371405900408
148. Galarza Vallejo A, Kroes M, Rey E, Acedo MV, Moratti S, Fernández G, Strange BA (2019) Propofol-induced deep sedation reduces emotional episodic memory reconsolidation in humans. Sci Adv 5(3):eaav3801. https://doi.org/10.1126/sciadv.aav3801
149. Raymundi AM, da Silva TR, Sohn J, Bertoglio LJ, Stern CA (2020) Effects of Δ9-tetrahydrocannabinol on aversive memories and anxiety: a review from human studies. BMC Psychiatry 20(1):420. https://doi.org/10.1186/s12888-020-02813-8
150. Soeter M, Kindt M (2013) High trait anxiety: a challenge for disrupting fear memory reconsolidation. PLoS One 8(11):e75239. https://doi.org/10.1371/journal.pone.0075239
151. Toth SL, Cicchetti D (2011) Frontiers in translational research on trauma. Dev Psychopathol 23(2):353–355. https://doi.org/10.1017/S0954579411000101
152. Daskalakis NP, Yehuda R, Diamond DM (2013) Animal models in translational studies of PTSD. Psychoneuroendocrinology 38(9):1895–1911. https://doi.org/10.1016/j.psyneuen.2013.06.006
153. Cohen H, Matar MA, Zohar J (2014) Maintaining the clinical relevance of animal models in translational studies of post-traumatic stress disorder. ILAR J 55(2):233–245. https://doi.org/10.1093/ilar/ilu006
154. Kroes MC, Schiller D, LeDoux JE, Phelps EA (2016) Translational approaches targeting reconsolidation. Curr Top Behav Neurosci 28:197–230. https://doi.org/10.1007/7854_2015_5008
155. Kida S (2020) Function and mechanisms of memory destabilization and reconsolidation after retrieval. Proc Jpn Acad Ser B Phys Biol Sci 96(3):95–106. https://doi.org/10.2183/pjab.96.008
156. Bustos SG, Maldonado H, Molina VA (2009) Disruptive effect of midazolam on fear memory reconsolidation: decisive influence of reactivation time span and memory age. Neuropsychopharmacology 34(2):446–457. https://doi.org/10.1038/npp.2008.75
157. Bustos SG, Giachero M, Maldonado H, Molina VA (2010) Previous stress attenuates the susceptibility to Midazolam's disruptive effect on fear memory reconsolidation: influence of pre-reactivation D-cycloserine administration. Neuropsychopharmacology 35(5):1097–1108. https://doi.org/10.1038/npp.2009.215
158. Marin FN, Franzen JM, Troyner F, Molina VA, Giachero M, Bertoglio LJ (2020) Taking advantage of fear generalization-associated destabilization to attenuate the underlying memory via reconsolidation intervention. Neuropharmacology 181:108338. https://doi.org/10.1016/j.neuropharm.2020.108338
159. Foa EB, Zinbarg R, Rothbaum BO (1992) Uncontrollability and unpredictability in post-traumatic stress disorder: an animal model. Psychol Bull 112(2):218–238. https://doi.org/10.1037/0033-2909.112.2.218
160. Yehuda R, Antelman SM (1993) Criteria for rationally evaluating animal models of post-traumatic stress disorder. Biol Psychiatry 33(7):479–486. https://doi.org/10.1016/0006-3223(93)90001-t
161. Pitman RK, Rasmusson AM, Koenen KC, Shin LM, Orr SP, Gilbertson MW, Milad MR, Liberzon I (2012) Biological studies of post-traumatic stress disorder. Nat Rev Neurosci 13(11):769–787. https://doi.org/10.1038/nrn3339
162. Aspesi D, Pinna G (2019) Animal models of post-traumatic stress disorder and novel treatment targets. Behav Pharmacol 30:130–150. https://doi.org/10.1097/FBP.0000000000000467
163. Verbitsky A, Dopfel D, Zhang N (2020) Rodent models of post-traumatic stress disorder: behavioral assessment. Transl Psychiatry 10(1):132. https://doi.org/10.1038/s41398-020-0806-x
164. Briscione MA, Jovanovic T, Norrholm SD (2014) Conditioned fear associated phenotypes as robust, translational indices of trauma-, stressor-, and anxiety-related behaviors. Front Psychiatry 5:88. https://doi.org/10.3389/fpsyt.2014.00088
165. Bali A, Jaggi AS (2015) Electric foot shock stress: a useful tool in neuropsychiatric

165. studies. Rev Neurosci 26(6):655–677. https://doi.org/10.1515/revneuro-2015-0015
166. Lissek S, van Meurs B (2015) Learning models of PTSD: theoretical accounts and psychobiological evidence. Int J Psychophysiol 98(3 Pt 2):594–605. https://doi.org/10.1016/j.ijpsycho.2014.11.006
167. Richter-Levin G, Stork O, Schmidt MV (2019) Animal models of PTSD: a challenge to be met. Mol Psychiatry 24(8):1135–1156. https://doi.org/10.1038/s41380-018-0272-5
168. Fanselow MS, Ponnusamy (2008) The use of conditioning tasks to model fear and anxiety. In: Blanchard RJ, Blanchard DC, Griebel G, Nutt D (eds) Handbook of anxiety and fear. Academic, Oxford, pp 29–48
169. Siegmund A, Wotjak CT (2006) Toward an animal model of posttraumatic stress disorder. Ann N Y Acad Sci 1071:324–334. https://doi.org/10.1196/annals.1364.025
170. Armstrong T, Federman S, Hampson K, Crabtree O, Olatunji BO (2021) Fear learning in veterans with combat-related PTSD is linked to anxiety sensitivity: evidence from self-report and pupillometry. Behav Ther 52(1):149–161. https://doi.org/10.1016/j.beth.2020.03.006
171. Miller MM, McEwen BS (2006) Establishing an agenda for translational research on PTSD. Ann N Y Acad Sci 1071:294–312. https://doi.org/10.1196/annals.1364.023
172. Cohen H, Richter-Levin G (2009) Toward animal models of post-traumatic stress disorder. In: Shiromani PJ, Keane TM, LeDoux JE (eds) Post-traumatic stress disorder: basic science and clinical practice. Humana Press, New York
173. Desmedt A, Marighetto A, Piazza PV (2015) Abnormal fear memory as a model for post-traumatic stress disorder. Biol Psychiatry 78(5):290–297. https://doi.org/10.1016/j.biopsych.2015.06.017
174. Maldonado NM, Martijena ID, Molina VA (2011) Facilitating influence of stress on the consolidation of fear memory induced by a weak training: reversal by midazolam pretreatment. Behav Brain Res 225(1):77–84. https://doi.org/10.1016/j.bbr.2011.06.035
175. Corley MJ, Caruso MJ, Takahashi LK (2012) Stress-induced enhancement of fear conditioning and sensitization facilitates extinction-resistant and habituation-resistant fear behaviors in a novel animal model of posttraumatic stress disorder. Physiol Behav 105(2):408–416. https://doi.org/10.1016/j.physbeh.2011.08.037
176. Olson VG, Rockett HR, Reh RK, Redila VA, Tran PM, Venkov HA, Defino MC, Hague C, Peskind ER, Szot P, Raskind MA (2011) The role of norepinephrine in differential response to stress in an animal model of posttraumatic stress disorder. Biol Psychiatry 70(5):441–448. https://doi.org/10.1016/j.biopsych.2010.11.029
177. Reznikov R, Diwan M, Nobrega JN, Hamani C (2015) Towards a better preclinical model of PTSD: characterizing animals with weak extinction, maladaptive stress responses and low plasma corticosterone. J Psychiatr Res 61:158–165. https://doi.org/10.1016/j.jpsychires.2014.12.017
178. Cabib S, Orsini C, Puglisi Allegra S (2019) Animal models of liability to post-traumatic stress disorder: going beyond fear memory. Behav Pharmacol 30(2 and 3-Spec Issue):122–129. https://doi.org/10.1097/FBP.0000000000000475
179. Daskalakis NP, Lehrner A, Yehuda R (2013) Endocrine aspects of post-traumatic stress disorder and implications for diagnosis and treatment. Endocrinol Metab Clin N Am 42(3):503–513. https://doi.org/10.1016/j.ecl.2013.05.004
180. Zoladz PR (2021) Animal models for the discovery of novel drugs for post-traumatic stress disorder. Expert Opin Drug Discovery 16(2):135–146. https://doi.org/10.1080/17460441.2020.1820982
181. Bremner JD, Randall P, Scott TM, Bronen RA, Seibyl JP, Southwick SM, Delaney RC, McCarthy G, Charney DS, Innis RB (1995) MRI-based measurement of hippocampal volume in patients with combat-related posttraumatic stress disorder. Am J Psychiatry 152(7):973–981. https://doi.org/10.1176/ajp.152.7.973
182. Gurvits TV, Shenton ME, Hokama H, Ohta H, Lasko NB, Gilbertson MW, Orr SP, Kikinis R, Jolesz FA, McCarley RW, Pitman RK (1996) Magnetic resonance imaging study of hippocampal volume in chronic, combat-related posttraumatic stress disorder. Biol Psychiatry 40(11):1091–1099. https://doi.org/10.1016/S0006-3223(96)00229-6
183. Stein MB, Koverola C, Hanna C, Torchia MG, McClarty B (1997) Hippocampal volume in women victimized by childhood sexual abuse. Psychol Med 27(4):951–959. https://doi.org/10.1017/s0033291797005242
184. Woon F, Hedges DW (2011) Gender does not moderate hippocampal volume deficits in adults with posttraumatic stress disorder: a meta-analysis. Hippocampus 21(3):243–252. https://doi.org/10.1002/hipo.20746

185. Kheirbek MA, Klemenhagen KC, Sahay A, Hen R (2012) Neurogenesis and generalization: a new approach to stratify and treat anxiety disorders. Nat Neurosci 15(12):1613–1620. https://doi.org/10.1038/nn.3262
186. Kasai K, Yamasue H, Gilbertson MW, Shenton ME, Rauch SL, Pitman RK (2008) Evidence for acquired pregenual anterior cingulate gray matter loss from a twin study of combat-related posttraumatic stress disorder. Biol Psychiatry 63(6):550–556. https://doi.org/10.1016/j.biopsych.2007.06.022
187. Kitayama N, Quinn S, Bremner JD (2006) Smaller volume of anterior cingulate cortex in abuse-related posttraumatic stress disorder. J Affect Disord 90(2–3):171–174. https://doi.org/10.1016/j.jad.2005.11.006
188. Shin LM, McNally RJ, Kosslyn SM, Thompson WL, Rauch SL, Alpert NM, Metzger LJ, Lasko NB, Orr SP, Pitman RK (1999) Regional cerebral blood flow during script-driven imagery in childhood sexual abuse-related PTSD: a PET investigation. Am J Psychiatry 156(4):575–584. https://doi.org/10.1176/ajp.156.4.575
189. Schuff N, Neylan TC, Fox-Bosetti S, Lenoci M, Samuelson KW, Studholme C, Kornak J, Marmar CR, Weiner MW (2008) Abnormal N-acetylaspartate in hippocampus and anterior cingulate in posttraumatic stress disorder. Psychiatry Res 162(2):147–157. https://doi.org/10.1016/j.pscychresns.2007.04.011
190. Kim SJ, Jeong DU, Sim ME, Bae SC, Chung A, Kim MJ, Chang KH, Ryu J, Renshaw PF, Lyoo IK (2006) Asymmetrically altered integrity of cingulum bundle in posttraumatic stress disorder. Neuropsychobiology 54(2):120–125. https://doi.org/10.1159/000098262
191. Liberzon I, Taylor SF, Amdur R, Jung TD, Chamberlain KR, Minoshima S, Koeppe RA, Fig LM (1999) Brain activation in PTSD in response to trauma-related stimuli. Biol Psychiatry 45(7):817–826. https://doi.org/10.1016/s0006-3223(98)00246-7
192. Pannu Hayes J, Labar KS, Petty CM, McCarthy G, Morey RA (2009) Alterations in the neural circuitry for emotion and attention associated with posttraumatic stress symptomatology. Psychiatry Res 172(1):7–15. https://doi.org/10.1016/j.pscychresns.2008.05.005
193. Shin LM, Handwerger K (2009) Is posttraumatic stress disorder a stress-induced fear circuitry disorder? J Trauma Stress 22(5):409–415. https://doi.org/10.1002/jts.20442
194. Bremner JD, Vermetten E, Schmahl C, Vaccarino V, Vythilingam M, Afzal N, Grillon C, Charney DS (2005) Positron emission tomographic imaging of neural correlates of a fear acquisition and extinction paradigm in women with childhood sexual-abuse-related post-traumatic stress disorder. Psychol Med 35(6):791–806. https://doi.org/10.1017/s0033291704003290
195. Rauch SL, Shin LM, Phelps EA (2006) Neurocircuitry models of posttraumatic stress disorder and extinction: human neuroimaging research – past, present, and future. Biol Psychiatry 60(4):376–382. https://doi.org/10.1016/j.biopsych.2006.06.004
196. Milad MR, Pitman RK, Ellis CB, Gold AL, Shin LM, Lasko NB, Zeidan MA, Handwerger K, Orr SP, Rauch SL (2009) Neurobiological basis of failure to recall extinction memory in posttraumatic stress disorder. Biol Psychiatry 66(12):1075–1082. https://doi.org/10.1016/j.biopsych.2009.06.026
197. Grossman R, Buchsbaum MS, Yehuda R (2002) Neuroimaging studies in posttraumatic stress disorder. Psychiatric Clin N Am 25(2):317–3vi. https://doi.org/10.1016/s0193-953x(01)00011-9
198. Bryant RA, Felmingham KL, Kemp AH, Barton M, Peduto AS, Rennie C, Gordon E, Williams LM (2005) Neural networks of information processing in posttraumatic stress disorder: a functional magnetic resonance imaging study. Biol Psychiatry 58(2):111–118. https://doi.org/10.1016/j.biopsych.2005.03.021
199. Karl A, Schaefer M, Malta LS, Dörfel D, Rohleder N, Werner A (2006) A meta-analysis of structural brain abnormalities in PTSD. Neurosci Biobehav Rev 30(7):1004–1031. https://doi.org/10.1016/j.neubiorev.2006.03.004
200. Liberzon I, Martis B (2006) Neuroimaging studies of emotional responses in PTSD. Ann N Y Acad Sci 1071:87–109. https://doi.org/10.1196/annals.1364.009
201. Etkin A, Wager TD (2007) Functional neuroimaging of anxiety: a meta-analysis of emotional processing in PTSD, social anxiety disorder, and specific phobia. Am J Psychiatry 164(10):1476–1488. https://doi.org/10.1176/appi.ajp.2007.07030504
202. Liberzon I, Sripada CS (2008) The functional neuroanatomy of PTSD: a critical review. Prog Brain Res 167:151–169. https://doi.org/10.1016/S0079-6123(07)67011-3

203. Simmons AN, Paulus MP, Thorp SR, Matthews SC, Norman SB, Stein MB (2008) Functional activation and neural networks in women with posttraumatic stress disorder related to intimate partner violence. Biol Psychiatry 64(8):681–690. https://doi.org/10.1016/j.biopsych.2008.05.027

204. Taber KH, Hurley RA (2009) PTSD and combat-related injuries: functional neuroanatomy. J Neuropsychiatry Clin Neurosci 21(1):1–4. https://doi.org/10.1176/jnp.2009.21.1.iv

205. Karl A, Werner A (2010) The use of proton magnetic resonance spectroscopy in PTSD research – meta-analyses of findings and methodological review. Neurosci Biobehav Rev 34(1):7–22. https://doi.org/10.1016/j.neubiorev.2009.06.008

206. Hayes JP, Hayes SM, Mikedis AM (2012) Quantitative meta-analysis of neural activity in posttraumatic stress disorder. Biol Mood Anxiety Disord 2:9. https://doi.org/10.1186/2045-5380-2-9

207. Patel R, Spreng RN, Shin LM, Girard TA (2012) Neurocircuitry models of posttraumatic stress disorder and beyond: a meta-analysis of functional neuroimaging studies. Neurosci Biobehav Rev 36(9):2130–2142. https://doi.org/10.1016/j.neubiorev.2012.06.003

208. Shucard JL, Cox J, Shucard DW, Fetter H, Chung C, Ramasamy D, Violanti J (2012) Symptoms of posttraumatic stress disorder and exposure to traumatic stressors are related to brain structural volumes and behavioral measures of affective stimulus processing in police officers. Psychiatry Res 204(1):25–31. https://doi.org/10.1016/j.pscychresns.2012.04.006

209. Scioli-Salter ER, Forman DE, Otis JD, Gregor K, Valovski I, Rasmusson AM (2015) The shared neuroanatomy and neurobiology of comorbid chronic pain and PTSD: therapeutic implications. Clin J Pain 31(4):363–374. https://doi.org/10.1097/AJP.0000000000000115

210. Akiki TJ, Averill CL, Abdallah CG (2017) A network-based neurobiological model of PTSD: evidence from structural and functional neuroimaging studies. Curr Psychiatry Rep 19(11):81. https://doi.org/10.1007/s11920-017-0840-4

211. Fitzgerald JM, DiGangi JA, Phan KL (2018) Functional neuroanatomy of emotion and its regulation in PTSD. Harv Rev Psychiatry 26(3):116–128. https://doi.org/10.1097/HRP.0000000000000185

212. Henigsberg N, Kalember P, Petrović ZK, Šečić A (2019) Neuroimaging research in posttraumatic stress disorder – focus on amygdala, hippocampus and prefrontal cortex. Prog Neuro-Psychopharmacol Biol Psychiatry 90:37–42. https://doi.org/10.1016/j.pnpbp.2018.11.003

213. Harnett NG, Goodman AM, Knight DC (2020) PTSD-related neuroimaging abnormalities in brain function, structure, and biochemistry. Exp Neurol 330:113331. https://doi.org/10.1016/j.expneurol.2020.113331

214. Kamiya K, Abe O (2020) Imaging of posttraumatic stress disorder. Neuroimaging Clin N Am 30(1):115–123. https://doi.org/10.1016/j.nic.2019.09.010

215. Kribakaran S, Danese A, Bromis K, Kempton MJ, Gee DG (2020) Meta-analysis of structural magnetic resonance imaging studies in pediatric posttraumatic stress disorder and comparison with related conditions. Biol Psychiatry Cogn Neurosci Neuroimag 5(1):23–34. https://doi.org/10.1016/j.bpsc.2019.08.006

216. Kunimatsu A, Yasaka K, Akai H, Kunimatsu N, Abe O (2020) MRI findings in posttraumatic stress disorder. J Magn Reson Imaging 52(2):380–396. https://doi.org/10.1002/jmri.26929

217. Nisar S, Bhat AA, Hashem S, Syed N, Yadav SK, Uddin S, Fakhro K, Bagga P, Thompson P, Reddy R, Frenneaux MP, Haris M (2020) Genetic and neuroimaging approaches to understanding post-traumatic stress disorder. Int J Mol Sci 21(12):4503. https://doi.org/10.3390/ijms21124503

218. Ross MC, Cisler JM (2020) Altered large-scale functional brain organization in posttraumatic stress disorder: a comprehensive review of univariate and network-level neurocircuitry models of PTSD. NeuroImage Clin 27:102319. https://doi.org/10.1016/j.nicl.2020.102319

219. Crombie KM, Ross MC, Letkiewicz AM, Sartin-Tarm A, Cisler JM (2021) Differential relationships of PTSD symptom clusters with cortical thickness and grey matter volumes among women with PTSD. Sci Rep 11(1):1825. https://doi.org/10.1038/s41598-020-80776-2

220. Neria Y (2021) Functional neuroimaging in PTSD: from discovery of underlying mechanisms to addressing diagnostic heterogeneity. Am J Psychiatry 178(2):128–135. https://doi.org/10.1176/appi.ajp.2020.20121727

221. Serra-Blasco M, Radua J, Soriano-Mas C, Gómez-Benlloch A, Porta-Casterás D,

Carulla-Roig M, Albajes-Eizagirre A, Arnone D, Klauser P, Canales-Rodríguez EJ, Hilbert K, Wise T, Cheng Y, Kandilarova S, Mataix-Cols D, Vieta E, Via E, Cardoner N (2021) Structural brain correlates in major depression, anxiety disorders and post-traumatic stress disorder: a voxel-based morphometry meta-analysis. Neurosci Biobehav Rev 129:269–281. https://doi.org/10.1016/j.neubiorev.2021.07.002

222. van Rooij S, Sippel LM, McDonald WM, Holtzheimer PE (2021) Defining focal brain stimulation targets for PTSD using neuroimaging. Depress Anxiety. https://doi.org/10.1002/da.23159

223. Shin LM, Orr SP, Carson MA, Rauch SL, Macklin ML, Lasko NB, Peters PM, Metzger LJ, Dougherty DD, Cannistraro PA, Alpert NM, Fischman AJ, Pitman RK (2004) Regional cerebral blood flow in the amygdala and medial prefrontal cortex during traumatic imagery in male and female Vietnam veterans with PTSD. Arch Gen Psychiatry 61(2):168–176. https://doi.org/10.1001/archpsyc.61.2.168

224. Richert KA, Carrion VG, Karchemskiy A, Reiss AL (2006) Regional differences of the prefrontal cortex in pediatric PTSD: an MRI study. Depress Anxiety 23(1):17–25. https://doi.org/10.1002/da.20131

225. Carrion VG, Weems CF, Richert K, Hoffman BC, Reiss AL (2010) Decreased prefrontal cortical volume associated with increased bedtime cortisol in traumatized youth. Biol Psychiatry 68(5):491–493. https://doi.org/10.1016/j.biopsych.2010.05.010

226. Gold AL, Shin LM, Orr SP, Carson MA, Rauch SL, Macklin ML, Lasko NB, Metzger LJ, Dougherty DD, Alpert NM, Fischman AJ, Pitman RK (2011) Decreased regional cerebral blood flow in medial prefrontal cortex during trauma-unrelated stressful imagery in Vietnam veterans with post-traumatic stress disorder. Psychol Med 41(12):2563–2572. https://doi.org/10.1017/S0033291711000730

227. Shin LM, Bush G, Milad MR, Lasko NB, Brohawn KH, Hughes KC, Macklin ML, Gold AL, Karpf RD, Orr SP, Rauch SL, Pitman RK (2011) Exaggerated activation of dorsal anterior cingulate cortex during cognitive interference: a monozygotic twin study of posttraumatic stress disorder. Am J Psychiatry 168(9):979–985. https://doi.org/10.1176/appi.ajp.2011.09121812

228. Aupperle RL, Allard CB, Grimes EM, Simmons AN, Flagan T, Behroozania M, Cissell SH, Twamley EW, Thorp SR, Norman SB, Paulus MP, Stein MB (2012) Dorsolateral prefrontal cortex activation during emotional anticipation and neuropsychological performance in posttraumatic stress disorder. Arch Gen Psychiatry 69(4):360–371. https://doi.org/10.1001/archgenpsychiatry.2011.1539

229. Knight LK, Naaz F, Stoica T, Depue BE, Alzheimer's Disease Neuroimaging Initiative (2017) Lifetime PTSD and geriatric depression symptomatology relate to altered dorsomedial frontal and amygdala morphometry. Psychiatry Res Neuroimaging 267:59–68. https://doi.org/10.1016/j.pscychresns.2017.07.003

230. Hinojosa CA, Kaur N, VanElzakker MB, Shin LM (2019) Cingulate subregions in posttraumatic stress disorder, chronic stress, and treatment. Handb Clin Neurol 166:355–370. https://doi.org/10.1016/B978-0-444-64196-0.00020-0

231. Kaldewaij R, Koch S, Hashemi MM, Zhang W, Klumpers F, Roelofs K (2021) Anterior prefrontal brain activity during emotion control predicts resilience to post-traumatic stress symptoms. Nat Hum Behav 5(8):1055–1064. https://doi.org/10.1038/s41562-021-01055-2

232. Weisholtz D, Silbersweig D, Pan H, Cloitre M, LeDoux J, Stern E (2021) Correlation between rostral dorsomedial prefrontal cortex activation by trauma-related words and subsequent response to CBT for PTSD. J Neuropsychiatry Clin Neurosci 33(2):116–123. https://doi.org/10.1176/appi.neuropsych.20030058

233. Alexandra Kredlow M, Fenster RJ, Laurent ES, Ressler KJ, Phelps EA (2022) Prefrontal cortex, amygdala, and threat processing: implications for PTSD. Neuropsychopharmacology 47(1):247–259. https://doi.org/10.1038/s41386-021-01155-7

234. Woon FL, Hedges DW (2009) Amygdala volume in adults with posttraumatic stress disorder: a meta-analysis. J Neuropsychiatry Clin Neurosci 21(1):5–12. https://doi.org/10.1176/jnp.2009.21.1.5

235. Aghajani M, Veer IM, van Hoof MJ, Rombouts SA, van der Wee NJ, Vermeiren RR (2016) Abnormal functional architecture of amygdala-centered networks in adolescent posttraumatic stress disorder. Hum Brain Mapp 37(3):1120–1135. https://doi.org/10.1002/hbm.23093

236. Ahmed-Leitao F, Spies G, van den Heuvel L, Seedat S (2016) Hippocampal and amygdala volumes in adults with posttraumatic stress disorder secondary to childhood abuse or maltreatment: a systematic review. Psychiatry

Res Neuroimaging 256:33–43. https://doi.org/10.1016/j.pscychresns.2016.09.008

237. Lieberman L, Gorka SM, DiGangi JA, Frederick A, Phan KL (2017) Impact of posttraumatic stress symptom dimensions on amygdala reactivity to emotional faces. Prog Neuro-Psychopharmacol Biol Psychiatry 79 (Pt B):401–407. https://doi.org/10.1016/j.pnpbp.2017.07.021

238. Badura-Brack A, McDermott TJ, Heinrichs-Graham E, Ryan TJ, Khanna MM, Pine DS, Bar-Haim Y, Wilson TW (2018) Veterans with PTSD demonstrate amygdala hyperactivity while viewing threatening faces: a MEG study. Biol Psychol 132:228–232. https://doi.org/10.1016/j.biopsycho.2018.01.005

239. Herzog JI, Thome J, Demirakca T, Koppe G, Ende G, Lis S, Rausch S, Priebe K, Müller-Engelmann M, Steil R, Bohus M, Schmahl C (2020) Influence of severity of type and timing of retrospectively reported childhood maltreatment on female amygdala and hippocampal volume. Sci Rep 10(1):1903. https://doi.org/10.1038/s41598-020-57490-0

240. Kang JI, Mueller SG, Wu G, Lin J, Ng P, Yehuda R, Flory JD, Abu-Amara D, Reus VI, Gautam A, PTSD Systems Biology Consortium, Hammamieh R, Doyle FJ 3rd, Jett M, Marmar CR, Mellon SH, Wolkowitz OM (2020) Effect of combat exposure and posttraumatic stress disorder on telomere length and amygdala volume. Biol Psychiatry Cogn Neurosci Neuroimag 5(7):678–687. https://doi.org/10.1016/j.bpsc.2020.03.007

241. Morey RA, Clarke EK, Haswell CC, Phillips RD, Clausen AN, Mufford MS, Saygin Z, VA Mid-Atlantic MIRECC Workgroup, Wagner HR, LaBar KS (2020) Amygdala nuclei volume and shape in military veterans with posttraumatic stress disorder. Biol Psychiatry Cogn Neurosci Neuroimag 5(3):281–290. https://doi.org/10.1016/j.bpsc.2019.11.016

242. Liu T, Ke J, Qi R, Zhang L, Zhang Z, Xu Q, Zhong Y, Lu G, Chen F (2021) Altered functional connectivity of the amygdala and its subregions in typhoon-related post-traumatic stress disorder. Brain Behav 11(1):e01952. https://doi.org/10.1002/brb3.1952

243. Sicorello M, Thome J, Herzog J, Schmahl C (2021) Differential Effects of Early adversity and posttraumatic stress disorder on amygdala reactivity: the role of developmental timing. Biol Psychiatry Cogn Neurosci Neuroimag 6(11):1044–1051. https://doi.org/10.1016/j.bpsc.2020.10.009

244. Bonne O, Brandes D, Gilboa A, Gomori JM, Shenton ME, Pitman RK, Shalev AY (2001) Longitudinal MRI study of hippocampal volume in trauma survivors with PTSD. Am J Psychiatry 158(8):1248–1251. https://doi.org/10.1176/appi.ajp.158.8.1248

245. Bremner JD, Vythilingam M, Vermetten E, Southwick SM, McGlashan T, Nazeer A, Khan S, Vaccarino LV, Soufer R, Garg PK, Ng CK, Staib LH, Duncan JS, Charney DS (2003) MRI and PET study of deficits in hippocampal structure and function in women with childhood sexual abuse and posttraumatic stress disorder. Am J Psychiatry 160(5):924–932. https://doi.org/10.1176/appi.ajp.160.5.924

246. Kitayama N, Vaccarino V, Kutner M, Weiss P, Bremner JD (2005) Magnetic resonance imaging (MRI) measurement of hippocampal volume in posttraumatic stress disorder: a meta-analysis. J Affect Disord 88(1):79–86. https://doi.org/10.1016/j.jad.2005.05.014

247. Smith ME (2005) Bilateral hippocampal volume reduction in adults with post-traumatic stress disorder: a meta-analysis of structural MRI studies. Hippocampus 15(6):798–807. https://doi.org/10.1002/hipo.20102

248. Bonne O, Vythilingam M, Inagaki M, Wood S, Neumeister A, Nugent AC, Snow J, Luckenbaugh DA, Bain EE, Drevets WC, Charney DS (2008) Reduced posterior hippocampal volume in posttraumatic stress disorder. J Clin Psychiatry 69(7):1087–1091. https://doi.org/10.4088/jcp.v69n0707

249. Wang Z, Neylan TC, Mueller SG, Lenoci M, Truran D, Marmar CR, Weiner MW, Schuff N (2010) Magnetic resonance imaging of hippocampal subfields in posttraumatic stress disorder. Arch Gen Psychiatry 67(3):296–303. https://doi.org/10.1001/archgenpsychiatry.2009.205

250. Logue MW, van Rooij S, Dennis EL, Davis SL, Hayes JP, Stevens JS, Densmore M, Haswell CC, Ipser J, Koch S, Korgaonkar M, Lebois L, Peverill M, Baker JT, Boedhoe P, Frijling JL, Gruber SA, Harpaz-Rotem I, Jahanshad N, Koopowitz S et al (2018) Smaller hippocampal volume in posttraumatic stress disorder: a multisite ENIGMA-PGC study: subcortical volumetry results from posttraumatic stress disorder consortia. Biol Psychiatry 83(3):244–253. https://doi.org/10.1016/j.biopsych.2017.09.006

251. Quidé Y, Andersson F, Dufour-Rainfray D, Descriaud C, Brizard B, Gissot V, Cléry H, Carrey Le Bas MP, Osterreicher S, Ogielska M, Saint-Martin P, El-Hage W (2018) Smaller hippocampal volume following sexual assault in women is associated with

252. Joshi SA, Duval ER, Kubat B, Liberzon I (2020) A review of hippocampal activation in post-traumatic stress disorder. Psychophysiology 57(1):e13357. https://doi.org/10.1111/psyp.13357
253. Dark HE, Harnett NG, Knight AJ, Knight DC (2021) Hippocampal volume varies with acute posttraumatic stress symptoms following medical trauma. Behav Neurosci 135(1):71–78. https://doi.org/10.1037/bne0000419
254. Li L, Pan N, Zhang L, Lui S, Huang X, Xu X, Wang S, Lei D, Li L, Kemp GJ, Gong Q (2021) Hippocampal subfield alterations in pediatric patients with post-traumatic stress disorder. Soc Cogn Affect Neurosci 16(3):334–344. https://doi.org/10.1093/scan/nsaa162
255. Weis CN, Webb EK, Huggins AA, Kallenbach M, Miskovich TA, Fitzgerald JM, Bennett KP, Krukowski JL, deRoon-Cassini TA, Larson CL (2021) Stability of hippocampal subfield volumes after trauma and relationship to development of PTSD symptoms. NeuroImage 236:118076. https://doi.org/10.1016/j.neuroimage.2021.118076
256. Fonzo GA, Simmons AN, Thorp SR, Norman SB, Paulus MP, Stein MB (2010) Exaggerated and disconnected insular-amygdalar blood oxygenation level-dependent response to threat-related emotional faces in women with intimate-partner violence posttraumatic stress disorder. Biol Psychiatry 68(5):433–441. https://doi.org/10.1016/j.biopsych.2010.04.028
257. Gong Q, Li L, Tognin S, Wu Q, Pettersson-Yeo W, Lui S, Huang X, Marquand AF, Mechelli A (2014) Using structural neuroanatomy to identify trauma survivors with and without post-traumatic stress disorder at the individual level. Psychol Med 44(1):195–203. https://doi.org/10.1017/S0033291713000561
258. Boukezzi S, El Khoury-Malhame M, Auzias G, Reynaud E, Rousseau PF, Richard E, Zendjidjian X, Roques J, Castelli N, Correard N, Guyon V, Gellato C, Samuelian JC, Cancel A, Comte M, Latinus M, Guedj E, Khalfa S (2017) Grey matter density changes of structures involved in Posttraumatic Stress Disorder (PTSD) after recovery following Eye Movement Desensitization and Reprocessing (EMDR) therapy. Psychiatry Res Neuroimaging 266:146–152. https://doi.org/10.1016/j.pscychresns.2017.06.009
259. Awasthi S, Pan H, LeDoux JE, Cloitre M, Altemus M, McEwen B, Silbersweig D, Stern E (2020) The bed nucleus of the stria terminalis and functionally linked neurocircuitry modulate emotion processing and HPA axis dysfunction in posttraumatic stress disorder. NeuroImage Clin 28:102442. https://doi.org/10.1016/j.nicl.2020.102442
260. Cwik JC, Vahle N, Woud ML, Potthoff D, Kessler H, Sartory G, Seitz RJ (2020) Reduced gray matter volume in the left prefrontal, occipital, and temporal regions as predictors for posttraumatic stress disorder: a voxel-based morphometric study. Eur Arch Psychiatry Clin Neurosci 270(5):577–588. https://doi.org/10.1007/s00406-019-01011-2
261. Harnett NG, Ference EW 3rd, Knight AJ, Knight DC (2020) White matter microstructure varies with post-traumatic stress severity following medical trauma. Brain Imag Behav 14(4):1012–1024. https://doi.org/10.1007/s11682-018-9995-9
262. Ju Y, Ou W, Su J, Averill CL, Liu J, Wang M, Wang Z, Zhang Y, Liu B, Li L, Abdallah CG (2020) White matter microstructural alterations in posttraumatic stress disorder: an ROI and whole-brain based meta-analysis. J Affect Disord 266:655–670. https://doi.org/10.1016/j.jad.2020.01.047
263. Sambuco N, Bradley MM, Lang PJ (2021) Trauma-related dysfunction in the frontostriatal reward circuit. J Affect Disord 287:359–366. https://doi.org/10.1016/j.jad.2021.03.043
264. Seidemann R, Duek O, Jia R, Levy I, Harpaz-Rotem I (2021) The reward system and post-traumatic stress disorder: does trauma affect the way we interact with positive stimuli? Chronic Stress 5:2470547021996006. https://doi.org/10.1177/2470547021996006
265. Yoshii T (2021) The role of the thalamus in post-traumatic stress disorder. Int J Mol Sci 22(4):1730. https://doi.org/10.3390/ijms22041730
266. Kang HJ, Yoon S, Lyoo IK (2015) Peripheral biomarker candidates of posttraumatic stress disorder. Exp Neurobiol 24(3):186–196. https://doi.org/10.5607/en.2015.24.3.186
267. Bandelow B, Baldwin D, Abelli M, Bolea-Alamanac B, Bourin M, Chamberlain SR, Cinosi E, Davies S, Domschke K, Fineberg N, Grünblatt E, Jarema M, Kim

YK, Maron E, Masdrakis V, Mikova O, Nutt D, Pallanti S, Pini S, Ströhle A et al (2017) Biological markers for anxiety disorders, OCD and PTSD: a consensus statement. Part II: neurochemistry, neurophysiology and neurocognition. World J Biol Psychiatry 18(3):162–214. https://doi.org/10.1080/15622975.2016.1190867

268. Geracioti TD Jr, Baker DG, Ekhator NN, West SA, Hill KK, Bruce AB, Schmidt D, Rounds-Kugler B, Yehuda R, Keck PE Jr, Kasckow JW (2001) CSF norepinephrine concentrations in posttraumatic stress disorder. Am J Psychiatry 158(8):1227–1230. https://doi.org/10.1176/appi.ajp.158.8.1227

269. Geracioti TD Jr, Baker DG, Kasckow JW, Strawn JR, Jeffrey Mulchahey J, Dashevsky BA, Horn PS, Ekhator NN (2008) Effects of trauma-related audiovisual stimulation on cerebrospinal fluid norepinephrine and corticotropin-releasing hormone concentrations in post-traumatic stress disorder. Psychoneuroendocrinology 33(4):416–424. https://doi.org/10.1016/j.psyneuen.2007.12.012

270. Norrholm SD, Jovanovic T, Smith AK, Binder E, Klengel T, Conneely K, Mercer KB, Davis JS, Kerley K, Winkler J, Gillespie CF, Bradley B, Ressler KJ (2013) Differential genetic and epigenetic regulation of catechol-O-methyltransferase is associated with impaired fear inhibition in posttraumatic stress disorder. Front Behav Neurosci 7:30. https://doi.org/10.3389/fnbeh.2013.00030

271. Ziegler C, Wolf C, Schiele MA, Feric Bojic E, Kucukalic S, Sabic Dzananovic E, Goci Uka A, Hoxha B, Haxhibeqiri V, Haxhibeqiri S, Kravic N, Muminovic Umihanic M, Cima Franc A, Jaksic N, Babic R, Pavlovic M, Warrings B, Bravo Mehmedbasic A, Rudan D, Aukst-Margetic B et al (2018) Monoamine oxidase A gene methylation and its role in posttraumatic stress disorder: first evidence from the South Eastern Europe (SEE)-PTSD study. Int J Neuropsychopharmacol 21(5):423–432. https://doi.org/10.1093/ijnp/pyx111

272. Perry BD, Giller EL Jr, Southwick SM (1987) Altered platelet alpha 2-adrenergic binding sites in posttraumatic stress disorder. Am J Psychiatry 144(11):1511–1512. https://doi.org/10.1176/ajp.144.11.1511a

273. Maes M, Lin AH, Verkerk R, Delmeire L, Van Gastel A, Van der Planken M, Scharpé S (1999) Serotonergic and noradrenergic markers of post-traumatic stress disorder with and without major depression. Neuropsychopharmacology 20(2):188–197. https://doi.org/10.1016/S0893-133X(98)00058-X

274. Hartwig CL, Sprick JD, Jeong J, Hu Y, Morison DG, Stein CM, Paranjape S, Park J (2020) Increased vascular α1-adrenergic receptor sensitivity in older adults with post-traumatic stress disorder. Am J Physiol Regul Integr Comp Physiol 319(6):R611–R616. https://doi.org/10.1152/ajpregu.00155.2020

275. Southwick SM, Krystal JH, Morgan CA, Johnson D, Nagy LM, Nicolaou A, Heninger GR, Charney DS (1993) Abnormal noradrenergic function in posttraumatic stress disorder. Arch Gen Psychiatry 50(4):266–274. https://doi.org/10.1001/archpsyc.1993.01820160036003

276. Morgan CA 3rd, Grillon C, Southwick SM, Nagy LM, Davis M, Krystal JH, Charney DS (1995) Yohimbine facilitated acoustic startle in combat veterans with post-traumatic stress disorder. Psychopharmacology 117(4):466–471. https://doi.org/10.1007/BF02246220

277. Bremner JD, Innis RB, Ng CK, Staib LH, Salomon RM, Bronen RA, Duncan J, Southwick SM, Krystal JH, Rich D, Zubal G, Dey H, Soufer R, Charney DS (1997) Positron emission tomography measurement of cerebral metabolic correlates of yohimbine administration in combat-related posttraumatic stress disorder. Arch Gen Psychiatry 54(3):246–254. https://doi.org/10.1001/archpsyc.1997.01830150070011

278. Blanchard EB, Kolb LC, Prins A, Gates S, McCoy GC (1991) Changes in plasma norepinephrine to combat-related stimuli among Vietnam veterans with posttraumatic stress disorder. J Nerv Ment Dis 179(6):371–373. https://doi.org/10.1097/00005053-199106000-00012

279. Mellman TA, Kumar A, Kulick-Bell R, Kumar M, Nolan B (1995) Nocturnal/daytime urine noradrenergic measures and sleep in combat-related PTSD. Biol Psychiatry 38(3):174–179. https://doi.org/10.1016/0006-3223(94)00238-X

280. Cordero MI, Sandi C (1998) A role for brain glucocorticoid receptors in contextual fear conditioning: dependence upon training intensity. Brain Res 786(1–2):11–17. https://doi.org/10.1016/s0006-8993(97)01420-0

281. Palma BD, Suchecki D, Tufik S (2000) Differential effects of acute cold and footshock on the sleep of rats. Brain Res 861(1):

97–104. https://doi.org/10.1016/s0006-8993(00)02024-2
282. Luine V, Martinez C, Villegas M, Magariños AM, McEwen BS (1996) Restraint stress reversibly enhances spatial memory performance. Physiol Behav 59(1):27–32. https://doi.org/10.1016/0031-9384(95)02016-0
283. Thompson BL, Erickson K, Schulkin J, Rosen JB (2004) Corticosterone facilitates retention of contextually conditioned fear and increases CRH mRNA expression in the amygdala. Behav Brain Res 149(2):209–215. https://doi.org/10.1016/s0166-4328(03)00216-x
284. Marchand AR, Barbelivien A, Seillier A, Herbeaux K, Sarrieau A, Majchrzak M (2007) Contribution of corticosterone to cued versus contextual fear in rats. Behav Brain Res 183(1):101–110. https://doi.org/10.1016/j.bbr.2007.05.034
285. Mourtzi N, Sertedaki A, Charmandari E (2021) Glucocorticoid signaling and epigenetic alterations in stress-related disorders. Int J Mol Sci 22(11):5964. https://doi.org/10.3390/ijms22115964
286. Meewisse ML, Reitsma JB, de Vries GJ, Gersons BP, Olff M (2007) Cortisol and post-traumatic stress disorder in adults: systematic review and meta-analysis. Br J Psychiatry J Ment Sci 191:387–392. https://doi.org/10.1192/bjp.bp.106.024877
287. Pan X, Wang Z, Wu X, Wen SW, Liu A (2018) Salivary cortisol in post-traumatic stress disorder: a systematic review and meta-analysis. BMC Psychiatry 18(1):324. https://doi.org/10.1186/s12888-018-1910-9
288. Yehuda R, Boisoneau D, Lowy MT, Giller EL Jr (1995) Dose-response changes in plasma cortisol and lymphocyte glucocorticoid receptors following dexamethasone administration in combat veterans with and without posttraumatic stress disorder. Arch Gen Psychiatry 52(7):583–593. https://doi.org/10.1001/archpsyc.1995.03950190065010
289. Yehuda R, Golier JA, Yang RK, Tischler L (2004) Enhanced sensitivity to glucocorticoids in peripheral mononuclear leukocytes in posttraumatic stress disorder. Biol Psychiatry 55(11):1110–1116. https://doi.org/10.1016/j.biopsych.2004.02.010
290. Yehuda R, Halligan SL, Golier JA, Grossman R, Bierer LM (2004) Effects of trauma exposure on the cortisol response to dexamethasone administration in PTSD and major depressive disorder. Psychoneuroendocrinology 29(3):389–404. https://doi.org/10.1016/s0306-4530(03)00052-0
291. Somvanshi PR, Mellon SH, Yehuda R, Flory JD, Makotkine I, Bierer L, Marmar C, Jett M, Doyle FJ 3rd (2020) Role of enhanced glucocorticoid receptor sensitivity in inflammation in PTSD: insights from computational model for circadian-neuroendocrine-immune interactions. Am J Phys Endocrinol Metab 319(1):E48–E66. https://doi.org/10.1152/ajpendo.00398.2019
292. Lehrner A, Bierer LM, Passarelli V, Pratchett LC, Flory JD, Bader HN, Harris IR, Bedi A, Daskalakis NP, Makotkine I, Yehuda R (2014) Maternal PTSD associates with greater glucocorticoid sensitivity in offspring of Holocaust survivors. Psychoneuroendocrinology 40:213–220. https://doi.org/10.1016/j.psyneuen.2013.11.019
293. Baker DG, Ekhator NN, Kasckow JW, Dashevsky B, Horn PS, Bednarik L, Geracioti TD Jr (2005) Higher levels of basal serial CSF cortisol in combat veterans with posttraumatic stress disorder. Am J Psychiatry 162(5):992–994. https://doi.org/10.1176/appi.ajp.162.5.992
294. Baker DG, West SA, Nicholson WE, Ekhator NN, Kasckow JW, Hill KK, Bruce AB, Orth DN, Geracioti TD Jr (1999) Serial CSF corticotropin-releasing hormone levels and adrenocortical activity in combat veterans with posttraumatic stress disorder. Am J Psychiatry 156(4):585–588. https://doi.org/10.1176/ajp.156.4.585
295. Deppermann S, Storchak H, Fallgatter AJ, Ehlis AC (2014) Stress-induced neuroplasticity: (mal)adaptation to adverse life events in patients with PTSD – a critical overview. Neuroscience 283:166–177. https://doi.org/10.1016/j.neuroscience.2014.08.037
296. Speer KE, Semple S, Naumovski N, D'Cunha NM, McKune AJ (2019) HPA axis function and diurnal cortisol in post-traumatic stress disorder: a systematic review. Neurobiol Stress 11:100180. https://doi.org/10.1016/j.ynstr.2019.100180
297. Castro-Vale I, Carvalho D (2020) The pathways between cortisol-related regulation genes and PTSD psychotherapy. Healthcare 8(4):376. https://doi.org/10.3390/healthcare8040376
298. Sarapultsev A, Sarapultsev P, Dremencov E, Komelkova M, Tseilikman O, Tseilikman V (2020) Low glucocorticoids in stress-related disorders: the role of inflammation. Stress 23(6):651–661. https://doi.org/10.1080/10253890.2020.1766020
299. Kim TD, Lee S, Yoon S (2020) Inflammation in post-traumatic stress disorder (PTSD): a review of potential correlates of PTSD with a

neurological perspective. Antioxidants 9(2): 107. https://doi.org/10.3390/antiox9020107

300. Yang JJ, Jiang W (2020) Immune biomarkers alterations in post-traumatic stress disorder: a systematic review and meta-analysis. J Affect Disord 268:39–46. https://doi.org/10.1016/j.jad.2020.02.044

301. Pan X, Kaminga AC, Wu Wen S, Liu A (2021) Chemokines in post-traumatic stress disorder: a network meta-analysis. Brain Behav Immun 92:115–126. https://doi.org/10.1016/j.bbi.2020.11.033

302. Morena M, Patel S, Bains JS, Hill MN (2016) Neurobiological interactions between stress and the endocannabinoid system. Neuropsychopharmacology 41(1):80–102. https://doi.org/10.1038/npp.2015.166

303. Ney L, Stone C, Nichols D, Felmingham K, Bruno R, Matthews A (2021) Endocannabinoid reactivity to acute stress: investigation of the relationship between salivary and plasma levels. Biol Psychol 159:108022. https://doi.org/10.1016/j.biopsycho.2021.108022

304. Hill MN, Bierer LM, Makotkine I, Golier JA, Galea S, McEwen BS, Hillard CJ, Yehuda R (2013) Reductions in circulating endocannabinoid levels in individuals with post-traumatic stress disorder following exposure to the World Trade Center attacks. Psychoneuroendocrinology 38(12):2952–2961. https://doi.org/10.1016/j.psyneuen.2013.08.004

305. Wilker S, Pfeiffer A, Elbert T, Ovuga E, Karabatsiakis A, Krumbholz A, Thieme D, Schelling G, Kolassa IT (2016) Endocannabinoid concentrations in hair are associated with PTSD symptom severity. Psychoneuroendocrinology 67:198–206. https://doi.org/10.1016/j.psyneuen.2016.02.010

306. Crombie KM, Leitzelar BN, Brellenthin AG, Hillard CJ, Koltyn KF (2019) Loss of exercise- and stress-induced increases in circulating 2-arachidonoylglycerol concentrations in adults with chronic PTSD. Biol Psychol 145: 1–7. https://doi.org/10.1016/j.biopsycho.2019.04.002

307. Dincheva I, Drysdale AT, Hartley CA, Johnson DC, Jing D, King EC, Ra S, Gray JM, Yang R, DeGruccio AM, Huang C, Cravatt BF, Glatt CE, Hill MN, Casey BJ, Lee FS (2015) FAAH genetic variation enhances fronto-amygdala function in mouse and human. Nat Commun 6:6395. https://doi.org/10.1038/ncomms7395

308. Crombie KM, Sartin-Tarm A, Sellnow K, Ahrenholtz R, Lee S, Matalamaki M, Almassi NE, Hillard CJ, Koltyn KF, Adams TG, Cisler JM (2021) Exercise-induced increases in Anandamide and BDNF during extinction consolidation contribute to reduced threat following reinstatement: preliminary evidence from a randomized controlled trial. Psychoneuroendocrinology 132:105355. https://doi.org/10.1016/j.psyneuen.2021.105355

309. Neumeister A, Normandin MD, Pietrzak RH, Piomelli D, Zheng MQ, Gujarro-Anton A, Potenza MN, Bailey CR, Lin SF, Najafzadeh S, Ropchan J, Henry S, Corsi-Travali S, Carson RE, Huang Y (2013) Elevated brain cannabinoid CB1 receptor availability in post-traumatic stress disorder: a positron emission tomography study. Mol Psychiatry 18(9):1034–1040. https://doi.org/10.1038/mp.2013.61

310. Hill MN, Campolongo P, Yehuda R, Patel S (2018) Integrating endocannabinoid signaling and cannabinoids into the biology and treatment of posttraumatic stress disorder. Neuropsychopharmacology 43(1):80–102. https://doi.org/10.1038/npp.2017.162

311. Hauer D, Kaufmann I, Strewe C, Briegel I, Campolongo P, Schelling G (2014) The role of glucocorticoids, catecholamines and endocannabinoids in the development of traumatic memories and posttraumatic stress symptoms in survivors of critical illness. Neurobiol Learn Mem 112:68–74. https://doi.org/10.1016/j.nlm.2013.10.003

312. Bailey CR, Cordell E, Sobin SM, Neumeister A (2013) Recent progress in understanding the pathophysiology of post-traumatic stress disorder: implications for targeted pharmacological treatment. CNS Drugs 27(3): 221–232. https://doi.org/10.1007/s40263-013-0051-4

313. Wimalawansa SJ (2014) Mechanisms of developing post-traumatic stress disorder: new targets for drug development and other potential interventions. CNS Neurol Disord Drug Targets 13(5):807–816. https://doi.org/10.2174/1871527313666140711091026

314. Fragkaki I, Thomaes K, Sijbrandij M (2016) Posttraumatic stress disorder under ongoing threat: a review of neurobiological and neuroendocrine findings. Eur J Psychotraumatol 7: 30915. https://doi.org/10.3402/ejpt.v7.30915

315. Sabban EL, Alaluf LG, Serova LI (2016) Potential of neuropeptide Y for preventing or treating post-traumatic stress disorder. Neuropeptides 56:19–24. https://doi.org/10.1016/j.npep.2015.11.004

316. Briscione MA, Michopoulos V, Jovanovic T, Norrholm SD (2017) Neuroendocrine underpinnings of increased risk for posttraumatic stress disorder in women. Vitam Horm

103:53–83. https://doi.org/10.1016/bs.vh.2016.08.003

317. DePierro J, Lepow L, Feder A, Yehuda R (2019) Translating molecular and neuroendocrine findings in posttraumatic stress disorder and resilience to novel therapies. Biol Psychiatry 86(6):454–463. https://doi.org/10.1016/j.biopsych.2019.07.009

318. Malikowska-Racia N, Salat K (2019) Recent advances in the neurobiology of posttraumatic stress disorder: a review of possible mechanisms underlying an effective pharmacotherapy. Pharmacol Res 142:30–49. https://doi.org/10.1016/j.phrs.2019.02.001

319. Ravi M, Stevens JS, Michopoulos V (2019) Neuroendocrine pathways underlying risk and resilience to PTSD in women. Front Neuroendocrinol 55:100790. https://doi.org/10.1016/j.yfrne.2019.100790

320. Torres-Berrio A, Nava-Mesa MO (2019) The opioid system in stress-induced memory disorders: from basic mechanisms to clinical implications in post-traumatic stress disorder and Alzheimer's disease. Prog Neuro-Psychopharmacol Biol Psychiatry 88:327–338. https://doi.org/10.1016/j.pnpbp.2018.08.011

321. Yoon S, Kim YK (2019) Neuroendocrinological treatment targets for posttraumatic stress disorder. Prog Neuro-Psychopharmacol Biol Psychiatry 90:212–222. https://doi.org/10.1016/j.pnpbp.2018.11.021

322. Seligowski AV, Harnett NG, Merker JB, Ressler KJ (2020) Nervous and endocrine system dysfunction in posttraumatic stress disorder: an overview and consideration of sex as a biological variable. Biol Psychiatry Cogn Neurosci Neuroimag 5(4):381–391. https://doi.org/10.1016/j.bpsc.2019.12.006

323. Thoma MV, Joksimovic L, Kirschbaum C, Wolf JM, Rohleder N (2012) Altered salivary alpha-amylase awakening response in Bosnian War refugees with posttraumatic stress disorder. Psychoneuroendocrinology 37(6):810–817. https://doi.org/10.1016/j.psyneuen.2011.09.013

324. Morris MC, Rao U (2013) Psychobiology of PTSD in the acute aftermath of trauma: integrating research on coping, HPA function and sympathetic nervous system activity. Asian J Psychiatr 6(1):3–21. https://doi.org/10.1016/j.ajp.2012.07.012

325. Busso DS, McLaughlin KA, Sheridan MA (2014) Media exposure and sympathetic nervous system reactivity predict PTSD symptoms after the Boston marathon bombings. Depress Anxiety 31(7):551–558. https://doi.org/10.1002/da.22282

326. Liberzon I, King AP, Ressler KJ, Almli LM, Zhang P, Ma ST, Cohen GH, Tamburrino MB, Calabrese JR, Galea S (2014) Interaction of the ADRB2 gene polymorphism with childhood trauma in predicting adult symptoms of posttraumatic stress disorder. JAMA Psychiatry 71(10):1174–1182. https://doi.org/10.1001/jamapsychiatry.2014.999

327. Nicholson EL, Bryant RA, Felmingham KL (2014) Interaction of noradrenaline and cortisol predicts negative intrusive memories in posttraumatic stress disorder. Neurobiol Learn Mem 112:204–211. https://doi.org/10.1016/j.nlm.2013.11.018

328. Keeshin BR, Strawn JR, Out D, Granger DA, Putnam FW (2015) Elevated salivary alpha amylase in adolescent sexual abuse survivors with posttraumatic stress disorder symptoms. J Child Adolesc Psychopharmacol 25(4):344–350. https://doi.org/10.1089/cap.2014.0034

329. Pietrzak RH, Sumner JA, Aiello AE, Uddin M, Neumeister A, Guffanti G, Koenen KC (2015) Association of the rs2242446 polymorphism in the norepinephrine transporter gene SLC6A2 and anxious arousal symptoms of posttraumatic stress disorder. J Clin Psychiatry 76(4):e537–e538. https://doi.org/10.4088/JCP.14l09346

330. Wingenfeld K, Whooley MA, Neylan TC, Otte C, Cohen BE (2015) Effect of current and lifetime posttraumatic stress disorder on 24-h urinary catecholamines and cortisol: results from the Mind Your Heart Study. Psychoneuroendocrinology 52:83–91. https://doi.org/10.1016/j.psyneuen.2014.10.023

331. Hendrickson RC, Raskind MA (2016) Noradrenergic dysregulation in the pathophysiology of PTSD. Exp Neurol 284:181–195. https://doi.org/10.1016/j.expneurol.2016.05.014

332. Gupta MA (2017) Recurrent hypersomnia and autonomic dysregulation in posttraumatic stress disorder. J Clin Sleep Med 13(12):1491. https://doi.org/10.5664/jcsm.6860

333. Park J, Marvar PJ, Liao P, Kankam ML, Norrholm SD, Downey RM, McCullough SA, Le NA, Rothbaum BO (2017) Baroreflex dysfunction and augmented sympathetic nerve responses during mental stress in veterans with post-traumatic stress disorder. J Physiol 595(14):4893–4908. https://doi.org/10.1113/JP274269

334. Silva L, Katayama PL (2017) Baroreflex-mediated sympathetic overactivation induced

335. Kanady JC, Maguen S, Neylan TC (2018) Further exploring the associations between sympathetic activation, fear of sleep, and insomnia symptoms in posttraumatic stress disorder. J Clin Sleep Med 14(12): 2095–2096. https://doi.org/10.5664/jcsm.7552

336. Naegeli C, Zeffiro T, Piccirelli M, Jaillard A, Weilenmann A, Hassanpour K, Schick M, Rufer M, Orr SP, Mueller-Pfeiffer C (2018) Locus coeruleus activity mediates hyperresponsiveness in posttraumatic stress disorder. Biol Psychiatry 83(3):254–262. https://doi.org/10.1016/j.biopsych.2017.08.021

337. Pan X, Kaminga AC, Wen SW, Liu A (2018) Catecholamines in post-traumatic stress disorder: a systematic review and meta-analysis. Front Mol Neurosci 11:450. https://doi.org/10.3389/fnmol.2018.00450

338. Ressler KJ (2018) Alpha-adrenergic receptors in PTSD – failure or time for precision medicine? N Engl J Med 378(6):575–576. https://doi.org/10.1056/NEJMe1716724

339. Cohen JR, Thomsen KN, Tu KM, Thakur H, McNeil S, Menon SV (2020) Cardiac autonomic functioning and post-traumatic stress: a preliminary study in youth at-risk for PTSD. Psychiatry Res 284:112684. https://doi.org/10.1016/j.psychres.2019.112684

340. Fonkoue IT, Marvar PJ, Norrholm S, Li Y, Kankam ML, Jones TN, Vemulapalli M, Rothbaum B, Bremner JD, Le NA, Park J (2020) Symptom severity impacts sympathetic dysregulation and inflammation in post-traumatic stress disorder (PTSD). Brain Behav Immun 83:260–269. https://doi.org/10.1016/j.bbi.2019.10.021

341. Fonkoue IT, Michopoulos V, Park J (2020) Sex differences in post-traumatic stress disorder risk: autonomic control and inflammation. Clin Auton Res 30(5):409–421. https://doi.org/10.1007/s10286-020-00729-7

342. Gupta MA (2020) Nightmare recurrence in patients with post-traumatic stress disorder is likely a primary feature of central sympathetic nervous activation. J Clin Sleep Med 16(11): 1995. https://doi.org/10.5664/jcsm.8782

343. Gupta MA (2020) Hypopneas with arousals: an important feature of central nervous system sympathetic activation in posttraumatic stress disorder. J Clin Sleep Med 16(2):335. https://doi.org/10.5664/jcsm.8198

344. Ross JA, Van Bockstaele EJ (2020) The role of catecholamines in modulating responses to stress: sex-specific patterns, implications, and therapeutic potential for post-traumatic stress disorder and opiate withdrawal. Eur J Neurosci 52(1):2429–2465. https://doi.org/10.1111/ejn.14714

345. Schneider M, Schwerdtfeger A (2020) Autonomic dysfunction in posttraumatic stress disorder indexed by heart rate variability: a meta-analysis. Psychol Med 50(12):1937–1948. https://doi.org/10.1017/S003329172000207X

346. Yoo JK, Badrov MB, Huang M, Bain RA, Dorn RP, Anderson EH, Wiblin JL, Suris A, Shoemaker JK, Fu Q (2020) Abnormal sympathetic neural recruitment patterns and hemodynamic responses to cold pressor test in women with posttraumatic stress disorder. Am J Physiol Heart Circ Physiol 318(5): H1198–H1207. https://doi.org/10.1152/ajpheart.00684.2019

347. Bicanic IA, Postma RM, Sinnema G, De Roos C, Olff M, Van Wesel F, Van de Putte EM (2013) Salivary cortisol and dehydroepiandrosterone sulfate in adolescent rape victims with post traumatic stress disorder. Psychoneuroendocrinology 38(3):408–415. https://doi.org/10.1016/j.psyneuen.2012.06.015

348. Dekel S, Ein-Dor T, Gordon KM, Rosen JB, Bonanno GA (2013) Cortisol and PTSD symptoms among male and female high-exposure 9/11 survivors. J Trauma Stress 26(5):621–625. https://doi.org/10.1002/jts.21839

349. Wahbeh H, Oken BS (2013) Salivary cortisol lower in posttraumatic stress disorder. J Trauma Stress 26(2):241–248. https://doi.org/10.1002/jts.21798

350. Walsh K, Nugent NR, Kotte A, Amstadter AB, Wang S, Guille C, Acierno R, Kilpatrick DG, Resnick HS (2013) Cortisol at the emergency room rape visit as a predictor of PTSD and depression symptoms over time. Psychoneuroendocrinology 38(11):2520–2528. https://doi.org/10.1016/j.psyneuen.2013.05.017

351. van Zuiden M, Kavelaars A, Geuze E, Olff M, Heijnen CJ (2013) Predicting PTSD: pre-existing vulnerabilities in glucocorticoid-signaling and implications for preventive interventions. Brain Behav Immun 30:12–21. https://doi.org/10.1016/j.bbi.2012.08.015

352. Labonté B, Azoulay N, Yerko V, Turecki G, Brunet A (2014) Epigenetic modulation of glucocorticoid receptors in posttraumatic

stress disorder. Transl Psychiatry 4(3):e368. https://doi.org/10.1038/tp.2014.3
353. Lian Y, Xiao J, Wang Q, Ning L, Guan S, Ge H, Li F, Liu J (2014) The relationship between glucocorticoid receptor polymorphisms, stressful life events, social support, and post-traumatic stress disorder. BMC Psychiatry 14:232. https://doi.org/10.1186/s12888-014-0232-9
354. Vukojevic V, Kolassa IT, Fastenrath M, Gschwind L, Spalek K, Milnik A, Heck A, Vogler C, Wilker S, Demougin P, Peter F, Atucha E, Stetak A, Roozendaal B, Elbert T, Papassotiropoulos A, de Quervain DJ (2014) Epigenetic modification of the glucocorticoid receptor gene is linked to traumatic memory and post-traumatic stress disorder risk in genocide survivors. J Neurosci 34(31): 10274–10284. https://doi.org/10.1523/JNEUROSCI.1526-14.2014
355. Kaminsky Z, Wilcox HC, Eaton WW, Van Eck K, Kilaru V, Jovanovic T, Klengel T, Bradley B, Binder EB, Ressler KJ, Smith AK (2015) Epigenetic and genetic variation at SKA2 predict suicidal behavior and post-traumatic stress disorder. Transl Psychiatry 5(8):e627
356. Steudte-Schmiedgen S, Stalder T, Schönfeld S, Wittchen HU, Trautmann S, Alexander N, Miller R, Kirschbaum C (2015) Hair cortisol concentrations and cortisol stress reactivity predict PTSD symptom increase after trauma exposure during military deployment. Psychoneuroendocrinology 59: 123–133. https://doi.org/10.1016/j.psyneuen.2015.05.00
357. Boks MP, Rutten BP, Geuze E, Houtepen LC, Vermetten E, Kaminsky Z, Vinkers CH (2016) SKA2 methylation is involved in cortisol stress reactivity and predicts the development of post-traumatic stress disorder (PTSD) after military deployment. Neuropsychopharmacology 41(5):1350–1356. https://doi.org/10.1038/npp.2015.286
358. Castro-Vale I, van Rossum EF, Machado JC, Mota-Cardoso R, Carvalho D (2016) Genetics of glucocorticoid regulation and posttraumatic stress disorder – what do we know? Neurosci Biobehav Rev 63:143–157. https://doi.org/10.1016/j.neubiorev.2016.02.005
359. Morris MC, Hellman N, Abelson JL, Rao U (2016) Cortisol, heart rate, and blood pressure as early markers of PTSD risk: a systematic review and meta-analysis. Clin Psychol Rev 49:79–91. https://doi.org/10.1016/j.cpr.2016.09.001
360. Cordero MI, Moser DA, Manini A, Suardi F, Sancho-Rossignol A, Torrisi R, Rossier MF, Ansermet F, Dayer AG, Rusconi-Serpa S, Schechter DS (2017) Effects of interpersonal violence-related post-traumatic stress disorder (PTSD) on mother and child diurnal cortisol rhythm and cortisol reactivity to a laboratory stressor involving separation. Horm Behav 90:15–24. https://doi.org/10.1016/j.yhbeh.2017.02.007
361. Olff M, van Zuiden M (2017) Neuroendocrine and neuroimmune markers in PTSD: pre-, peri- and post-trauma glucocorticoid and inflammatory dysregulation. Curr Opin Psychol 14:132–137. https://doi.org/10.1016/j.copsyc.2017.01.001
362. Straub J, Klaubert LM, Schmiedgen S, Kirschbaum C, Goldbeck L (2017) Hair cortisol in relation to acute and post-traumatic stress symptoms in children and adolescents. Anxiety Stress Coping 30(6):661–670. https://doi.org/10.1080/10615806.2017.1355458
363. Li Y, Seng JS (2018) Child maltreatment trauma, posttraumatic stress disorder, and cortisol levels in women: a literature review. J Am Psychiatr Nurses Assoc 24(1):35–44. https://doi.org/10.1177/1078390317710313
364. McNerney MW, Sheng T, Nechvatal JM, Lee AG, Lyons DM, Soman S, Liao CP, O'Hara R, Hallmayer J, Taylor J, Ashford JW, Yesavage J, Adamson MM (2018) Integration of neural and epigenetic contributions to posttraumatic stress symptoms: the role of hippocampal volume and glucocorticoid receptor gene methylation. PLoS One 13(2): e0192222
365. Szeszko PR, Lehrner A, Yehuda R (2018) Glucocorticoids and hippocampal structure and function in PTSD. Harv Rev Psychiatry 26(3):142–157. https://doi.org/10.1097/HRP.0000000000000188
366. Dunlop BW, Wong A (2019) The hypothalamic-pituitary-adrenal axis in PTSD: pathophysiology and treatment interventions. Prog Neuro-Psychopharmacol Biol Psychiatry 89:361–379. https://doi.org/10.1016/j.pnpbp.2018.10.010
367. Schumacher S, Niemeyer H, Engel S, Cwik JC, Laufer S, Klusmann H, Knaevelsrud C (2019) HPA axis regulation in posttraumatic stress disorder: a meta-analysis focusing on potential moderators. Neurosci Biobehav Rev 100:35–57. https://doi.org/10.1016/j.neubiorev.2019.02.005
368. van den Heuvel LL, Wright S, Suliman S, Stalder T, Kirschbaum C, Seedat S (2019)

Cortisol levels in different tissue samples in posttraumatic stress disorder patients versus controls: a systematic review and meta-analysis protocol. Syst Rev 8(1):7. https://doi.org/10.1186/s13643-018-0936-x

369. Hadad NA, Schwendt M, Knackstedt LA (2020) Hypothalamic-pituitary-adrenal axis activity in post-traumatic stress disorder and cocaine use disorder. Stress 23(6):638–650. https://doi.org/10.1080/10253890.2020.1803824

370. Metz S, Duesenberg M, Hellmann-Regen J, Wolf OT, Roepke S, Otte C, Wingenfeld K (2020) Blunted salivary cortisol response to psychosocial stress in women with posttraumatic stress disorder. J Psychiatr Res 130:112–119. https://doi.org/10.1016/j.jpsychires.2020.07.014

371. Pan X, Kaminga AC, Wen SW, Wang Z, Wu X, Liu A (2020) The 24-hour urinary cortisol in post-traumatic stress disorder: a meta-analysis. PLoS One 15(1):e0227560. https://doi.org/10.1371/journal.pone.0227560

372. Sheerin CM, Lind MJ, Bountress KE, Marraccini ME, Amstadter AB, Bacanu SA, Nugent NR (2020) Meta-analysis of associations between hypothalamic-pituitary-adrenal axis genes and risk of posttraumatic stress disorder. J Trauma Stress 33(5):688–698. https://doi.org/10.1002/jts.22484

373. van den Heuvel LL, Stalder T, du Plessis S, Suliman S, Kirschbaum C, Seedat S (2020) Hair cortisol levels in posttraumatic stress disorder and metabolic syndrome. Stress 23(5):577–589. https://doi.org/10.1080/10253890.2020.1724949

374. Almeida FB, Pinna G, Barros H (2021) The role of HPA axis and allopregnanolone on the neurobiology of major depressive disorders and PTSD. Int J Mol Sci 22(11):5495. https://doi.org/10.3390/ijms22115495

375. Danan D, Todder D, Zohar J, Cohen H (2021) Is PTSD-phenotype associated with HPA-axis sensitivity? Feedback inhibition and other modulating factors of glucocorticoid signaling dynamics. Int J Mol Sci 22(11):6050. https://doi.org/10.3390/ijms22116050

376. Fischer S, Schumacher T, Knaevelsrud C, Ehlert U, Schumacher S (2021) Genes and hormones of the hypothalamic-pituitary-adrenal axis in post-traumatic stress disorder. What is their role in symptom expression and treatment response? J Neural Transm 128(9):1279–1286. https://doi.org/10.1007/s00702-021-02330-2

377. Koumantarou Malisiova E, Mourikis I, Darviri C, Nicolaides NC, Zervas IM, Papageorgiou C, Chrousos GP (2021) Hair cortisol concentrations in mental disorders: a systematic review. Physiol Behav 229:113244. https://doi.org/10.1016/j.physbeh.2020.113244

378. Schaffter N, Ledermann K, Pazhenkottil AP, Barth J, Schnyder U, Znoj H, Schmid JP, Meister-Langraf RE, von Känel R, Princip M (2021) Serum cortisol as a predictor for post-traumatic stress disorder symptoms in post-myocardial infarction patients. J Affect Disord 292:687–694. https://doi.org/10.1016/j.jad.2021.05.065

379. Lu AT, Ogdie MN, Järvelin MR, Moilanen IK, Loo SK, McCracken JT, McGough JJ, Yang MH, Peltonen L, Nelson SF, Cantor RM, Smalley SL (2008) Association of the cannabinoid receptor gene (CNR1) with ADHD and post-traumatic stress disorder. Am J Med Genet 147B(8):1488–1494. https://doi.org/10.1002/ajmg.b.30693

380. Hauer D, Schelling G, Gola H, Campolongo P, Morath J, Roozendaal B, Hamuni G, Karabatsiakis A, Atsak P, Vogeser M, Kolassa IT (2013) Plasma concentrations of endocannabinoids and related primary fatty acid amides in patients with post-traumatic stress disorder. PLoS One 8(5):e62741. https://doi.org/10.1371/journal.pone.0062741

381. Schaefer C, Enning F, Mueller JK, Bumb JM, Rohleder C, Odorfer TM, Klosterkötter J, Hellmich M, Koethe D, Schmahl C, Bohus M, Leweke FM (2014) Fatty acid ethanolamide levels are altered in borderline personality and complex posttraumatic stress disorders. Eur Arch Psychiatry Clin Neurosci 264(5):459–463. https://doi.org/10.1007/s00406-013-0470-8

382. Mota N, Sumner JA, Lowe SR, Neumeister A, Uddin M, Aiello AE, Wildman DE, Galea S, Koenen KC, Pietrzak RH (2015) The rs1049353 polymorphism in the CNR1 gene interacts with childhood abuse to predict posttraumatic threat symptoms. J Clin Psychiatry 76(12):e1622–e1623. https://doi.org/10.4088/JCP.15l10084

383. Neumeister A, Seidel J, Ragen BJ, Pietrzak RH (2015) Translational evidence for a role of endocannabinoids in the etiology and treatment of posttraumatic stress disorder. Psychoneuroendocrinology 51:577–584. https://doi.org/10.1016/j.psyneuen.2014.10.012

384. Berardi A, Schelling G, Campolongo P (2016) The endocannabinoid system and

post traumatic stress disorder (PTSD): from preclinical findings to innovative therapeutic approaches in clinical settings. Pharmacol Res 111:668–678. https://doi.org/10.1016/j.phrs.2016.07.024

385. Spagnolo PA, Ramchandani VA, Schwandt ML, Kwako LE, George DT, Mayo LM, Hillard CJ, Heilig M (2016) FAAH gene variation moderates stress response and symptom severity in patients with posttraumatic stress disorder and comorbid alcohol dependence. Alcohol Clin Exp Res 40(11):2426–2434. https://doi.org/10.1111/acer.13210

386. Ney LJ, Matthews A, Bruno R, Felmingham KL (2018) Modulation of the endocannabinoid system by sex hormones: implications for posttraumatic stress disorder. Neurosci Biobehav Rev 94:302–320. https://doi.org/10.1016/j.neubiorev.2018.07.006

387. Sloan ME, Grant CW, Gowin JL, Ramchandani VA, Le Foll B (2019) Endocannabinoid signaling in psychiatric disorders: a review of positron emission tomography studies. Acta Pharmacol Sin 40(3):342–350. https://doi.org/10.1038/s41401-018-0081-z

388. Lohr JB, Chang H, Sexton M, Palmer BW (2020) Allostatic load and the cannabinoid system: implications for the treatment of physiological abnormalities in post-traumatic stress disorder (PTSD). CNS Spectr 25(6):743–749. https://doi.org/10.1017/S1092852919001093

389. Mayo LM, Asratian A, Lindé J, Holm L, Nätt D, Augier G, Stensson N, Vecchiarelli HA, Balsevich G, Aukema RJ, Ghafouri B, Spagnolo PA, Lee FS, Hill MN, Heilig M (2020) Protective effects of elevated anandamide on stress and fear-related behaviors: translational evidence from humans and mice. Mol Psychiatry 25(5):993–1005. https://doi.org/10.1038/s41380-018-0215-1

390. Navarrete F, García-Gutiérrez MS, Jurado-Barba R, Rubio G, Gasparyan A, Austrich-Olivares A, Manzanares J (2020) Endocannabinoid system components as potential biomarkers in psychiatry. Front Psychiatry 11:315. https://doi.org/10.3389/fpsyt.2020.00315

391. Ney LJ, Crombie KM, Mayo LM, Felmingham KL, Bowser T, Matthews A (2022) Translation of animal endocannabinoid models of PTSD mechanisms to humans: where to next? Neurosci Biobehav Rev 132:76–91. https://doi.org/10.1016/j.neubiorev.2021.11.040

392. Gold PE, Van Buskirk RB (1975) Facilitation of time-dependent memory processes with posttrial epinephrine injections. Behav Biol 13(2):145–153. https://doi.org/10.1016/s0091-6773(75)91784-8

393. Izquierdo I, Dias RD (1983) Effect of ACTH, epinephrine, beta-endorphin, naloxone, and of the combination of naloxone or beta-endorphin with ACTH or epinephrine on memory consolidation. Psychoneuroendocrinology 8(1):81–87. https://doi.org/10.1016/0306-4530(83)90043-4

394. Izquierdo I, Netto CA (1985) Factors that influence test session performance measured 0, 3, or 6 h after inhibitory avoidance training. Behav Neural Biol 43(3):260–273. https://doi.org/10.1016/s0163-1047(85)91606-1

395. Liang KC, Juler RG, McGaugh JL (1986) Modulating effects of posttraining epinephrine on memory: involvement of the amygdala noradrenergic system. Brain Res 368(1):125–133. https://doi.org/10.1016/0006-8993(86)91049-8

396. Introini-Collison IB, Baratti CM (1986) Opioid peptidergic systems modulate the activity of beta-adrenergic mechanisms during memory consolidation processes. Behav Neural Biol 46(2):227–241. https://doi.org/10.1016/s0163-1047(86)90710-7

397. Introini-Collison IB, McGaugh JL (1986) Epinephrine modulates long-term retention of an aversively motivated discrimination. Behav Neural Biol 45(3):358–365. https://doi.org/10.1016/s0163-1047(86)80024-3

398. Introini-Collison IB, McGaugh JL (1987) Naloxone and beta-endorphin alter the effects of post-training epinephrine on memory. Psychopharmacology 92(2):229–235. https://doi.org/10.1007/BF00177921

399. Introini-Collison IB, McGaugh JL (1988) Modulation of memory by post-training epinephrine: involvement of cholinergic mechanisms. Psychopharmacology 94(3):379–385. https://doi.org/10.1007/BF00174693

400. Netto CA, Maltchik M (1990) Distinct mechanisms underlying memory modulation after the first and the second session of two avoidance tasks. Behav Neural Biol 53(1):29–38. https://doi.org/10.1016/0163-1047(90)90763-v

401. Introini-Collison I, Saghafi D, Novack GD, McGaugh JL (1992) Memory-enhancing effects of post-training dipivefrin and epinephrine: involvement of peripheral and central adrenergic receptors. Brain Res 572(1–2):81–86. https://doi.org/10.1016/0006-8993(92)90454-h

402. Costa-Miserachs D, Portell-Cortés I, Aldavert-Vera L, Torras-García M, Morgado-Bernal I (1993) Facilitation of a distributed shuttlebox conditioning with post-training epinephrine in rats. Behav Neural Biol 60(1):75–78. https://doi.org/10.1016/0163-1047(93)90755-7

403. Williams CL, McGaugh JL (1993) Reversible lesions of the nucleus of the solitary tract attenuate the memory-modulating effects of posttraining epinephrine. Behav Neurosci 107(6):955–962

404. Costa-Miserachs D, Portell-Cortés I, Aldavert-Vera L, Torras-García M, Morgado-Bernal I (1994) Long-term memory facilitation in rats by posttraining epinephrine. Behav Neurosci 108(3):469–474. https://doi.org/10.1037//0735-7044.108.3.469

405. Torras-Garcia M, Portell-Cortés I, Costa-Miserachs D, Morgado-Bernal I (1997) Long-term memory modulation by posttraining epinephrine in rats: differential effects depending on the basic learning capacity. Behav Neurosci 111(2):301–308. https://doi.org/10.1037/0735-7044.111.2.301

406. Torras-Garcia M, Costa-Miserachs D, Portell-Cortés I, Morgado-Bernal I (1998) Posttraining epinephrine and memory consolidation in rats with different basic learning capacities. The role of the stria terminalis. Exp Brain Res 121(1):20–28. https://doi.org/10.1007/s002210050432

407. Nordby T, Torras-Garcia M, Portell-Cortés I, Costa-Miserachs D (2006) Posttraining epinephrine treatment reduces the need for extensive training. Physiol Behav 89(5):718–723. https://doi.org/10.1016/j.physbeh.2006.08.010

408. Tuon L, Comim CM, Petronilho F, Barichello T, Izquierdo I, Quevedo J, Dal-Pizzol F (2008) Memory-enhancing treatments reverse the impairment of inhibitory avoidance retention in sepsis-surviving rats. Crit Care 12(5):R133. https://doi.org/10.1186/cc7103

409. Roozendaal B, Mirone G (2020) Opposite effects of noradrenergic and glucocorticoid activation on accuracy of an episodic-like memory. Psychoneuroendocrinology 114:104588. https://doi.org/10.1016/j.psyneuen.2020.104588

410. Gold PE, Van Buskirk R (1976) Enhancement and impairment of memory processes with post-trial injections of adrenocorticotrophic hormone. Behav Biol 16(4):387–400. https://doi.org/10.1016/s0091-6773(76)91539-x

411. Flood JF, Vidal D, Bennett EL, Orme AE, Vasquez S, Jarvik ME (1978) Memory facilitating and anti-amnesic effects of corticosteroids. Pharmacol Biochem Behav 8(1):81–87. https://doi.org/10.1016/0091-3057(78)90127-2

412. Cabib S, Castellano C, Patacchioli FR, Cigliana G, Angelucci L, Puglisi-Allegra S (1996) Opposite strain-dependent effects of post-training corticosterone in a passive avoidance task in mice: role of dopamine. Brain Res 729(1):110–118

413. Roozendaal B, McGaugh JL (1996a) Amygdaloid nuclei lesions differentially affect glucocorticoid-induced memory enhancement in an inhibitory avoidance task. Neurobiol Learn Mem 65(1):1–8. https://doi.org/10.1006/nlme.1996.0001

414. Roozendaal B, McGaugh JL (1996b) The memory-modulatory effects of glucocorticoids depend on an intact stria terminalis. Brain Res 709(2):243–250. https://doi.org/10.1016/0006-8993(95)01305-9

415. Quirarte GL, Roozendaal B, McGaugh JL (1997) Glucocorticoid enhancement of memory storage involves noradrenergic activation in the basolateral amygdala. Proc Natl Acad Sci U S A 94(25):14048–14053. https://doi.org/10.1073/pnas.94.25.14048

416. Setlow B, Roozendaal B, McGaugh JL (2000) Involvement of a basolateral amygdala complex-nucleus accumbens pathway in glucocorticoid-induced modulation of memory consolidation. Eur J Neurosci 12(1):367–375. https://doi.org/10.1046/j.1460-9568.2000.00911.x

417. Zorawski M, Killcross S (2002) Posttraining glucocorticoid receptor agonist enhances memory in appetitive and aversive Pavlovian discrete-cue conditioning paradigms. Neurobiol Learn Mem 78(2):458–464. https://doi.org/10.1006/nlme.2002.4075

418. Hui GK, Figueroa IR, Poytress BS, Roozendaal B, McGaugh JL, Weinberger NM (2004) Memory enhancement of classical fear conditioning by post-training injections of corticosterone in rats. Neurobiol Learn Mem 81(1):67–74. https://doi.org/10.1016/j.nlm.2003.09.002

419. Venturella R, Lessa D, Luft T, Roozendaal B, Schwartsmann G, Roesler R (2005) Dexamethasone reverses the memory impairment induced by antagonism of hippocampal gastrin-releasing peptide receptors. Peptides 26(5):821–825. https://doi.org/10.1016/j.peptides.2004.12.010

420. Roozendaal B, Hui GK, Hui IR, Berlau DJ, McGaugh JL, Weinberger NM (2006) Basolateral amygdala noradrenergic activity mediates corticosterone-induced enhancement of auditory fear conditioning. Neurobiol Learn Mem 86(3):249–255. https://doi.org/10.1016/j.nlm.2006.03.003

421. Miranda MI, Quirarte GL, Rodriguez-Garcia G, McGaugh JL, Roozendaal B (2008) Glucocorticoids enhance taste aversion memory via actions in the insular cortex and basolateral amygdala. Learn Mem 15(7):468–476. https://doi.org/10.1101/lm.964708

422. Roozendaal B, Barsegyan A, Lee S (2008) Adrenal stress hormones, amygdala activation, and memory for emotionally arousing experiences. Prog Brain Res 167:79–97. https://doi.org/10.1016/S0079-6123(07)67006-X

423. Abrari K, Rashidy-Pour A, Semnanian S, Fathollahi Y, Jadid M (2009) Post-training administration of corticosterone enhances consolidation of contextual fear memory and hippocampal long-term potentiation in rats. Neurobiol Learn Mem 91(3):260–265. https://doi.org/10.1016/j.nlm.2008.10.008

424. Campolongo P, Roozendaal B, Trezza V, Hauer D, Schelling G, McGaugh JL, Cuomo V (2009) Endocannabinoids in the rat basolateral amygdala enhance memory consolidation and enable glucocorticoid modulation of memory. Proc Natl Acad Sci U S A 106(12):4888–4893. https://doi.org/10.1073/pnas.0900835106

425. Kaouane N, Porte Y, Vallée M, Brayda-Bruno L, Mons N, Calandreau L, Marighetto A, Piazza PV, Desmedt A (2012) Glucocorticoids can induce PTSD-like memory impairments in mice. Science 335(6075):1510–1513. https://doi.org/10.1126/science.1207615

426. Liao Y, Shi YW, Liu QL, Zhao H (2013) Glucocorticoid-induced enhancement of contextual fear memory consolidation in rats: involvement of D1 receptor activity of hippocampal area CA1. Brain Res 1524:26–33. https://doi.org/10.1016/j.brainres.2013.05.030

427. Zalachoras I, Houtman R, Atucha E, Devos R, Tijssen AM, Hu P, Lockey PM, Datson NA, Belanoff JK, Lucassen PJ, Joëls M, de Kloet ER, Roozendaal B, Hunt H, Meijer OC (2013) Differential targeting of brain stress circuits with a selective glucocorticoid receptor modulator. Proc Natl Acad Sci U S A 110(19):7910–7915. https://doi.org/10.1073/pnas.1219411110

428. Yang C, Liu JF, Chai BS, Fang Q, Chai N, Zhao LY, Xue YX, Luo YX, Jian M, Han Y, Shi HS, Lu L, Wu P, Wang JS (2013) Stress within a restricted time window selectively affects the persistence of long-term memory. PLoS One 8(3):e59075. https://doi.org/10.1371/journal.pone.0059075

429. McReynolds JR, Holloway-Erickson CM, Parmar TU, McIntyre CK (2014) Corticosterone-induced enhancement of memory and synaptic Arc protein in the medial prefrontal cortex. Neurobiol Learn Mem 112:148–157. https://doi.org/10.1016/j.nlm.2014.02.007

430. Kashefi A, Rashidy-Pour A (2014) Effects of corticosterone on contextual fear consolidation in intact and ovariectomized female rats. Neurobiol Learn Mem 114:236–241. https://doi.org/10.1016/j.nlm.2014.06.013

431. Souza RR, Dal Bó S, de Kloet ER, Oitzl MS, Carobrez AP (2014) Paradoxical mineralocorticoid receptor-mediated effect in fear memory encoding and expression of rats submitted to an olfactory fear conditioning task. Neuropharmacology 79:201–211. https://doi.org/10.1016/j.neuropharm.2013.11.017

432. Xiong H, Cassé F, Zhou Y, Zhou M, Xiong ZQ, Joëls M, Martin S, Krugers HJ (2015) mTOR is essential for corticosteroid effects on hippocampal AMPA receptor function and fear memory. Learn Mem 22(12):577–583. https://doi.org/10.1101/lm.039420.115

433. Siller-Pérez C, Fuentes-Ibañez A, Sotelo-Barrera EL, Serafín N, Prado-Alcalá RA, Campolongo P, Roozendaal B, Quirarte GL (2019) Glucocorticoid interactions with the dorsal striatal endocannabinoid system in regulating inhibitory avoidance memory. Psychoneuroendocrinology 99:97–103. https://doi.org/10.1016/j.psyneuen.2018.08.021

434. Santucci AC, Kanof PD, Haroutunian V (1989) Effect of physostigmine on memory consolidation and retrieval processes in intact and nucleus basalis-lesioned rats. Psychopharmacology 99(1):70–74. https://doi.org/10.1007/BF00634455

435. Young SL, Bohenek DL, Fanselow MS (1995) Scopolamine impairs acquisition and facilitates consolidation of fear conditioning: differential effects for tone vs context conditioning. Neurobiol Learn Mem 63(2):174–180. https://doi.org/10.1006/nlme.1995.1018

436. Castellano C, Cabib S, Puglisi-Allegra S, Gasbarri A, Sulli A, Pacitti C, Introini-Collison IB, McGaugh JL (1999) Strain-dependent involvement of D1 and D2 dopamine receptors in muscarinic cholinergic influences on memory storage. Behav Brain Res 98(1):17–26. https://doi.org/10.1016/s0166-4328(98)00046-1

437. Castellano C, Cestari V, Cabib S, Puglisi-Allegra S (1991) Post-training dopamine receptor agonists and antagonists affect memory storage in mice irrespective of their selectivity for D1 or D2 receptors. Behav Neural Biol 56(3):283–291. https://doi.org/10.1016/0163-1047(91)90439-w

438. Puglisi-Allegra S, Cestari V, Cabib S, Castellano C (1994) Strain-dependent effects of post-training cocaine or nomifensine on memory storage involve both D1 and D2 dopamine receptors. Psychopharmacology 115(1–2):157–162. https://doi.org/10.1007/BF02244766

439. Castellano C, Zocchi A, Cabib S, Puglisi-Allegra S (1996) Strain-dependent effects of cocaine on memory storage improvement induced by post-training physostigmine. Psychopharmacology 123(4):340–345. https://doi.org/10.1007/BF02246644

440. Cestari V, Castellano C (1996) Caffeine and cocaine interaction on memory consolidation in mice. Arch Int Pharmacodyn Ther 331(1):94–104

441. Flood JF, Cherkin A (1987) Fluoxetine enhances memory processing in mice. Psychopharmacology 93(1):36–43. https://doi.org/10.1007/BF02439584

442. Montezinho LP, Miller S, Plath N, Jensen NH, Karlsson JJ, Witten L, Mørk A (2010) The effects of acute treatment with escitalopram on the different stages of contextual fear conditioning are reversed by atomoxetine. Psychopharmacology 212(2):131–143. https://doi.org/10.1007/s00213-010-1917-5

443. Zhang G, Ásgeirsdóttir HN, Cohen SJ, Munchow AH, Barrera MP, Stackman RW Jr (2013) Stimulation of serotonin 2A receptors facilitates consolidation and extinction of fear memory in C57BL/6J mice. Neuropharmacology 64(1):403–413. https://doi.org/10.1016/j.neuropharm.2012.06.007

444. Charlier Y, Tirelli E (2011) Differential effects of histamine H(3) receptor inverse agonist thioperamide, given alone or in combination with the N-methyl-d-aspartate receptor antagonist dizocilpine, on reconsolidation and consolidation of a contextual fear memory in mice. Neuroscience 193:132–142. https://doi.org/10.1016/j.neuroscience.2011.07.034

445. Charlier Y, Brabant C, Serrano ME, Lamberty Y, Tirelli E (2013) The prototypical histamine H3 receptor inverse agonist thioperamide improves multiple aspects of memory processing in an inhibitory avoidance task. Behav Brain Res 253:121–127. https://doi.org/10.1016/j.bbr.2013.07.016

446. Brabant C, Charlier Y, Tirelli E (2013) The histamine H3-receptor inverse agonist pitolisant improves fear memory in mice. Behav Brain Res 243:199–204. https://doi.org/10.1016/j.bbr.2012.12.063

447. Kim DH, Lee Y, Lee HE, Park SJ, Jeon SJ, Jeon SJ, Cheong JH, Shin CY, Son KH, Ryu JH (2014) Oroxylin A enhances memory consolidation through the brain-derived neurotrophic factor in mice. Brain Res Bull 108:67–73. https://doi.org/10.1016/j.brainresbull.2014.09.001

448. Wetzel W, Matthies H (1982) Effect of substance P on the retention of a brightness discrimination task in rats. Acta Biol Med German 41(7–8):647–652

449. Fulginiti S, Cancela LM (1983) Effect of naloxone and amphetamine on acquisition and memory consolidation of active avoidance responses in rats. Psychopharmacology 79(1):45–48. https://doi.org/10.1007/BF00433015

450. Castellano C, Pavone F (1984) Effects of DL-allylglycine, alone or in combination with morphine, on passive avoidance behaviour in C57BL/6 mice. Arch Int Pharmacodyn Ther 267(1):141–148

451. Castellano C, Introini-Collison IB, Pavone F, McGaugh JL (1989) Effects of naloxone and naltrexone on memory consolidation in CD1 mice: involvement of GABAergic mechanisms. Pharmacol Biochem Behav 32(2):563–567. https://doi.org/10.1016/0091-3057(89)90197-4

452. del Cerro S, Borrell J (1990) Dynorphin1-17 can enhance or impair retention of an inhibitory avoidance response in rats. Life Sci 47(16):1453–1462. https://doi.org/10.1016/0024-3205(90)90524-u

453. Castellano C, Ventura R, Cabib S, Puglisi-Allegra S (1999) Strain-dependent effects of anandamide on memory consolidation in mice are antagonized by naltrexone. Behav Pharmacol 10(5):453–457. https://doi.org/10.1097/00008877-199909000-00003

454. Leri F, Nahas E, Henderson K, Limebeer CL, Parker LA, White NM (2013) Effects of post-training heroin and d-amphetamine on

consolidation of win-stay learning and fear conditioning. J Psychopharmacol 27(3): 292–301. https://doi.org/10.1177/0269881112472566
455. Brioni JD, McGaugh JL (1988) Post-training administration of GABAergic antagonists enhances retention of aversively motivated tasks. Psychopharmacology 96(4):505–510. https://doi.org/10.1007/BF02180032
456. Castellano C, Populin R (1990) Effect of ethanol on memory consolidation in mice: antagonism by the imidazobenzodiazepine Ro 15-4513 and decrement by familiarization with the environment. Behav Brain Res 40(1):67–72. https://doi.org/10.1016/0166-4328(90)90044-f
457. Castellano C, Cestari V, Cabib S, Puglisi-Allegra S (1993) Strain-dependent effects of post-training GABA receptor agonists and antagonists on memory storage in mice. Psychopharmacology 111(2):134–138. https://doi.org/10.1007/BF02245514
458. Kopf SR, Melani A, Pedata F, Pepeu G (1999) Adenosine and memory storage: effect of A(1) and A(2) receptor antagonists. Psychopharmacology 146(2):214–219. https://doi.org/10.1007/s002130051109
459. Viu E, Zapata A, Capdevila J, Skolnick P, Trullas R (2000) Glycine(B) receptor antagonists and partial agonists prevent memory deficits in inhibitory avoidance learning. Neurobiol Learn Mem 74(2):146–160. https://doi.org/10.1006/nlme.1999.3947
460. Gould TJ, McCarthy MM, Keith RA (2002) MK-801 disrupts acquisition of contextual fear conditioning but enhances memory consolidation of cued fear conditioning. Behav Pharmacol 13(4):287–294. https://doi.org/10.1097/00008877-200207000-00005
461. Kruk-Slomka M, Biala G (2016) CB1 receptors in the formation of the different phases of memory-related processes in the inhibitory avoidance test in mice. Behav Brain Res 301: 84–95. https://doi.org/10.1016/j.bbr.2015.12.023
462. Morena M, Berardi A, Peloso A, Valeri D, Palmery M, Trezza V, Schelling G, Campolongo P (2017) Effects of ketamine, dexmedetomidine and propofol anesthesia on emotional memory consolidation in rats: consequences for the development of post-traumatic stress disorder. Behav Brain Res 329:215–220. https://doi.org/10.1016/j.bbr.2017.04.048
463. Morena M, Colucci P, Mancini GF, De Castro V, Peloso A, Schelling G, Campolongo P (2021) Ketamine anesthesia enhances fear memory consolidation via noradrenergic activation in the basolateral amygdala. Neurobiol Learn Mem 178:107362. https://doi.org/10.1016/j.nlm.2020.107362
464. Campolongo P, Roozendaal B, Trezza V, Cuomo V, Astarita G, Fu J, McGaugh JL, Piomelli D (2009) Fat-induced satiety factor oleoylethanolamide enhances memory consolidation. Proc Natl Acad Sci U S A 106(19):8027–8031. https://doi.org/10.1073/pnas.0903038106
465. Ratano P, Petrella C, Forti F, Passeri PP, Morena M, Palmery M, Trezza V, Severini C, Campolongo P (2018) Pharmacological inhibition of 2-arachidonoilglycerol hydrolysis enhances memory consolidation in rats through CB2 receptor activation and mTOR signaling modulation. Neuropharmacology 138:210–218. https://doi.org/10.1016/j.neuropharm.2018.05.030
466. Frye CA, Lacey EH (2001) Posttraining androgens' enhancement of cognitive performance is temporally distinct from androgens' increases in affective behavior. Cogn Affect Behav Neurosci 1(2):172–182. https://doi.org/10.3758/cabn.1.2.172
467. Arteni NS, Lavinsky D, Rodrigues AL, Frison VB, Netto CA (2002) Agmatine facilitates memory of an inhibitory avoidance task in adult rats. Neurobiol Learn Mem 78(2): 465–469. https://doi.org/10.1006/nlme.2002.4076
468. Janezic EM, Uppalapati S, Nagl S, Contreras M, French ED, Fellous JM (2016) Beneficial effects of chronic oxytocin administration and social co-housing in a rodent model of post-traumatic stress disorder. Behav Pharmacol 27(8):704–717. https://doi.org/10.1097/FBP.0000000000000270
469. Scavuzzo CJ, Rakotovao I, Dickson CT (2020) Differential effects of L- and D-lactate on memory encoding and consolidation: potential role of HCAR1 signaling. Neurobiol Learn Mem 168:107151. https://doi.org/10.1016/j.nlm.2019.107151
470. Camera K, Mello CF, Ceretta AP, Rubin MA (2007) Systemic administration of polyaminergic agents modulate fear conditioning in rats. Psychopharmacology 192(4):457–464. https://doi.org/10.1007/s00213-007-0734-y
471. Bazin MA, El Kihel L, Boulouard M, Bouët V, Rault S (2009) The effects of DHEA, 3beta-hydroxy-5alpha-androstane-6,17-dione, and 7-amino-DHEA analogues on short term and long term memory in the mouse. Steroids 74(12):931–937. https://doi.org/10.1016/j.steroids.2009.06.010

472. Hauer D, Ratano P, Morena M, Scaccianoce S, Briegel I, Palmery M, Cuomo V, Roozendaal B, Schelling G, Campolongo P (2011) Propofol enhances memory formation via an interaction with the endocannabinoid system. Anesthesiology 114(6):1380–1388. https://doi.org/10.1097/ALN.0b013e31821c120e

473. Amiri S, Jafari-Sabet M, Keyhanfar F, Falak R, Shabani M, Rezayof A (2020) Hippocampal and prefrontal cortical NMDA receptors mediate the interactive effects of olanzapine and lithium in memory retention in rats: the involvement of CAMKII-CREB signaling pathways. Psychopharmacology 237(5): 1383–1396. https://doi.org/10.1007/s00213-020-05465-4

474. Kopf SR, Opezzo JW, Baratti CM (1993) Glucose enhancement of memory is not state-dependent. Behav Neural Biol 60(3): 192–195. https://doi.org/10.1016/0163-1047(93)90333-d

475. Quartermain D, Hawxhurst A, Ermita B, Puente J (1993) Effect of the calcium channel blocker amlodipine on memory in mice. Behav Neural Biol 60(3):211–219. https://doi.org/10.1016/0163-1047(93)90390-4

476. Boccia MM, Acosta GB, Baratti CM (2001) Memory improving actions of gabapentin in mice: possible involvement of central muscarinic cholinergic mechanism. Neurosci Lett 311(3):153–156. https://doi.org/10.1016/s0304-3940(01)02181-4

477. Baldi E, Bucherelli C, Schunack W, Cenni G, Blandina P, Passani MB (2005) The H3 receptor protean agonist proxyfan enhances the expression of fear memory in the rat. Neuropharmacology 48(2):246–251. https://doi.org/10.1016/j.neuropharm.2004.09.009

478. Navarrete A, Flores-Machorro FX, Téllez-Ballesteros RI, Alfaro-Romero A, Balderas JL, Reyes A (2014) Study on action mechanism of 1-(4-methoxy-2-methylphenyl)piperazine (MMPP) in acquisition, formation, and consolidation of memory in mice. Drug Dev Res 75(2):59–67. https://doi.org/10.1002/ddr.21094

479. Al Abed AS, Ducourneau EG, Bouarab C, Sellami A, Marighetto A, Desmedt A (2020) Preventing and treating PTSD-like memory by trauma contextualization. Nat Commun 11(1):4220. https://doi.org/10.1038/s41467-020-18002-w

480. Kolodziejczyk MH, Fendt M (2020) Corticosterone treatment and incubation time after contextual fear conditioning synergistically induce fear memory generalization in neuropeptide S receptor-deficient mice. Front Neurosci 14:128. https://doi.org/10.3389/fnins.2020.00128

481. Lesuis SL, Brosens N, Immerzeel N, van der Loo RJ, Mitrić M, Bielefeld P, Fitzsimons CP, Lucassen PJ, Kushner SA, van den Oever MC, Krugers HJ (2021) Glucocorticoids promote fear generalization by increasing the size of a dentate gyrus engram cell population. Biol Psychiatry 90(7):494–504. https://doi.org/10.1016/j.biopsych.2021.04.010

482. Zhu RT, Liu XH, Shi YW, Wang XG, Xue L, Zhao H (2018) Propranolol can induce PTSD-like memory impairments in rats. Brain Behav 8(2):e00905. https://doi.org/10.1002/brb3.905

483. De Bundel D, Zussy C, Espallergues J, Gerfen CR, Girault JA, Valjent E (2016) Dopamine D2 receptors gate generalization of conditioned threat responses through mTORC1 signaling in the extended amygdala. Mol Psychiatry 21(11):1545–1553. https://doi.org/10.1038/mp.2015.210

484. Lynch JF, Winiecki P, Gilman TL, Adkins JM, Jasnow AM (2017) Hippocampal GABAB (1a) receptors constrain generalized contextual fear. Neuropsychopharmacology 42(4): 914–924. https://doi.org/10.1038/npp.2016.255

485. Vanvossen AC, Portes M, Scoz-Silva R, Reichmann HB, Stern C, Bertoglio LJ (2017) Newly acquired and reactivated contextual fear memories are more intense and prone to generalize after activation of prelimbic cortex NMDA receptors. Neurobiol Learn Mem 137:154–162. https://doi.org/10.1016/j.nlm.2016.12.002

486. Bayer H, Bertoglio LJ (2020) Infralimbic cortex controls fear memory generalization and susceptibility to extinction during consolidation. Sci Rep 10(1):15827. https://doi.org/10.1038/s41598-020-72856-0

487. Kritman M, Maroun M (2013) Inhibition of the PI3 kinase cascade in corticolimbic circuit: temporal and differential effects on contextual fear and extinction. Int J Neuropsychopharmacol 16(4):825–833. https://doi.org/10.1017/S1461145712000636

488. Adamec R, Muir C, Grimes M, Pearcey K (2007) Involvement of noradrenergic and corticoid receptors in the consolidation of the lasting anxiogenic effects of predator stress. Behav Brain Res 179(2):192–207. https://doi.org/10.1016/j.bbr.2007.02.001

489. Atsak P, Roozendaal B, Campolongo P (2012) Role of the endocannabinoid system in regulating glucocorticoid effects on

memory for emotional experiences. Neuroscience 204:104–116. https://doi.org/10.1016/j.neuroscience.2011.08.047

490. Akirav I (2013) Cannabinoids and glucocorticoids modulate emotional memory after stress. Neurosci Biobehav Rev 37(10 Pt 2): 2554–2563. https://doi.org/10.1016/j.neubiorev.2013.08.002

491. Busquets-Garcia A, Gomis-González M, Srivastava RK, Cutando L, Ortega-Alvaro A, Ruehle S, Remmers F, Bindila L, Bellocchio L, Marsicano G, Lutz B, Maldonado R, Ozaita A (2016) Peripheral and central CB1 cannabinoid receptors control stress-induced impairment of memory consolidation. Proc Natl Acad Sci U S A 113(35):9904–9909. https://doi.org/10.1073/pnas.1525066113

492. Zerbes G, Kausche FM, Müller JC, Wiedemann K, Schwabe L (2019) Glucocorticoids, noradrenergic arousal, and the control of memory retrieval. J Cogn Neurosci 31(2):288–298. https://doi.org/10.1162/jocn_a_01355

493. Bahtiyar S, Gulmez Karaca K, Henckens MJAG, Roozendaal B (2020) Norepinephrine and glucocorticoid effects on the brain mechanisms underlying memory accuracy and generalization. Mol Cell Neurosci 108: 103537. https://doi.org/10.1016/j.mcn.2020.103537

494. Warren WG, Papagianni EP, Stevenson CW, Stubbendorff C (2022) In it together? The case for endocannabinoid-noradrenergic interactions in fear extinction. Eur J Neurosci 55(4):952–970. https://doi.org/10.1111/ejn.15200

495. Berardi A, Trezza V, Palmery M, Trabace L, Cuomo V, Campolongo P (2014) An updated animal model capturing both the cognitive and emotional features of post-traumatic stress disorder (PTSD). Front Behav Neurosci 8:142. https://doi.org/10.3389/fnbeh.2014.00142

496. Flandreau EI, Toth M (2018) Animal models of PTSD: a critical review. Curr Top Behav Neurosci 38:47–68. https://doi.org/10.1007/7854_2016_65

497. Hefner K, Whittle N, Juhasz J, Norcross M, Karlsson RM, Saksida LM, Bussey TJ, Singewald N, Holmes A (2008) Impaired fear extinction learning and cortico-amygdala circuit abnormalities in a common genetic mouse strain. J Neurosci 28(32): 8074–8085. https://doi.org/10.1523/JNEUROSCI.4904-07.2008

498. Andero R, Heldt SA, Ye K, Liu X, Armario A, Ressler KJ (2011) Effect of 7,8-dihydroxyflavone, a small-molecule TrkB agonist, on emotional learning. Am J Psychiatr 168(2):163–172. https://doi.org/10.1176/appi.ajp.2010.10030326

499. Ortiz V, Giachero M, Espejo PJ, Molina VA, Martijena ID (2015) The effect of midazolam and propranolol on fear memory reconsolidation in ethanol-withdrawn rats: influence of d-cycloserine. Int J Neuropsychopharmacol 18(4):pyu082. https://doi.org/10.1093/ijnp/pyu082

500. Espejo PJ, Ortiz V, Martijena ID, Molina VA (2016) Stress-induced resistance to the fear memory labilization/reconsolidation process. Involvement of the basolateral amygdala complex. Neuropharmacology 109:349–356. https://doi.org/10.1016/j.neuropharm.2016.06.033

501. Deslauriers J, Toth M, Der-Avakian A, Risbrough VB (2018) Current status of animal models of posttraumatic stress disorder: behavioral and biological phenotypes, and future challenges in improving translation. Biol Psychiatry 83(10):895–907. https://doi.org/10.1016/j.biopsych.2017.11.019

Chapter 14

Loudness Dependence of Auditory Evoked Potentials: A Promising Pre-treatment Predictor of Selective Serotonin Reuptake Inhibitor Response

Suzanne L. Pineles, Shivani Pandey, Rachel Shor, Ronnie F. Abi-Raad, Matthew O. Kimble, and Scott P. Orr

Abstract

Selective serotonin reuptake inhibitors are the most commonly prescribed class of medications for patients with PTSD. However, many patients are not responsive to SSRIs and there is no rapid, uncomplicated way to determine who will benefit from this family of medications. Loudness dependence of auditory evoked potentials, a measure derived from auditory event-related potentials to a series of increasingly loud tones, appears to be strongly influenced by brain serotonin level and thereby holds considerable promise as an indicator of the brain's potential responsiveness to SSRIs. The overarching goal of this chapter is to (a) describe the scientific rationale supporting the use of the LDAEP task in research advancing precision medicine approaches to the treatment of PTSD and (b) provide specific guidance to enable readers to administer this task in their own labs. Information regarding the requisite equipment is provided, as well as detailed instructions regarding the procedures for data collection, cleaning, and scoring.

Key words Electrophysiology, event-related potentials, P2, Loudness dependence of auditory evoked potentials, LDAEP, Post-traumatic stress disorder, PTSD, Selective serotonin reuptake inhibitors, SSRIs

Abbreviations

Cz	Electrode placement at the central region of the cap on the midline
EBA	Electro-Based Adapter
EEG	Electroencephalogram
ERP	Event-Related Potentials
Fz	Electrode placement at the frontal region of the cap on the midline
HEOR/HEOL	Horizontal Electro Oculogram Right/Left
LDAEP	Loudness Dependence of Auditory Evoked Potentials
M1/M2	Left Mastoid/Right Mastoid
Pz	Electrode placement at the posterior region of the cap on the midline

REF	Reference
SNRI	Selective Norepinephrine Reuptake Inhibitor
SSRI	Selective Serotonin Reuptake Inhibitor
VEOU/VEOL	Vertical Electro Oculogram Upper/Lower

1 Introduction

Selective serotonin reuptake inhibitors (SSRIs) constitute a class of medications commonly used to treat a range of mood, anxiety, and trauma-related disorders with approximately 109 million SSRIs prescribed annually [1]. However, many, if not most, patients are not responsive to SSRIs [2, 3]. Currently, there is no way to predict whether a particular patient will benefit from an SSRI. Determining an SSRI's effectiveness for a patient is accomplished through trial and error over several weeks or months. This can lead the patient to feel frustrated and hopeless as their provider works to find a medication, or a combination of medications, to alleviate their symptoms.

This chapter describes a pre-treatment auditory event-related potential (ERP) procedure that has the potential to inform how likely a person will be to benefit from an SSRI. This electrophysiological procedure can be conducted in a lab or office and entails the participant listening to a series of tones ranging in intensity (i.e., loudness dependence of auditory evoked potentials, LDAEP) while brain electrical activity is recorded through an electroencephalogram (EEG). The slope of P2 ERP component amplitudes in response to the four different tone intensities appears to be strongly influenced by brain serotonin level. Thus, the P2 ERP task holds considerable promise as a potential indicator of the brain's potential responsiveness to SSRIs.

1.1 LDAEP and Serotonergic Transmission

The P2 component of an ERP represents a positive ("P") electrophysiological response that occurs at approximately 200 ("2") msec following the onset of a stimulus. It is thought to reflect the sensitivity of a gating mechanism that regulates sensory input to the cortex [4]. Using a four-tone, stimulus-intensity-modulation (i.e., augmenting-reducing) paradigm, the slope of the P2 response amplitudes across a series of increasingly loud sound intensity levels (74, 84, 94, and 104 dB) is calculated and represents the LDAEP [5]. A reduction in the amplitude of the P2 component at higher tone intensity levels produces a shallow LDAEP. This ERP response pattern of decreased intensity dependence, or "reducing," may reflect a protectively tuned sensory system that protects the organism from sensory overload. In contrast, the opposite pattern (i.e., an increasing or "augmenting" LDAEP with enhanced intensity dependence) is thought to reflect a cortex that has a heightened sensitivity to increases in stimulus intensity [6–8]. LDAEP is calculated by different research groups in one of two ways, that is, by using P2 amplitudes alone or by computing the ratio of the ERP N1 and P2 (N1/P2) components. As findings from studies using these methods largely parallel each other, we will refer to them both simply as "LDAEP." LDAEP has excellent test-retest reliability (r_{tt}s > 0.76 for test-retest intervals of 1 week to 1 year [6–8]) in the absence of intervention.

There is substantial evidence from human studies and rodent models demonstrating that aberrantly strong or weak LDAEP intensity dependence reflects inverse abnormalities in central serotonergic (5-HT) transmission [9], which are also thought to play a key role in the pathophysiology of anxiety and depression [10, 11]. Specifically, increased LDAEP appears to reflect low 5-HT neurotransmission in the primary auditory cortex, whereas decreased LDAEP appears to reflect high 5-HT neurotransmission in this brain region. Direct support for this position comes from animal studies [12, 13], while human studies provide indirect support [9–11].

Abnormalities in LDAEP have been found in several neuropsychiatric disorders. Specifically, increased LDAEP has been observed in individuals with bipolar affective disorder [14], histrionic disorder [15], fibromyalgia [16], and migraine conditions [17]. Furthermore, abstinent users of "ecstasy" (methylenedioxymethamphetamine), a drug with neurotoxic effects on central serotonergic terminals, show increased LDAEP, compared to normal controls [18, 19]. In contrast, decreased LDAEP has been reported in generalized anxiety disorder [20]. Mixed findings have been found regarding whether individuals with unipolar depression diagnoses differ from healthy controls with regard to LDAEP, particularly when medication use is not controlled [14, 21]. In PTSD, some studies have found increased LDAEP [4, 22, 23], while others found decreased LDAEP [24, 25]. It is possible that the heterogeneity in LDAEP across PTSD studies represents biological, and potentially genetically based, variations in PTSD endophenotypes.

More compelling evidence for an inverse relationship between 5-HT neurotransmission and LDAEP is derived from several studies demonstrating a link between increased pre-treatment LDAEP and a favorable response to SSRIs in depressed individuals [21, 26–34]. Notably, this association was demonstrated despite mixed findings regarding whether LDAEP differed between groups of individuals with and without depression [21]. LDAEP may be selectively related to the unique mechanisms of action of SSRIs, as LDAEP has been found to have an inverse relationship with selective norepinephrine reuptake inhibitor (SNRI) treatment outcome, that is, reduced LDAEP predicts a more positive outcome in response to reboxetine, an SNRI [27, 30, 35]. As suggested by

Linka et al. [30], reduced LDAEP could be used as an indicator for the use of a noradrenergic, rather than serotonergic, antidepressant; whereas increased LDAEP may indicate treatment with an SSRI. In addition, a case report suggests that a low LDAEP is an indicator of who should *not* be given an SSRI; a patient diagnosed with a major depressive disorder who showed a reduced LDAEP experienced significant side effects in response to SSRI treatment, but not in response to tianeptine, a selective serotonin re-uptake enhancer [36].

Although there is growing evidence for the potential usefulness of LDAEP as a predictor of SSRI response, it is important to note several limitations of the work to date. First, the research has mostly been limited to studies of depression and has not included populations with primary diagnoses of PTSD or anxiety disorders. Second, most of these studies have contrasted the top half of their sample (i.e., stronger LDAEP) with the bottom half [33]. Consequently, the mid-point used to form the high and low groups is study specific, and the results cannot be translated into clinical cutoff scores for use in predicting the benefit from SSRIs in individual patients. Third, with few exceptions [28], most of this research has focused on LDAEP as a predictor of SSRI response and has not examined whether LDAEP normalizes as a function of successful SSRI treatment. Such information may increase our understanding of potential mechanisms underlying clinical response. Thus, more research is needed before LDAEP can be utilized as a clinical tool to inform treatment decisions.

2 Equipment

Figure 1 shows an example of a typical electrophysiology laboratory schematic. An ERP laboratory requires two computers—one for presenting the stimuli (i.e., stimulus presentation computer) and one for recording EEG (i.e., data acquisition computer). There also needs to be a way for event codes to be sent from the stimulus presentation system to the amplifier. These event codes will be used as reference points ("markers") for averaging the EEG signals for each type of stimulus presented (i.e., 74, 84, 94, 104 dB tones). The EEG signal is detected through electrodes placed on the scalp, typically via an electrode cap (*see* **Note 1**). Most modern EEG amplifier systems have a headbox that interfaces with the electrode cap. The amplifier is then connected to the data acquisition PC for display, storage, and further data processing. If the lab includes a control room and a participant room, then it is recommended that there be two additional monitors in the participant's room that are connected to the stimulus presentation computer and data acquisition/digitization computer, respectively. One of these monitors is used to display the impedance levels, which allows study staff to

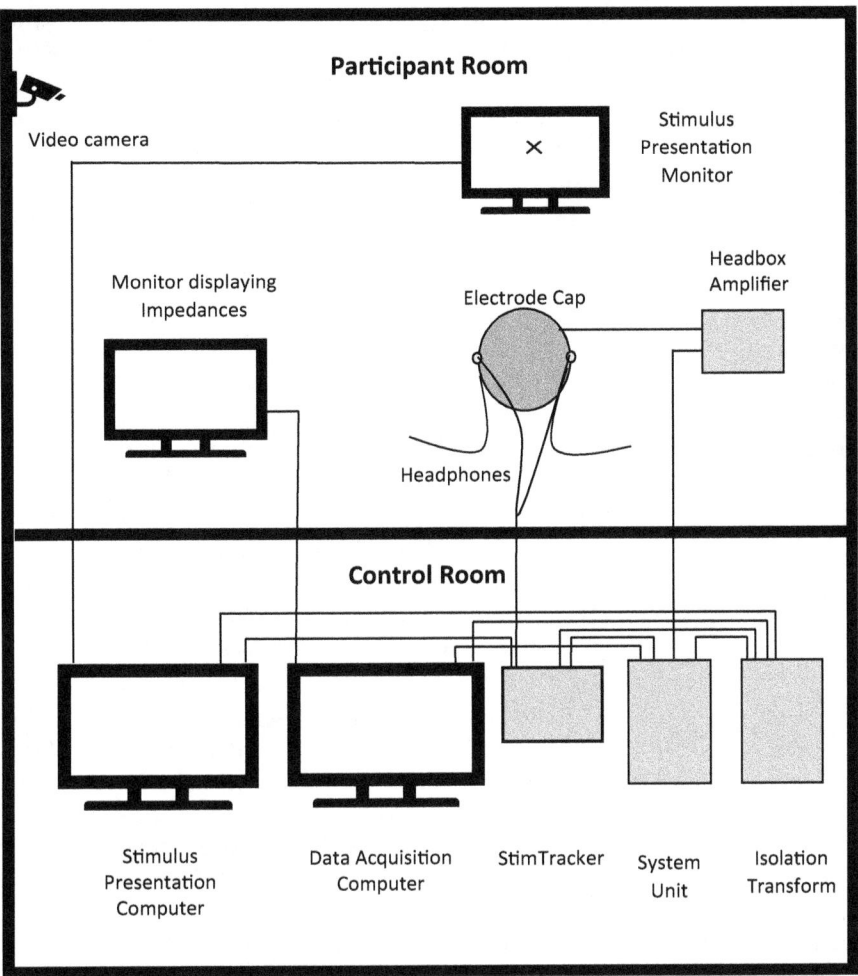

Fig. 1 Example of an electrophysiology laboratory set up

adjust electrodes to ensure the optimal quality of the EEG recordings. The other monitor will present an image (e.g., an X) where participants are instructed to focus their eyes during the LDAEP task (*see* **Note 7**).

In our laboratory, we use a Compumedics-Neuroscan SynAmps RT 64 channel amplifier (Charlotte, NC, USA) and two PCs for data acquisition and stimulus presentation. The stimulus presentation PC uses Stim2 software (Compumedics-Neuroscan, Charlotte, NC, USA) and is factory-calibrated alongside a 2G StimTracker DUO (Cedrus Corporation, San Pedro, California, USA), which provides accurate timing of event marker delivery to the data acquisition computer. The system is also factory-calibrated for accurate sound amplitude delivery through a specialized Sound Blaster card (Creative Labs, Inc., Milpitas, CA, USA). Curry 8 software (Compumedics-Neuroscan, Charlotte, NC, USA) is used for data acquisition and processing. The raw data are acquired in DC mode using a 1KHz sampling rate, a low-pass filter setting of 400 Hz and a fixed amplifier gain of 10. See Subheading 3.2 for offline filter settings. For EEG electrodes, we use 21Channel QuickCaps with embedded sintered Ag/AgCl electrodes (Compumedics-Neuroscan, Charlotte, USA) and a proprietary electro-board adapter (EBA) that interfaces the cap with the headbox.

3 Methods

3.1 Overview and Description of the LDAEP Task

For measurement of LDAEP, the participant listens to a series of tones ranging in intensity while the EEG signal is recorded. The auditory stimuli are 50 millisecond, 780 Hz tones gated with rise and fall times of 25 milliseconds. The tones are presented at four intensities (74, 84, 94, and 104 dB SPL) in four blocks of 16 tones, which are repeated four times in a Latin-Square design for a total of 256 tones (*see* **Note 2**). While some participants might experience some of the louder tones as unpleasant, they are safe and not damaging to the ear. Using these parameters, the LDAEP task lasts approximately 13 min, with interstimulus interval (ISI) ranging from 2 to 4 s ($ISI_{mean} = 3$ s). The EEG is amplified by a factor of 10 and digitized for storage and offline processing as continuous voltage measurements. As discussed above, the electrophysiological components of interest are N1, P2, and the N1/P2 ratio. See below for data processing and scoring of LDAEP slope.

3.2 Procedures

The procedures below are presented in three sections: (a) procedures for ensuring quality EEG data collection, (b) procedures for running the LDAEP data acquisition protocol, and (c) data cleaning, processing, and analysis. The respective steps are described for a Compumedics-Neuroscan SynAmps RT system with Curry 8 data acquisition software and Stim2 stimulus presentation software. However, these procedures can be adapted to any electrophysiological system that has the capability to collect ERP data in response to auditory stimuli.

(a) **Procedures for Ensuring Quality EEG Data Collection:** As is true with all electrophysiological research, the importance of attending to the accurate placement of electrodes and minimizing impedance levels cannot be overstated. The procedures described will maximize the likelihood of collecting data with the greatest possible signal-to-noise ratio. Signal noise reduction techniques used with individual participants will focus on attending to possible causes of high impedance levels. *See* **Note 2** for a discussion of ways to increase signal detection by increasing the number of trials and **Note 3** for a discussion of reducing sources of environmental noise. High impedance can lead to decreased common-mode rejection (i.e., the ability of the amplifiers to subtract away environmental noise) and increased skin potentials [37]. This, in turn, can result in the exclusion of large blocks of data at the data processing stage, and/or difficulty detecting the typical ERP response patterns. Ensuring a clean scalp prior to any EEG session is the primary strategy for reducing the naturally high impedance of the skin.

(a1) Exclusion criteria that may impact successful data collection and procedures to prepare the scalp and skin before placing the electrode cap: Electrophysiology studies with auditory stimuli typically exclude participants who use hearing aids or with significant hearing loss at the frequency of tones presented (e.g., inability to hear tones below 35 dB in either ear). In addition, individuals with severe head injuries, epilepsy, or other neurological diagnoses are often excluded. There are also several practices that can help reduce impedance levels before the electrode cap is placed on the participant's head.

1. The electrophysiology laboratory should be quiet and set at a comfortable temperature. Hot rooms or rooms that get hot because of equipment cause sweating that increases impedance levels and makes it difficult to detect an ERP response pattern. Uniformity of temperature across participants should be confirmed.

2. Before coming to the lab, participants should wash their hair with plain shampoo; no conditioner or other hair products should be used. *See* **Note 4** for additional considerations for people who wear braids, extensions, or hair weaves.

3. Once in the lab, the participant should use a brush or dandruff-control scalp massager to abrade their scalp. This will decrease debris buildup at the hair follicles from dirt or hair product. Because this step focuses on the scalp rather than the hair, this step should be also completed for bald participants.

4. The experimenter should use a large, cotton-tipped applicator to apply NuPrep (Aurora, CO, USA) using small, circular motions, down the midline of the participant's scalp. NuPrep is a mildly abrasive gel used to prepare the skin for electrode placement. Application to the center line of the scalp will facilitate the reduction of impedance levels for the midline electrodes of the participant's scalp (i.e., those used to measure LDAEP).

5. The sites for the face and mastoid electrodes should be cleaned with alcohol wipes or NuPrep applied in a circular motion using a cotton-tipped applicator. As the skin on the face is sensitive, the application of the alcohol wipes or NuPrep should be fairly gentle. It is useful to complete this step before placing the electrode cap, which makes it difficult to access these sites, and to prepare a relatively large area at each site. *See* Fig. 2 for the placement of these electrodes.

(a2) **Placement of the electrode cap and auxiliary electrodes:**

1. Electrode caps come in different sizes. To determine the correct size cap for a participant, the experimenter should use a cloth measuring tape to measure the circumference of the participant's head at its largest point, that is, just above the participant's eyebrows. The manual for the electrode caps provides guidance regarding circumference ranges and suggested cap sizes. If on the cusp of sizes, the smaller size should be used so as to provide a more secure fit.

2. Use the cloth measuring tape to measure the distance between the inion (the bump located at the base of the skull) and the nasion, the location between the participant's eyebrows where the forehead frontal bone and nasal bone meet. To ensure the correct placement of the cap, the Cz electrode will be located at the halfway point of this measurement (*see* Subheading a2, **Step 8**).

3. When first applying the electrode cap, it can be helpful to ask the participant to hold the tip of the cap close to the middle of their eyebrows while the experimenter pulls the cap down on the back and sides. The exact placement will be adjusted later (*see* Subheading a2, **Step 8**).

4. The QuikCap has several electrodes attached via wires that will be placed on the face and mastoids. These electrodes should be secured with 1″ wide pieces of Hypafix (County Bluffs, Iowa, USA) tape placed on the back of the electrode (i.e., the side not against the skin). For each electrode, there is a hole that is used to fill the electrode cup with electrolyte gel. *Before* placing the electrode on the participant, the experimenter should use a syringe with a blunted needle to poke a hole in the tape so that this hole is accessible (*see* **Note 5**).

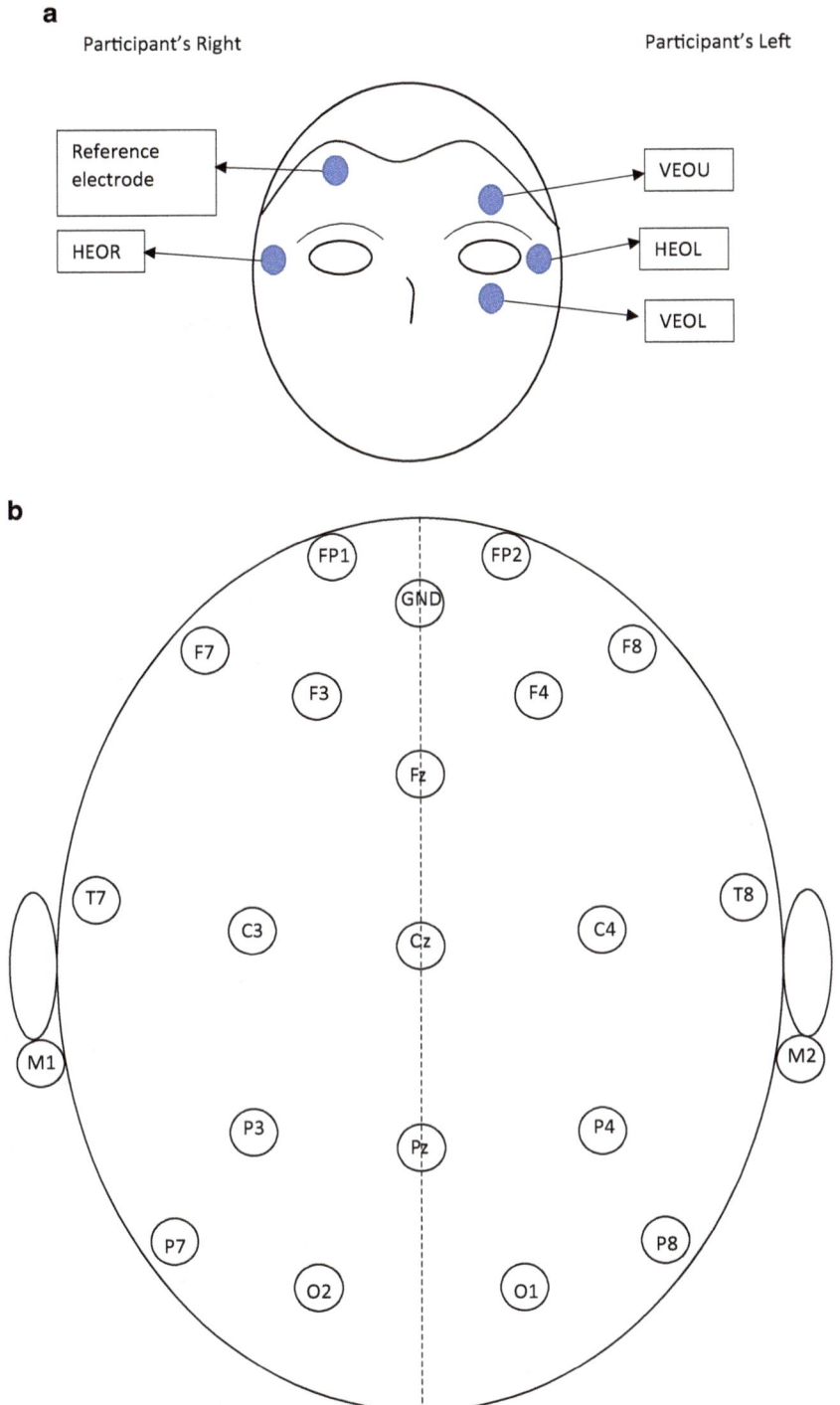

Fig. 2 (**a**) Placement of facial electrodes and (**b**) map of electrodes included on a 21-channel cap

5. The hanging electrodes to be placed include: "VEOU/VEOL," "HEOR/HEOL," "M1/M2," and "REF" (*see* Fig. 2 for placement of these electrodes).

    (a). VEOU and VEOL measure eyeblinks and are placed above and below the left eye, respectively. These electrodes should align vertically with the pupil when the participant looks straight ahead.

    (b). HEOR and HEOL measure horizontal eye movements and are placed on the right and left temple, horizontally aligned with the pupil.

    (c). M1 and M2 are placed on the mastoids, that is, the bony protrusion behind each ear. These electrodes will be designated as the reference electrodes during data processing (*see* Subheading d, **Step 2**).

    (d). REF is the reference electrode that is used during EEG data acquisition. The exact placement of this electrode is not critical, provided it does not interfere with the placement of another electrode. Our lab uses the upper right side of the forehead. This placement ensures that when the cap is pulled down and secured, the REF will not be directly underneath one of the other electrodes.

6. After the hanging electrodes are placed and secured, our lab fills the electrodes with gel because it is difficult to access the mastoid and reference electrodes once the chin strap is in place. However, other labs may choose to verify the correct placement of the cap and secure the chinstrap before this step (**Steps 7** and **8** below).

    The experimenter should use blunt-tipped syringes to fill the electrode cups on the participant's face and behind their ears with conductive electrolyte gel (e.g., Signa Gel, Fairfield, NJ, USA). The thumb and forefinger should be used to secure the electrode against the skin when filling it. A small bead of gel should appear at the top of the electrode's hole when the electrode cup is sufficiently filled. Overfilling these electrodes may cause the gel to seep under the Hypafix tape and lead to loss of adhesiveness (this will not happen if the previous step of applying gentle pressure against the electrode is properly executed). *See* **Notes 5** and **6** for tips on applying facial electrodes.

7. Next, the experimenter should carefully insert the back end of the cotton swab into the electrode hole and gently rotate the applicator against the participant's skin. This will help abrade the underlying skin and ensure the conductive gel is in contact with the skin. The pressure level used on the facial electrodes is much gentler than will be used on the scalp due to the sensitivity of the face and the lack of interference from hair.

8. After the facial and mastoid electrodes have been applied, the cap should be adjusted to ensure correct placement. Correct placement involves: (a) alignment of the midline of electrodes between the eyes and (b) ensuring the Cz electrode is half the distance of the calculation described above (*see* Subheading a2, **Step 2**). Some of the facial electrodes (such as the reference and VEOU) may be covered by the cap.

9. Next, the chin strap should be secured and adjusted to ensure that the participant's ears are accessible to position the intra-aural earphone inserts that will be used for the LDAEP task (*see* Subheading b, **Step 2**). If it hasn't previously been connected, this is a good time to connect the electrode cap to the EBA on the headbox, as the experimenter will now begin to focus on reducing the impedances of the electrodes embedded in the cap.

(a3) **Ensuring low impedance levels for electrodes:** The next phase of the study preparation is focused on ensuring that the electrodes have low impedance values. This is important to minimize the likelihood of artifact in data.

1. To help guide this process, one can use the Curry 8 software to view the impedance levels of each electrode in real time. Within the Curry 8 program, click on "start amplifier" and then the Ohm (Ω) icon to view an electrode map with impedance levels for each electrode (*see* **Note 7**). The experimenter should aim for impedance levels ≤5 kOhms, with an upper limit of 10 kOhms, if necessary.

2. As with the facial and mastoid electrodes, syringes are used to fill the electrodes embedded in the cap with electrode gel. The thumb and finger should be used to secure the electrode flush with the scalp when filling the electrodes. After filling the electrode, rotate the back end of the cotton-tipped applicator in the electrode hole against the scalp to help ensure that the conductive gel is in contact with the participant's

skin. This also abrades the scalp and helps move aside any hair that is interfering with a good connection. It is important to avoid overfilling the electrode well as excessive gel can cause bridging between electrodes which will result in poor spatial resolution in the recording.

3. As the experimenter works to improve the impedance levels for each electrode (*see* **Note 1**), it is most useful to attend to the ground and reference electrodes first. If impedances are not below 10 kOhms, strategies to improve impedance levels of these electrodes and the remaining electrodes involve gently swirling the back end of the cotton-tipped applicator in the electrode cup, followed by injecting a small amount of additional gel into the cup (*see* **Note 8**).

4. As a final step, Surgilast netting (Integra Life Science Corp., Princeton, NJ) can be used to further secure the electrode cap. An approximately 4″ strip of the netting can be cut and tied at one end; we use Surgilast—working stretch (sizes 5.5 and 6 for adults). The untied end of the strip is stretched and placed over the electrode cap. This step should be completed only if the cap does not have a snug fit onto the subject's head. The Surgilast will exert pressure when applied, causing discomfort in prolonged applications of 2 h or more.

5. After the electrode cap is in place and the impedance levels are acceptable, the participant is now ready to undergo the LDAEP task.

(b) **Procedures for running the LDAEP task:**

1. The experimenter should read the LDAEP instructions aloud. Sample instructions are: "In this session you will hear a series of tones. The tones will vary in loudness from soft to very loud. You do not have to respond to the tones, but you should stay alert and pay attention to the tones while remaining relaxed. We ask that you try to keep your body and especially your eyes as still as possible. We have placed a cross in front of you as a place to focus your eyes. Use the cross as a place to look to help keep your eyes from wandering. Please keep your eyes open throughout the task. Do you have any questions?"

2. After the instructions are read, the participant should place the earphone inserts into both ear canals. With the Neuroscan system, intra-aural earphone inserts are used.

3. The experimenter can now initiate the LDAEP task from the laboratory computers (*see* **Note 9**).

(a) To initiate data acquisition using Curry 8 software, first de-select the "Ω" icon. This will launch a screen depicting the EEG with a "not recording" message in red. Initiate the data recording button shortly *before* initiating the task using the Stim2 program on the stimulus presentation computer (*see* **Note 10**).

(b) To initiate the stimulus presentation of the LDAEP task using the STIM2 software, select the Gentask tab and select the LDAEP task. Use the mouse to select the green play button after data recording has been initiated through Curry 8.

(c) When data collection is complete, use the mouse to select the blue square to stop recording in the Curry 8 software.

(c) **Post-study cleaning procedures for the electrode cap:**

1. The electrode cap should be cleaned as soon as possible after use. The cap should be soaked in lukewarm water for 30 min; a water pic is useful to remove the gel from the electrode cups. It is important to use care when washing the cap: (a) ensure that the electrode cups are not scratched and (b) the cap-to-headbox connector does not get wet. The connector should not be immersed in liquid and does not need to be cleaned with water or any liquid.

2. The electrode cap should be placed in a disinfecting solution for 10 min, such as Control III laboratory germicide (Maril Products, Inc., Tustin, CA, USA). After disinfection, the cap should again be rinsed with lukewarm water and hung to dry, being careful to avoid water dripping onto the cap-to-headbox connector. *See* **Note 11**.

(d) **Processing and scoring the electrophysiological data:** As discussed above, the electrophysiological data are amplified, digitized and filtered, and stored as a set of continuous voltage measurements on the data acquisition computer. Additional steps are necessary to extract the ERPs from the EEG data; these include digital filtering, identifying and removing trials with artifacts, and averaging across trials.

When processing electrophysiological data, many programs have scripts that allow for automated processing and scoring. These scripts should be used with caution. At a minimum, it is useful to "hand score" a portion of the data to ensure the automated processing scripts yield comparable results. Our group uses the following parameters in Curry 8 to clean and score the electrophysiological data for use in calculating a LDAEP score.

1. Baseline correction: A baseline correction is used to compensate for signal drift; for this we use the "Constant" setting. The constant baseline is calculated page-wise with each page consisting of a 2^n number of samples. For each channel, all samples within the "page" are averaged, and this average is then subtracted from all samples. This means that while paging through ongoing data, a separate baseline is computed for each page.

2. Re-referencing: With the Neuroscan SynAmps RT system, the online reference during data acquisition is determined by the hang-down electrode location, which has been placed on the forehead. However, using this electrode can lead to distortions of the ERP waveform and misinterpretations of the ERP signal [37]. Thus, in the data processing stage, we use the re-referencing functionality in Curry 8 to change the reference to the linked mastoids. The average of the mastoids was chosen as the reference because it is convenient, not biased toward one hemisphere, and has been previously used in LDAEP research [26, 37, 38].

3. Digital filtering: The high- and low-pass filters are set at 0.1 and 30 Hz, respectively. The low-pass filter of 30 Hz is used to correct for environmental noise (e.g., AC line noise or video noise). Because ERPs of interest are mostly composed of frequencies under 30 Hz, this filter setting will not likely impact the ERP signal of interest. The high-pass filter setting serves to remove very slow voltage changes, such as those that occur as a result of changes in skin potentials from sweating.

4. Artifact detection and reduction: It is useful to start with a visual inspection of the data and mark bad blocks of data (i.e., data with a lot of artifacts) for exclusion. This might include motor or movement artifacts of the head, jaw, and mouth that might overwhelm the smaller brain potentials. Following that, an artifact detection and reduction procedure is used to remove eyeblinks that may contaminate the EEG signal. A threshold method is used to detect ocular artifact in the VEO channel. Covariance analysis between each EEG channel and the VEO channel is used to reduce the artifact. The voltage threshold setting is individually determined for each participant after visually inspecting the raw EEG data, balancing the need to detect eye blinks while not excluding data resulting from smaller eyelid fluctuations.

5. Signal averaging: Within each of the four different sound intensity levels (i.e., 74 dB, 84 dB, 94 dB, and 104 dB), data are averaged across trials. First, EEG epochs are aligned with the stimulus and averaged together in a point-by-point manner. This averaging technique is based on the assumption that noise will average out across trials allowing the neural activity evoked by the stimulus to emerge [37]. The first step in this

process is auto-aligning the Stim2 triggers with those generated by the Stimtracker to correct for the lag between when the system initiates the auditory stimulus and the presentation of the stimulus. Next, the data are epoched, which captures the neural response in the selected latency range for the P2 ERP. Our group uses a -200 msec to 800 msec epoch for the LDAEP.

6. Identifying peak amplitude and latency: For the LDAEP task, the electrodes of interest are those located at the midline frontal, central, and parietal sites (Fz, Cz, and Pz). Thus, peak amplitude and latency measures are determined for each stimulus intensity for each of these electrode sites. P2 is defined as the most positive point between 140 and 230 msec post-stimulus onset relative to the 100 msec pre-stimulus baseline. N1 is defined as the most negative point between 70 and 170 msec post-stimulus onset relative to the 100 msec pre-stimulus baseline.

7. Plotting ERP data: Publications should include a graphical depiction of averaged ERP waveforms to illustrate the LDAEP phenomenon. For an example, *see* Fig. 3 taken from Metzger et al.'s [4] published work. Plots of ERP waveforms should include both the voltage scale and time scale to highlight both the amplitude and latencies of the ERP components.

8. Computing LDAEP score: As discussed above, the slope of the change in P2 response amplitude across a series of increasing sound intensity levels (74, 84, 94, and 104 dB) is calculated and represents the LDAEP [38]. *See* Fig. 4 for a graphic depiction of exemplar LDAEP slopes from Metzger et al.'s [4] published work. Log transformations of P2 amplitude for each sound intensity level at each electrode site are used before creating the slopes to minimize overdependence on the lowest and highest sound Intensity levels. Similar slopes can also be computed for N1 and N1/P2 ratio.

4 Notes

1. Electrode caps vary in terms of how many electrode sites they include. Because the LDAEP is assessed using only electrodes along the midline (Fz, Cz, and Pz), it is best to use an electrode cap with the fewest electrodes possible. A cap with fewer electrodes will minimize the time needed to ensure low impedance levels across all of the electrode sites. Our lab uses a 21-channel cap with a floating hang-down reference that can be placed anywhere on the subject's head.

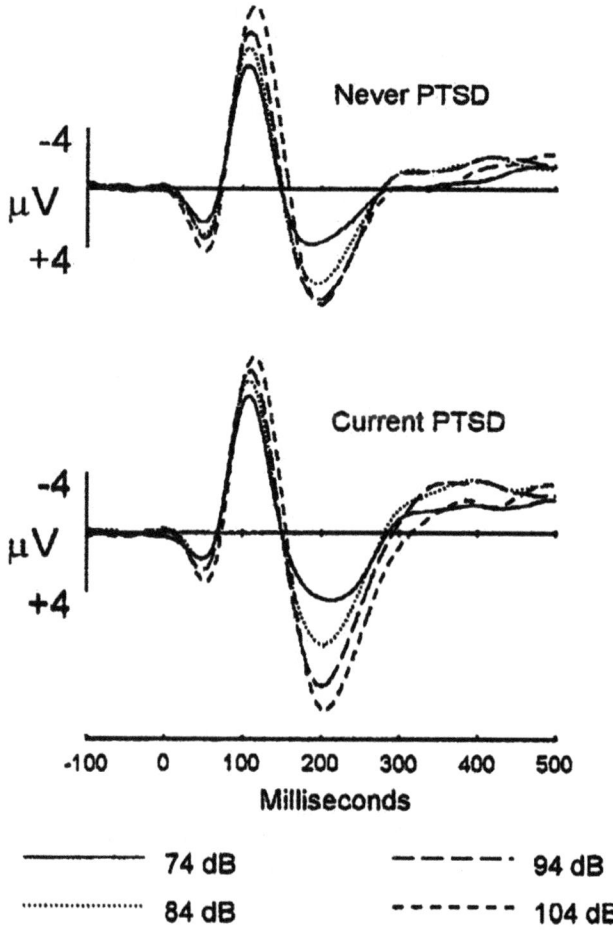

Fig. 3 Example of group grand average waveforms to tones of increasing intensities in a previously published study by Ref. [4] investigating PTSD group differences in performance on an LDAEP task

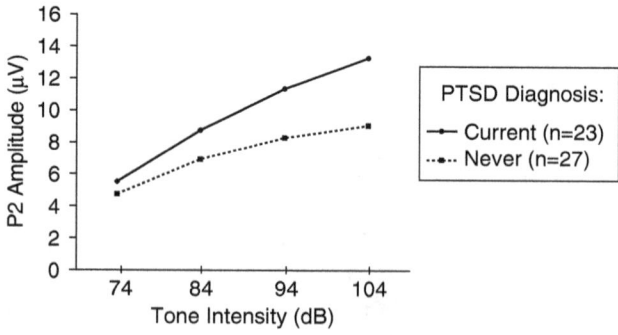

Fig. 4 Example of mean P2 amplitudes measured at Cz, plotted as a function of tone intensity for the current PTSD and never PTSD groups in a previously published study by Ref. [4]

2. Including a large number of trials in the LDAEP protocol is important for signal detection. A large number of trials can compensate for noise from non-EEG biological signals such as skin potentials and from environmental EEG noise sources. In our version of the LDAEP task, we use 256 trials (i.e., 64 trials for each stimulus intensity level). This number of trials was chosen in an effort to strike a balance between a sufficient number of trials to obtain quality ERPs while not overburdening the participant or causing habituation to multiple trials.

3. There can be sources of noise and aberrant electrical signals in the electrophysiology lab. An electrophysiological chamber can be helpful but is not always necessary if members of the lab engage in due diligence to minimize sources of electrical noise in the laboratory. This may include temporarily turning off equipment such as air conditioners, air purifiers, and sound machines. Our group had an experience where we had an unknown source of an aberrant electrical signal that we eventually determined to be from a nearby cell tower. The only way to rectify that situation was to move the lab to a different room in a nearby facility that had concrete walls. To minimize distractions and auditory noise in the laboratory, a "do not disturb" sign should be used during testing. We recommend the insert earphones as the foam inserts minimize extraneous noise.

4. In addition to instructions about hair care prior to the appointment (e.g., washing hair, not using hair products), we recommend orienting potential participants to the electrode cap application procedures such as the application of NuPrep and electrolyte gel to the face and scalp during the informed consent procedures and/or before scheduling the ERP session. This is important for all potential participants, but particularly relevant to participants with braids or weaves.

5. Throughout the procedures, it is helpful for the experimenter to narrate the procedures as they occur. For example, "Now I'm about to put gel in the electrode by your left eye. Let me know if you feel the gel or if it is uncomfortable." This can both speed up the process by ensuring gel is being adequately placed and also ease the discomfort of participants. In general, most participants prefer to see the blunt-tipped syringes, observe that the tip is blunt, and hear an explanation of how they are used. For participants who are afraid of needles, it may be helpful to clarify that Q-tips, rather than the blunt-tipped syringe, are used to abrade the skin.

6. When placing facial electrodes, it is recommended that the electrodes be placed to prevent their wires from interfering with the participant's field of vision. Additional strips of tape can be used on the wire for this purpose as well. The

experimenter should also avoid applying the electrodes and tape over eyelashes or eyebrows. At the end of the task, participants may opt to remove facial electrodes themselves as they are placed on sensitive areas of the skin. Before the participants remove the electrodes, the experimenter should instruct them to pull from the electrode cup itself and avoid tugging on the wires.

7. If your electrophysiology laboratory has separate control and participant rooms, it is helpful to have a monitor housed in the participant room and cloned to the monitor of the data acquisition computer. This will allow the experimenter to view the electrode map with impedance levels while working to lower impedance levels on individual electrodes. This can easily be achieved with an inexpensive video splitter if the video card has a single output. The monitor in the participant room should be turned off while the participant is engaging in the LDAEP task.

8. Trouble-shooting tips for achieving acceptable impedance levels: (1) Make sure that the reference and ground electrodes are properly engaging the scalp; (2) add a small amount of gel and use the back end of the Q-tip to further abrade the skin for the problematic electrodes; and (3) check that the Surgilast net is enhancing (and not interfering with) the connection between the electrode and the scalp.

9. A camera (e.g., a portable video baby monitor) in the participant room with two-way audio is helpful to ensure that participants are attending to the task and not making too many movements. The two-way audio will allow the experimenter to hear the participant at all times and also enable the experimenter to give instructions without entering the room. Alternatively, a video can be recorded and saved in a synchronized manner with the EEG for offline examination.

10. Before initiating the LDAEP task and after its completion, it is useful to save the impedance values in the data acquisition file. These data can serve as a record of each participant's starting and ending impedance levels. With Curry 8 software, this is done by briefly displaying the impedance screen (selecting and deselecting the Ohm icon).

11. It may take 3–4 h for an EEG cap to dry after being cleaned. Depending on the frequency of participants being run in the lab, it may be necessary to purchase multiple caps in sizes that are frequently used. Alternatively, a desktop blow fan can also be used to speed up the cap-drying process.

Acknowledgments

Support for this work was provided by the Department of Veterans Affairs, Clinical Sciences R&D Service, Merit Award Program (CX001627-01A1; PI: Pineles).

Disclosures Dr. Abi-Raad is affiliated with Compumedics USA Inc., the manufacturer and distributor of the hardware and software described in this report. Other authors report no conflicts of interest related to this study disclosure.

References

1. Grohol J (2016) Top 25 psychiatric medication prescriptions for 2013. Psych Central. http://psychcentral.com/lib/top-25-psychiatric-medication-prescriptions-for-2013/. Accessed 23 Sept 2016
2. Brady K, Pearlstein T, Asnis GM et al (2000) Efficacy and safety of sertraline treatment of posttraumatic stress disorder: a randomized controlled trial. JAMA 283(14):1837–1844
3. Davidson JR, Rothbaum BO, van der Kolk BA et al (2001) Multicenter, double-blind comparison of sertraline and placebo in the treatment of posttraumatic stress disorder. Arch Gen Psychiatry 58(5):485–492
4. Metzger LJ, Carson MA, Paulus LA et al (2002) Event-related potentials to auditory stimuli in female Vietnam nurse veterans with posttraumatic stress disorder. Psychophysiology 39(1):49–63
5. Buchsbaum M, Stevens S (1971) Neural events and psychophysical law. Science 172(3982):502–502
6. Carrillo-De-La-Pena MT (2001) One-year test–retest reliability of auditory evoked potentials (AEPs) to tones of increasing intensity. Psychophysiology 38(03):417–424
7. Hensch T, Herold U, Diers K et al (2008) Reliability of intensity dependence of auditory-evoked potentials. Clin Neurophysiol 119(1):224–236
8. Tenke CE, Kayser J, Pechtel P et al (2017) Demonstrating test-retest reliability of electrophysiological measures for healthy adults in a multisite study of biomarkers of antidepressant treatment response. Psychophysiology 54(1):34–50
9. Hegerl U, Juckel G (1993) Intensity dependence of auditory evoked potentials as an indicator of central serotonergic neurotransmission: a new hypothesis. Biol Psychiatry 33(3):173–187
10. Davis LL, Suris A, Lambert MT et al (1997) Post-traumatic stress disorder and serotonin: new directions for research and treatment. J Psychiatry Neurosci 22(5):318–326
11. Eison MS (1990) Serotonin: a common neurobiologic substrate in anxiety and depression. J Clin Psychopharmacol 10(3):26S–30S
12. Juckel G, Molnár M, Hegerl U et al (1997) Auditory-evoked potentials as indicator of brain serotonergic activity first evidence in behaving cats. Biol Psychiatry 41(12):1181–1195
13. Juckel G, Hegerl U, Molnár M et al (1999) Auditory evoked potentials reflect serotonergic neuronal activity—a study in behaving cats administered drugs acting on 5-HT1A autoreceptors in the dorsal raphe nucleus. Neuropsychopharmacology 21(6):710–716
14. Brocke B, Beauducel A, John R et al (2000) Sensation seeking and affective disorders: characteristics in the intensity dependence of acoustic evoked potentials. Neuropsychobiology 41(1):24–30
15. Wang W, Wang Y, Fu X et al (2006) Cerebral information processing in personality disorders: I. intensity dependence of auditory evoked potentials. Psychiatry Res 141(2):173–183
16. Carrillo-de-la-Pena M, Vallet M, Perez MI et al (2006) Intensity dependence of auditory-evoked cortical potentials in fibromyalgia patients: a test of the generalized hypervigilance hypothesis. J Pain 7(7):480–487
17. Siniatchkin M, Kropp P, Neumann M et al (2000) Intensity dependence of auditory evoked cortical potentials in migraine families. Pain 85(1):247–254

18. Croft RJ, Klugman A, Baldeweg T et al (2001) Electrophysiological evidence of serotonergic impairment in long-term MDMA ("ecstasy") users. Am J Psychiatr 158(10):1687–1692
19. Tuchtenhagen F, Daumann J, Norra C et al (2000) High intensity dependence of auditory evoked dipole source activity indicates decreased serotonergic activity in abstinent ecstasy (MDMA) users. Neuropsychopharmacology 22(6):608–617
20. Senkowski D, Linden M, Zubrägel D et al (2003) Evidence for disturbed cortical signal processing and altered serotonergic neurotransmission in generalized anxiety disorder. Biol Psychiatry 53(4):304–314
21. Jaworska N, Protzner A (2013) Electrocortical features of depression and their clinical utility in assessing antidepressant treatment outcome. Can J Psychiatry 58(9):509–514
22. McPherson WB, Newton JE, Ackerman P et al (1997) An event-related brain potential investigation of PTSD and PTSD symptoms in abused children. Integr Physiol Behav Sci 32(1):31–42
23. Metzger LJ, Pitman RK, Miller GA et al (2008) Intensity dependence of auditory P2 in monozygotic twins discordant for Vietnam combat: associations with posttraumatic stress disorder. J Rehabil Res Dev 45(3):437–449
24. Lewine JD, Thoma RJ, Provencal SL et al (2002) Abnormal stimulus-response intensity functions in posttraumatic stress disorder: an electrophysiological investigation. Am J Psychiatr 159(10):1689–1695
25. Paige SR, Reid GM, Allen MG et al (1990) Psychophysiological correlates of posttraumatic stress disorder in Vietnam veterans. Biol Psychiatry 27(4):419–430
26. Gallinat J, Bottlender R, Juckel G et al (2000) The loudness dependency of the auditory evoked N1/P2-component as a predictor of the acute SSRI response in depression. Psychopharmacology 148(4):404–411
27. Juckel G, Pogarell O, Augustin H et al (2007) Differential prediction of first clinical response to serotonergic and noradrenergic antidepressants using the loudness dependence of auditory evoked potentials in patients with major depressive disorder. J Clin Psychiatry 68(8):1206–1212
28. Lee BH, Park YM, Lee SH et al (2015) Prediction of long-term treatment response to selective serotonin reuptake inhibitors (SSRIs) using scalp and source loudness dependence of auditory evoked potentials (LDAEP) analysis in patients with major depressive disorder. Int J Mol Sci 16(3):6251–6265
29. Lee TW, Yu YW, Chen TJ et al (2005) Loudness dependence of the auditory evoked potential and response to antidepressants in Chinese patients with major depression. J Psychiatry Neurosci 30(3):202–205
30. Linka T, Müller BW, Bender S et al (2005) The intensity dependence of auditory evoked ERP components predicts responsiveness to reboxetine treatment in major depression. Pharmacopsychiatry 38(03):139–143
31. Mulert C, Juckel G, Brunnmeier M et al (2007) Prediction of treatment response in major depression: integration of concepts. J Affect Disord 98(3):215–225
32. Paige SR, Fitzpatrick DF, Kline JP et al (1994) Event-related potential amplitude/intensity slopes predict response to antidepressants. Neuropsychobiology 30(4):197–201
33. Wade EC, Iosifescu DV (2016) Using EEG for treatment guidance in major depressive disorder. Biol. Psychiatry Cogn Neurosci Neuroimagin 1:411–422
34. Yoon S, Kim Y, Lee SH (2021) Does the loudness dependence of auditory evoked potential predict response to selective serotonin reuptake inhibitors?: a meta-analysis. Clin Pschopharmacol Neurosci 19(2):254
35. Linka T, Sartory G, Wiltfang J et al (2009) Treatment effects of serotonergic and noradrenergic antidepressants on the intensity dependence of auditory ERP components in major depression. Neurosci Lett 463(1):26–30
36. Park YM, Lee SH, Park EJ (2012) Usefulness of LDAEP to predict tolerability to SSRIs in major depressive disorder: a case report. Psychiatry Investig 9(1):80–82
37. Luck SJ (2014) An introduction to the event-related potential technique. MIT Press, Cambridge, MA
38. Nathan PJ, Segrave R, Phan KL et al (2006) Direct evidence that acutely enhancing serotonin with the selective serotonin reuptake inhibitor citalopram modulates the loudness dependence of the auditory evoked potential (LDAEP) marker of central serotonin function. Hum Psychopharmacol Clin Exp 21(1):47–52

Chapter 15

Preclinical Methods of Neurosteroid-Induced Facilitation of Fear Extinction and Fear Extinction Retention

Luca Spiro Santovito and Graziano Pinna

Abstract

Post-traumatic stress disorder and other stress-related mood disorders are a major public health burden. Psychotherapy and SSRIs, the only pharmacological treatment currently approved for PTSD, are only partial treatments. Validated PTSD animal models could provide a better understanding of PTSD neurobiology via the discovery of new pharmacologic targets to facilitate resilience after trauma. However, the complexity of PTSD makes the development of adequate animal models a challenge. To mimic some of the phenotypes underlying the pathophysiology of PTSD, several trauma-focused rodent models have been assessed. The protracted social isolation paradigm used in our laboratory results in a time-dependent reduction in the synthesis of neurosteroids, such as allopregnanolone and its isomer, pregnanolone, which is consistent with PTSD clinical findings. Neurosteroids act at membrane receptors to regulate neuronal excitability and the stress response. Neurosteroids, allopregnanolone and pregnanolone, are potent positive allosteric modulators of extrasynaptic $GABA_A$ receptors. Several alterations in the GABAergic system, both in GABA levels and $GABA_A$ receptor subunit composition, have been shown in PTSD. Following social isolation, changes in GABAergic receptor sensitivity and receptor conformation have been observed in corticolimbic areas that correlate with receptor pharmacology changes and a relative lack of sensitivity to anxiolytic benzodiazepines. Allopregnanolone plays a crucial role in the pathophysiology of PTSD and depression, mainly by potentiating GABAergic neurotransmission. Several agents have recently been shown to mimic the pharmacology of GABAergic neurosteroids and offer novel treatments for PTSD. In a translational approach, allopregnanolone, allopregnanolone analogs, SSRIs at low doses, and the endocannabinoid-like molecule, N-palmitoylethanolamine (PEA), have been shown to elevate endogenous allopregnanolone levels and to be effective in reducing PTSD-like behavioral alterations in rodent stress models and, thus, provide potential candidates for clinical testing.

Key words Allopregnanolone, GABA-A receptors, PPAR-alpha, N-palmitoyl ethanolamide, Mood disorders, Biomarkers, Rapid-acting-antidepressants

1 Introduction

Post-traumatic stress disorder (PTSD) is "an anxiety disorder that develops about an event which creates psychological trauma in response to actual or threatened death, serious injury, or sexual violation" [1], as defined by *The Diagnostic and Statistical Manual of Mental Disorders* (5th edition; DSM-5). Although only a minority of individuals develop PTSD after a traumatic event, its prevalence has increased over the last two decades and it represents a major burden on the health system, affecting 8.3% of the US population, and costing billions [2].

Intrusive flashbacks, avoidance of trauma-associated stimuli, negative mood, impaired cognition surrounding the trauma, and hyperarousal/hypervigilance all characterize PTSD [3]. These symptoms are principally mediated by impairment of fear learning, and lack of retaining extinction learning. Fear is defined as "an unpleasant often strong emotion caused by anticipation or awareness of danger, an instance of this emotion, a state marked by this emotion" [4, 5]. The traumatic event that triggers PTSD can be conceived as a brief and long-lasting form of fear acquisition [6]. According to the Pavlovian paradigm, the threat to one's safety is named unconditioned stimulus (US) and it elicits a psychobiological fear response without any prior learning. When environmental cues (named conditioned stimuli, or CS) such as smells, sounds, or sights or other sensory stimuli are associated with the traumatic event, they can trigger fear responses and an individual who was previously exposed to trauma can acquire persistent fear in response to these cues. The association of the CS with the US is the typical paradigm in a classic translational fear conditioning model, where a neutral stimulus is regularly paired with an aversive outcome [5].

Several psychotherapy approaches have proven effective in the treatment of PTSD, based on the Pavlovian paradigm. One is trauma-focused cognitive behavioral therapy (TF-CBT), which includes prolonged exposure therapy (PE) [5, 7]. PE consists of re-experiencing the traumatic event through both engaging continuously with the traumatic memories of the event (so-called imaginal exposure) and everyday reminders (the in vivo exposure), rather than avoiding triggers [5, 8–10]. Lately, virtual reality exposure therapy (VRET) has been implemented [11–13]. VRET uses virtual reality to expose clients to their fears; it is based on the hypothesis that the fear network can be changed by integrating new and incompatible information into the network after being faced with feared stimuli [14]. VRET not only has been proven to be as effective as an in vivo exposure, but it also allows the therapist to adjust the intensity, frequency, and pace of the exposure, based on the clients' needs [11, 15, 16]. It also allows the recording of physiological parameters (skin conductance and heart rate), and it can be performed in a safe environment [17]. Another therapy that uses a trauma-focused approach is eye movement desensitization and reprocessing therapy (EMDR). Clients are subjected to a continual recall of stressful images during the therapy sessions while receiving sensory inputs. On the contrary, cognitive processing therapy (CPT) and present-centered therapy (PCT) are less focused on the traumatic event, the former highlighting the correct attribution of faults (PTSD clients tend to see the world as uncontrollable and dangerous) and the latter focusing mostly on current work and relationships. In addition, group and family therapies, where subjects with PTSD share their traumatic experiences and find support from the group, have been proven effective [10].

Selective serotonin reuptake inhibitors (SSRIs), including sertraline and paroxetine, are currently the only class of drugs approved for PTSD [5, 18], even though recent evidence suggests that serotonin and norepinephrine reuptake inhibitors (SNRI), such as venlafaxine, may have higher effectiveness than SSRIs [19]. Despite benzodiazepines being commonly used in clinical practice for the treatment of anxiety disorders, evidence suggests their lack of effectiveness in PTSD treatment.

In addition to their well-known serotonergic effects [20], SSRIs may also increase the biosynthesis of the GABAergic neurosteroid, allopregnanolone, as well as upregulate the expression and signaling of brain-derived neurotrophic factor (BDNF)—both of which have been found to be decreased in subjects with PTSD [21].

Allopregnanolone and its equipotent GABAergic stereoisomer, pregnanolone, have been a topic of recent interest. Allopregnanolone is a potent positive allosteric modulator of GABA's action at the $GABA_A$ receptor. It is synthesized from progesterone by glutamatergic pyramidal neurons and is involved in the regulation of emotional behavior via GABAergic pathways [22, 23]. Corticolimbic levels of this neurosteroid are markedly decreased in two well-recognized rodent stress models of PTSD, the protracted social isolation model and the single prolonged stress model. These models are associated with depressive-like phenotypes and increased contextual fear conditioning responses. The socially isolated (SI) mouse also shows increased aggressive behavior, which is an important marker of suicidality risk in humans.

Results from these animal behavior models have corroborated clinical phenotypes observed in human subjects with PTSD [24]. Namely, downregulated neurosteroid biosynthesis, as well as reduced allopregnanolone and pregnanolone levels in corticospinal fluid (CSF), plasma, and serum have been demonstrated in both animal and human studies. Moreover, an inverse relationship was found between levels of allopregnanolone and pregnanolone with the severity of PTSD symptoms in human clients [25, 26].

Thus, allopregnanolone's central role in the neurobiology of PTSD and other stress-related mood disorders, combined with its involvement in fear learning and fear extinction, make it not only a promising biomarker candidate but also a potential target for the treatment of these devastating psychiatric disorders.

2 Role of the GABA$_A$ Receptor and GABA Levels in PTSD

Five glycoproteic subunits surrounding a chloride-permeable channel comprise the structure of the GABA$_A$ receptor. Two α, two β, and one γ subunit represent the most prominent synaptic configuration in the brain. Opening of the channel facilitates an influx of chloride which results in hyperpolarization and, hence, imparts an inhibitory effect on the cell membrane [27]. GABA promotes receptor activation by binding between the α and β subunits, where two different GABA binding sites are located [28], while benzodiazepines bind the receptor at the interface of the α and γ2 subunits [29]. Neurosteroids bind the GABA$_A$ receptor at two specific binding sites, an activation site between the α and β subunits, and a potentiation site found within a cavity in the α subunit [30]. The most common configuration of the GABA$_A$ receptor is $2\alpha_n 2\beta_n \gamma_n$ [31]; however, different configurations with different physiological and pharmacological properties have also been described. For instance, different α subunits drive different responses to benzodiazepine binding and pharmacodynamics. Benzodiazepines have anticonvulsant, sedative, and amnesic properties when the α_1 subunit is expressed [32], anxiolytic properties when the α_2 subunit is involved [33], or they can act as muscle relaxants when the α_3 subunit is expressed [34]. Importantly, GABA$_A$ receptors formed by α_4 and α_6 subunits conversely elicit no response to benzodiazepines [35].

GABA$_A$ receptors can be classified as synaptic or extrasynaptic. Synaptic GABA$_A$ receptors (mainly α/β/γ configuration) are responsible for inhibitory phasic currents, and while highly sensitive to benzodiazepines, have low sensitivity to GABA and neurosteroids. Extrasynaptic GABA$_A$ receptors (mainly α/β/δ configuration), which facilitate inhibitory tonic currents, conversely show high sensitivity to neurosteroids, low sensitivity to GABA, and no sensitivity to benzodiazepines [31, 36].

Several important alterations to the GABAergic system have been associated with stress-related disorders. Decreased concentrations of GABA and altered expression of GABA$_A$ receptors were observed in the brains of PTSD clients, and low plasma levels of GABA were proposed as a risk factor for PTSD. On the other hand, another recent study observed an increase in plasma GABA levels in PTSD-affected individuals. In this study, the plasma level of GABA was measured in a large cohort of military personnel, predominantly male, pre-deployment or 1- to 6-month post-deployment. They showed that an increase in plasma GABA was associated with an increase in depressive and/or PTSD symptoms [37]. These results corroborate a study from Arditte Hall and colleagues, showing that PTSD dysphoria, avoidance, and total PTSD symptom severity positively correlated with plasma GABA levels [38]. Stress-related mood disorders are also characterized by changes in the structure and expression of the GABA$_A$ receptor in different brain regions such as the hippocampus, frontal cortex, and hypothalamus. In PTSD patients, diminished overall GABA$_A$ receptor activity has been observed, whereas in depressed or suicidal patients a dysregulation of the genes encoding for several receptor subunits (i.e., α1, α3, α4, α5, γ2, and δ) has been reported [39–42]. Moreover, GABAergic involvement in stress-related mood disorders has also been described in several preclinical studies and involves reduced expression of the α1 subunit after repeated forced swim tests [43]. Interestingly, the balance between synaptic and extrasynaptic GABA$_A$ receptor subtypes is modified by protracted social isolation stress in male rodents and is characterized by an increased expression of extrasynaptic GABA$_A$ receptor subunits (α4, α5, and δ subunits) and a reduction of synaptic receptor subunit composition (α1, α2, and γ2 subunits) [44–46]. Accordingly, benzodiazepines that have a high affinity for synaptic GABA$_A$ receptor subunits, such as diazepam and zolpidem, exhibit altered pharmacological effects in these SI mice [45, 47]. From a translational perspective, it is worth noting the SI mouse model shows several similarities with PTSD pathophysiology, such as changes in GABA$_A$ receptor distribution and subunit composition in the cortex, thalamus, and hippocampus. Such changes may alter the effectiveness of benzodiazepine therapy in PTSD [27]. Collectively, these data indicate that protracted stress may cause an imbalance between synaptic and extrasynaptic GABA$_A$ receptor subtypes and, hence, between tonic and phasic inhibitory currents, making the GABAergic system a major target candidate for the development of novel neurosteroid-based pharmacologic agents in preclinical and clinical studies of PTSD.

3 Role of GABAergic Neurosteroids in PTSD

Allopregnanolone acts at membrane receptor level and regulates neural excitability by increasing the influx of chloride seven- to ten-fold in response to GABA binding [22]. They were named neurosteroids by Baulieau et al. after the groundbreaking discovery that the enzymatic machinery needed for their biosynthesis is expressed in the brain and neurosteroids, including allopregnanolone, reach higher concentrations in the brain than in peripheral organs and blood [48]. Neurosteroids are directly produced *ex novo* by glutamatergic neurons, but also in GABAergic long-projecting neurons. Glial cells synthesize pregnenolone, the precursor of all neurosteroids, via cholesterol and the action of several enzymes, including CYP11A1 [49–51], which is found in the inner membrane of mitochondria. Pregnenolone is further metabolized into progesterone by 3β-hydroxysteroid dehydrogenase and then into 5α-dihydro-progesterone (5α-DHP) by 5α-reductase type 1 or into 5β-dihydro-progesterone (5β-DHP), by 5β-reductase. At the end of this process, 5α-DHP and 5β-DHP can be reduced by 3α-hydroxysteroid dehydrogenase (3α-HSD) into allopregnanolone and pregnanolone (together termed ALLO), respectively [27].

Other than their neuroprotective and neurotrophic effects, neurosteroids also act as anxiolytic, sedative, analgesic, and antidepressant neuromodulators. Additionally, their levels fluctuate during pregnancy, the menstrual cycle, development, and acute or chronic stress, for example.

Numerous studies have established a connection between defective neurosteroidogenesis and neuropsychiatric disorders, especially stress-related mood disorders such as PTSD and depression. A significant decrease in CSF allopregnanolone and pregnanolone levels was described in women with PTSD during the follicular phase of the menstrual cycle, along with a reduction in the ALLO/5α-DHP ratio [52]. These findings were confirmed by Pineles' team, who measured a reduced ALLO/5α-DHP ratio in the plasma of women with PTSD during different phases of the menstrual cycle. Interestingly, this ratio failed to increase in women with PTSD in response to a stressful fear conditioning task, while it strongly increased in healthy women in response to the same task, suggesting an enzymatic deficit in the conversion of 5α-DHP into allopregnanolone [53]. Furthermore, decreased levels of neurosteroids in CSF show a strong negative correlation with PTSD symptoms in male patients. Decreased levels of allopregnanolone have also been observed in the serum, plasma, and CSF of women affected by major depression or during pregnancy and postpartum [25].

Since major depression and anxiety are frequently comorbid with anorexia nervosa (AN) and overweight/obesity (OW/OB), Dichtel et al. investigated GABAergic neurosteroids in these clients. They found lower allopregnanolone levels in women with AN and OW/OB when compared with healthy controls. Allopregnanolone levels were negatively associated with the severity of anxiety and depression symptoms, independently of BMI. These findings suggest a dysregulation of GABAergic neurosteroid metabolism and/or synthesis, and thus of the GABAergic system, in AN and OW/OB women with concomitant affective symptoms [54].

Very few studies have measured neurosteroid levels in the human postmortem brain from PTSD subjects. In an intriguing study using gas chromatography-tandem mass spectrometry preceded by high-performance liquid chromatography, Cruz and colleagues measured levels of several neurosteroids (allopregnanolone, pregnenolone, pregnanolone, epiallopregnanolone, epipregnanolone, tetrahydrodesoxycorticosterone [THDOC], and androsterone) in the medial orbitofrontal cortex of the post-mortem brain of PTSD clients. They showed decreased levels of androsterone and allopregnanolone in male PTSD clients when compared with healthy controls. Interestingly, they also measured decreased levels of allopregnanolone in male PTSD clients when compared with healthy individuals. Moreover, androsterone was decreased in PTSD males in the same brain area. By contrast, females with PTSD had increased levels of pregnanolone and pregnenolone relative to healthy female individuals. However, this study failed to record menstrual cycles in this cohort, leaving results difficult to interpret [55].

Together, these results suggest a potential role of neurosteroids in the pathophysiology of PTSD, both as biomarkers and potential treatment targets for PTSD symptoms, including a role in facilitating fear extinction and fear extinction retention.

4 Studying Fear Extinction in Animal Models of PTSD

Disorders that originate from the exaggerated feeling of fear, such as PTSD, are characterized by the inability to suppress inappropriate fear responses [5, 56], thus, a common approach to treating stress-related maladaptive fear responses is trauma-focused psychotherapy, such as CBT and PE therapy. PE aims to prevent avoidance of fear-linked cues by repeatedly exposing clients to these cues in a safe environment with the goal of causing the extinction of fear responses [10, 57, 58]. The Pavlovian model allows for the replication of this method in research laboratories, both with humans and animals, linking the fear with a previously innocuous stimulus (CS), and then attempting to decrease the fear by presenting the CS alone. Fear extinction is not obliteration of the original fear but an inhibition of the originally acquired fear. Indeed, fear extinction is not a permanent phenomenon; it can re-emerge spontaneously because of a change in the experimental environment (renewal shift), or if the subject is exposed to an aversive unconditioned stimulus after the extinction process (reinstatement effect) [5, 59–63]. The wide range of PTSD symptoms combined with the complexity of stress-related disorders represents a challenge in creating an adequate and complete rodent model. This variety in symptoms may be due to varying individual susceptibility and resilience to the disorder. Different types of traumas and repeated exposures to them could result in PTSD subtypes as well. Several rodent models utilize exposure to various traumatic procedures to induce PTSD phenotypes that could be useful in investigating mechanistic aspects of the disorder [64–66]. One of the most widely used rodent fear learning models is the inescapable shock model. This involves administering one or more electric shocks lasting 0.5–10 s to the rodent's paw or tail [64]. Another rodent model of fear learning is the predator-stress model, which exposes the rodents to a potential predator or predator scent. This in turn causes hyperarousal, avoidance, exaggerated fear response, impaired fear extinction, and anxiety-like behaviors, which are more prominent with direct predator exposure, and intermediate with exposure to the predator's scent [64, 67]. The exposure to three different stressors in succession (2 h of restraint-immobilization stress, 20 min of forced swimming, and exposure to diethyl ether until loss of consciousness) characterizes another common and well-accepted PTSD rodent model: the single prolonged stress model. This experimental paradigm increases hyperarousal and contextual freezing, and compromises fear extinction retention [64, 68, 69]. Of note, the 129S1/SvlmJ mouse strain provides an interesting investigation approach for the genetic mechanisms underlying fear extinction. Indeed, the 129S1/SvlmJ mice exhibit overgeneralization of fear cues, and when safety signals are presented, they show an impaired inhibition of fear [64, 70].

4.1 The Social Isolation Model and Impairment of GABAergic Signal Transduction

The SI mouse model originates from evidence that protracted social isolation in humans (perceived social isolation or loneliness) underlies several neurological and psychiatric conditions, from Alzheimer's disorder to major depression. During the COVID-19 pandemic, quarantine has been efficacious to contain the spread of the disease. While social distancing was necessary to mitigate the effects of the pandemic, it has increased the prevalence of social isolation (an objective state in which an individual is alone) and loneliness (the subjective feeling of inadequacy of social connections). Social isolation and loneliness are known risk factors for stress-related mood disorder exacerbations and suicidal behavior. The subjective feeling of being alone was associated with an increased risk of suicidal thoughts and attempts, as demonstrated by the Quebec Health Survey and the previous SARS pandemic in 2010. An increased suicide rate was seen during this time, and social isolation was associated with one-third of these events [71–73].

In rodents, protracted social isolation is a distressing event that elicits increased fear response deficits and anxiety-like and aggressive behavior that mimic neurobiological alteration encountered in humans with PTSD. Most studies that use this animal model are conducted in male mice, while in female mice social isolation is limited to exploring depressive-like behavior [64].

4.1.1 Fear Conditioning Responses, Fear Extinction, and Fear Extinction Retention in the socially isolated Mouse Paradigm

In our laboratory, social isolation comprises isolation of the animals in individual cages for 3–4 weeks post-weaning. Contextual fear conditioning response is evaluated using a Pavlovian model with the administration of conditioned (acoustic tone) and unconditioned (foot shock) stimuli in a novel context. SI mice subjected to this paradigm show increased freezing (indicator of elevated fear response) 24 h post-training session (Fig. 1). The response to fear conditioning increases in a time-dependent fashion during 4 weeks

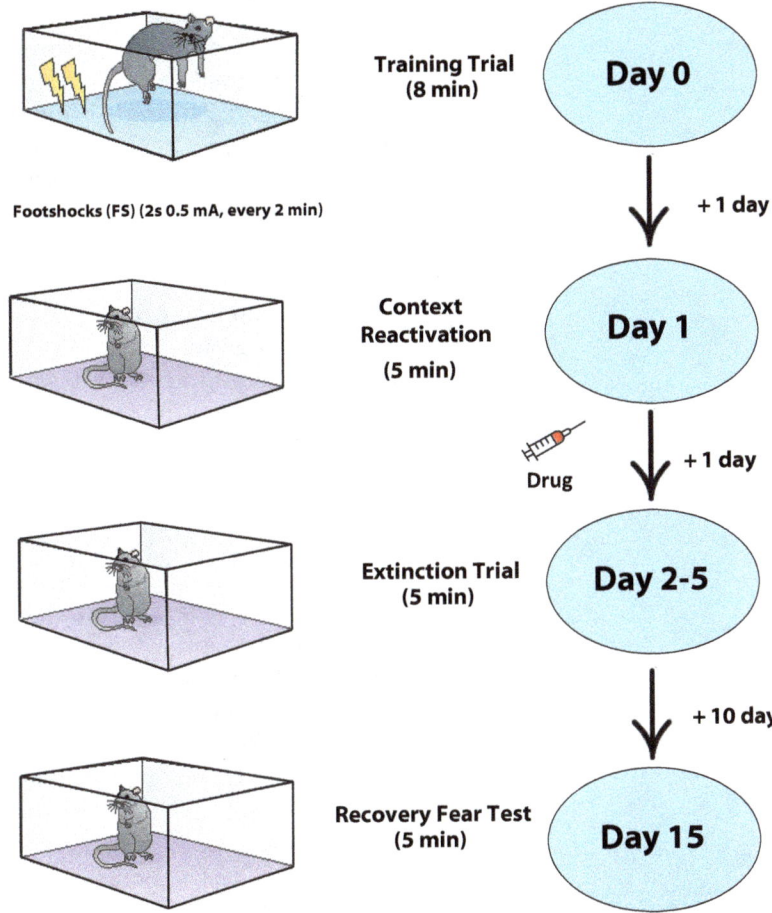

Fig. 1 Schematic presentation of the experimental procedure to determine contextual fear conditioning responses in a rodent model of PTSD. Contextual fear conditioning responses, fear extinction, and extinction retention are measured in mice after four weeks of social isolation. During a training section, mice are allowed to explore the training chamber for 8 min. During this time, mice receive an electric foot shock (2 s; 0.5 mA) (the unconditioned stimulus), that will be presented three times every 2 min. The duration of freezing behavior is measured 24 h after training during a re-activation session in absence of a footshock presentation. Contextual fear extinction is measured by placing the mouse in the contextual chamber 24 h after the re-activation session for five consecutive days for 5 min each day. After an interval of 10 days, extinction retention is assessed by placing the mice in the contextual chamber for 5 min without footshock presentation. Aversive memories disruption is achieved through a reconsolidation blockade by administering pharmacological agents immediately after a contextual reactivation session

of isolation, reaching a plateau between 4 and 6 weeks when neural allopregnanolone level decline is maximal [24, 74–78].

The test is performed in a transparent acrylic chamber (25 cm wide, 18 cm high, and 21 cm deep) in which the floor is made of stainless steel rods linked to an electric shock generator, and the top wall of the enclosure is a small fan. Sixteen infrared photo beams organized in a frame surround the chamber. Delivery of the electric shock, the auditory stimuli, and freezing time (defined as beam interruptions and latencies to beam interruptions) are computer-assisted, regulated, and recorded. The test is divided into three portions: the training trial, the contextual reactivation, and the extinction trial, including a recall section. The training test consists of placing the mouse into the training chamber and allowing it to explore for 2 min. The conditioned stimulus, a sound presented for 30 s, is then presented three times every 2 min paired with the unconditioned stimulus (electric foot shock, 2 s, 0.5 mA), which is administered during the last 2 s of the CS. The mice are allowed to explore the chamber for an additional two minute after the last shock, with a total test time of 8 min (Fig. 1). Twenty-four hours post-training, mice are placed into the contextual chamber and freezing behavior is assessed for 5 min; without presenting foot shock. During the extinction phase, mice are placed in the contextual cage 24 h post context reactivation and again for five consecutive days. Fear extinction retention can be evaluated by placing mice back into the contextual chamber 10 days from the last day of the extinction trial (San Diego instruments; freeze monitor system, San Diego instruments) [24]. Freezing behavior, defined as "the absence of any movement except for those related to respiration while the animal is in a stereotypical crouching posture" [74], is then measured for 5 min while the mice are placed in the contextual cage, in the absence of tone or foot shock (Fig. 1).

4.1.2 Behavioral Responses of Contextual Fear Conditioning in SI Mice

Male mice that have been socially isolated for 3–4 weeks post-weaning and then exposed to a fear conditioning paradigm display elevated freezing time, a proxy for fear response, 24 h post-training. In 4-week-long social isolation studies, freezing time tends to increase in a time-dependent manner until reaching a plateau after week 3. Moreover, SI mice show impaired fear extinction learning and defective fear extinction retention after the passage of time. SI mice express several other behavioral deficits, including increased aggression toward a same-sex intruder and anxiety-like and depressive-like phenotypes [24, 74, 76].

In this model, social isolation can be considered a prolonged stressor that is then paired with an acute traumatic event (the foot shock) that leads to impaired stress adaptation, precipitating the onset of PTSD-like behavioral deficits.

4.1.3 Fear Paradigm in SI Mice and Role of Allopregnanolone

Prolonged stress induced by the social isolation model results in a time-dependent reduction of allopregnanolone biosynthesis, due to lowered expression of the rate-limiting enzyme in allopregnanolone production, 5α-reductase type I [77–82]. This results in a 50–70% reduction in allopregnanolone and its equipotent GABAergic isomer, pregnanolone, in SI mice [76, 83]. The downregulation of allopregnanolone biosynthesis in the olfactory bulb and corticolimbic areas involved in fear acquisition and extinction, such as the amygdala, hippocampus, and frontal cortex, is strongly associated with an increased fear response as well as diminished contextual fear extinction and extinction retention [24, 74, 84]. These behavioral changes observed in rodent stress models are similar to the behavioral deficits described in human clients who suffer from PTSD and anxiety spectrum disorders. Decreased ALLO levels in these crucial corticolimbic areas are associated with impairment of GABAergic neurotransmission, which leads to an altered fear response. Notably, in situ immunohistochemistry has been used to study the expression and neuronal localization of 5α-reductase type I. These studies revealed that enzymatic loss occurs in specific corticolimbic neurons in SI mice, including pyramidal-like neurons of the basolateral amygdala, layers V/VI pyramidal neurons in the medial prefrontal cortex, hippocampal CA3 pyramidal neurons, and granule cells in the dentate gyrus (Fig. 2). SI-induced downregulation of enzyme expression in these specific neuronal populations may explain the increased contextual fear conditioning responses observed in this PTSD mouse model. Conversely, decreased fear extinction can also be obtained by normalizing cortico-hippocampal and amygdaloid circuitry by pharmacologically upregulating $GABA_A$ receptor neurotransmission (Fig. 2) [23, 85, 86].

5 Pharmacological Blockade of Reconsolidation Processes Using Neurosteroid-Based Molecules

There are few studies that investigate the role of allopregnanolone and GABAergic neurosteroids in non-aversive learning. In one of these studies, allopregnanolone was administered at doses of 3.2, 10, and 17 mg/kg in mice to assess non-spatial memory performance. Allopregnanolone was administered 15 min before exposing mice to two sample objects in a novel object recognition test. After 24 h, allopregnanolone-treated mice showed a dose-dependent impairment in encoding and consolidation of memory, with a reduction in time spent exploring the novel object that was presented together with a familiar one. These investigators also tested allopregnanolone's effect on hippocampus-specific non-aversive learning. First, male mice were administered either

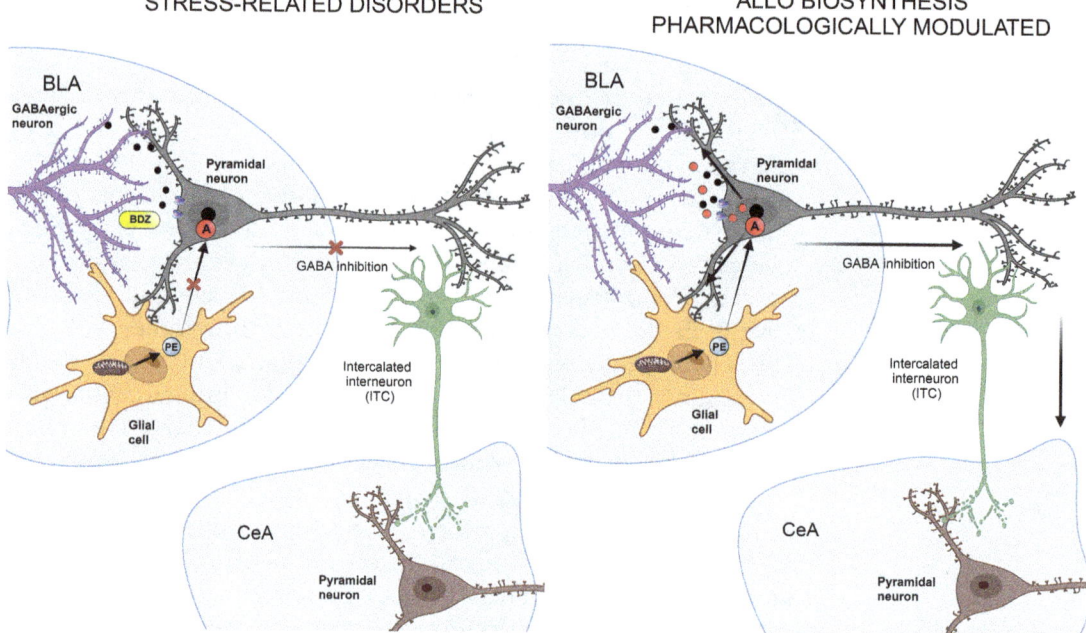

Fig. 2 Stress-induced downregulation of allopregnanolone (A, red solid circle) impairs GABAergic signaling resulting in dramatic behavioral effects. Stressful event exposure impairs the cortical inhibitory projections to the basolateral amygdala (BLA) unleashing exaggerated hyperactivity and fear responses, which have been observed in PTSD animal models and clients with PTSD. In PTSD, lower GABA (solid black circle) and plasma and CSF Allo levels, and altered $GABA_A$ receptor subunit expressions have been observed, which constitute possible biomarkers of stress-induced disorders, including PTSD. In socially isolated (SI) mice, a rodent stress model of PTSD, cortico-hippocampal projections to the BLA show lower allopregnanolone biosynthesis in association with deficits in fear responses. Decreased allopregnanolone levels in glutamatergic neurons may underlie the impairment of cortico-hippocampal-amygdaloid circuits and result in exaggerated fear responses. Neurosteroidogenic agents that normalize allopregnanolone levels also normalize cortical and hippocampal neuronal connectivity by affecting extrasynaptic $GABA_A$ receptor-mediated neurotransmission, which modulates emotional processing and PTSD behavioral deficits. Abbreviations: CeA, central amygdaloid nucleus; PE, pregnenolone, BDZ, benzodiazepine

allopregnanolone (10 mg/kg) or vehicle for 10 min before exposing them to a novel recognition chamber for 5 min. Twenty-four hours later, vehicle was administered to all mice 10 min before a fear conditioning task, consisting of a 30 s long, 90 dB tone co-terminating with a 1 s, 15 mA foot shock presented three times over a 5-minute period, was performed. They found that mice treated with allopregnanolone before pre-exposure showed no difference in fear response to the tone but manifested diminished overall contextual fear response [87].

The role of allopregnanolone in modulating contextual fear conditioning responses is further confirmed by pharmacologic PTSD animal model studies. When allopregnanolone or S-norfluoxetine (a selective brain steroidogenic stimulant) was administered before the training session, the increased contextual fear conditioning response expressed by SI mice was reversed. Moreover, pre-injection of a potent 5α-reductase inhibitor (SKF 105,111), which causes the decline of allopregnanolone levels by 90% in about 1 h, precipitated an enhanced fear response when administered to group-housed mice during the fear conditioning task [80, 88–90]. When analyzed together, these results suggest the behavioral deficits observed in SI isolated mice could be the result of a downregulation in allopregnanolone and subsequent impairment of GABAergic neurotransmission. These deficits could underlie modifications in corticolimbic circuits that regulate emotional responses. The role of the amygdala in mediating fear responses is crucial. Inhibitory spiny GABAergic interneurons (ITC) control the connection between the basolateral amygdala (BLA) and the central amygdaloid nucleus (CeA). ITCs have an inhibitory function, regulated by the amygdala's pyramidal-like neurons. Amygdala hyperactivity is one of the best characterized PTSD mechanisms underlying emotional instability [23, 24]. This hyperactivity is due to functional alterations in glutamatergic neurons of the prefrontal cortex and hippocampus that project to GABAergic amygdaloid neurons, thus inhibiting these amygdaloid neurons [91]. This inhibition of the amygdala nuclei can be suppressed or weakened during a maladaptive response to a traumatic event, resulting in an inappropriate and exaggerated fear response, and impaired fear extinction due to amygdala hyperactivity [92]. Evidently, GABAergic ITC outputs play a crucial role in emotion regulation following a stressful event by directly affecting fear extinction learning and regulation of CeA output, key response mediators to conditioned fear [93]. In addition, several findings showed that ITC lesions hinder fear extinction retention, while their activation hastens extinction learning. CeA and ITC GABAergic neurons also project to the hypothalamus and brainstem to attenuate fear responses and extinction after a stressful event (Fig. 2) [94, 95]. Taken together, these corticolimbic circuits express decreased levels of allopregnanolone and pregnanolone [96] in SI mice and they may be responsible for emotional responses such as altered fear response and aggressive behavior (Fig. 3).

These findings suggest that supplementation with allopregnanolone, analogs thereof, or neurosteroidogenic agents may improve PTSD symptoms. To test this hypothesis, the allopregnanolone analog, ganaxolone (3.75–30 mg/kg), was administered subcutaneously before testing social isolation–induced behavior in mice. Ganaxolone decreased aggressive behavior during a resident-intruder test in a dose-dependent manner, resulting in an EC_{50} of 9.7 mg/kg. Of note, when the contextual fear response was assessed in both SI and group-housed mice, we observed a decreased freezing time in ganaxolone-treated SI mice compared with vehicle-treated SI mice. No differences were found in group-housed mice who received ganaxolone. Moreover, a single 10 mg/

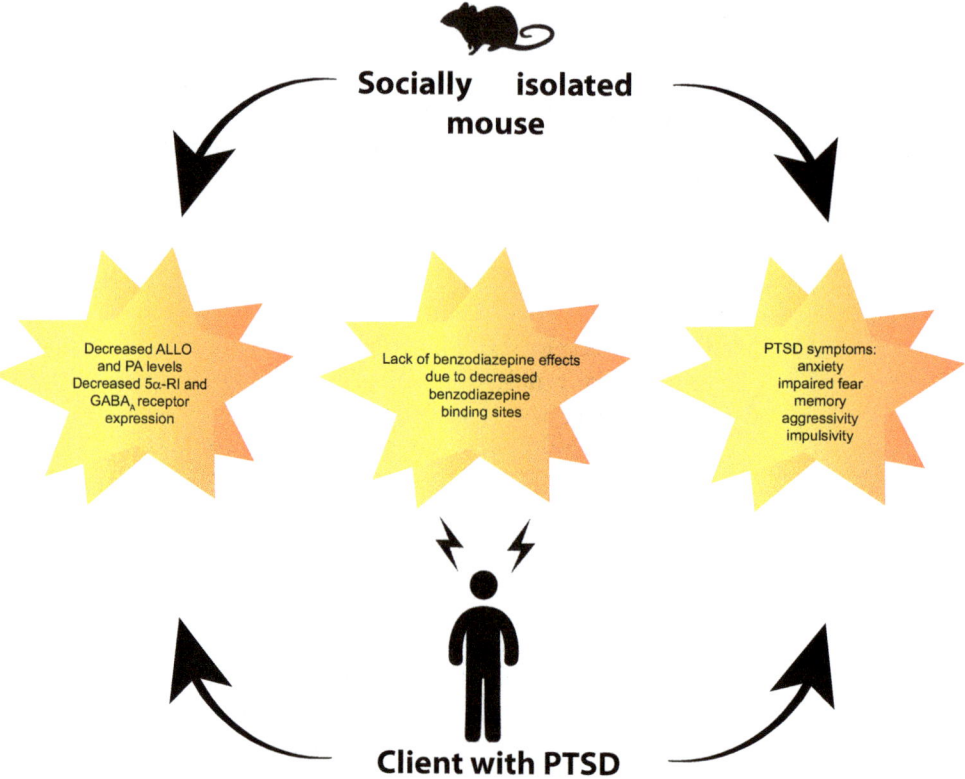

Fig. 3 PTSD-like behavioral phenotypes in socially isolated (SI) mice. SI mice express several behavioral phenotypes that mimic the neurobiology of PTSD in humans. Behaviorally, SI mice show increased aggressive behavior, anxiety-like behavior, and contextual fear extinction deficits. These deficits are hallmarks of PTSD. Neurobiological deficits in SI mice also include changes in $GABA_A$ receptor subunit composition and downregulation of neurosteroid biosynthesis, including allopregnanolone and pregnanolone. Like in PTSD clients, SI mice fail to respond to the anxiolytic effects of benzodiazepines. 5α-RI, 5α-reductase type I; ALLO, allopregnanolone; PA, pregnanolone

kg dose of ganaxolone administered immediately after a contextual fear reactivation section improved fear extinction memory and facilitated fear extinction retention via a reconsolidation blockade in SI mice. Additionally, ganaxolone treatment proved effective in decreasing anxiety-like behavior in SI mice when tested in an elevated plus-maze test. Ganaxolone demonstrated dose-dependent anxiolysis, with doses ranging from 3.75 mg/kg to 15 mg/kg. The anxiolytic effect was determined by the ratio of time spent in the open arm versus closed arm, and rest time versus total time. Interestingly, the highest effective dose of ganaxolone injected also had an anxiolytic effect in the group-housed mice that did not exhibit decreased levels of allopregnanolone [97]. Taken together, these findings corroborate the hypothesis that allopregnanolone biosynthesis and GABAergic neurotransmission deficiencies underlie the neurobiological mechanisms of stress-related disorders, and support ganaxolone as a candidate for the treatment of mood disorders in clinical practice. Additionally, several allopregnanolone analogs, such as BR351 and BR297, showed an anti-aggression effect in SI mice. These drugs were administered 60 min before a resident-intruder test, both to early and late adolescent SI mice. BR351, administered at the doses of 1, 2.5, and 5 mg/kg, showed an anti-aggressive effect at the dose of 5 mg/kg. This drug appeared to have higher potency in late (EC_{50} = 3.75 mg/kg) than in early (EC_{50} = 4.5 mg/kg) adolescent mice. BR297 was administered at doses of 0.3125, 0.625, and 2.5 mg/kg. BR297 showed a strong anti-aggression effect at all tested doses in both early and late adolescent SI mice, with equal potency (EC_{50} = 0.25 mg/kg). Importantly, the non-response rate of mice treated with BR351 or BR297 was lower when compared to mice treated with a higher dose of S-fluoxetine. Additionally, the anti-aggression effect of a single dose of BR297 was long-lasting, requiring 7 days in early adolescent SI mice and 9 days in late adolescent SI mice to revert to basal levels of aggression. These results suggest that these molecules could be tested in a fear-response paradigm [98].

N-palmitoylethanolamine (PEA) is also a promising potential treatment agent for stress-related mood conditions. PEA, an endogenous fatty acid ethanolamide with neuromodulator properties, stimulates the peroxisome proliferator-activated receptor (PPAR)-α, which is highly expressed in the brain and has demonstrated anti-inflammatory properties [99, 100]. It has been hypothesized that PEA may also act as a weak activator of the endocannabinoid receptor type 1 (CB1) and 2 (CB2), and be part of the endocannabinoid system, which has been implicated in the pathophysiology of stress-related mood disorders [101–103]. Reports observed that PEA stimulates the biosynthesis of allopregnanolone by binding to PPAR-α [104, 105]; moreover, it has been shown that PPAR-α is largely distributed in glutamatergic corticolimbic neurons where allopregnanolone is also abundantly found [100]. Recently, Locci and Pinna observed that by treating wild-type male Swiss-Webster mice or PPAR-α knockout male mice with PEA (5–20 mg/kg IP) 1 h before exposure to the elevated plus maze, forced swim test (FST), and tail suspension test (TST) resulted in improvement of affective behavior in Swiss-Webster but not in PPAR-α KO mice. PEA at a dose of 5 mg/kg was also administered immediately after a contextual fear conditioning reactivation session to block the reconsolidation processes. PEA treatment showed a dose-dependent anxiolytic effect, which was determined by the lower time spent in and the number of entries to the open arms during the elevated plus-maze test in SI mice. Furthermore, PEA treatment induced an antidepressant effect in SI-isolated mice by reducing the time of immobility during the FST and the TST. When injected as a single dose immediately after a contextual fear conditioning reactivation session, PEA facilitated contextual fear extinction during the first 3 days of the extinction trial and blocked the re-appearance of contextual fear responses after the passage of time (14 days after the training trial). Importantly, neurosteroid levels measured using state-of-the-art gas chromatography-mass spectrometry (GC-MS) were increased in the brain. Specifically, allopregnanolone and its precursors were found elevated in corticolimbic areas such as the olfactory bulb, prefrontal cortex, hippocampus, and amygdala. Of note, PEA treatment had no effect on PPAR-α KO mice. The specificity of the effect of PEA at PPAR-α was confirmed by analyzing the effects of PPAR-α agonists and antagonists. While PPAR-α agonists such as the synthetic GW7647 (G7) and fenofibrate showed anxiolytic effects, the PPAR-α antagonist, GW6471, prevented the anxiolytic effect elicited by administration with PEA. Moreover, when G7 was administered, allopregnanolone levels increased in corticolimbic areas. These authors demonstrated that the expression of several proteins involved in neurosteroidogenesis, such as steroidogenic acute regulatory protein (StAR), cholesterol side-chain cleavage enzyme (CYP11A1), and 5α-RI, were downregulated in SI mice. To understand the mechanisms through which PEA stimulates the biosynthesis of allopregnanolone, expression of the aforementioned proteins and enzymes was measured by Western Blot after the administration of PEA at an ED_{50} dose of 10 mg/kg. Experiments showed the expression level of these proteins was normalized by a PEA treatment [84]. Taken together, these findings suggest PEA and PPAR-α agonists to be promising new clinical approaches for PTSD and stress-related mood disorder treatment.

6 Reconsolidation Blockade and Fear Extinction in Humans

Non-aversive episodic memory (a conscious long-term memory of past experiences together with their context), especially autobiographical episodic memory (episodic memory that recalls individuals' specific events) and source memory (a type of episodic memory that recalls the origin of how we have built our past information), are crucial in the pathophysiology of PTSD and stress-related anxiety disorders [7, 106, 107]. To date, the role of allopregnanolone and $GABA_A$ receptor activation on episodic memory is poorly investigated, particularly in humans. Clinical studies of PTSD have quantified ALLO and/or GABA levels in serum, plasma, and CSF; however, only a few have measured them in the post-mortem brain. Kask and colleagues [108] investigated how intravenous injection of allopregnanolone affects memory processes in healthy women. When administered, their episodic memory was impaired. They used free verbal recall to assess episodic memory. The participants were instructed to memorize

12 common, unrelated words presented one by one at 2 s intervals using a tape recorder. The participants recalled as many words as possible, in any order, as soon as the words had been presented. The same test was repeated two more times, during the subjects' follicular phase of their menstrual cycle, to ensure that endogenous hormonal levels would not have influenced the results. Interestingly, the concomitant administration of 3β-20β-dihydroxy-5-α-pregnane (UC1011), a known inhibitor of allopregnanolone in vitro, blocked the allopregnanolone-induced learning impairment [109, 110]. This cognitive effect of allopregnanolone seems to be dependent on the serum concentrations achieved after the treatment dose (0.07 mg/kg) is administered; indeed, a lower dose of allopregnanolone enhanced both working and long-term memory [111]. Interestingly, in accordance with the aforementioned study, memory processing and recall were found impaired during pregnancy and in the first 8 months post-partum by different studies which evaluated the effect of pregnancy on cognition, especially those which investigated the third trimester, when allopregnanolone reaches its peak (about 6–10 times higher levels compared to pre-pregnancy baseline levels), and early motherhood [112, 113]. Another longitudinal study showed that third-trimester pregnant women scored lower when evaluated for their learning and memory retrieval ability by the visual verbal word learning task (WLT). This test consists of presenting a set of 15 monosyllabic meaningful words that occur frequently, are acquired early in life, and evoke a mental image in three trials. A free recall followed each trial. To test their long-term memory, the authors of the study asked the participants to reproduce the set of words 20 min after the last trial [114].

Allopregnanolone's role in episodic memory confirms the hypothesis of its beneficial effect on extinction retention in PTSD. For example, higher levels of 17β-estradiol, an upregulator of 3-α-HSD (an enzyme involved in downstream allopregnanolone synthesis), resulted in better extinction retention in healthy women [115]. In accordance with these findings, traumatized women without PTSD showed optimum extinction retention during the mid-luteal phase, when 17β-estradiol, allopregnanolone, and progesterone are all present at high levels [116]. Interestingly, extinction retention was similar between trauma-exposed women with and without PTSD during the early follicular phase, while it was considerably different during the mid-luteal phase in PTSD-affected women when compared with healthy trauma-exposed women. Additionally in PTSD-affected women, there was a strong association between resting plasma levels of allopregnanolone and extinction retention [117]. Notably, during the early follicular phase, extinction retention correlated with the plasma ratio of allopregnanolone and pregnanolone/DHEA rather than allopregnanolone or pregnanolone alone. DHEA is an allosteric antagonist of GABA$_A$ receptors, and its plasma levels are comparable to those of allopregnanolone during the early follicular phase [7, 117]. This may suggest that the two neurosteroids compete at GABA$_A$ receptors located in neuronal networks modulating emotions, and influence extinction retention. Moreover, exposure to pharmacologic agents and environmental toxins that disrupt GABAergic neurosteroid function and/or synthesis may worsen PTSD symptoms, chronicity, and comorbidity. For example, chronic use of tobacco was associated with PTSD severity in a large population of post-9/11 deployed veterans. Veterans that used tobacco had higher rates of comorbid conditions such as major depressive disorder and chronic pain [118]. Chronic nicotine exposure disrupts the adrenal gland activity and alters the timing between stressors and stress-hormones response, altering the normal process of learning and retention of new associations. Indeed, nicotine may alter the release of adrenal hormones that are crucial in long-term potentiation and long-term depression, such as allopregnanolone and DHEA [26, 119]. In addition, rodent models of chronic intermittent ethanol administration showed altered memory functions, assessed with the Morris water maze, in association with decreased neurosteroid levels in the hippocampus [120]. A better understanding of the role of allopregnanolone and thus of all the other GABAergic neurosteroids on fear extinction and extinction retention is pivotal in developing biomarkers for neuropathology related to fear extinction deficits and for discovering new therapeutics that can remediate this condition.

7 Conclusions and Future Directions

Several clinical and preclinical studies have established a link between learning and memory mechanisms that are fundamental in the pathophysiology of stress-related mood disorders, and the role played by GABAergic neurotransmission and GABAergic neurosteroids within these mechanisms [87, 108, 114] (Fig. 3). Modifications of GABA$_A$ receptor signal transduction in key brain areas involved in memory consolidation and emotion processing underscore the importance of investigating how GABAergic neurosteroids are involved in modulating aversive fear memories and, thus, in regulating PTSD risk as well as enhancing PTSD recovery and resilience. Furthermore, exploring mechanisms of GABAergic neurosteroids on GABAergic neurotransmission in the context of PTSD comorbidities such as major depression, alcohol and substance abuse, chronic pain, and neurodegenerative disorders will be crucial for the full recovery of individuals suffering from these debilitating conditions.

Elucidating the role these neurosteroids have in PTSD pathophysiology and that of neurosteroidogenic agents in alleviating PTSD symptoms may be impactful for clinical practice, in addition to trauma-focused therapy, to accelerate and prolong recovery. Several molecules show promise to be effective in PTSD treatment by acting at different neurosteroidogenic targets. TSPO ligands may increase neurosteroidogenesis by enhancing cholesterol entry into the inner mitochondrial membrane [27, 64]. The administration of allopregnanolone's precursor, pregnenolone, is another treatment strategy that may improve emotional behavior by increasing allopregnanolone biosynthesis and restoring it to physiological levels [27, 64]. SSRIs are among the most used pharmacologic treatments for PTSD. SSRIs, administered at a non-serotoninergic dose, act as selective brain steroidogenic stimulants (SBSS) by stimulating the enzyme 3α-HSD and increasing the conversion of 5α-DHP into allopregnanolone (although there are several contradictory findings regarding this mechanism) [27, 64]. Moreover, the direct injection of allopregnanolone analogs, such as ganaxolone [97], BR351, and BR297 [98], has proven effective in ameliorating fear extinction and other emotional behaviors, including aggression and anxiety-like behavior, in SI mice.

Altogether, these findings corroborate the hypothesis that GABAergic neurosteroids and GABAergic neurotransmission are crucial in PTSD pathophysiology and that they are promising treatment targets for the development of future pharmacologic therapies.

Conflict of Interest

G.P. is a paid consultant to PureTech Health (Boston, MA, USA), GABA Therapeutics, and NeuroTrauma Sciences (Alpharetta, GA, USA). He has two patent applications, one on *N*-palmitoylethanolamine (PEA) and peroxisome proliferator-activated receptor alpha (PPAR-α) agonists US20180369171A1, pending, and one on allopregnanolone analogs US11266663B2 granted on March 8, 2022, in the treatment of neuropsychiatric disorders. The other authors declare no conflict of interest.

References

1. American Psychiatric Association (2013) Diagnostic and statistical manual of mental disorders, 5th edn. American Psychiatric Association, Arlington
2. Kilpatrick DG et al (2013) National estimates of exposure to traumatic events and PTSD prevalence using DSM- IV and DSM- 5 criteria. J Trauma Stress 26:537–547
3. Brewin CR (2001) A cognitive neuroscience account of posttraumatic stress disorder and its treatment. Behav Res Ther 39:373–393

4. Merriam-Webster (2017) https://www.merriam-webster.com/dictionary/fear
5. Raber J et al (2019) Current understanding of fear learning and memory in humans and animal models and the value of a linguistic approach for analyzing fear learning and memory in humans. Neurosci Biobehav Rev 105:136–177
6. Rothbaum BO, Davis M (2003) Applying learning principles to the treatment of post-trauma reactions. Ann NY Acad Sci 1008: 112–121
7. Rasmusson AM et al (2021) A role for deficits in GABAergic neurosteroids and their metabolites with NMDA receptor antagonist activity in the pathophysiology of posttraumatic stress disorder. J Neuroendocrinol 34(2): e13062
8. Bentz D et al (2010) Enhancing exposure therapy for anxiety disorders with glucocorticoids: from basic mechanisms of emotional learning to clinical applications. J Anxiety Disord 24:223–230
9. Richard D, Lauterbach D (2011) Handbook of exposure therapies. Academic Press, New York
10. Boland R, Verdiun M, Ruiz P (2022) Trauma and stress-related disorders. In: Kaplan and Sadock's synopsis of psychiatry, 12th edn. Wolters Kluver, Alphen aan den Rijn
11. Freeman D et al (2017) Virtual reality in the assessment, understanding, and treatment of mental health disorders. Psychol Med 47:1–8
12. Opris D et al (2012) Virtual reality exposure therapy in anxiety disorders: a quantitative meta-analysis. Depress Anxiety 25:85–93
13. Powers M et al (2008) Virtual exposure therapy for anxiety disorders: a meta-analysis. J Anxiety Disord 22:561–569
14. Parsons TD, Rizzo AA (2008) Affective outcomes of virtual reality exposure therapy for anxiety and specific phobias: a meta-analysis. J Behav Ther Exp Psychiatry 39:250–261
15. Anderson PL, Rothbaum BO, Hodges LF (2003) Virtual reality exposure in the treatment of social anxiety. Cogn Behav Pract 10: 240–247
16. Emmelkamp P (2005) Technological innovations in clinical assessment and psychotherapy. Psychother Psychosom 74:336–343
17. Diemer J et al (2014) Virtual reality exposure in anxiety disorders: impact on psychophysiological reactivity. World J Biol Psychiatr 15: 427–442
18. Locci A, Pinna G (2017) Neurosteroids biosynthesis downregulation and changes in GABA receptors subunit composition: a biomarker axis in stress-induced cognitive and emotional impairment. Br J Pharmacol 174:3226–3241
19. Lee D et al (2016) Psychotherapy versus pharmacotherapy for posttraumatic stress disorder: systemic review and metanalyses to determine first-line therapy treatments. Depress Anxiety 33:792–806
20. Pinna G (2015) Fluoxetine. Pharmacology, mechanisms of action and potential side effects. Nova Biomedical, Hauppauge
21. Pinna G (2015) The neurosteroidogenic action of fluoxetine unveils the mechanism for the anxiolytic property of SSRIs. In: Pittman J (ed) Fluoxetine. Pharmacology, mechanisms of action and potential side effects, 1st edn. Nova Biomedical, Hauppauge, pp 25–42
22. Puia G et al (1990) Neurosteroids act on recombinant human $GABA_A$ receptors. Neuron 4:759–765
23. Agis-Balboa RC et al (2007) Downregulation of neurosteroid biosynthesis in corticolimbic circuits mediates social isolation-induced behavior in mice. Proc Natl Acad Sci U S A 104(47):18736–18741
24. Pinna G (2019) Animal models of PTSD: the socially isolated mouse and the biomarker role of allopregnanolone. Front Behav Neurosci 11(13):114. eCollection 2019
25. Rasmusson AM, Pinna G, Pineles SJ (2022) Pleiotropic Endophenotypic and phenotype effects of GABAergic Neurosteroid synthesis deficiency in posttraumatic stress disorder. Curr Opin Endocrin Metab Res 25:100359
26. Rasmusson AM, Pineles SJ (2018) Neurotransmitter, peptide, and steroid hormone abnormalities in PTSD: biological endophenotypes relevant to treatment. Curr Psychiatry Rep 20(52):52
27. Locci A et al (2019a) Neurosteroid-based biomarkers and therapeutic approaches to facilitate resilience after trauma. In: Pinna G, Izumi T (eds) Facilitating resilience after PTSD. A translational approach. Nova Biomedical, Hauppauge, pp 199–236
28. Smith GB, Olsen RW (1995) Functional domains of $GABA_A$ receptors. Trends Pharmacol Sci 16(5):162–168
29. Siegel E (2002) Mapping of the benzodiazepine recognition site on GABA(A) receptors. Curr Top Med Chem 2(8):833–839
30. Hosie AM et al (2006) Endogenous neurosteroids regulate GABAA receptors through two discrete transmembrane sites. Nature 444(7118):486–489

31. Nusser Z, Mody I (2002) Selective modulation of tonic and phasic inhibitions in dentate gyrus granule cells. J Neurophysiol 87(5): 2624–2628
32. McKernan RM et al (2000) Sedative but not anxiolytic properties of benzodiazepines are mediated by GABA(A) receptor alpha1 subtype. Nat Neurosci 3(6):587–592
33. Low K et al (2000) Molecular and neuronal substrate for the selective attenuation of anxiety. Science 290(5489):131–134
34. Whiting PG (1999) The GABA-A receptor gene family: new targets for therapeutic intervention. Neurochem Int 34(5):387–390
35. Wisden W et al (1991) Cloning, pharmacological characteristics and expression pattern of the rat GABAA receptor alpha 4 subunit. FEBS Lett 289(2):227(230)
36. Wohlfarth KM, Bianchi MT, MacDonald RL (2002) Enhanced neurosteroid potentiation of ternary GABA(A) receptors containing the delta subunit. J Neurosci 22(5):1541–1549
37. Shur RR et al (2016) Development of psychopathology in deployed armed forces in relation to plasma GABA levels. Psychoneuroendocrinology 73:263–270
38. Harditte Hall KA et al (2021) Plasma gammaamminobutyric acid (GABA) levels and posttraumatic stress disorders symptoms in trauma-exposed women; a preliminary report. Psychopharmacology 238:1541–1552
39. Merali Z et al (2004) Dysregulation in the suicide brain: mRNA expression of corticotropin-releasing hormone receptors and GABA(a) receptor subunits in frontal cortical brain region. J Neurosci 24(6): 1478–1485
40. Sequeria A et al (2009) Global brain gene expression analysis links glutamatergic and GABAergic alterations to suicide and major depression. PLoS One 4(8):e6585
41. Geuze E et al (2008) Reduced GABAA benzodiazepine receptor binding in veterans with posttraumatic stress disorder. Mol Psychiatry 13(1):74–83
42. Bremner JD et al (2000) Decreased benzodiazepine receptor binding in prefrontal cortex in combat-related posttraumatic stress disorder. Am J Psychiatry 157(7):1120–1126
43. Montpied P et al (1993) Repeated swim-stress reduces GABAA receptor alpha subunit mRNAs in the mouse hippocampus. Brain Res 18(3):267–272
44. Matzumoto K et al (2007) GABA(A) receptor neurotransmission dysfunction in a mouse model of social isolation-induced stress: possible insights into a non-serotonergic mechanism of action of SSRIs in mood and anxiety disorders. Stress 10(1):3–12
45. Pinna G et al (2006) Imidazenil and diazepam increase locomotor activity in mice exposed to protracted social isolation. Proc Natl Acad Sci U S A 103(11):4275–4280
46. Serra M et al (2006) Social isolation-induced increase in alpha and delta subunit gene expression is associated with a greater efficacy of ethanol on steroidogenesis and GABA receptor function. J Neurochem 98(1): 122–133
47. Costa E et al (2002) GABAA receptors and benzodiazepines: a role for dendritic resident subunit mRNAs. Neuropharmacology 43(6): 925–937
48. Corpechot C et al (1981) Characterization and measurement of dehydroepiandrosterone sulfate in rat brain. Proc Natl Acad Sci U S A 78:4704–4707
49. Ukena K et al (1998) Cytochrome P450 side-chain cleavage enzyme in the cerebellar Purkinje neuron and its neonatal change in rats. Endocrinology 139(1):137–147
50. Le Goascogne C et al (1987) Neurosteroids: cytochrome P-450scc in rat brain. Science 237(4819):1212–1215
51. Hu ZY et al (1987) Neurosteroids: oligodendrocyte mitochondria convert cholesterol to pregnenonole. Proc Natl Acad Sci U S A 84(23):8215–8219
52. Rasmusson AM et al (2006) Decreased cerebrospinal fluid allopregnanolone levels in women with posttraumatic stress disorder. Biol Psychiatry 60:704–713
53. Pineles SL et al (2018) PTSD in women is associated with a block of conversion of progesterone to the GABAergic neurosteroids allopregnanolone and pregnanolone measured in plasma. Psychoneuroendocrinology 93:133–141
54. Dichtel L et al (2018) Neuroactive steroids and affective symptoms in women across the weight spectrum. Neuropsychopharmacology 43(6):1436–1444
55. Cruz DA et al (2019) Neurosteroids levels in the orbitofrontal cortex of subjects with PTSD and controls: a preliminary report. Chron Stress 3:1–10
56. Rosen JB, Shulkin J (1998) From normal fear to pathological anxiety. Psychol Rev 105: 325–350
57. Hoffman SG (2008) Cognitive processes during fear acquisition and extinction in animals and humans: implications for exposure therapy of anxiety disorders. Clin Psychol Rew 28: 199–210

58. Foa EB (2011) Prolonged exposure therapy: past, present, and future. Depress Anxiety 28: 1043–1047
59. Bouton ME (2004) Context and behavioral processes in extinction. Learn Mem 11:458–494
60. Bouton ME (2002) Context, ambiguity, and unlearning: sources of relapse after behavioral extinction. Biol Psychiatry 52:976–986
61. Meyer K, Davis M (2007) Mechanisms of fear extinction. Mol Psychiatry 12:120–150
62. Vansteenwegen D et al (2005) Return of fear in human differential conditioning paradigm caused by a return to the original acquisition context. Behav Res Ther 43:323–336
63. Hermans D et al (2006) Extinction in humans fear conditioning. Biol Psychiatry 60:361–368
64. Aspesi D, Pinna G (2019) Animal model of traumatic stress disorder and novel treatment targets. Behav Pharmacol 30:130–150
65. Locci A, Pinna G (2019b) Social isolation as a promising animal model of PTSD comorbid suicide: neurosteroids and cannabinoids as possible treatment options. Prog Neuro-Psychopharmacol Biol Psychiatry 92:243–259
66. Torok B et al (2019) Modelling posttraumatic stress disorders in animals. Prog Neuro-Psychopharmacol Biol Psychiatry 90:117–133
67. Adamec R, Walling S, Burton P (2004) Long-lasting, selective, anxiogenic effects of feline predator stress in mice. Physiol Behav 83: 401–410
68. Liberzon I, Krstov M, Young EA (1997) Stress-restress: effects on ACTH and fast feedback. Psychoneuroendocrinology 22:443–453
69. Yamamomto S et al (2009) Single prolonged test: toward an animal model of posttraumatic stress disorder. Depress Anxiety 26:1110–1117
70. Camp MC et al (2012) Genetic strain differences in learned fear inhibition associated with variation in neuroendocrine, autonomic, and amygdala dendritic pheno-types. Neuropsychopharmacology 37:1534–1547
71. Wilkialis L et al (2021) Social isolation, loneliness and generalized anxiety: implications and associations during the Covid-19 quarantine. Brain Sci 11(12):1620
72. Sher L (2020) The impact of the Covid pandemic on suicide rates. QJM 113(10): 707–712
73. Stravynski A, Boyer R (2001) Loneliness in relation to suicide ideation and parasuicide: a population-wide study. Suicide Life Threat Behav 31(1):32–40
74. Pibiri F et al (2008) Decreased corticolimbic allopregnanolone expression during social isolation enhances contextual fear: a model relevant for posttraumatic stress disorder. Proc Natl Acad Sci U S A 105:5567–5572
75. Rau V, DeCola JP, Fanselow MS (2005) Stress-induced enhancement of fear learning: an animal model of posttraumatic stress disorder. Neurosci Biobehav Rev 29:1207–1223
76. Pinna G, at al. (2003) In socially isolated mice, the reversal of brain allopregnanolone down-regulation mediates the anti-aggressive action of fluoxetine. Proc Natl Acad Sci U S A 100:2035–2040
77. Guidotti A et al (2001) The socially-isolated mouse: a model to study the putative role of allopregnanolone and 5-α-dihydroprogesterone in psychiatric disorders. Brain Res Rev 37:110–115
78. Pinna G et al (2008) Neurosteroid biosynthesis regulates sexually dimorphic fear and aggressive behavior in mice. Neurochem Res 33:1990–2007
79. Bortolato M et al (2011) Isolation rearing induced reduction of brain 5α-reductase expression: relevance to dopaminergic impairments. Neuropharmacology 60:1301–1308
80. Matsumoto K et al (1999) Permissive role of brain allopregnanolone content in the regulation of pentobarbital-induced righting reflex loss. Neuropharmacology 38:955–963
81. Matsumoto K et al (2007) GABAA receptor neurotransmission dysfunction in a mouse model of social isolation-induced stress: possible insights into a non-serotonergic mechanism of action of SSRIs in mood and anxiety disorders. Stress 10:3–12
82. Serra M, Pisu MG, Littera M, Papi G, Sanna E, Tuveri F et al (2000) Social solation-induced decreases in both the abundance of neuroactive steroids and GABAA receptor function in rat brain. J Neurochem 75:732–740
83. Dong E et al (2001) Brain 5-α-dihydroprogesterone and allopregnanolone synthesis in a mouse model of protracted social isolation. Proc Natl Acad Sci U S A 98:2849–2854
84. Locci A, Pinna G (2019c) Stimulation of peroxisome proliferator-activation of PPAR-α by N-palmitoylethanolamine engages allopregnanolone biosynthesis to modulate emotional behavior. Biol Psychiatry 85(12):1036–1045

85. Agís-Balboa RC et al (2006) Characterization of brain neurons that express enzymes mediating neurosteroid biosynthesis. Proc Natl Acad Sci U S A 103:14602–14607
86. Agis-Balboa RC, Guidotti A, Pinna G (2014) 5α-reductase type I expression is downregulated in the prefrontal cortex/Brodmann's area s9 (BA9) of depressed patients. Psychopharmacology 231:3569–3580
87. Rabinowitz A et al (2014) The neurosteroid allopregnanolone impairs object memory and contextual fear memory in male C57BL/6J mice. Horm Behav 66:238–246
88. Pinna G (2003) In socially isolated mice, the reversal of brain allopregnanolone downregulation mediates the anti-aggressive action of fluoxetine. Proc Natl Acad Sci USA 100(4): 2035–2040
89. Pinna G et al (2000) Brain allopregnanolone regulates the potency of the GABA(A) receptor agonist muscimol. Neuropharmacology 39(3):440–448
90. Pinna G, Costa E, Guidotti A (2004) Fluoxetine and norfluoxetine stereospecifically facilitate pentobarbital sedation by increasing neurosteroids. Proc Natl Acad Sci USA 101(16):6222–62225
91. Akirav I, Maroun M (2007) The role of the medial prefrontal cortex-amygdala circuit in stress effects on the extinction of fear. Neural Plast 2007:30873
92. Liberzon I, Sripada CS (2008) The functional neuroanatomy of PTSD: a critical review. Prog Brain Res 167:151–169
93. Pare D, Quirk GJ, Ledoux JE (2004) New vistas on amygdala networks in conditioned fear. J Neurophysiol 92:1–9
94. Jüngling K et al (2008) Neuropeptide S-mediated control of fear expression and extinction: role of intercalated GABAergic neurons in the amygdala. Neuron 59:298–310
95. Likhtik E et al (2008) Amygdala intercalated neurons are required for expression of fear extinction. Nature 454:642–645
96. Matrisciano F, Pinna G (2021) PPAR-α hypermethylation in the hippocampus of mice exposed to social isolation stress is associated with enhanced neuroinflammation and aggressive behavior. Int J Mol Sci 22(19): 10678
97. Pinna G, Rasmusson AM (2014) Ganaxolone improves behavioral deficits in a mouse model of post-traumatic stress disorder. Front Cell Neurosci 11(8):256
98. Locci A et al (2017) Social isolation in early versus late adolescent mice is associated with persistent behavioral deficits that can be improved by neurosteroid-based treatment. Front Cell Neurosci 11:208
99. Petrosino S, Di Marzo V (2017) The pharmacology of palmitoyletha-nolamide and first data on the therapeutic efficacy of some of its new formulations. Br J Pharmacol 174:1349–1365
100. Moreno S, Farioli-Vecchioli CMP (2004) Immunolocalization of peroxisome proliferator-activated receptors and retinoid X receptors in the adult rat CNS. Neuroscience 123:131–145
101. Musella A et al (2017) A novel crosstalk within the endocannabinoid system controls GABA transmission in the striatum. Sci Rep 7: 7363
102. Häring M, Guggenhuber S, Lutz B (2012) Neuronal populations mediating the effects of endocannabinoids on stress and emotionality. Neuroscience 204:145–158
103. Neumeister A et al (2014) Translational evidence for a role of endocannabinoids in the etiology and treatment of posttraumatic stress disorder. Psychoneuroendocrinology 51: 577–584
104. Sasso O et al (2010) Palmitoylethanolamide modulates pentobarbital-evoked hyp-notic effect in mice: involvement of allopregnanolone biosynthesis. Eur Neuropsychopharmacol 20:195–206
105. Raso GM et al (2011) Palmitoylethanolamide stimulation induces allopregnanolone synthesis in C6 cells and primary astrocytes: involvement of peroxisome-proliferator activated receptor-a. J Neuroendocrinol 23:591–600
106. Mitchell KJ, Johnson MK (2009) Source monitoring 15 years later: what have we learned from fMRI about the neural mechanisms of source memory? Psychol Bull 135: 638–677
107. Golier J et al (1997) Source monitoring in PTSD. Ann N Y Acad Sci 21:472–475
108. Kask K et al (2008) Allopregnanolone impairs episodic memory in healthy women. Psychopharmacology 199:161–168
109. Turkmen S et al (2004) 3beta-20beta-dihydroxy-5alpha-pregnane (UC1011) antagonism of the GABA potentiation and the learning impairment induced in rats by allopregnanolone. Eur J Neurosci 20:1604–1612
110. Stromberg J et al (2006) Neurosteroid modulation of allopregnanolone and GABA effect on the GABA-A receptor. Neuroscience 143: 73–81
111. Wang MD, Backstrom T, Landgren S (2000) The inhibitory effects of allopregnanolone

and pregnanolone on the population spike, evoked in the rat hippocampal CA1 stratum pyramidale in vitro, can be blocked selectively by epiallopregnanolone. Acta Physiol Scand 169:333–341

112. Casey P et al (1999) Memory in pregnancy. II. Implicit, incidental, explicit, semantic, short-term, working and prospective memory in primigravid, multigravid and postpartum women. J Psychosom Obstet Gynaecol 20:158–164

113. Keenan PA et al (1998) Explicit memory in pregnant women. Am J Obstet Gynecol 179:731–737

114. de Groot RH et al (2006) Differences in cognitive performance during pregnancy and early motherhood. Psychol Med 36:1023–1032

115. Milad MR et al (2010) The influence of gonadal hormones on conditioned fear extinction in healthy humans. Neuroscience 168:652–658

116. Pineles SL et al (2016) Extinction retention and the menstrual cycle: different associations for women with posttraumatic stress disorder. J Abnorm Psychol 125:349–355

117. Pineles SL et al (2020) Associations between PTSD-related extinction retention deficits in women and plasma steroids that modulate brain GABAA and NMDA receptor activity. Neurobiol Stress 13:100225

118. Fonda JR et al (2019) Tobacco dependence is associated with increased risk for multimorbid clustering of posttraumatic stress disorder, depressive disorder, and pain among post-9/11 deployed veterans. Psychopharmacology 236:1729–1739

119. Rasmusson AM, Vythilingam M, Morgan CA (2003) The neuroendocrinology of posttraumatic stress disorder: new directions. CNS Spectr 8:651–667

120. Cagetti E et al (2004) Chronic intermittent alcohol (CIE) administration in rats decreases levels of neurosteroids in hippocampus, accompanied by altered behavioral responses to neurosteroids and memory function. Neuropharmacology 46:570–579

INDEX

A

Actigraphy .... viii, 52, 53, 67–72, 75, 77
Acute stress .... ix, 77, 193
Adolescence ....viii, 6, 15, 29, 46
Allopregnanolone .... x, 327, 329–331, 334–343
Amygdala 2–6, 15, 17, 46, 61, 62, 110,
  111, 118, 119, 126, 130, 205–208, 219, 237,
  239, 248, 251, 266, 267, 271, 273, 335–337, 340
Animal models... v, vii–xi, 2, 17,
  23, 31, 32, 37, 46, 61, 136, 138, 139, 142, 174,
  191, 192, 204, 207, 215–217, 219, 221, 222,
  224, 225, 253–255, 266, 271, 274, 332, 336
Anxiety... 15, 24, 37–39, 46,
  57, 61, 74, 101, 102, 121, 143, 197, 204, 222,
  236, 238, 252, 253, 306, 307, 330, 335
Anxiety disorders................................. viii, 2, 57, 60, 112,
  118, 142, 232, 252, 307, 308, 325, 327, 340
Apolipoprotein E 170–185

B

Behaviors ....vii, viii, 2, 3, 7, 9, 10, 12,
  15, 16, 26, 28–32, 38, 40, 45, 53, 54, 56, 62, 74,
  75, 117, 118, 128, 136, 142, 143, 145, 154, 157,
  170, 174, 175, 184, 194, 195, 204–211, 250,
  251, 266, 274, 327, 332–334, 337–339, 343
Biomarkers... vii, viii, x, xi, 53, 59,
  128, 138, 255, 266, 267, 327, 331, 336, 342
Brain imaging .... viii, 62, 83, 126, 219

C

Children .... viii, 2, 14–17, 24, 37–47, 203
Cognitive enhancers.................................. 206, 207, 211
Conditioned fear .... vii, viii, 23–32,
  37, 38, 40, 42, 44, 52, 98, 118, 119, 125, 126,
  207, 208, 210, 219–222, 337
Conditioned stimulus (CS) .... vii, 4, 7, 9,
  12, 13, 37, 41–46, 79, 99, 103, 104, 106, 108,
  109, 111, 118, 122, 192, 193, 195, 198, 218–
  220, 250, 326, 331, 334
Conditioning paradigm ....viii, 3, 4,
  117, 122, 142, 219–221, 334
Cortisol 59, 60, 80, 81, 138, 170, 224, 268

D

Development .... v, vii–xi, 2, 3, 7, 17,
  23, 29, 31, 38, 39, 52, 56, 57, 59, 74, 75, 84, 98,
  112, 118, 122, 128, 137, 138, 185, 215, 216,
  221, 224, 252, 253, 267, 269, 329, 330, 343
Dim light melatonin onset
  (DLMO) 60, 81, 82

E

Early life adversities .... viii, 14, 23–25, 224
Electrophysiology..63, 85, 307,
  309, 311, 321, 322
Event related potential (ERP) .... x, 44, 306,
  307, 310, 311, 317–319, 321
Extinction .... vii–x, 6, 23–32,
  37–47, 52, 56, 57, 61, 78–80, 83, 98–112, 117,
  118, 122–126, 128, 198, 199, 204–211, 219,
  222–224, 249, 252–255, 269, 271–274, 276,
  326, 331, 333–335, 337, 340–342

F

Fear circuit...6, 110
Fear conditioning .... viii–x, 1–17, 27,
  30–31, 38, 39, 41–43, 45, 46, 52, 56, 61, 78–80,
  83, 100, 103, 104, 108, 110, 111, 117, 119,
  120, 122, 124, 126, 128, 142, 192–194, 197,
  198, 200, 204–207, 215–225, 250, 251, 253,
  255, 259–265, 273, 274, 276, 326, 327, 330,
  332–337, 339, 340
Fear extinction...vii–xi, 24, 37, 38,
  41, 56, 79, 80, 98, 99, 125, 126, 191–200, 204,
  219, 224, 235–237, 249, 276, 325–343
Fears...vii–x, 2–6, 12, 14,
  23, 24, 26, 28, 30, 31, 37–47, 57, 61, 62, 74,
  78–80, 97–112, 117, 118, 122, 124–128, 142,
  192, 194, 197–200, 204–206, 216–224,
  234–237, 239, 248–255, 266–269, 274, 276,
  326, 327, 331, 332, 334–338, 340, 342
5-hydroxytryptamine
  (5-HT) .... ix, x, 204, 206, 307
Functional MRI (fMRI) 42–46, 61,
  82, 83, 98–100, 102, 106, 107, 109, 240

G

GABA-A receptor 271

H

Heart rate variability (HRV) 57, 59, 80, 83
Hippocampus ....ix, x, 2, 4, 6, 17, 46, 61, 110, 111, 124, 127, 204–208, 237, 239, 248, 249, 251, 267, 271, 273, 329, 335, 337, 340, 342

I

Immobilization .... ix, 191–200, 255
Individual variability .... ix, x, 216, 222, 224, 276
Infants .... viii, 1–17
Inflammation 136

L

Loudness dependence of auditory evoked potentials (LDAEP) .... x, 306–322

M

Memory consolidation 30, 56, 237, 239, 248–251, 255, 266, 269, 342
Mice .... ix, x, 3, 4, 6, 9, 10, 12, 30, 39, 142, 170, 174–176, 182–185, 192, 193, 195–199, 220, 221, 223, 250, 256–264, 269–273, 317, 327, 329, 332–340, 343
Microbiomes 154
Mood disorders .... x, 327–330, 332, 339, 340, 342

N

Neurobiology .... vii, x, 15, 128, 199, 250, 251, 255, 327, 338
Nightmares .... viii, 51–53, 57, 60, 61, 63, 74, 76–78, 85, 203, 216, 252
Non-invasive brain stimulation (NIBS) .... x, 232–240
N-palmitoylethanolamine 339, 343

O

Oxysterol 171, 174, 176–179, 182–185

P

P2 .... 28, 306, 307, 310, 319, 320
Pavlovian conditioning .... vii, ix, 250
Polysomnography (PSG) .... viii, 51–54, 56, 62–68, 72, 74, 75, 77, 78, 80, 82–84
Post-traumatic stress disorder (PTSD) and related syndromes .... vii–xi, 1–17, 23, 24, 26, 32, 37–39, 45, 51–83, 85, 97–112, 117–130, 135–157, 169–185, 199, 203, 204, 207, 208, 210, 211, 215–225, 231, 232, 234, 235, 237–240, 248, 250, 252–269, 274, 276, 307, 308, 320, 325–338, 340–343
PPAR-alpha 339, 340, 343
Psychophysiology 102, 120
PTSD models 254, 255, 266, 271

R

Rat 3, 8, 13, 15, 17, 25, 26, 28, 29, 138, 141, 143, 146, 148, 151, 153, 205, 263, 264
Resilience .... vii, ix, x, 144, 172, 216, 224, 331, 342

S

Selective serotonin reuptake inhibitor (SSRI) .... x, 206, 208, 210, 211, 306–322, 327, 343
Serotonin .... x, 5, 270, 306, 308, 327
7-ketocholesterol 171, 174, 177–180, 182–185
Sex as a biological variable .... ix, 191, 200
Skin conductance response (SCR) .... ix, 39, 40, 42, 44–46, 78, 99, 101, 102, 106, 108–112, 118, 121, 123–128, 235, 236, 238
Sleep .... viii, 25, 46, 51–85, 101, 136, 150, 152, 154, 252
Social buffering 4, 6, 12, 184
Social isolation .... viii, x, 23–32, 184, 327, 329, 332–335, 337
Stress .... viii–xi, 4, 7, 14, 15, 23–26, 28, 30–32, 41, 46, 59, 78, 111, 121, 135–157, 170, 175, 191–193, 195–197, 199, 200, 204, 206–209, 216, 217, 219, 222, 231–240, 249, 250, 252, 268, 269, 327, 329, 330, 332, 334–336

T

Threat .... viii, 2, 98, 99, 102–104, 110, 138, 218, 231, 239, 248, 253, 326
Translational .... viii, x, 2, 23, 31, 32, 38, 117–130, 136–138, 144, 145, 253–255, 266, 269, 326, 329
Translational research .... xi, 110, 141, 144, 254
Trauma .... vii–ix, 1, 2, 14–17, 39, 47, 51–57, 59–61, 63, 67, 74, 76, 97, 101, 118, 120, 136, 138, 142, 169–171, 191, 203, 216–219, 221, 222, 224, 231, 252, 253, 267, 268, 276, 325, 326
Traumatic stress 191, 200, 219, 220

V

Virtual reality exposure 231–240, 326

CPSIA information can be obtained
at www.ICGtesting.com
Printed in the USA
LVHW020413140623
749744LV00007B/64